Developing Successful ICT Strategies:
Competitive Advantages in a Global Knowledge-Driven Society

Hakikur Rahman
SDNP, Bangladesh

Information Science REFERENCE

INFORMATION SCIENCE REFERENCE
Hershey · New York

Acquisitions Editor:	Kristin Klinger
Development Editor:	Kristin Roth
Senior Managing Editor:	Jennifer Neidig
Managing Editor:	Sara Reed
Copy Editor:	Joy Langel
Typesetter:	Jeff Ash
Cover Design:	Lisa Tosheff
Printed at:	Yurchak Printing Inc.

Published in the United States of America by
Information Science Reference (an imprint of IGI Global)
701 E. Chocolate Avenue, Suite 200
Hershey PA 17033
Tel: 717-533-8845
Fax: 717-533-8661
E-mail: cust@igi-pub.com
Web site: http://www.igi-pub.com/reference

and in the United Kingdom by
Information Science Reference (an imprint of IGI Global)
3 Henrietta Street
Covent Garden
London WC2E 8LU
Tel: 44 20 7240 0856
Fax: 44 20 7379 0609
Web site: http://www.eurospanonline.com

Library of Congress Cataloging-in-Publication Data

Developing successful ICT strategies : competitive advantages in a global knowledge-driven society / M. Hakikur Rahman, editor.

p. cm.

Summary: "This book collects an authoritative core of research investigating the notion that information communication technologies (ICTs) have the potential to improve the lives of people and contribute to enhancing social conditions in developing countries through such concepts as the Knowledge Society, open education, and e-governance"--Provided by publisher.

Includes bibliographical references and index.

ISBN-13: 978-1-59904-654-9 (hardcover)

ISBN-13: 978-1-59904-656-3 (ebook)

1. Information technology--Developing countries. 2. Technological innovations--Developing countries. 3. Information technology--Economic aspects--Developing countries. 4. Technological innovations--Economic aspects--Developing countries. 5. Knowledge management. I. Rahman, Hakikur, 1957-

HC59.72.I55D48 2008

303.48'33091724--dc22

2007016962

British Cataloguing in Publication Data
A Cataloguing in Publication record for this book is available from the British Library.

Table of Contents

Section II
Education and Research

Section III
E-Concepts

Section IV
Strategic ICT Applications

Detailed Table of Contents

As nations continue to conceptualize, adopt, implement, and monitor their respective information and communications technologies (ICTs) policies and e-strategies, it is vital that a critical assessment is undertaken on their effective progress. This includes analysis of strategies, methodologies, and best practices while weighing the impact, effectiveness, and efficiency of these policies and strategies. However, the demands, motivations, and incentives of various governments for adopting, adapting, and initiating ICTs varies from country to country, with diverse economic, social, cultural, and political environments. Researches in this aspect recognized the necessity of a cohesive ICT policy and strategic framework in each country for socio-economic development at their grass roots. The execution of these policies and strategies should include concrete national commitment and strong political will at the highest levels of government, and an enabling environment that promotes stakeholder involvement in setting agendas and implementing plans and programs. This chapter provides insights into various national implementations in promoting related activities, tries to establish an analytical approach that would assist in formulating ICT policies and strategies by identifying different ICT indicators. Furthermore, this chapter focuses on critical aspects of different strategic national level policies with short-medium-long term visions that targeted both the immediate needs of the populace and long-term needs of nations by integrating ICTs. Finally, this chapter has recommended that via coherent and complimentary policies that engage both the private sector and civil society organizations, nations can move forward towards creating a knowledge society and at the same time by leveraging capabilities of ICT can address the social, economic and political issues on the ground.

Trade globalization and advancement in ICT may be considered as the two major forces that will be directing economic growth of a country in the coming years. Policy makers have realized the importance of ICT in achieving national goals in the digital era. The developing and underdeveloped countries generally lack research capacity and standard policy development processes that are critical for developing a successful ICT policy. Cause of policy failure in these countries can most of the time, be traced to flaws in policy development process rather than environmental complexity. The chapter provides an insight on policy development process from strategic management perspective to highlight critical issues, which are sometimes overlooked by policy makers. The purpose is to assist the policy makers in developing successful ICT Policy for their countries.

The focus of this chapter is on mapping the environment. It will show the main aspects of the model for industry analysis and explain the process of forming strategic groups and pressure maps.

The expansion of the social influence of the Internet has led both general interest and national public broadcasters to a structural challenge. The new online paradigm demands a full version of the public enterprise, which include all the available tools to achieve its mission. The Internet would be the ideal complement to the public service. Before this, public broadcasters should define its core business to attend the public demand and to collaborate in the creation of the public sphere. Finally, the clear definition of the public activities and the separation of public and commercial finance will be a main point. To confirm the benefits of the intervention of the public broadcasters the cases of Spain and United Kingdom will be studied.

This chapter examines the Internet and law relationship. Using rules and decisions on ISP liability for civil defamation to illustrate, the chapter attempts to identify and proffer solutions to the challenges posed by the Internet to law. It suggests legal recognition of the distinctive nature of the Internet by policy makers and administrators as the key to addressing the legal issues continually raised by the Internet with the result that appropriate legal strategies would be applied to the Internet. The chapter also advocates international legislative action on the Internet issues. Law, including its substantive and procedural contents, interpretation and administration, is an important factor for encouraging or hindering continued Internet development.

In this chapter, it is argued that one of the benefits of IT is its role in enabling the disintermediation of education to provide an environment in which all can participate. In the context of this chapter however, the focus is on the role of the Internet as a disintermediating technology. There are many ramifications to this but the focus is upon the implications as far as education is concerned. The controlling mechanisms of education have been cost and geographical access. The Internet has changed both of these through its disintermediating function, potentially allowing access to all. The purpose of this chapter is to theorise this in considering the prognosis for development in various parts of the world.

This chapter introduces actor-network theory (ANT) as an approach to the analysis and improvement of the use of ICT for development (ICT4D). It argues that ANT helps to conceptualize ICT beyond the technological systems of the conventional "digital divide." ANT supports thinking about the sociotechnical networks that incorporate humans, hardware, institutions, texts, and policies, and so forth, into ICT networks. It also supports the inclusion of marginal actors, helps to address development problems from the perspective of those populations, and traces the networks of power that supports their participation or exclusion. The author hopes that the ideas in this chapter will promote further discussion on the topic and the refinement of an analytical framework for ICT4D.

Instructional systems are products of the cultures in which they are developed. Culture, which we define here as "the collective programming of the mind which distinguishes the members of one group or category of people from another" (Hofstede, 2001), has a pervasive influence on instructional systems, regardless of whether these systems involve self-paced e-learning, synchronous or asynchronous computer-based learning activities, or online communities of learners. The issue of culture's impact on instructional systems becomes most relevant and challenging where such systems are transferred across cultural boundaries, or developed for multiple cultures. This is currently happening in many large, globally dispersed organizations that use e-learning technologies to support the learning of their staff around the world. Theories of learning and of cultural dimensions suggest that the effectiveness of certain types of learning systems will be affected where they are used in culturally diverse environments. The aim of this paper is to highlight the issues that designers of a wide range of e-learning experiences face when designing e-learning for culturally diverse learner groups. We provide some models to support learning practitioners, focusing in particular on the importance of a conscious, culturally-informed selection of instructional strategies as the most critical part of the design and development process.

This chapter presents the strategies of higher education institutions and how they can be described using the balanced scorecard approach. The pedagogical ICT strategy describes the virtual learning and e-entrepreneurship in higher education. Strategic themes are presented to describe what management believes must be done to succeed and achieve the desired outcomes in virtual learning and e-entrepreneurship. Strategy maps are used to describe the strategy in a graphical representation. In addition, the study presents an example of the cooperation between a higher education institution and a spin-off company. This chapter helps the educational administrators to better describe and implement strategies for virtual learning and e-entrepreneurship.

As student numbers and lecturer workloads increase, traditional methods of assessment make it rather difficult to undertake effective assessment and provide students with in-depth, relevant, specific, and speedy feedback. It is thus incumbent on tertiary academic institutions, more especially distance education providers, to consider alternative ways, means, and strategies towards the incorporation of computer-based assessment (CBA) methods. This automation can be easily realised by the ever-growing field of computer-aided assessment (CAA) software that is currently permeating the educational landscape of the world. While some individuals may argue that CBA software has been proven in only limited situations and that it is very costly to develop, it nevertheless remains a viable alternative technology that is beginning to challenge many traditional and out-dated modes of assessment. In this chapter the rationale for developing CBA has been discussed and thereafter pertinent issues around the types of CBA, advantages of CBA, proprietary, and non proprietary software have been analysed. The chapter concludes by firstly, highlighting strategic planning issues for consideration in engaging computer based assessment and secondly it provides modalities for future research directions within the arena of computer based assessment. These two components are key in designing ICT strategies for continuous academic evaluation.

This chapter addresses collaborative business networks at the level of industry/cluster networks, which is important and relevant from the strategic management perspective in several industries. This chapter examines two current "hot topics": the emergence of a new Internet driven business group, and its

Foreword

ICTs have great potential to improve the lives of common people and contribute to accelerating the pace of social development in developing countries. They have proved effective in providing access to information to millions of people even in remote areas, facilitating health care, assisting in distance education and learning, enhancing rural productivity through access to market information, improving service delivery of governments, and empowering grass roots communities by enabling their increased participation in national development efforts.

During the last decade, many governments around the world have formulated national ICT policies and strategies, where ICT is treated as a *thrust sector*. Donors and aid agencies responded by funding a variety of ICT applications for specific sectors or stakeholders, by including ICT components in developing projects, by dealing with telecommunications infrastructure as a free-standing sector, and by carrying out assessments of country-wide e-readiness.

This book aims to improve the understanding of the challenges and opportunities of the ICT revolution and their implications for development policy and strategies. It examines why developing countries should try to adapt and harness ICT in support of economic and social development. It tries to analyze the risk factors of the ICT revolution, and its potential impact on productivity, markets, organizations, and education. It also outlines the rationale for designing national strategies for e-development (or e-enhanced development), objectives, and parameters related to create major thrusts for such strategies in support of economic growth, poverty reduction, and achieving the Millennium Development Goals (MDGs).

A book featuring all the above aspects can be of immense benefit to the policy initiators, development partners, and researchers in developing substantial national ICT strategies for their economies and make use of their competitive advantages in a global knowledge driven society. Those who understand the inherent processes can direct their efforts towards learning the new practices and may find a route to leaping forward and catching up.

Professor Jamilur Reza Choudhury
Vice Chancellor, BRAC University
Dhaka, Bangladesh

Preface

The world is undergoing through challenges of an all-embracing technological revolution. Although this revolution has taken various names, but in deep sense it is a revolution of information technology, or an innovation in techno-economic paradigm, brought about by a set of neoteric information and communication technologies (ICTs). The enduring revolution of information technology combined with the forces of globalization has incited the aspirations and fears of countries at all levels of development, to leapfrog to this new economy. As a result many governments have formulated national ICT policies and strategies, where ICT is treated mainly as a thrust sector. Donors and aid agencies responded by piloting a variety of ICT applications for specific sectors or stakeholders, by including its components in developing projects, by dealing with telecommunications infrastructure as a free-standing sector, and foremost, by carrying out assessments of country-wide e-readiness (Rodriguez & Wilson, 2000; European Parliament, 2001). However, in the current status quo while mainstream development practitioners, including national leaders continue to ignore the inherent roles of ICT imposes serious vulnerability in development effectiveness. They must strategically assess their inherent knowledge resources and human capabilities, and try to establish their knowledge strategy to sustain competitive advantages. Therefore, the strategic significance of ICT for enabling national development and poverty reduction strategies must be perceptibly understood and operationalized by countries striving to gain global competency in a knowledge-driven society.

ICT's pervasive impact on competency and all aspects of life in advanced economies and its potential influence on social and economic development cannot be denied. ICTs have the potential to improve the lives of common people, contribute to social developments, and creation of a knowledge-based society. They have proved effective in disseminating information en masse in raising awareness; facilitating health care services; assisting distance education and learning; enhancing rural productivity through access to market information; facilitating environmental activism; improving service delivery of governments by emending monitoring utilities and simplifying response systems; and empowering grass roots communities by enabling their increased participation in national and international policy making processes (Government of Namibia, 2005; Government of Kenya, 2006; United Nations, 2003).

Countries that are enjoying greater technological progress have created better ambiance of democratic rights and civil liberties conducive to innovation, adaptation, and utilization of ICTs (Sayo, Chacko, & Pradhan, 2004). They have established respect for the rule of law and security of property rights; increased investment in human capital; and noticed low levels of government deformations. Following these, almost all developing countries, even the poorest, are trying to improve their access to and use of contemporary ICTs. However, the gap between the rich OECD countries and the poor developing countries is growing (Rodriguez & Wilson, 2000), both in terms of quality ICT products as well as in terms of technology based earnings. Although these new technologies appear to be improving overall economic performance and welfare among the user populations, the link between ICTs and society-wide economic progress has been more elusive. There is a lack of association between homogeneous economic growth and proper utilization of ICTs.

Concurrently, there is no scientific measurement to predict whether the ICT gap is not growing internally within countries between the rich and the poor; nor it can be forecasted whether ICTs are contributing to greater equality of incomes at national levels (United Nations, 2005). Knowledgeable sources indicate that information technologies can even cause substantial increase in inequality. But, no further research works have been found that can compute whether this effect will be ultimately reversed in the longer run.

Evidence indicates that ICTs can be enormously beneficial to individual communities and countries, and under right circumstances ICTs can improve education, health, job creation, governance, and skill development. But, to achieve these, informed policy decisions are indispensable that may escalate economic and social equity impacts across the communities. Countries with similar levels of per capita incomes and economic structures exhibit wide variation in their ICT performances (IMF, 2001). Some developing countries are surging ahead while others are falling behind. It needs to identify the pro-ICT policies that appear to be causing these differences in outcomes.

The impact of ICTs on development primarily depends on the existing economic, social, and cultural fabric into which they are composed. But, in formulating a revised, improved, extended, intelligent, and pragmatic ICT policy, among many, a few parameters require thorough review; like how contributions of ICTs relate to economic growth, how ICTs influence competitiveness, what can public policy do to affect ICTs contribution to growth and competitiveness?

Furthermore, despite the massive prospects of ICT in assisting communities to increase their overall well-being through community development, there are relatively a very few examples of sustained community networks that have built around ICT, even in the developed countries where the technology has been increasingly available for more than 20 years. Hence, the critical factors related to economic viability of communities and countries that are retrospective to ICT initiations are need to refocused, revisited, refurbished, redesigned, and re-designated to make rational and comprehensive national ICT policies.

A national development strategy that attempts to position an economy by taking advantage of the ongoing technology revolution must take a comprehensive view to the enabling character of ICT. Intrinsic parameters related to development of ICTs must be intricately furbished. Economic environment conducive to investment and a climate of civil liberties conducive to research, extension of knowledge base, and expansion of communications must be prevailed (Njuguna, 2006; UNESCO, 2005). Measurable indicators must be introduced to apprehend wastage of newly evolved technology dynamics that should properly routed to the grass roots.

These demand a public knowledge source, where situation analysis and cases are being reviewed including success cases, upholding successful ICT strategies. This book has tried to feature these aspects that may act as a handbook, learning kit, and guide for the policy initiators, researchers, and development partners in developing substantial national ICT strategies for their economies and make use of their competitive advantages in a global knowledge driven society. Those who recognize the inherent processes can direct their efforts towards learning the new practices and may find a route to leaping forward and catching up. This will involve a great degree of learning and understanding the impacts of ICT on markets, organizations, competitive strategies, innovation as well as the implications for services, employment, education and learning, regional and spatial development, and poverty reduction.

WHERE THE BOOK STANDS

ICTs are both critical for economic development and highly disruptive to existing employment opportunities. Maximizing the potential of ICTs requires both promoting their use and addressing the inequality and dislocation they may cause. New technologies and the human and institutional capacity building in the public and private sectors, coupled with trends towards a greater emphasis on development initiatives at the state level, have resulted in a paradigm shift in the roles of the private sector and governments in ICT.

Maximizing the potential of ICT requires both promoting their use and addressing the inequality and dislocation they may cause. New technologies and the human and institutional capacity building in the public and private sectors, coupled with trends towards a greater emphasis on development initiatives at the state level, have resulted in a paradigm shift in the roles of the private sector and governments in ICT. Despite the massive prospects of ICT in assisting communities to increase their overall well being through community development, there are relatively a very few examples of sustained community networks that have built around ICT. This book features these aspects in developing substantial national ICT strategies for their economies and makes use of their competitive advantages in a global knowledge driven society. This book also includes policies and perspectives of similar natures, including success cases around the world to synthesize and assist the reader in acquiring comprehensive knowledge from various analyses.

ORGANIZATION OF CHAPTERS

Looking at the practicality of the subject area, the book has been divided into four sections; policy initiation, education and research, e-concepts, and strategic ICT applications. Policy initiation deals with strategic ICT implementation issues, its management perspectives, and related regulatory aspects. This section has five chapters. The second section, education and research, comprise of five chapters and it talks about research aspects and networking theory to promote establishment of knowledge societies by utilizing successful ICT strategies in education and research. E-concepts is the third section and it has four chapters. This section illustrates new concept of ICT based applications in the context of economic development, governance, and learning. Finally, strategic ICT based applications has put forward four case studies contemplating strategy management processes and policy implications to uphold grass roots ICT developments.

Altogether the book has 17 chapters and a brief description of each of the chapter is given below:

Chapter I provides a synopsis on various national implementations in promoting ICT based development activities, and tries to establish an analytical approach that would assist in formulating ICT policies and strategies by identifying different ICT indicators. This chapter also focuses on critical aspects of different strategic national level policies with short-medium-long term visions targeting the immediate needs of the populace and long-term needs of nations by integrating ICTs. It has made a few recommendations on developing ICT strategies for private sector and civil society organizations.

Chapter II offers an insight on policy development process from strategic management perspective to highlight critical issues that are sometimes overlooked by policy makers. The purpose of the chapter is to provide an insight on ICT policy-development-process for the policy makers, so as to assist them in developing successful ICT policy for their country. The chapter approaches ICT strategy development as an integrated process.

In Chapter III the authors claimed that developing a good competitive strategy largely depends on mapping the overall environment and constant monitoring. The main focus of this chapter is on mapping the environment. It shows the main aspects of the model for industry analysis and explains the process of forming strategic groups and pressure maps.

Chapter IV focuses on Internet challenges for public service broadcasters. As the expansion of social influence on the Internet have led both general interest and national public broadcasters to a structural challenge, the authors deduce that the new online paradigm demands a full version of the public enterprise that include all the available tools to achieve its mission. In this context, it is stated that the Internet would be the ideal complement to the public service. But, to achieve this, public broadcasters should define its core business to attend the public demand and to collaborate in the creation of the public sphere. Along this, the clear definition of the public activities and the separation of public and commercial finance are also to be taken care of. To confirm the benefits of the initiatives carried out by the public broadcasters, this chapter has studied different cases of Spain and United Kingdom.

Chapter V examines the Internet and its relationship with law. Using rules and verdicts on ISP liability for civil defamation to illustrate, the chapter attempts to identify and suggest proper solutions to the challenges posed by the Internet to law. It suggests legal recognition of the distinctive nature of the Internet by policy makers, as well as administrators as the key to address the legal issues continually rose by the Internet with the result that appropriate legal strategies would be applied to the Internet. The chapter also tries to advocate international legislative action on the Internet issues. As law, including its substantive and procedural contents, interpretation and administration, is an important factor for encouraging or hindering continued Internet development; this chapter addresses the question on the liability of Internet service providers (ISPs) for defamatory materials created by third parties. However, the discussion is restricted to civil defamation only and does not extend to criminal defamation. The objective of this chapter is to reveal the difficulties with the current regime on Internet defamation and the concern is with the appropriate relationship between legal rules and the Internet.

In Chapter VI the authors argued that one of the benefits of IT is its role in enabling the disintermediation of education to provide an environment in which all can participate. In this context they focus on the role of the Internet as a disinter mediating technology. There are many ramifications to this but this chapter focuses upon the implications as far as education is concerned. It has been found that the controlling mechanisms of education have been cost and geographical access. The Internet has changed both of these through its disinter mediating function, potentially allowing access to all. The purpose of this chapter is to theorize this in considering the prognosis for development around the globe.

Chapter VII introduces actor-network theory (ANT) as an approach to the analysis and improvement of the use of ICT for development (ICT4D). The chapter argues that ANT helps to conceptualize ICT beyond the technological systems of the conventional "digital divide." ANT supports thinking about the socio-technical networks that incorporate humans, hardware, institutions, texts, and policies, and so forth, into ICT networks. Moreover, it supports the inclusion of marginal actors, helps to address development problems from the perspective of those populations, and traces the networks of power that supports their participation or exclusion. The author hopes that the ideas in this chapter will promote further debate on the topic and the refinement of an analytical framework for ICT4D. In this chapter, the author wishes to call for an ongoing and focused dialogue around the problem of ICT4D analysis, and made an additional contribution to the effort, which she believes has the potential to provide coherence and association to the efforts in this aspect.

Chapter VIII highlights various issues that designers of a wide range of e-learning experiences face when designing e-learning for culturally-diverse learner groups. The authors provide some models to support learning practitioners, focusing in particular on the importance of a conscious, culturally informed selection of instructional strategies as the most critical part of the design and development process for effective e-learning.

Chapter IX presents the strategies of higher education institutions and how they can be described using the balanced scorcard approach. The pedagogical ICT strategy describes the virtual learning and e-entrepreneurship in higher education. Strategic themes are presented to described what management believes must be done to succeed and achieve the desired outcomes in virtual learning and e-entrepreneurship. Authors use strategy maps to describe the strategy in a graphical representation. Furthermore, the study presents an example of the cooperation between a higher education institution and a spin-off company. The chapter will assist the educational administrators to better understand and implement strategies for virtual learning and e-entrepreneurship.

In Chapter X rationale for developing computer-based assessment (CBA) has been discussed and in this context, pertinent issues around the types of computer-aided assessment (CAA), advantages of CAA, proprietary and non-proprietary software have been analyzed. The chapter concludes by highlighting strategic planning issues that would complement in preparing ICT strategies for academic evaluation at national level.

Chapter XI examines two current topics: the emergence of a new Internet driven business group and its construction through a series of intentionally developed *collaborative networks*. This chapter aims to show how an intentional networking can benefit a focal actor in developing a new business group based on emerging technologies, and thus, create competitive advantage among different stakeholders. This study is based on a longitudinal case study in the ICT sector and it takes the viewpoint of a focal actor. The author looks at how new technologies (i.e., xDSL, multimedia, mobile technology, Internet) and the combination of traditional technologies and businesses are transformed into a viable business during the time period of 1990-2003. This study also examines how the focal actor tries intentionally to construct this new business group by developing a series of interlinked strategic alliances and networks of organizations.

Chapter XII introduces the term e-governance and puts on view that e-governance is not simply the computerization of government processes but a government process re-engineering (GPR). E-governance is not only about implementing technologies and supplying services but also provides solutions. Therefore, initiatives taken in this regard should not only remain limited to informational and transactional level but also reach complete transformational level. Some success stories of e-governance initiatives have been reported in this chapter with their impact and value addition in citizen's services. At the same time, it is argued that technical issues are equally imperative to be addressed along with socio-economic and political issues for implementing interoperable e-governance solutions. An attempt has been taken to identify such technical issues and provide solution so as to go for successful ICT strategy. In this context, this chapter deals with some technical issues and challenges in implementing e-governance at national level. It also discusses different stages of implementations of e-governance system and recommends developing successful ICT strategies incorporating comprehensive e-governance action plans.

Chapter XIII describes past developments and current challenges in governing IT portfolio of IT applications, IT development, IT operations, and IT platforms. Based upon the lessons, the author has set key objectives to move beyond 'descriptives' and discuss how organizations can diagnose and design IT governance architecture for future performance improvement and sustained business growth. Finally, this chapter provided a thorough understanding and holistic picture of effective IT governance practices and presents a new organizing logic for IT governance.

Chapter XIV analyses the relevance and limits of the use of e-commerce as a paradigm in government. For this purpose it starts out by distinguishing between e-government and e-democracy. In the next step the chapter discusses which factors have led to the success of e-commerce and might, therefore, be applicable as parts of the paradigm. It then discusses the strengths and the weaknesses of e-commerce as applied to government. In conclusion, the author justified the reasons to use the commercial paradigm in e-government and e-democracy.

Chapter XV looks at the role of language and the community consultation process in overcoming the digital divide by facilitating the advantages of ICTs in small regional communities in Australia. It focuses on one of the 'telecentre' programs funded by State and Federal governments in Australia—the Community Technology Centre at New South Wales (CTC@NSW) program. The authors look at some of the key issues that emerged in communities that applied for CTC@NSW when a more culturally relevant consultation process was developed and implemented. The chapter looks at the consultation process and methodology used in achieving successful ICT strategies and outcomes for the regional communities. It also highlights the importance of using simple English in the consultation process and in dealing with the underlying resentment that may exist in various stakeholder groups when their needs are misunderstood or inadequately addressed.

Chapter XVI explores the diffusion of ICT in the country of Jamaica, and within the context of the national strategic plans for development of information technology. Further, it utilizes the concept of institutional contribution on policies and initiatives undertaken in support of the plans. Within this framework, the impacts of managed initiations via collaborative projects between government and international agencies on the adoption of Internet technologies by small and medium-sized enterprises are also examined. The aim of the authors is to point out that as far as developing countries are concerned, governments can affect the diffusion and adoption of ICT through policymaking but more positive outcomes can be realized through managed initiatives.

Finally, Chapter XVII provides policy makers with a comprehensive framework for developing national digital government strategies. This framework raises the importance of technical and economic situations, cross-country comparison of laws and institutions, and the necessity of considering political contexts. To illuminate the utility and application of the framework through examination of two cases from Poland and Taiwan, this chapter yields insights into specific considerations for designing and improving digital government.

CONCLUSION

By bringing down impediments to trade and communications and lowering entry barriers, ICTs seem to be introducing a powerful force towards convergence of global wages. At the same time, ICT strategies should focus on human development by establishing information driven policies in a knowledge-driven society. Although the potential benefits from advances in ICTs appear to be clear, how they will be distributed is not. Well-founded fears exist that the poor are being left behind by the information revolution. Access to ICTs requires education, infrastructure, and institutions, three basic resources that many developing countries lack. Without them, it is increasingly likely that the poor may be on the losing side of this revolution. This book will find its standpoint for its readers among multi-sectoral research bases. Apart from academics, researchers, and think tanks, the book will be extremely handy to the international financial institutions, national governments in LDCs, non-governmental organizations, donor agencies, development partners, and private sector entrepreneurs operating in ICT4D arena.

REFERENCES

Civi, E. (2000). Knowledge management as a competitive asset: a review. *Journal of Marketing intelligence and Planning*, *18*(4), 166-174.

European Parliament (2001, March). *Developing countries and the ICT revolution: Final study, Working document for the STOA Panel.* Directorate General for Research, European Parliament, Luxembourg.

Government of Namibia (2005). *Education and Training Sector Improvement Programme: Palling for a learning nation.* Programme Document: Phase I (2006-2011), Working Draft Edition of October 2005, Government of Namibia.

Government of Kenya (2006). *National Information and Communications Technology Policy.* Ministry of Information and Communications, Government of Kenya, January 2006.

IMF (2001). *World Economic Outlook: The Information Technology Revolution, A Survey by the Staff of the International Monetary Fund.* International Monetary Fund, October 2001.

Njuguna, E. (2006). ICT Policy in Developing Countries: Understanding the Bottlenecks, a *Proceedings of PTC'06.*

Rodriguez, F., & Wilson, E.J. III (2000). *Are poor countries losing the information revolution?* University of Maryland, College Park.

Sayo, P., Chacko, J.G., & Pradhan, G. (2004). *ICT Policies and e-Strategies in the Asia-Pacific: A critical assessment of the way forward.* United Nations Development Programme-Asia Pacific Development Information Programme (UNDP-APDIP), Elsevier, New Delhi, India.

UNESCO (2005). *From the information society to knowledge societies.* UNESCO World Report, Paris: UNESCO Publishing.

United Nations (2003). *Regional road map towards an information society in Asia and the Pacific.* Economic and Social Commission for Asia and the Pacific. United Nations, New York.

United Nations (2005). *Measuring ICT: The global status of ICT indicators. Partnership on measuring ICT for development.* United Nations ICT Task Force, New York.

Acknowledgment

The editor would like to acknowledge the assistance from all involved in the entire accumulation and thorough review process of the book, without whose support the project could not have been amply completed. Reviewers who provided the most comprehensive, significant and productive comments include: Arla Juntunen of Helsinki School of Economics, Suleman Aziz Lodhi of National College of Business Administration and Economics, and Sheryl Thompson of Lancaster University Management School. Thanks to my close friends at UNDP and colleagues at SDNP for their encouragements during the whole process.

Special thanks also go to the publishing team at IGI Global. Particularly to Kristin Roth, Ross Miller, and Meg Stocking for their continuous suggestions and supports via e-mail for keeping the project on schedule, and to Mehdi Khosrow-Pour and Jan Travers for their enduring professional supports. Finally, I would like to thank my wife and daughter for their love and support throughout this period.

Hakikur Rahman, PhD
SDNP, Bangladesh
May 2007

Section I
Policy Initiation

Chapter I
An Overview on Strategic ICT Implementations Toward Developing Knowledge Societies

Hakikur Rahman
Sustainable Development Networking Progamme (SDNP), Bangladesh

ABSTRACT

As nations continue to conceptualize, adopt, implement, and monitor their respective information and communications technologies (ICTs) policies and e-strategies, it is vital that a critical assessment is undertaken on their effective progress. This includes analysis of strategies, methodologies, and best practices while weighing the impact, effectiveness, and efficiency of these policies and strategies. However, the demands, motivations, and incentives of various governments for adopting, adapting, and initiating ICTs varies from country to country, with diverse economic, social, cultural, and political environments. Researches in this aspect recognized the necessity of a cohesive ICT policy and strategic framework in each country for socio-economic development at their grass roots. The execution of these policies and strategies should include concrete national commitment and strong political will at the highest levels of government, and an enabling environment that promotes stakeholder involvement in setting agendas and implementing plans and programs. This chapter provides insights into various national implementations in promoting related activities, tries to establish an analytical approach that would assist in formulating ICT policies and strategies by identifying different ICT indicators. Furthermore, this chapter focuses on critical aspects of different strategic national level policies with short-medium-long term visions that targeted both the immediate needs of the populace and long-term needs of nations by integrating ICTs. Finally, this chapter has recommended that via coherent and complimentary policies that engage both the private sector and civil society organizations, nations can move forward towards creating a knowledge society and at the same time by leveraging capabilities of ICT can address the social, economic and political issues on the ground.

INTRODUCTION

During 1970 to 2000, the largest users of foreign direct investment (FDI) flows were generally in Asia (China alone had 33.2%), while portfolio flows were more concentrated in Latin America (Brazil, Mexico, and Argentina added up to 26.9%, Singapore and Malaysia obtained 10%, and rest of the developing world had 29.9%). An important point can be noted that, despite low FDI, Singapore and Malaysia have successes in economic development due to the implementation of successful ICT strategies in their countries.

In essence in many countries, ICTs provide the means to accelerate their progress or even leapfrog into the new phase of development and to enable their integration into the global economy. Particularly, in developed countries the knowledge-based economy allows further specialization, improvements in productivity, and the achievement of sustainable growth. By virtue knowledge capital is the only asset that can grow without limits; and new knowledge increases the efficient use of resources that are in finite supply (Léautier, 2005). Figure 1 gives a global view on how knowledge-based economy has been adopted in many countries, since 1995. It also shows that a few countries have been achieved almost highest ranking in terms of knowledge economy index (KEI) in recent years[1].

With an estimated span of about 50 years of Information Age (1950-2010)[2], the world has been passing through revolution of changes and witnessing astounding development in ICTs. Today, the Second World[3] has about 30% of the world's population and about 20% of its wealth. The Third World[4] accounts for some 60% of the population but enjoys less than 10% of its wealth. These classifications relate primarily to economic indicators such as GDP, but they can also be equated to the differing rates of adoption of industrial technologies. In this context, information technology presents an opportunity to the Second and Third World countries to improve their economies. However, the First World[5] countries were among the leading developers and early adopters of Information Age technologies. Therefore, Second and Third World countries

Figure 1. Global view: Knowledge economy index by countries and regions (Adapted from Dahlman, Routti, & Ylä-Anttila, 2005)

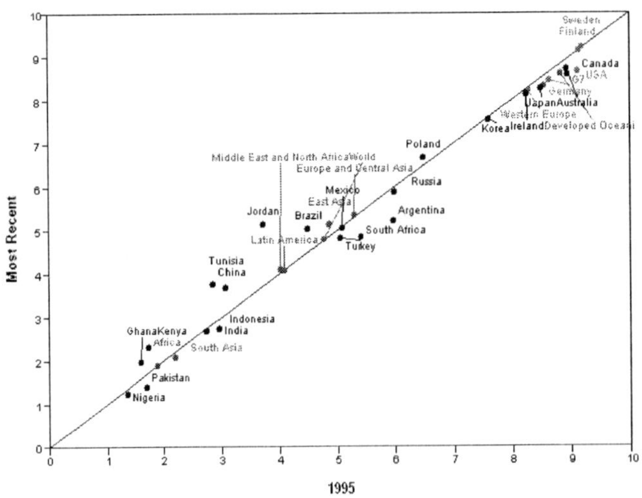

have an opportunity by embracing ICT to equal these information economies (UN, 2003).

This chapter has analytically observed some of the ICT indicators in a few countries of the world, synthesized their approaches of overall development, tried to find out any synergy among them that can be replicated elsewhere, and endeavored on critical assessment of potential ICT parameters that are being used to represent a nation's digital presence, including status of their economic development. It has been observed that a few countries in Europe and Asia have reached an acceptable level of development through implementation of strategic ICT policies. Keeping this in mind, brief review on transformation of digital governance in Finland and Ireland have been incorporated, while Turkey has been discussed due to its transitional nature in the European Union that may be treated as a success case among many nations. In Asia, Japan, Singapore, and Thailand are leading in many parameters of digital access index (DAI)[6], and at the same time, a few countries of Africa have been included in the synthesis with specific focus to Kenya.

BACKGROUND

Technology planning is the process of strategically deciding what technologically should be implemented to best poise a nation's vision, mission, strategy, activities, outputs, and outcome. It involves a candid assessment of the current scenario and an analysis of what can be improved upon with technology tools to enhance their effectiveness. It is also useful in identifying innovative program delivery opportunities that could be available through technology. Where most nations employ technology in a reactive fashion, responding to current short-term measures, strategic technology planning encourages a more long-term focus, making a realistic assessment of both the economic value and the cost of implementing technology. Although technology planning is a process, at the same time it also results in a document-the Strategic Technology Plan, which serves as a snapshot of current technology adoption and a roadmap for integrating technology more effectively (Uganda-Africa, 2005).

Strategy is to be defined as a plan to reach the goals in terms of systematic actions to be taken by different actors at all levels. In addition, an action plan goes beyond by clarifying measures to be taken—what should be accomplished, by whom, and when?[7] Currently, all strategic planning on ICTs emphasizes on building knowledge societies apart from establishing networks, contents, and learning. Knowledge is fueling economic growth and social development in all regions of the world. Along the way, the forces of globalization such as migration, travel, trade, foreign investment, and communications are speeding the dissemination and utilization of information across boundaries. Novel ideas and innovation are spreading faster than ever. Furthermore, knowledge-based growth and development offer opportunities for both developed and developing economies, and advancement in ICTs has been a necessary precondition for these new developments in the global economy (Léautier, 2005).

The world has seen remarkable achievement in terms of Internet penetration and broadband penetration as many countries have adopted successful ICT strategies in their policies and economies. Table 1 shows the top five countries of the world with highest Internet penetration[8] in 2005, Table 2 shows the top five countries with highest broadband penetration in 2005, Table 3 shows the top five countries with highest technology achievement index (TAI) in 2001[9] (UNDP, 2001), and Table 4 shows the top five countries with highest ICT diffusion index (ICTDI) in 2004.

From Table 1 it can be noted that top 30 countries (with 13.74% of world population) have 66.2% penetration in terms of their percentage of population, while the rest of the world has a penetration rate of 8.8%. At the same time from Tables 1, 2, 3, and 4 it has been found that Hong

Table 1. Top 5 countries with the highest Internet penetration rate

#	Country or Region	Penetration (% Population)	Internet Users Latest Data	Population (2006 Est.)	Source and Date of Latest Data
1	New Zealand	76.3 %	3,200,000	4,195,729	ITU - Sept/05
2	Iceland	75.9 %	225,600	297,072	ITU - Sept/05
3	Sweden	74.9 %	6,800,000	9,076,757	ITU - Oct/05
4	Denmark	69.4 %	3,762,500	5,425,373	ITU - Sept/05
5	Hong Kong	69.2 %	4,878,713	7,054,867	Nielsen//NR Feb./05
TOP 30 in Penetration		66.2 %	519,461,313	785,179,437	IWS - Mar/06
Rest of the World		8.8 %	503,401,994	5,714,517,623	IWS - Mar/06
World Total Users		15.7 %	1,022,863,307	6,499,697,060	IWS - Mar/06

Table 2. Top 5 economies by broadband penetration (Source: ITU statistics)

Year	Country	Per 100 inhabitants	Country	Per 100 inhabitants	Country	Per 100 inhabitants	Country	Per 100 inhabitants	Country	Per 100 inhabitants
2002	Korea (Rep.)	21.3	Hong Kong, China	14.6	Canada	11.5	Taiwan, China	9.4	Iceland	8.6
2003	Korea (Rep.)	23.3	Hong Kong, China	18.0	Canada	14.8	Iceland	14.5	Taiwan, China	13.4
2004	Korea (Rep.)	24.8	Hong Kong, China	22.0	Netherlands	19.8	Denmark	19.1	Iceland	18.8
2005	Iceland	25.0	Korea (Rep.)	23.6	Netherlands	23.5	Denmark	23.4	Liechtenstein	23.3

Table 3. Top 5 countries with technology achievement index (TAI) in 2001 (Source: UNDP, 2004)

Country	TAI Value
Finland	0.744
USA	0.733
Sweden	0.703
Japan	0.698
Korea (R.)	0.666

Table 4. Top 5 countries with ICT diffusion index (ICTDI) in 2004 (Source: UN, 2006)

Rank	Country	Access Index	Connectivity Index	ICT Diffusion Index[10]
1	Luxembourg	0.928	0.703	0.815
2	USA	0.833	0.754	0.794
3	Iceland	0.854	0.706	0.780
4	Sweden	0.836	0.700	0.768
5	Denmark	0.828	0.667	0.748

Kong, Korea (R.), Iceland, Sweden, and USA have accomplished as such in developing their ICT sector to remain at the top in terms of Internet penetration, broadband penetration, TAI, or ICTDI.

In terms of Internet penetration, the European Union is leading the world with 49.8% of penetration rate, comprising 230,396,996 users with a growth rate of 147% between 2000 and 2005. Though growth rate in Africa rises to 424% between 2000 and 2005, the Internet penetration rate is only 2.6% of its population. Similarly, Asia has shown a growth rate of 219% with 9.9% penetration and America has a growth rate of 143%, but Internet penetration rate is much higher at 34.7%. Tables 5, 6, 7, and 8 show other parameters in Europe, Africa, Asia, and America.

To show the trend on settling down of Internet penetration in the developed world, a survey conducted by the Oxford Internet Institution (OII) can be highlighted. OII has been found that Internet penetration is growing slowly in UK with 3% between 2003 and 2005. However, Internet users are skeptical about the Internet's ability to help develop social relations. Sixty-five percent disagree ('disagree strongly' and. 'disagree') that it is easier to meet people online than in person, while 73% agree ('agree' and 'agree strongly') that the Internet is a toll of benefit in a time-starved world that saves time; and 63% agree ('agree and 'agree strongly') that technology is making things better for people (OII, 2005).

Subsequently, case of a small country, Macedonia is being highlighted that is thriving for ICT development. ICT Assessment Report of Municipalities in FYR Macedonia 2004 and Recommendations and E-model of Municipalities summarizes that the 6.75% of budget spent on ICT and its development is graded very low, with 1.73% of the total budget being spent on IT infrastructure and 5.02% on communications. It follows that municipalities do not assign sufficient importance to ICT usage and do not recognize the work-related benefits of the new technologies. The existing municipality profile is without a strategy for development of ICT usage. Nearly all municipality decision-making persons know little of ICT and no ICT initiatives are implemented, with the exception of a few sponsored projects and donations. The level of computer literacy of employees and members of the municipality council is graded very low, with a quality measure of 0.75 on a scale from 0 to 4, where value 1 signifies basic IT skills and value 2 moderate IT skills, which scores very low in comparison to the surprisingly high educational level (UNDP, 2004).

More importantly, in recent years many researchers, including world reputed agencies have been showing particular interest in ICTs. According to the Committee on Science and Technology at the Service of Development, ICTs "will become crucially important for sustainable

Table 5. Internet usage in Europe (Source: http://www.internetworldstats.com/stats4.htm)

EUROPE	Population (2006 Est.)	% Pop. of World	Internet Users, Latest Data	Penetration (% Population)	% Usage of World	Use Growth (2000-2005)
European Union	462,371,237	7.1 %	230,396,996	49.8 %	22.5 %	147.3 %
EU Candidate Countries	110,206,019	1.7 %	19,055,671	17.3 %	1.9 %	450.7 %
Rest of Europe	234,711,764	3.6 %	42,148,231	18.0 %	4.1 %	397.5 %
TOTAL EUROPE	807,289,020	12.4 %	291,600,898	36.1 %	28.5 %	177.5 %
Rest of World	5,692,408,040	87.6 %	730,462,384	12.8 %	71.5 %	185.5 %
TOTAL WORLD	6,499,697,060	100.0 %	1,022,063,282	15.7 %	100.0 %	183.1 %

Table 6. Internet users and population statistics for Africa (Source: http://www.internetworldstats.com/stats1.htm)

AFRICA REGION	Population (2006 Est.)	Pop. % in World	Internet Users, Latest Data	Penetration (% Population)	% Users in World	Use Growth (2000-2005)
Total for Africa	915,210,928	14.1 %	23,649,000	2.6 %	2.3 %	423.9 %
Rest of World	5,584,486,132	85.9 %	999,214,307	17.9 %	97.7 %	180.3 %
WORLD TOTAL	6,499,697,060	100.0 %	1,022,863,307	15.7 %	100.0 %	183.4 %

Table 7. Internet users and population statistics for Asia (Source: http://www.internetworldstats.com/stats3.htm)

ASIA REGION	Population (2006 Est.)	% Pop. of World	Internet Users, Latest Data	Penetration (% Population)	% Usage of World	Use Growth (2000-2005)
Asia Only	3,667,774,066	56.4 %	364,270,713	9.9 %	35.8 %	218.7 %
Rest of the World	2,831,922,994	43.6 %	653,786,676	23.1 %	64.2 %	165.0 %
WORLD TOTAL	6,499,697,060	100.0 %	1,018,057,389	15.7 %	100.0 %	182.0 %

Table 8. Internet users and population statistics for the Americas (Source: http://www.internetworldstats.com/stats2.htm)

AMERICA	Population (2006 Est.)	% Pop. America	Internet Users, Latest Data	% Population (Penetration)	% Usage America	Use Growth (2000-2005)
Central America	144,550,714	16.3 %	20,021,900	13.9 %	6.5 %	522.3 %
South America	370,118,282	41.8 %	55,629,500	15.0 %	18.1 %	289.2 %
The Caribbean	39,239,636	4.4 %	4,298,409	11.0 %	1.4 %	670.7 %
SUB-TOTAL	553,908,632	62.6 %	79,962,809	14.4 %	26.0 %	342.5 %
North America	331,473,276	37.4 %	227,303,680	68.6 %	74.0 %	110.3 %
TOTAL AMERICAS	885,381,908	100.0 %	307,266,489	34.7 %	100.0 %	143.5 %

development in developing countries." (Credé & Mansell 1998: ix) For the past two decades, most developed countries have witnessed significant changes that can be traced to ICTs. These multidimensional changes (technical, financial and economic, cultural, social, and political) have been observed in almost all facets of life: economics; education; health; communications; leisure; and travel (Thioune, 2003). The World Summit on Sustainable Development in 2002 and subsequent sessions of the Commission of Sustainable Development (CSD) encouraged further work on indicators for sustainable development by countries in line with national conditions and priorities and invited the international community to support these efforts (UN, 2005). Therefore, a number of areas, including sustainable tourism, marine environment, national strategies for sustainable development, and good governance, require further work to identify appropriate indicators. In the next section, various ICT indicators have been critically analyzed and appropriate measures at national levels have been closely followed, that have created positive environment in formulating successful ICT strategies.

Furthermore, the changes observed in these countries have led to what is now referred to as "the knowledge society." ICTs have made it possible to find fast access to, and distribution of, information as well as new ways of communication in real time at a cheaper cost. However, still a considerable gap exists between developing countries and developed ones in terms of the contribution of ICTs to the creation of wealth. The gap has tended to widen among developed countries, the technology suppliers, and the receiving developing countries. At the same time, the gap between the privileged and the grassroots communities within these developing countries is also expanding in terms of their access to ICTs. If measures are not taken to make ICTs both affordable and usable, mere access to them will be insignificant (Thioune, 2003).

Finally, technologies can play a crucial role in determining how networks can be used, whether they will be centralized or decentralized, who can own them, who can set them up, and whether they can be adapted to specific needs of individuals and communities. Certain technological developments and choices can decentralize networks and offer local communities a role in shaping the services and applications that are achievable. Conversely, any restricted choices and configurations may leave entire communities completely outside of the networks (Siochru & Girard, 2005).

MAIN THRUST: ANALYSIS OF ICT INDICATORS

This section has tried to identify some leading countries across the globe that have developed in terms of ICTs indicators, successful implementations of their ICT strategies in achieving human development, and have transformed into knowledge societies. Firstly, a few of ICT indicators have been discussed with specific focus on their developments, and consequently some of the parameters have been critically analyzed to build some form of inference. Secondly, a few of ICT parameters have been highlighted to observe their follow up in different economies, and finally ICT strategies in a few countries have been closely monitored, focusing their regional contribution.

For this purpose, one of the ICT indicators, the penetration of main telephone lines in top five countries during 2002 to 2005 have been highlighted in Tables 9 to 12. Penetration of main telephone lines per 100 inhabitants has been considered and it has been found that Bermuda remains at the top. This gives the indication of Bermuda's growth of basic information infrastructure and dependency on communication media.

In 2005, Vietnam had the highest CAGR[11] of 42.5% per 100 inhabitants from 2000-2005 (ITU, 2006). The lowest CAGR was –16.6 in D.R. Congo. In 2004 Somalia had the highest CAGR of 46.6% per 100 inhabitants from 1999-2004, and the lowest CAGR was –14.0 in D.R. Congo, while Vietnam maintained CAGR of 35.6. In 2003 Sudan had the highest CAGR of 38.5% per 100 inhabitants from 1998-2003, with the lowest CAGR of –24.0 in D.R. Congo, and Vietnam kept the CAGR at 19.2. In 2002 Sudan had the highest CAGR of 39.0% per 100 inhabitants from 1997-2002, while the lowest CAGR was –5.7 in Solomon Island, but Vietnam had a CAGR of 22.7.

Main Telephone Lines

Due to non-availability of continuous data, two small countries in Europe, Guernsey, and Jersey could not be put in the table, but Guernsey had over 90 main telephone lines per 100 inhabitants in 2002 and 2003, while Jersey had over 80 main telephone lines per 100 inhabitants in 2002 and 2003. At the same time, a country can be put into focus, which is Vietnam. That country has been consistently increasing its number of land telephone per 100 inhabitants throughout the period. This has been possible due to Vietnam's recent ICT reformations. Table 13 shows various stages of ICT strategies that have been taken in Vietnam.

Vietnam will boost its digital content industry through streamlining its legal environment and upgrading its broadband network by 2006. In this context, a master plan on digital content industry drafted by MPT will be submitted to the government later in 2006. The country hopes that this industry will generate USD400 million by 2010 with an annual growth rate of 50%. Meanwhile, Vietnam will provide skills development training to around 300,000 digital content industry experts. Digital content industry will include e-learning, games for computer, online game, interactive game, game for mobile handset, Internet e-content (online newspaper, Web sites, searching, library), and value-added services for mobile (logo, ring tones, digital movies, television, and cartoon). Furthermore, Vietnam's ICT development strategy aims to reach annual ICT revenue of USD6-7 billion by 2010 including USD3 billion in the computer hardware industry, USD2 billion in electronics production, and USD1.2 billion in the software/outsourcing[12].

Next, compound annual growth rate of cellular phones have been considered in Tables 14 to 17 and found that Hong Kong, Italy, Luxembourg, and Sweden remain at the top, though Lithuania

Table 9. Penetration of main telephone lines in order of per 100 inhabitants – 2005

Country	CAGR (%) 2000-2005	Per 100 inhabitants	CAGR (%) per 100 inhabitants
Bermuda	-1.4	86.15	-0.5
Luxembourg	2.8	79.75	1.8
Sweden	-1.1	71.54	-1.4
Switzerland	-0.4	68.66	-1.1
Germany	1.9	66.57	1.7

Table 10. Penetration of main telephone lines in order of per 100 inhabitants – 2004

Country	CAGR (%) 1999-2004	Per 100 inhabitants	CAGR (%) per 100 inhabitants
Bermuda	0.6	86.15	0.2
Luxembourg	3.7	79.75	2.6
Sweden	-0.2	71.54	0.1
Switzerland	0.8	70.97	0.1
Germany	2.5	66.15	2.4

Table 11. Penetration of main telephone lines in order of per 100 inhabitants – 2003

Country	CAGR (%) 1998-2003	Per 100 inhabitants	CAGR (%) per 100 inhabitants
Bermuda	1.0	86.15 (2002)	0.6
Luxembourg	-5.8	79.75	-4.5
Sweden	0.5	72.89	0.2
Switzerland	1.7	72.74	1.2
Germany	3.1	65.71	3.0

Table 12. Penetration of main telephone lines in order of per 100 inhabitants – 2002 (Source: ITU statistics)

Country	CAGR (%) 1997-2002	Per 100 inhabitants	CAGR (%) per 100 inhabitants
Bermuda	1.6	86.15	1.2
Luxembourg	4.9	79.68	3.7
Switzerland	2.8	73.99	2.3
Sweden	1.0	73.57	0.8
Denmark	2.1	68.86	1.7

Table 13. Actions taken by the Vietnam's Government for the improvement of ICT (Source: Thanh, 2003; Soan, 2006; ThuHue, 2005; MPT, 2003; World Bank, 2005)

Action Taken	Year	Remarks
National Program on IT (NPIT) initiated	1996-2000	Implementing major application projects in state management and in some socio-economic areas
Directive No. 58-CT/TW	2000	Articulated the national strategy plan and policies for ICT development, and pushed IT towards the forefront of the country's industrialization drive for the period 2000-2010
Decision No. 81/2001/QD-TTg	May 2001	Approval of the Action Program to implement Directive No. 58-CT/TW
Internet Policies initiated	2001	Amended in 2003-2004
	February 2002	Ordinance on Posts and Telecommunications adopted
Decision No. 95/2002/QD-TTg	July 2002	Approval of the IT Application and Development Master Plan in Vietnam by 2005
Decree No. 90/2002/ND-CP	November 2002	A new Ministry of Posts and Telematics (MPT) created to develop ICT Strategy along with other ministries
ICT Strategy initiated	January 2003	The Vietnam ICT development strategy to 2010 and forward to 2020 approved
ICT Strategy approved	2005	
e-transaction law drafted	2005	
ICT Law	2006	Drafted by MPT and to be approved by end of 2006

had the highest CAGR of with 127.10 per 100 inhabitants and highest percent of total telephone subscriber in 2005. It has also been found that Nigeria had the highest CAGR of 261.8% between 2000 and 2005. Furthermore, D.R. Congo had the highest percent of total cell phone subscribers as indicated in Table 18. Previously it was observed that D.R. Congo had the lowest CAGR in main telephone lines (Tables 9 to 12) and this reflect their dependency on the cellular phone, in lack of any robust information infrastructure. It is also to be noted that Nigeria had the highest CAGR of 261.8% from 2000-2005 in terms of cellular phone subscribers.

Cellular Subscriber

Afterwards, Internet penetration has been considered and found that USA remains at the top consequently in 2002, 2003, and 2004 (see Tables 17 to 19), with the highest Internet hosts per 10,000 inhabitants. But, Iceland and Korea (R.) took the lead in terms of Internet users per 100 inhabitants. This supports towards their efforts in implementing successful ICT strategies in those countries for creating enabling environments.

Table 14. Compound annual growth rate of cellular phone 2002-2005 in order of per 100 inhabitants

Country	CAGR (%) 2000-2005	Per 100 inhabitants	As % of total telephone subscribers
Lithuania	52.7	127.10	84.5
Italy	11.1	123.14	74.1
Hong Kong, China	9.7	122.65	69.5
Macao, China	30.4	115.82	75.3
Czech Republic	22.1	115.22	78.5

Table 15. Compound annual growth rate of cellular phone 1999-2004 in order of per 100 inhabitants

Country	CAGR (%) 1999-2004	Per 100 inhabitants	As % of total telephone subscribers
Luxembourg	24.8	138.17	63.8
Hong Kong, China	14.0	118.77	68.6
Sweden	13.8	108.47	60.3
Italy	15.7	108.19	70.7
Czech Republic	40.9	105.64	75.9

Table 16. Compound annual growth rate of cellular phone 1998-2003 in order of per 100 inhabitants

Country	CAGR (%) 1998-2003	Per 100 inhabitants	As % of total telephone subscribers
Luxembourg	32.8	119.38	60.0
Taiwan, China	40.4	114.14	65.9
Hong Kong, China	18.3	107.92	65.9
Italy	22.6	98.07	68.1
Sweden	16.5	98.05	57.4

Table 17. Compound annual growth rate of cellular phone 1997-2002 in order of per 100 inhabitants

Country	CAGR (%) 1997-2002	Per 100 inhabitants	As % of total telephone subscribers
Taiwan, China	74.9	108.30	65.1
Luxembourg	47.7	106.05	57.1
Italy	35.8	95.99	66.6
Israel	30.5	95.45	67.8
Hong Kong, China	23.5	94.25	62.5

Table 18. Highest and lowest growth rate of total cell phone subscribers 2002-2005 (Source: ITU statistics www.itu.int/ITU-D/ict/statistics.htm

2002		2003		2004		2005	
Country	**%**	**Country**	**%**	**Country**	**%**	**Country**	**%**
D.R. Congo	98.2	D.R. Congo	99.0	D.R. Congo	99.5	D.R. Congo	97.3
Turkmenistan	2.1	Turkmenistan	2.4	Turkmenistan	11.8	Turkmenistan	11.8

Information Technology

Lastly, before performing a critical analysis of an ICT indicator, basic indicators of ICT during 2002 to 2005 have been portrayed in Tables 22 to 25. It has been found that, despite high population density of over 6,500 per Km^2 Hong Kong has maintained to keep higher number of total telephone subscriber per 100 inhabitants, with over 23,000 USD per capita. This reflects the national commitment in implementing successful ICT strategies for their economy.

Table 19. Internet hosts per 10,000 inhabitants of top 5 countries in 2004

Country	Internet hosts per 10,000 inhabitants	Internet users per 100 inhabitants	PCs per 100 inhabitants	Country	Internet users per 100 inhabitants
USA	6645.16	63.00	76.22	Iceland	77.00
Iceland	4758.60	77.00	48.22	Greenland	66.32
Netherlands	3334.42	61.63	68.47	Korea (R.)	65.68
Denmark	2681.94	60.41	65.48	Australia	65.28
Finland	2215.16	63.00	48.22	Guernsey	64.52

Table 20. Internet hosts per 10,000 inhabitants of top 5 countries in 2003

Country	Internet hosts per 10,000 inhabitants	Internet users per 100 inhabitants	PCs per 100 inhabitants	Country	Internet users per 100 inhabitants
USA	5577.84	55.58	62.44	Iceland	67.47
Iceland	3789.65	67.47	46.37	Sweden	63.00
Finland	2436.55	49.05	46.08	Korea (R.)	61.07
Denmark	2312.67	56.21	61.40	Guernsey	59.14
Netherlands	2162.66	52.19	50.83	Faroe Island	58.74

Table 21. Internet hosts per 10,000 inhabitants of top 5 countries in 2002 (Source: ITU statistics www.itu.int/ITU-D/ict/statistics.htm)

Country	Internet hosts per 10,000 inhabitants	Internet users per 100 inhabitants	PCs per 100 inhabitants	Country	Internet users per 100 inhabitants
USA	4004.25	50.10	62.44	Iceland	64.79
Iceland	2370.17	64.79	45.14	Sweden	57.31
Finland	2343.12	48.57	44.17	Korea (R.)	55.17
Tonga	1964.21	2.92	2.02	Guernsey	53.57
Netherlands	1937.14	50.63	46.66	Faroe Island	53.23

Basic Indicators

All these indicators are providing with facts that countries/communities will emerge as successful if and only if, successful ICT strategies can be implemented properly through longer-term strategies. Else, the ultimate result will not be visible in terms of their outcomes. Table 26 shows a critical analysis on one of ICT indicators (number of Internet users per 100 populations):

Table 22. Basic ICT indicators in 2005 in order of total telephone subscribers per 100 inhabitants

Country	Total telephone subscribers per 100 inhabitants	Per capita USD (2004)	Population density (per Km^2)
Luxembourg	199.13	69,027	180
Sweden	180.02	38,850	20
Hong Kong, China	176.54	23,960	6,630
Iceland	169.34	41,765	3
Denmark	162.39	45,059	126

Table 23. Basic ICT indicators in 2004 in order of total telephone subscribers per 100 inhabitants

Country	Total telephone subscribers per 100 inhabitants	Per capita USD (2003)	Population density (per Km^2)
Luxembourg	199.13	59,626	177
Sweden	180.02	36,738	20
Guernsey	177.31	32,428	858
Hong Kong, China	173.19	23,245	6,512
Iceland	164.01	35,973	3

Table 24. Basic ICT indicators in 2003 in order of total telephone subscribers per 100 inhabitants

Country	Total telephone subscribers per 100 inhabitants	Per capita USD (2002)	Population density (per Km^2)
Luxembourg	199.13	47,255	175
Guernsey	173.79	32,428	858
Taiwan, China	173.22	12,453	628
Sweden	170.94	26,864	20
Hong Kong, China	163.81	23,566	6,413

Table 25. Basic ICT indicators in 2002 in order of total telephone subscribers per 100 inhabitant (Source: ITU statistics www.itu.int/ITU-D/ict/statistics.htm)

Country	Total telephone subscribers per 100 inhabitants	Per capita USD (2001)	Population density (per Km^2)
Luxembourg	185.74	44,857	172
Taiwan, China	166.47	12,553	626
Guernsey	164.82	32,428	862
Sweden	162.45	24,626	20
Iceland	155.71	26,687	57

In this context, to make pragmatic steps, indicators on national sustainable development strategy and implementation of ratified global agreements have been modified to establish good global governance. At the same time, indicator #47[14] and 48[15] of MDGs have been made less dense by extending to 1000 inhabitants, instead of 100 (Table 27). They seem realistic in sense, that many countries will not be able to achieve MDGs by 2015, unless they keep forcing on improving indicators on knowledge building and increase expenditures on research and development as a percent of their GDP.

Before proceeding further, ICT strategies of a few countries in regional aspect have been highlighted. They are Finland, Ireland, and Turkey from European Union; Japan, Singapore, and Thailand from Asia Pacific and Kenya from Africa.

European Union

An international cooperation dimension integrated with the activities of each of the thematic priority fields of these specific programmes, "Integrating and strengthening the European Research Area," "Specific activities covering a wider field of research," and "Structuring the European Research Area" allowed the European research community to benefit from the knowledge and expertise of third countries and institutions through the participation of researchers, teams, and institutions from those countries. The general objective under the 6th Framework programme is to help open up the European research area to the world focusing mutual benefits of the community. The overall objective is, therefore, for the European Union to develop strong scientific partnerships with developing countries in order to contribute to their sustainable development (EC, 2004), by means of human capital development, mobility, and institution-building (EC, 2005, 2006). It is thus important that the ownership of the knowledge and technology developed in partnership under this activity be equitably shared. The results will be used to contribute to improved regional cooperation strategies, the elaborations of concepts aimed at sustainable development, and societal innovation.

It has been observed that, over a quarter of EU GDP growth and around 50% of productivity growth is due to improvement in ICT. In recent years ICT services, skills development, media and content services, and their usage in other economic sectors, are growing fast within the economy and society. However, many Europeans are yet to reap a few benefits from ICT. As such, 57% of individuals living in the EU did not regularly use the Internet in 2005; only 10% of persons over 65 used Internet against 68% of those with age

Table 26. A critical analysis on one of ICT indicators (Adapted from UN, 2005)

Parameter: Number of Internet users per 100 populations

Purpose: The number of Internet users is a measure of Internet access and its use

Relevance to building knowledge society: As an information distribution system, the Internet and its usage provide opportunities for bringing education and information within the reach of all. It can significantly shorten time lags as well as opening up a new range of information resources and creating a knowledge society by enabling them to take informed decisions.

Relevance to Sustainable/Unsustainable Development: Telecommunications and social, economic, and institutional development are closely linked. Modern communications is considered to be relatively favorable to the development processes. There is unlikely to be sustainable development without a well-developed communications infrastructure. Communication is critical to support sustainable development.

International conventions and agreements: The four-year strategic Istanbul Action Plan adopted by the ITU World Telecommunication Development Conference in 2002 provides a six-point action plan that address the key elements needed to bridge the digital divide. Similarly G8 Okinawa charter on Global Information Society in 2000; two phases of World Summit on the Information Society (WSIS), first phase in Geneva from December 10-12, 2003 and second phase in Tunis from November 16-18, 2005 endorsed by the UN General Assembly Resolution 56/183 [13] (December 21, 2001); and WSSD in Johannesburg August 26-September 4, 2002, including many follow up meetings are providing dynamic guidance to the global communities.

International targets/ Recommended standards: The benefits of new technologies, especially ICTs should be available to all, UN Millennium Declaration #55/2.

Linkages to Other Indicators: There are also other variables (e.g., Internet hosts and subscribers), which provide a measure of how many people are accessing the Internet. This indicator is also related to other telecommunication indicators (e.g., main telephone lines, mobile cellular subscribers), as well as income and education indicators.

Table 27. Modified MDG indicators (Adapted from UN, 2005)

CSD[16] Indicators	MDG Indicators	New Indicators
National Sustainable Development Strategy		
Implementation of Ratified Global Agreements		Good governance indicators
Number of Internet Subscribers per 1000 inhabitants	48. Personal computers in use per 100 population and Internet users per 100 population	
Main Telephone Lies per 1000 inhabitants	47. Telephone lines and cellular subscribers per 100 population	
Expenditure on R&D as a percent of GDP		

between 16 and 24; only 24% of persons with low education used the Internet against 73% of those with high education; and only 32% of unemployed persons used the Internet against 54% of those with employment (ICT Riga, 2006).

Finland

Finland is a unique example of how a country can rapidly transform itself from a resource-based economy into one based on innovative and intellectual technologies. In Finland, the specialization of production, trade, and R&D in more knowledge-intensive products and services coincided with the gradual opening of the economy and deregulation of capital flows. A peculiarity of the Finnish case is the atypical pattern of industrial renewal from essentially natural-resource-based industries toward machinery, engineering, electronics, and ICT. However, there are a few other examples of natural-resource-abundant countries that have managed to transform their industrial structures toward higher knowledge intensity and value added so rapidly and successfully as Finland. The origins of the Finnish knowledge economy can be traced back to user-producer linkages between the forest-based industries as early users of high technology, and the emerging engineering, electronics, and ICT industries in the 1960s and 1970s (see Figure 2).

The Finnish experience in the 1990s is an example of how knowledge becomes the driving force in economic transformation and growth. Although as recently as in the 1970s Finland was relying mainly on resource-intensive industries, it has transformed to be the most ICT specialized economy in the world. In this aspect, rapid structural transformation beginning in the mid nineties coincided with equally rapid improvements in macro balances, and by the end of the decade the country's macroeconomic implementation was one of the strongest in Europe. Although Finland has many characteristics that cannot easily be replicated by other countries, much of its experience in designing knowledge-based economic and social strategies are highly relevant. However, a key lesson is the importance of flexibility in responding to change and the critical role of a responsive education system. Finland's experience also highlights the importance of developing a vision, a process for consensus building and accomplishment (Léautier, 2005).

Figure 2. Finland's stages of industrial and economic development (Adapted from Dahlman, Routti, & Ylä-Anttila, 2005)

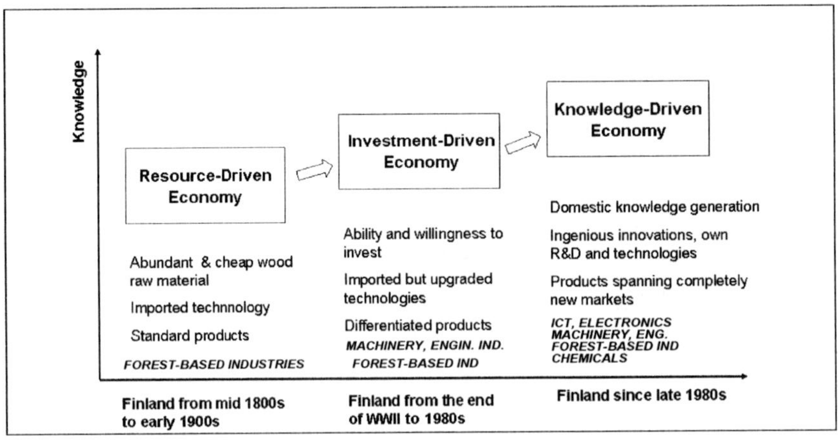

Ireland

ICTs lead a transition in the Irish economy to get away from low-end manufacturing to hardware manufacturer and software developer. Over the past decade, Ireland has become one of the strongest growing partners in the EU and ranked nineteenth in the Networked Readiness Index[17]. Heavy government spending on technical and tertiary institutions in the early 1980s gave the country to provide steady supply of highly skilled manpower and at the same time creation of tax and trade regulators gave it a better environment for ICT investors.

The Irish economy escalated during the 1990s, with GDP per capita growing by an annual average of 7.1% between 1990 and 2000, compared to an OECD average of 2.6%[18]. A few of the strategies are being illustrated in Table 28:

Turkey

Since the founding by Mustafa Ataturk in 1923, the government of the Republic of Turkey has been moving to integrate the country into Europe. To achieve this long term aim the country in recent years has been investing aggressively in networked readiness. Turkey ranks forty-first in overall Networked Readiness[23]. Digital access index (DAI) of Turkey in 2002 was 0.48, that falls under medium access (under a category of high access, upper access, medium access, and low access, respectively)[24]. Table 29 shows the broadband subscribers per 100 inhabitants, by technology in December 2005:

Asia Pacific

The world economy is in the midst of a profound transformation spurred by globalization and sustained by the rapid development of ICT that is accelerating the transmission of information and knowledge in every part of the globe, including the countries and regions in the Asia Pacific. ICT has a great potential to facilitate the achievement of the UN Millennium Development Goals, particularly to combat poverty, hunger, disease, illiteracy, environmental degradation, and gender inequality (Hak-Su, 2003).

Table 28. Various steps taken by the Irish government to promote ICT for knowledge development[19;20;21]

Country	Policies taken	Date of implementation	DAI Ranking in 2002
Ireland	Higher Education Authority Act	1971	0.69 (upper access)
	Strategy for Education Technology	October 1997	
	The Education Act	1998	
	Government Action Plan on the Information Society	January 1999	
	Policy for the Acceptable use of the Internet in Schools	1999	
	Classroom 2000	2000	
	Establishment of Information Society Commission	2000	
	Information Society Fund	2000	
	Establishment of Media Lab Europe	2000-2005	
	Inclusion of Research Technological Development and Innovation Priority (RTDI) in the National Development Plan 2000-2006	2000	
	Establishment of Technology Foresight Fund	February 2000	
	Establishment of Irish Research Council for Science, Engineering and Technology	2001	
	Establishment of Cabinet Committee on the Information Society	December 2001	
	New Connections- A Strategy to Realise the Potential of the Information Society	March 2002	
	Inclusion of Programme for research in Third Level Institutions (PRTLI)	March 2002	
	New Connections: the Second Government Action Plan	April 2002	
	Establishment of Cabinet Committee on the Information Society		
	The Empowering Schools Strategy	2003	
	Strategic Innovation Fund	2006	
	Institutes of technology Act	2006	
	New Government Action Plan for 2006-2008[22]	2006	

Table 29. Broadband subscribers per 100 inhabitants of Turkey (Source: www.oecd.org/ict/braodband accessed on August 16, 2006

DSL	Cable	Other	Total	Rank	Total subscribers
2.1	0.0	0.0	2.1	29	1,530,000

At the same time, the Asia Pacific region is also the most dynamic and fastest growing region of the world and it is expected to account for more than half of the world trade in the twenty-first century. Current trade estimates among the three continents is given in Figure 3, and it has been observed that it is highest between Asia and North America.

Particularly, the economic and technological levels in East Asian countries are not uniform, and the bases and environments for network economic development are also diverse. However, the growth tendency for the network economy in East Asia as a whole is very noticeable. For the average growth rate of Internet users, in 2000, it was 150% in the East Asia region, higher than 130% in Europe and 51% in North America. Similarly, network economy is becoming the new motive force for the development of Asian economy[25].

Asia Pacific homes more than 60% of the world's population, among them more than 75% are below the poverty line, despite being the most dynamic and fastest growing region of the world. However, due to its diverse nature of characteristics comprising the world's richest and the poorest nations, it represents more than half of the world trade. At the same time, it establishes a unique feature in this region to promote growth of ICT, as the region dwells orient cultures that have been centers of learning and have created huge intellectual property in several languages. GDP per capita grew from USD120 in the 1900s to USD1600+ in the 2000s (Japan has the highest GDP per capita of US $37890 in 2005)[26] despite the fact that only 11 economies in this region has over 25% of teledensity and Internet penetration rate ranges between 10 and 50 (UN, 2003; ITU, 2005)[27]. Currently, Japan has 21,304,292 (2005) Internet hosts and 86.3 million (2005) Internet users[28] that shows their ICT development, and at the same time, Japan's KDDI has launched an unlimited access mobile Internet tariff at a price equivalent to entry-level digital subscriber line (DSL) that gives their competency in establishing local knowledge repository[29]. Another country, with more than 95% of internal content, Korea (R.) is currently ranked highest in the world in terms of the development of information infrastructure[30]. Korea (R.) also enjoys the highest penetration of broadband Internet with over 70% of total households and all schools of that country have broadband Internet. Korea (R.) ranked fourth in DAI 2002 with high access of 0.82[31].

Figure 3. Amount of trade estimates in three continents (Adapted from UN, 2003)

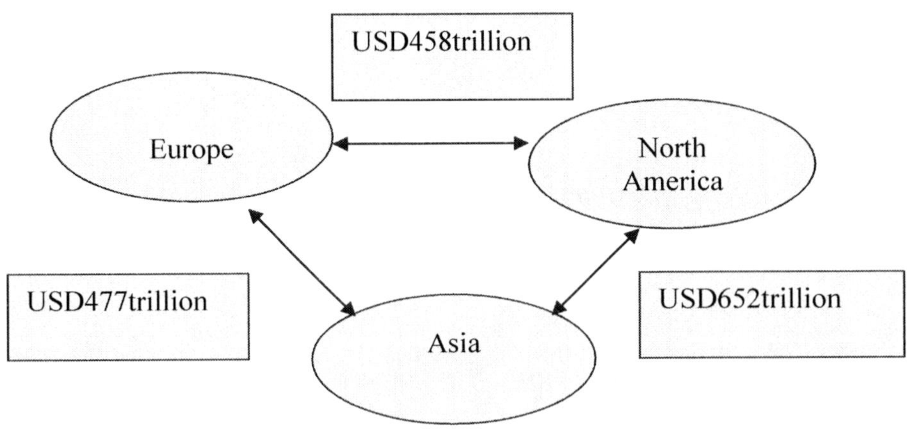

Table 30. Various steps taken by the Japan government to promote a knowledge society

Country	Policies taken	**Period of implementation**	**DAI Ranking in 2002**
Japan (MIC, 2004; 2005)	National ICT Policy	1994-1999	0.75 (high access)
	Formation of the Advanced Information and telecommunications Society Promotion Headquarters	2000-2001	
	Strategic Headquarters for the Promotion of an Advanced Information and telecommunications Network Society	2001-2002	
	e-Japan Strategy-I	2001-2003	
	e-Japan Priority Policy Program (e-Japan Strategy-II)	2003-2005	
	u-Japan Initiative	2004->	

Table 31. E-strategies of Singapore and Thailand

Country	**Policies taken**	**Period of implementation**	**DAI Ranking in 2002**
Singapore	National Computerization Plan	1980-1985	0.75 (high access)
	National IT Plan	1986-1991	
	IT 2000	1992-1999	
	InfoComm 21	1999-2003	
	Connected Singapore	2003-onwards	
Thailand	National Science and Technology Act	1991	0.48 (medium access)
	National IT Committee	1992	
	National IT Policy: IT 2000	1996-2000	
	e-Thailand initiative	2000	
	IT 2010 Policy	2001-2010	
	National ICT Master Plan	2002-2006	

Japan

Initially, at the first meeting of the IT Strategic Headquarters in January 2001 "e-Japan Strategy" was launched to establish the goal for Japan 'to become the world's most advanced IT nation within 5 years (by the end of 2005). To achieve this goal the public and private sectors together have been expending great efforts in implementing various measures, such as the development of communications infrastructure, promotion of e-government, and e-commerce. As a result, stunning achievements have been made in the area of ICT. For example, Internet service in Japan is the fastest and cheapest in the world, e-government instrument is expanding (online procedures operate 24 hours a day for 365 days and handle more than 100,000 applications annually) and its e-commerce market is growing in scale to become the second largest in the world after the United States (the United States ranked at the top with 9.0%, followed by Japan with 7.1%) (ECOM, 2003). At the 29th meeting of the IT Strategic Headquarters, the "IT Policy Package 2005: Towards the Realization of the World's Most Advanced IT Nation" was announced. It is desired that the "IT Policy Package 2005" will further enhance measures mainly in the areas closely related to the peoples' daily lives, including administrative service, medical care, and education. The IT Strategic Headquarters will work to realize a knowledge society in which people can experience the change and enjoy the benefits brought upon by ICT (Government of Japan, 2005).

If observed closely, one will comment that these have been achieved not just in a few days or months, but years of relentless efforts and endeavors to fulfill the mission of a nation. Table 30 shows a chronological overview of various steps taken by the Japan government along this prospect.

Table 31 provides a comparison on e-strategies of Singapore and emerging Thailand who may take a leading role in South East Asia within a few years.

Africa

Africa is confronted with numerous challenges. But, one of the most important of these challenges is to integrate the continent into the information society. Furthermore, reducing the digital divide and facilitating the continent's absorption into the global information society could overcome Africa's isolation. In an avant-garde approach, concerned authorities have been convinced that research through the production of learning and applicable knowledge could contribute significantly to a better development. For this purpose, the International Development Research Centre (IDRC) launched a program known as Acacia to contribute to the production of an essentially African body of knowledge on the role of ICTs in the economic and social development process. This focus on learning and the production and sharing of knowledge is reflected in the systematic documentation and evaluation of Acacia's experiences in sub-Saharan Africa (Olukoshi & O'Neil, 2003).

For the past few decades, the international community has noted a growing digital gap between developed countries and Africa, on the one hand, and within African countries, between the elites and the poorer and underprivileged populations, on the other hand. New information and communication technologies can serve as a development lever to speed up the economic development of Africa and its poor communities. However, the details of these transformations and the degree and pace of such changes in poor communities have yet to be fully grasped. Equally, the ways in which ICTs might best serve development are still relatively unknown.

Therefore, it is critical to determine the implications that these changes will have on the poor communities that ICTs are supposed to transform. Studies and investigations are needed to generate new hypotheses that can be tested. The results of such research must also be shared rapidly and effectively to keep up with the pace at which ICTs evolve (Adera & Camara, 2003). To illustrate on how ICT strategies have been implemented in Africa and how indicators of ICT are being developed, Tables 32-35 are showing the basic ICT indicators for the five top countries, in terms of total telephone subscribers. From these tables, it has been observed that Reunion, Seychelles, and Mauritius are at the top with robust information infrastructure, while South Africa and Botswana have been driving towards the achievement with increased GDP per capita.

Table 32. Basic ICT indicators in 2005

Country	Total telephone subscribers per 100 inhabitants	Per capita USD (2004)	Population density (per Km^2)
Reunion	98.65	-	313
Seychelles	96.80	8,348	200
Mauritius	86.13	5,146	668
Tunisia	69.06	2,821	61
Botswana	54.11	4,957	3

Table 33. Basic ICT indicators in 2004

Country	Total telephone subscribers per 100 inhabitants	Per capita USD (2003)	Population density (per Km^2)
Reunion	98.65	1,893	306
Seychelles	86.94	8,348	200
Mauritius	70.06	4,606	661
Tunisia	47.97	2,527	61
South Africa	46.76	2,293	38

Table 34. Basic ICT indicators in 2003

Country	Total telephone subscribers per 100 inhabitants	Per capita USD (2002)	Population density (per Km^2)
Reunion	98.65	1,893	301
Seychelles	85.07	8,348	205
Mauritius	55.22	3,957	655
South Africa	46.76	2,293	39
Botswana	37.19	2,939	3

Table 35. Basic ICT indicators in 2002 (Source: ITU statistics (www.itu.int/ITU-D/ict/statistics.htm)

Country	Total telephone subscribers per 100 inhabitants	Per capita USD (2001)	Population density (per Km^2)
Reunion	98.65	1,893	296
Seychelles	78.81	7,603	207
Mauritius	55.79	3,771	649
South Africa	40.80	2,562	38
Botswana	33.57	2,921	3

Table 36. Various steps taken by the Kenyan government to promote ICT

Country	Policies taken	Period of implementation	DAI Ranking in 2002
Kenya	The Kenya Information and Communications Bill 2006	February 2006	0.19 (low access)
	National ICT Policy	January 2006	
	Freedom of Information Act 2005	2005	
	The Kenya Communication Act, 1998	October 1998	
	The Science and Technology Act	1980	

Basic Indicators

Kenya

Despite advances in private sector initiatives, the government has been found to be reluctant to embrace ICTs for social and economic development for some time[32]. However, Kenya's e-strategies have been best poised through the approval of its national ICT policy by the Ministry of Information and Communication in January 2006. Table 36 shows the steps of the Kenyan government towards establishment of a knowledge society through initiating relevant ICT strategies:

The next section puts forward a few recommendations for successful implementation of ICT strategies by upholding the concept of creating knowledge societies.

RECOMMENDATIONS

Creation of knowledge societies demand high initial investment in sectors that are seemingly looked as unproductive at the beginning, commitment at all levels of the policy initiations, and mass awareness at the end users level. Furthermore, increasing investments in R&D during these periods of high unemployment demand great political wisdom and courage, when an easier path may seem to generate immediate employment rather than build up longer-term strengths. At the same time, increasing R&D must be supported by national industrial and innovation strategies, which should be properly communicated by governments. These national strategies should focus on consensus building; such as, by organizing economic policy programs attended by practically all members of the Parliament and other decision makers from the public and private sectors, media, and labor market organizations (Dahlman, Routti & Ylä-Anttila, 2005).

There should be a triangular incorporation of partnerships, as shown in figure 4[33].

Consequently, it is almost impossible to identify the embedded ICT component due to non-availability of the financial data, and for that reason the aggregate figure cannot be calculated. However, figures for ICT-specific projects/programmes are easier to produce and some reported figures (including some non-ODA) have been highlighted, and it also shows various approaches taken by governments across the globe through their direct commitments in establishing knowledge societies:

- Canada estimates a minimum expenditure of $33 million per annum
- European Commission has a commitment of €250 million for multi-year ICT-specific programmes in addition to €110 million from the European Development Fund and €750 million from the European Investment Bank (1999–2003)
- France committed about €40 million (2002–2005) to global programmes over and above its country programmes and other facilities
- Germany supports at present ICT applications with approximately €180 million;
- Japan launched its Comprehensive Cooperation Package for bridging the "digital divide" —Japan's commitment for ICT support, announced at the G8 Kyushyu-Okinawa Summit in 2000–which consists of non-ODA and ODA funding with a total of $15 billion over 5 years (2000–2005)
- Sweden spent approximately $18 million in 2003
- the United Kingdom currently has multi-year ICT-specific programmes and projects, mounting to approximately a total of $83 million
- the United States estimates its spending of ICTs for development at more than $200 million in 2003, and through leveraged or matching outside resources a further $240 million was mobilized (OECD, 2005).

Figure 4. Transformation to knowledge society demands tri-partite involvements

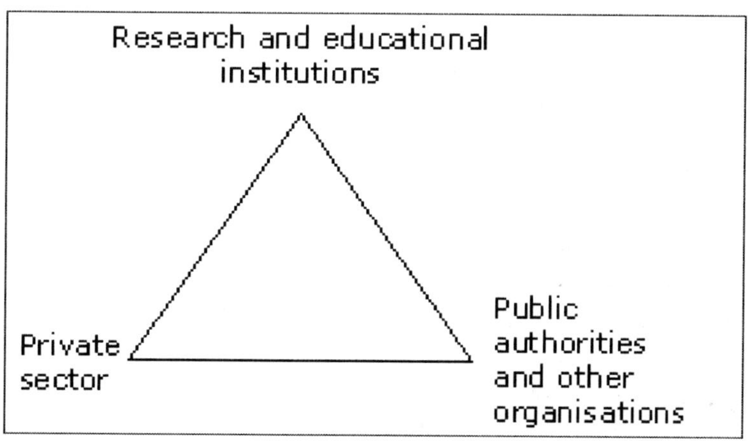

Taking all these facts into consideration, the following recommendations may be adopted for successful implementation of ICT strategies in each country:

- Formulation of appropriate policies for promoting investment in ICT sector and establishing a transparent and consistent legal and regulatory framework to foster ICT development
- Ensure investment to improve access and network infrastructure, establish investment friendly climate by setting up policy, regulatory, and legal framework and by providing fiscal incentives, induction of new less expensive and cost-effective broadband wireless technology, and optical fiber communication transport backbone infrastructure (UN, 2003)
- Provide a framework for establishing national, regional, and international partnerships and help in coordinating programmes and activities aimed at specific ICT application areas with well defined objectives, milestones, and outcomes
- Formulate action plan for implementation at national and regional levels aimed at achieving the MDGs by mainstreaming ICTs
- Supporting national actions aimed at harnessing the potential of ICT for development through providing necessary assistance
- Serving as a common global forum to accelerate and promote universal access to knowledge and information, contribute to the development of norms and standards in a transparent, meaningful, and participatory way, taking into account development priorities and incentives for innovations and contribute to addressing issues like cultural diversity, information ethics, information rights, intellectual property rights, privacy, security, and cyber crime
- Contributing to a more systematic approach, review, and dissemination of ICT expertise, distance learning programmes, information and reliable data on ICT, case studies, best practices and successful models, thus becoming and important 'knowledge bank' in this field
- Emphasizing universal access to knowledge and information for promoting grass roots development
- Taking actions in bridging the digital divide and promoting digital opportunities and adopting a coherent ICT strategy that would ensure cooperation and synergy among programmes and activities of the involved organizations or agencies so as to transform into a knowledge based society (UN, 2003)

The author feels that many nations are yet to cross the boundaries of implanting ICTs for fostering their economic outgrowths, despite substantive themes were addressed by various agencies including the Commission of Sustainable Development (CSD) since 1995. The box shows the themes of CSD with years as a ready reference and it can be observed that, including 2001 (focus was on biotechnology), all the themes were on capacity development or related to it.

- Technology for small-scale economic activities to address the basic needs of low-income populations; the gender implications of science and technology; and the contributions of science and technology to an integrated approach to land management (1995)
- Information and communication technologies for development (1997)
- Science and technology partnerships and networking for national capacity building (1999)
- National capacity-building in biotechnology (2001)
- Technology development and capacity building for competitiveness in a digital society (2003)

- Promoting the application of science and technology to meet the development goals contained in the United Nations Millennium Declaration (2004)
- Science and technology promotion, advice and application for the achievement of the internationally agreed development goals contained in the United Nations Millennium Declaration (2005)
- Bridging the technology gap between and within nations (2006)

Furthermore, nations can endorse a two-prong approach to promote successful ICT strategies. One may be in the technology context and the other one in the management perspective. The box (Figure 5) shows different action plans; a nation can take in these aspects.

An integrated approach is desirable in the implementation process along with the design and development phases, including critical assessment of the implementation strategies. Notwithstanding technologies are dynamically adjustable and varied in nature, but countries with success cases can easily be adopted without much duplication. This would save time, effort, and resources for many economies. Rather, efforts can be concentrated to localize the technology, content, and the change management processes through recursive simplifications. A circular modality can be followed as shown in Figure 6.

Finally, a context diagram is being developed that may be adopted in implementing national ICT strategies (see Figure 7).

Figure 5. Parameters of the two-prong approach

Technology parameters

Initiation of Strategic Implementation Plan
Uniform Network Deployment
Adoption of Definitive Economic Indicators
Implementation of Recursive Algorithm for Content Management
Implementation of Adaptive Algorithm for Bandwidth Management
Establishment of Tools for Monitoring and Evaluation
Taking Corrective Measures

Management parameters

Development of Strategies
Transformation of Strategies into Plans and Actions
Correlations among Implementing Partners
Coherence among Stakeholders

Figure 6. An ideal strategic implementation process

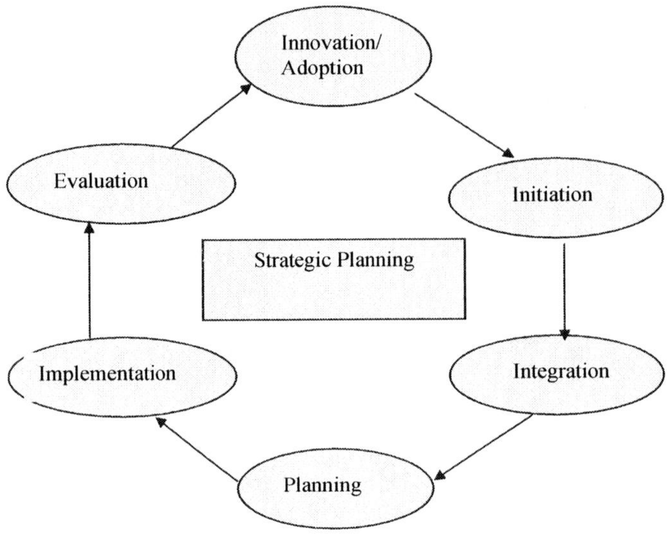

Figure 7. Context Diagram of activities for a knowledge society

At local level:

- Awareness creation
- Infrastructure
- Stakeholders group
- Local committee
- Capacity development

At national level:

- Development of a monitoring and evaluation system on the readiness to information society
- Programme formulation
- Enabling environment
- Human resource development
- Access to information and knowledge

At regional level:

- Policy, regulatory, and legal framework
- Regional broadband network
- Human resource development
- Research and development
- Building partnerships

WAY AHEAD AND CONCLUSION

National regulators, financial institutions, and market participants have shown that the global financial system can continue to function smoothly even under a difficult and totally unanticipated form of extreme constraints. Globalization of the world economy has been driven by a variety of forces: rising trade in goods, increasing international capital flows, enhanced technological spillovers, and emergent labor mobility.

A second channel has evolved in recent years through which capital account liberalization can have a positive impact is technology spillovers. These spillovers are most noticeable in the case of foreign direct investment (FDI), especially through foreign firms incorporating new technologies in their subsidiaries. As novel technologies are generally developed and adapted by firms in industrial countries, FDI may be the most efficient way for developing economies to gain access to them. In addition, this knowledge may become more widely available in the country over a longer time, as employees with experience in the techniques used in foreign companies switch to other firms. Furthermore, foreign investment could increase competition in the host-country industry, and hence force local firms to become more productive by adopting more efficient methods or by investing in human and or physical resources (IMF, 2001).

Many countries have already released various strategic plans over the last decade. The worldwide technology revolution makes such strategies much more imperative. However, unlike the industrial revolution, the knowledge revolution encompasses all aspects of society, so e-strategies must go beyond technical goals in addressing the challenge of universal access to, and the usage of ICTs. The responsibility for adopting appropriate e-strategies rests not just with government, but also with civil society and the private sector. Moreover, much depends on how open governments want to be with their people in formulating and implementing such policies (Ulrich, Chacko & Says, 2004).

Along this aspect, the UN regional roadmap adopts the common vision and the key principles consistent with the WSIS Inter-sessional meeting recommendations that envisage the Information Society as a new and higher form of social organization, where highly-developed ICT networks, equitable and ubiquitous access to information, appropriate content in accessible formats, and effective communication must enable all the people to achieve their full potential, promote sustainable economic and social development, improve quality of life and therefore, alleviate poverty and hunger (UN, 2003).

Similarly, the UNESCO declaration urged the governments, parliaments, and other agencies to establish the legislative and political framework for reform and further development of higher education, taking into account of the fact that education and research are two closely related elements in establishing knowledge economy. Innovation schemes of collaboration among education institutions and different sectors of society might ensure that education and research programs effectively contribute to local, national, regional, and global development (Rossman, 2005).

Furthermore, the World Summit on the Information Society (WSIS) held in Tunis in November 2005 requested ECOSOC[34] to oversee the system-wide follow-up of its Geneva and Tunis outcomes. To this end, it requested ECOSOC, at its substantive session of 2006, to appraise the mandate, agenda, and composition of the Commission on Science and Technology for Development (CSTD), including considering the strengthening of the Commission, taking into account the multi-stakeholder approach. This mandate was immediately endorsed by the General Assembly in its resolution (A/60/252) adopted by consensus on March 27, 2006 (ECOSOC, 2006).

In these contexts, if the Internet penetration rate were taken as an indicator of successful implementation of ICT strategies, then it has been observed that during 1994-2000 it reaches to about 7%, while till 2003 the rate has increased to about 11%. However, mobile penetration rate was around 13% in 2000 and in 2003 it was about 23%. Hence, it can be seen that during 2000-2003 Internet penetration rate has increased by 4%, while mobile penetration rate has increased by 10%. If Internet growth is to recover, it is important to reach out to the growing number of users who have a mobile phone but no fixed-line telephone. Handheld Internet enabled devices could open up the information gateway in a new and exciting manner—through that one could help further the goals of universal access while challenging manufactures and service providers to meet different users' need across the globe. Table 37 synthesizes the growth rate of Internet penetration and mobile penetration across the globe during 1994 to 2003.

Technologies, such as Bluetooth, ZigBee, and RFID allow low-power—short-distance connectivity within a range of 30 meters. Medium-range technologies can communicate at least 150 meters from a hotspot (e.g., Wi-Fi or IEEE 802.11b) and up to several kilometers, depending on environmental, geographical, and regulatory factors. Similarly, long-range technologies such as WiMAX (IEEE 802.16) and IMT-2000 (3G) have ranges that may extend up to 50 kilometers from a base station, and to near-nationwide coverage through a networked service. Furthermore, in this category fit solutions based on high or low-altitude platform stations (HAPS/LAPS), LMDS, and MMDS that can serve a whole town or city, and finally satellite that can serve a whole region (ITU, 2004).

Academic institutions can play greater role in helping design and evaluate ICT projects that may involve technically challenging research. In addition, their corporate research counterparts could become active in developing standards that are revolutionizing the spread and use of ICT: from open source software and the next-generation Internet IPv6, to wireless LAN (IEEE 802.11b or 'Wi-Fi'), the longer-reaching standard of IEEE802.16 (or Wi-Max'), and the IEEE802.20 (or 'Wi-Mobile'). At the one end of the academic spectrum are scholars involved in cutting-edge innovations designed at leading research institutions and higher education institutions (HEIs); at the other are those toiling on the front lines of basic education. Each stage of the continuum has a stake in national e-strategies and input to provide into their own design.

Developing a cadre of digitally literate workers is necessary for a country to participate in the ICT revolution. However, such a process takes time—in fact, education and the invest-

Table 37. Internet and mobile penetration rates (Source: ITU World Telecommunication Indicator Database)

Year	Internet penetration rate	Mobile penetration rate
1994	0	0
1996	3	5
1998	4	7
2000	7	13
2002	9	19
2003	11	23

ment in human capital can have a longer horizon for payback than almost any other investments. In 1980, when Singapore first began looking at computerization, the island city-state had just 850 computer professionals among its three million people. In 2004 it reaches to over 100,000. Similarly, India's policy of putting resources into tertiary ICT education took over two decades to bear fruit in creating global centers for outsourcing software development and business processes for multinational firms (call centers, etc.).

North American Companies' outsourcing of software development, processing of health-insurance claims, and maintenance of customer databases has generated 3000 jobs long back in 1998 in the Barbados, which was the same number of people employed in sugar cane farming[35]. After about 8 years in the age of ICT, still those sectors deserve the same attention from many nations who need to adopt ICT in their knowledge economy.

Mexico's GDP per capita in 1960 was about 2.5 times higher than that of Korea (R.), but by 2003 Korea (R.)'s GDP per capita was more than double than Mexico's. Without the contribution of knowledge, Korea (R.)'s actual GDP per capita would have been lower than Mexico (World Bank, 2006).

No matter how undeveloped they are, virtually all countries recognize that ICT development is the key to future prosperity, and ICT has to be embedded into the daily life with an enabling environment (Capacity dot org, 2004). Furthermore, without knowledge workers, nations cannot move up to higher rank of GDP, and more remunerative links in the global value chain. Hence, in a few countries (some indicated in this chapter) and elsewhere, digital literacy has become the highest priority, emphasizing math, basic science, technical skills, and for better or worse, English, as the prevailing lingua franca of international business and the Internet (Ulrich, Chacko & Says, 2004).

FUTURE ICT STRATEGIES

National development strategies calls for the creation of a high-skill, high-value added economy to increase the GDP and help bridge the wealth gap among cities and between country and the rest of the world. Skills development can be seen as moving a country eventually into the knowledge-based economy. Hence, the vision of the strategy should be dynamic, inclusive, and sustainable depending on world-class competitive business and skilled motivated people (National Assembly for Wales, 2001). From these aspects, national ICT strategies of a country should act as catalytic agents in reinforcing the overall development through innovative application of technologies and promotion of ubiquitous access to information.

Development of ICT needs assessment and strategic planning; policy development and implementation; information management; project development and implementation for e-services in health, education, environment, and other economic sectors; private sector entrepreneurship; awareness, education, and capacity building; e-government strategic planning; information infrastructure improvement; enterprise information resources integration; information systems design and deployment; business process reengineering and optimization; grass-roots and national institutional reforms and change management; and foremost improved regulatory environment.

For successful strategic ICT development, the following steps may be initiated at national level:

- Formulation and realization of ICT-enabling policy, institutional and regulatory framework and its incorporation into the overall strategy of the national social and economic development
- Formulation and implementation of sectoral ICT strategies and policies in health, education, environment, SMEs, and other economic sectors

- Advisory services to central and local governments on policies to raise awareness and encourage application of ICT to strengthen the economic and social position of the poor masses and ensure their enhanced participation in decision-making processes
- Advisory services on e-governance including strengthening of e-government road-map, strategies and policies, and implementation framework, e-government systems architecture improvement, strategic planning for information integration, e-service planning and research, information security strategies, knowledge, and content management
- Advisory services on ICT infrastructure advancement, ICT manpower development, institutional restructuring, change management and business process re-engineering
- Sponsor training courses, capacity building and awareness creation workshops on ICT policy, institutional and legislative issues, e-government, ICT standards, information security, and data protection (UNESCAP, 2006)

Figure 8 shows the various steps needed to develop strategic ICT implementations in a country.

As described earlier, in recent years many governments have undertaken, as part of the Millennium Declaration, to develop and put into effect strategies to give common people a real change to find decent and productive work. The Heads of State and governments have also resolved to 'ensure that the benefits of new technologies, especially information and communication technologies are available to all' (UN General Assembly, 2000). Furthermore, as developing countries join the global information infrastructure, each country will need to find competent ways of maximizing the benefits and reducing the risks from ICTs. This will involve synchronized action through national ICT strategies encompassing the technologies and services as well as many aspects of the institutional settings. Strategies are needed to build the necessary scientific, technical, and engineering knowledge as well as the management techniques so that the social and economic institutions that are consistent with creativity

Figure 8. Steps towards successful ICT strategy implementation

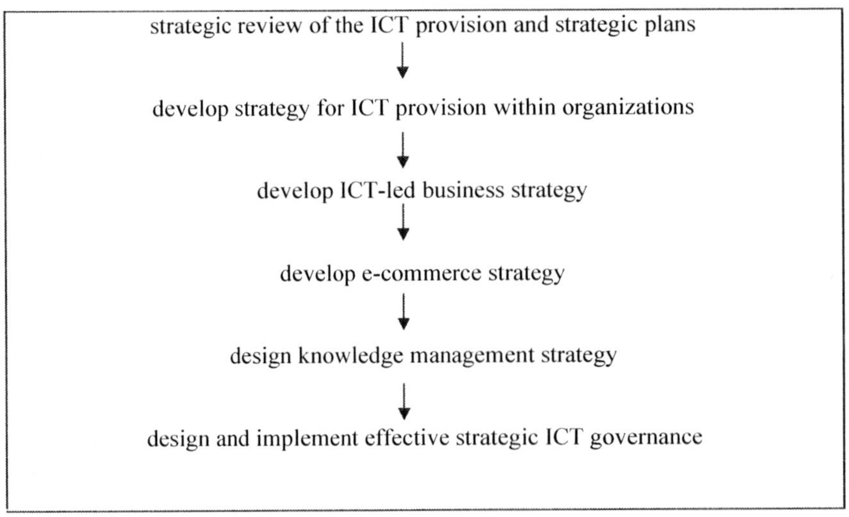

using ICTs may reap the potential social and economic benefits. Eventually, effective national ICT strategies are most likely to materialize through creatively combining competencies in scientific, technological, and management fields related to ICTs in each country and recombining these strengths with externally available technologies, information, and knowledge resources (Mansell, R. & When, U., 1998).

Figure 9 illustrates an ideal environment to develop successful ICT strategies. The core parameters are hardware, software and access to ICT services, and the basic preconditions are to improve appropriate infrastructure, enhance human capacity, incorporate supportive public policy, establish supportive framework for enterprises, and strengthen local content and applications.

In terms of hardware, ICT refers to a wide range of products and service activities. These include computers, related equipment, and accessories. In recent years, it also includes the fiber optics, satellites, sub-marine cables, voice over internet protocol, cellular phones (3Gs, 4Gs), high speed communication equipment, high definition television, and wifi/wimax equipment. ICT strategy should also encompass the software and related services industries, as such computer programming services, pre-packaged software, wholesale trade of software, retail trade of software, computer-integrated system design, computer processing, data generation, data security and information retrieval services, computer application and services management, computer rental and leasing, as well as computer maintenance and repair (Curtin, 2001).

REFERENCES

Adera, E., & Camara, A. (2003). Opportunities and Challenges for Community Development. Preface. In Ramata Malo Thioune (Ed.), *Information and Communication Technologies for Development in Africa: Vol. 1.* International Development Research Centre, Canada, 2003. ISBN 2-86978-114-8.

Capacity dot org (2004). *Embedding ICT in development*, Issue 23, October 2004, ISSN 1566-6603.

Crede, A., & Mansell, R. (1998). *Knowledge Societies in a Nutshell.* Information Technologies for Sustainable Development. IDRC, Ottawa, Canada.

Figure 9. Preconditions for a successful ICT strategy

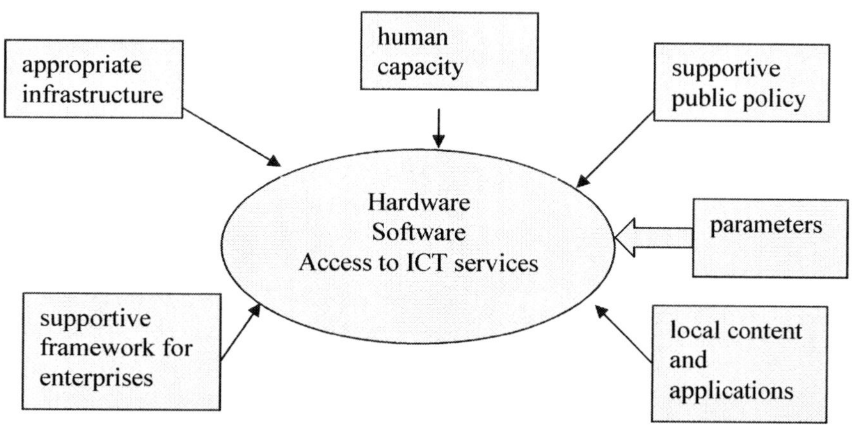

Curtin, R. (2001). *Promoting youth employment through information and communication technologies (ICT): Best practices examples in Asia and the Pacific.* Report prepared for ILO/Japan Tripartite Regional Meeting on Youth Employment in Asia and the Pacific, Bangkok.

Dahlman, C.J., Routti, J., & Ylä-Anttila, P. (Eds.). (2005). A prepublication version of the Overview of the book *Finland as a Knowledge Economy: Elements of Success and Lessons Learned,* published in January 2006. The International Bank for Reconstruction and Development / The World Bank, Washington, 2005.

ECOM International Relations Group announced its 2002 Survey on the State of E-commerce in Other Countries. (2003). ECOM Newsletter, 34.Retrieved from http://www.ecom.jp/ecom_e/latest/newsletter_no34.htm

EC (2004). *Specific programme for research, technological development and demonstration: Integrating and strengthening the European research Area.* 2004 Work Programme (SP1-10), European Commission, 2004.

EC (2005). *Specific programme for research, technological development and demonstration: Integrating and strengthening the European research Area.* 2005 Work Programme (SP1-10), European Commission, 2005.

EC (2006). *Specific programme for research, technological development and demonstration: Integrating and strengthening the European research Area.* 2006 Work Programme (SP1-10), European Commission, 2006.

ECOSOC (2006). Informal background note on possible options for consideration by ECOSOC in the reform of the CSTD.

Government of Japan (2005, February). *IT Policy Package 2005: Towards the Realization of the World's Most Advanced IT Nation.* The 29th Meeting of the IT Strategic Headquarters.

Hak-Su, K. (2003). Forward. *Regional Road Map towards an Information Society in Asia and the Pacific, Economic and Social Commission for Asia and the Pacific.* United Nations, New York.

ICT Riga (2006, June). *Ministerial Declaration.* Latvia.

IMF (2001, October). *World Economic and Financial Surveys, World Economic Outlook: The Information Technology Revolution.* International Monetary Fund, Washington, ISBN 1-58906-073-3.

ITU (2004, September). *The Portable Internet.* ITU Internet Reports 2004, Switzerland. ISBN 92-61-10741-9.

ITU (2005). *World Telecommunications Development Report 2005*, ITU, Geneva.

ITU (2006). *World Telecommunication/ICT Indicators*, ITU, Geneva.

Léautier, F.A. (2005). Forward. In Dahlman, C.J., Routti, J., & Ylä-Anttila, P. (Eds.) A prepublication version of the Overview of the book *Finland as a Knowledge Economy: Elements of Success and Lessons Learned,* published in January 2006. The International Bank for Reconstruction and Development / The World Bank, Washington, 2005.

Mansell, R., & When, U. (Eds.). (1998). *Information Technology for Sustainable Development.* Oxford: Oxford University Press.

MIC (2004). Ministry of Internal Affairs and Communications. *MIC Communications News. 15*,12.

MIC (2005). Ministry of Internal Affairs and Communications, *MIC Communications News.16*,1.

MPT (2003, December). *Info-Communication Country Report in Vietnam, Ministry of Posts and Telematics (MPT).* The 29th Asian Info-communication Council conference, Tokyo, Japan.

National Assembly for Wales (2001). *Information and Communication Technology for Education.* Education and Life-Long Learning Committee. Final Report, Cardiff, UK.

OECD (2005). *Financing ICTs for Development: Efforts of DAC Members,*

Review of Recent Trends of ODA and its Contribution, Report to the UN Task Force on Financial Mechanisms for ICT for Development (TFFM), OECD Publications, France, 2005.

OII (2005, May). *The Internet in Britain.* The Oxford Internet Survey (OxIS), Oxford Internet Institute, University of Oxford, Oxford.

Olukoshi, A., & O'Neil, M. (2003). In Ramata Malo Thioune (Ed.), *INFORMATION AND COMMUNICATION TECHNOLOGIES FOR DEVELOPMENT IN AFRICA: Vol.1., Opportunities and Challenges for Community Development.* International Development Research Centre, Canada. ISBN 2-86978-114-8.

Rossman, P. (2005). *TRANSFORMING HUMAN SOCIETY: The Future of Higher (Lifelong) Education: A Vision of Job Training and Education for Everyone on the Planet during this Century, Planning For all Worldwide, a Holistic View.* Retrieved June 12, 2006, from http://ecolecon.missouri.edu/globalresearch/

Siochru, Sean O., & Girard, Bruce (2005). *Community-based Networks and Innovative Technologies: New models serve and empower the poor.* UNDP, New York, 2005.

Soan (2006). *Digital sector comes into sharp focus.* Retrieved July 12, 2006, from http://english.vietnamnet.vn/tech/2006/06/582289/

ThuHue, N. (2005, July). *Internet governance for socio-economic development.* Case study from Vietnam, a case study in GIPI, Geneva.

Thanh, P.X. (2003, March). *ICT Policy Report.* Vietnam, Paper presented at the Interregional Seminar and Regional Demonstration Workshop on Accessible Information and Communications Technologies and Persons with Disabilities, Manila, Philippines.

Thioune, R.M. (Ed.) (2003). *Information and Communication Technologies for Development in Africa, Volume 1.* Opportunities and Challenges for Community Development, IDRC, Ottawa, Canada, 2003.

UN General Assembly (2000). *United nations Millennium Declaration 55/2,* available at http://www.un.org/millennium/declaration/ares552e.htm), para. 20.

UN (2003). *Regional Road Map towards an Information Society in Asia and the Pacific, Economic and Social Commission for Asia and the Pacific,* United Nations, New York.

UN (2005). *United Nations for Sustainable Development, Expert Group Meeting on Indicators of Sustainable Development,* New York.

UN (2006). *The Digital Divide Report: ICT Diffusion Index 2005,* United Nations Conference on Trade and Development, United nations, New York and Geneva.

UNDP (2001). *Human Development Report 2001: Making New technologies Work for Human Development,* Oxford University Press, New York.

UNDP (2004, May). *Local e-Government in FYR Macedonia 2004.* ICT Assessment Report of Municipalities, Skopje, UNDP FYR Macedonia.

UNESCAP (2006). Retrieved February 15, 2007, from http://www.unescap.org/pmd/03_tc/pmd_dp_tc_ra_ictps.asp

Ulrich, P., Chacko, J.G., & Says, P, (2004). Overview of ICT Policies and e-strategies in the Asia-Pacific Region.In Sayo, P., Chacko, J.G., & Pradhan, G. (Eds.), *ICT Policies and e-Strategies in the Asia-Pacific: A critical assessment of the way forward,* Elsevier, New Delhi. ISBN: 81-8147-513-5.

Uganda-Africa (2005). Retrieved June 06, 2006 from http://www.ungana-afrika.org/resources/techplanning/tech_planning_guide.pdf

World Bank (2005). *Project Appraisal Document and Project Information Document*, World Bank, Retrieved July 12, 2006, from http://web.worldbank.org/external/projects/main?pagePK=104231&.htm

World Bank (2006). *Korea as a Knowledge Economy: Evolutionary Process and Lessons Learned, The World Bank,* Washington, 2006.

Additional Readings

Ang, J., & Pavri, F. (1994). A survey and critique of impacts of information technology, *International Journal of Information Management, 14,* 122-133.

Balzan, R., & Vella, G. (1997). *A review of the National Strategy for Information Technology.* Office of the Prime Minister, Government of Malta.

Bennett, I. (May 9, 1995). *National industrial policy information technology sector.* (Draft Report), Planning Institute of Jamaica.

Bessette, G. (1996). *Empowering people through information and communication technology: Lessons from experience.* Paper submitted to the International Development Research Centre, Ottawa, Canada.

Bourdeal, J., Vazquez-Abad, J., & Winer, L. (January, 1996). *Information and communication technologies for generating and disseminating know-how.* (Draft report). Department of Scientific and Technological Affairs, Organization of American States.

Byron, L., & Gagliardi, R. (1996). *Communities and the information society: The role of information and communication technologies in education.* Report prepared for International Development Research Centre, Ottawa, UNESCO International Bureau of Education (IBE).

Camilleri, J. (October, 1994). *A national strategy for information technology for Malta.* Office of the National Strategy for Information technology, University of Malta.

Chataway, B., & Cooke, A. (1996). *Measuring the impact of information on development: Related literature: 1993-1995.* Available at http://www.idrc.ca/books/focus/783/chataway.htm

Cooper, C. (1998). An outline for policy analysis on access to information technology. In C. Cooper (Ed.), *Information technology policy and national economic development.* London: Routledge and Tokyo: UNU Press.

Correa, C.M. (1996). Strategies for software exports from developing countries. *World Development, 24*(1), 171-182.

Curtain, R. (2000). *Identifying the basis for a youth employment strategy aimed at transitional and developing countries.* United Nations Social Development Division. Available at http://www.un.org/esa/socdev/youthemployment/research.html

D'Orville, H. (April 30, 1996). *Technology revolution study: Communications and knowledge-based technologies for sustainable human development.* Final report, United Nations Development Programme, New York.

Drake, W.J. (Ed.). (1995). *The new information infrastructure: Strategies for US policy.* New York: Twentieth Century Fund Press.

Dutton, W.H. (Ed.). (1997). *Information and communication technologies: Vision and realities.* Oxford: Oxford University Press.

Eliasson, G. (1990). *The knowledge-based information economy.* Stockholm: Almquist & Wiksell International.

Friedman, A.L. (1990). Four phases of information technology: Implications for forecasting IT work. *Future, 22*(8), 787-800.

Hafkin, N., & Taggart, N. (2001, June). *Gender, information technology and developing countries: An analytic study.* For the Office of Women in Development, Bureau for Global Programs, Field Support and Research, United States Agency for International Development. Available at http://www.usaid.gov/wid/pubs/it01.htm

Institute on Governance (1996). *Information and communication technologies (ICTs) and governance: Linkages and challenges.* International Development Research Centre, Ottawa, Canada.

Kahin, B., & Wilson III, E.J. (1997). *National information infrastructure Initiatives: Vision and policy design.* Cambridge MA: MIT Press.

Kenyon, P., & White, S. (2001). *Enterprise-based youth employment policies, strategies and programmes.* International Labour Office, Geneva. Available at http://www.ilo.org/public/english/employment/skills/youth/publ/

Lundvall, B.A. (1996). Information technology in the learning economy-challenges for development strategy. In a background paper for the *UNCSTD Working Group on IT and Development.*

Mahan, A.K., & Melody, W.H. (2005). *Stimulating investment in network development: Roles for regulators.* Case studies and research from World Dialogue on Regulation Cycle 2, Technical University of Denmark, Lyngby.

National Computerization Agency (2003). *Informatization: White Paper.* e-Korea, National Computerization Agency, Seoul, Rep. of Korea.

OECD (1997). *Information infrastructure: Their impact and regulatory requirements.* Committee for Information, Computer and Communications Policy, Paris, OECD/GD(97)18.

Surman, M., & Reilly, K. (2003). Appropriating the Internet for social change: Towards the strategic use of networked technologies by transnational civil society organizations. A report for Social Sciences research Council.

The Digital Opportunity Task Force (DOT Force) (2001). Digital opportunities for all: Meeting the challenge: Including a proposal for a Genoa Plan of Action, May 11, Genoa. Available at http://www.dotforce.org/reports/

Valantin, R. (1995). Information and communication technologies for development: Some principles and lessons from IDRC experience. *Information Technology for Development, 10,* 540-544.

ENDNOTES

1. *Source:* World Bank–Knowledge Assessment Methodology. www.worldbank.org/kam. *Note:* The Knowledge Economy Index (KEI) comprises of 80 structural or qualitative variables with benchmark performance from more than 128 countries. KEI is an aggregate of all variables normalized from 0 (worst) to 10 (best).
2. J. Bradiford De Long. http://econ161.berkley.edu
3. Eastern Europe, South America, and more developed parts of Asia
4. Rural Asia, Central America, and Africa
5. North America, Western Europe, Japan, Australia, and New Zealand
6. DAI measures the overall ability of individuals in a country to access and use new ICTs. The DAI is built around four fundamental vectors that impact a country's ability to access ICTs: infrastructure, affordability, knowledge and quality, and actual usage of ICTs. http://www.ictlogy.net/wiki/index.php?title=Digital_Access_Index accessed on August 02, 2006.
7. Retrieved June 15, 2006 from http://www.barents2010.net/default.asp?ML=2663

8 Retrieved July 14, 2006, from http://www.internetworldstats.com/top25.htm

9 TAI is a measure of how a country is creating and diffusing technology and building a human skill base, reflecting capacity to participate in the technological innovations at the network age.

10 ICTDI: Scores are derived as an index relative to the maximum and minimum achieved by countries in any indicator, as such; Index score= (Value – Minimum)/(Maximum – Minimum).

11 The compound annual growth rate (CAGR) is computed by the formula: $[(P_v/P_0)^{(1/n)}]-1$, where P_v = Present value, P_0 = Beginning value and n = Number of periods. The result is multiplied by 100 to obtain a percentage.

12 http://english.vietnamnet.vn/tech/2006/06/582289 accessed July 12, 2006

13 http://www.itu.int/wsis/docs/background/resolutions/56_183_unga_2002.pdf

14 47: MDG- Telephone lines and cellular subscribers per 100 population

15 48: MDG- Personal computers in use per 100 population and Internet users per 100 population

16 Commission on Sustainable Development

17 The Global Information Technology Report 2001-2002: Readiness for the Networked World, Center for International Development, Harvard University, Oxford University Press, Oxford, 2002.

18 Country Note: Ireland, OECD Review of Career Guidance Policies, OECD, April 2002.

19 http://insight.eun.org/ww/en/pub/insight/misc/country_report.cfm accessed on August 16, 2006

20 http://www.american.edu/initeb/ty5746a/policies.htm accessed on August 08, 2006

21 http://www.taoiseach.gov.ie/index.asp?docID=764; 229; 557 accessed on August 08, 2006

22 Knowledge Society News, Issue 1, August 03, 2006 available online from http://newsweaver.ie/

23 The Global Information Technology Report 2001-2002: Readiness for the Networked World, Center for International Development, Harvard University, Oxford University Press, Oxford, 2002.

24 ITU Digital Access Index: World's First Global ICT Ranking, ITU 2002.

25 http://www.ecdc.net.cn/newindex/chinese/page/sitemap/reports/IT_report/english/02/03.htm, accessed on June 12, 2006.

26 http://www.mfat.govt.nz/foreign/regions/northasia/country/japanpaper.html, accessed on June 14, 2006 http://www.mfat.govt.nz/foreign/regions/northasia/country/japanpaper.html, accessed on June 14, 2006

27 http://www.ecom.jp/ecom_e/latest/newsletter_no34.htm, accessed on June 12, 2006

28 http://www.cia.gov/cia/publications/factbook/geos/ja.html, accessed on June 14, 2006

29 http://reports.tmgtelecom.com/apmmo/TMG%20APMMO%20datasheet.pdf accessed on June 20, 2006

30 http://www.digital-review.org/03_Korea.htm; http://www.digital-review.org/05_Korea.htm accessed on August 16, 2006

31 ITU Digital Access Index: World's First Global ICT Ranking, ITU 2002

32 http://www.uneca.org/aisi/NICI/Kenya/kenya.htm, accessed on August 01, 2006

33 Adapted from Strategy and action plan for the Barents Region until 2010, Retrieved June 15, 2006 from http://www.barents2010.net/files/llloRhbb.pdf

34 Economic and Social Council: United Nations organ facilitating international cooperation on standards- making and problem-solving in economic and social issues

[35] http://www.idrc.ca/en/ev-28856-201-1-DO_TOPIC.htm accessed on August 05, 2006

Chapter II
Strategy Development:
A Focus on Process

Suleman Aziz Lodhi
National College of Business Administration and Economics, Pakistan

Munir Ahmad
National College of Business Administration and Economics, Pakistan

ABSTRACT

Trade globalization and advancement in ICT may be considered as the two major forces that will be directing economic growth of a country in the coming years. Policy makers have realized the importance of ICT in achieving national goals in the digital era. The developing and underdeveloped countries generally lack research capacity and standard policy development processes that are critical for developing a successful ICT policy. Cause of policy failure in these countries can most of the time, be traced to flaws in policy development process rather than environmental complexity. The chapter provides an insight on policy development process from strategic management perspective to highlight critical issues, which are sometimes overlooked by policy makers. The purpose is to assist the policy makers in developing successful ICT Policy for their countries.

INTRODUCTION

Transformation of world trade under World Trade Order (WTO) framework and advancement in information and communication technology (ICT) may be considered as two major forces that will be directing economic growth of countries in the coming years. The new scenario has brought ICT policy of a country to center of discussion on trade development and economic sustainability. The chapter approaches ICT strategy development as an integrated process (see Figure 1, p. 49). References are made from strategic management domain to highlight critical points, which are sometimes overlooked in policy making.

Causes of policy failure in developing countries can, most of the time, be traced to flaws in policy development process rather than environmental complexity. A policy based on inadequate research and makeshift procedures is bound to cause more problems than it is designed to solve. It is no doubt difficult to develop a strategy in the complex environment that presently people are living in, but following a framework will certainly increase chances of success. The underdeveloped and developing countries generally lack the research capacity and integration in policy making processes that are critical for developing policies to compete in global knowledge driven society. Policy makers are often guided by insufficient data and background research for economic planning, and as a result of that enormous problems arise when a policy is being implemented. Some policies are stopped in the middle of implementation due to lack of directions. This gives impression that the government organizations in these countries are moving in circles. Drawing inspiration from Beer's (1994) research on team synergy and especially on development of viable system model, one would comment that the policies (in underdeveloped and developing countries) fail because they are designed to fail, or in other words the policies developed contain inherent faults, which finally lead to their failure.

The purpose of the chapter is to provide an insight on ICT policy—development-process for the policy makers, so as to assist them in developing successful ICT policy for their country.

BACKGROUND

Traditional competitive advantages in manufacturing industry, like availability of raw material or technical skills are now diminishing as the price of raw material, power, and even labour are becoming globally equivalent. Major cause for this shift is due to implications of WTO framework and a shift in world exports from manufacturing to knowledge-based or intangible products over the last few years (Learner, 1995). The increase in world trade in intangibles is growing at a much higher rate than other sectors. These changes have produced socio-economic impact in developing as well as developed countries. With ICT playing a focal role in trade and economic development, new jobs of many different categories are emerging in developing countries (Kalim & Lodhi, 2004). This new business dynamics developing around the ICT are termed by many as Digital Economy. Rappa (2002) identifies eight basic Internet-based business models for the digital economy and some additions are made by Turban et al. (2002) to include *name your price*, *find best price*, *dynamic brokering*, *affiliate marketing*, *group purchasing*, *electronic tendering system*, *online auction*, *customization and personalization*, *electronic market place*, and *collaborative commerce*. These models have no doubt brought fundamental changes in the business (Gosh, 1998) and as such has changed the way supply chains are managed (Anderson & Lee, 2000) ICT developments have been taking place incrementally over the years and it has now reached a maturity level where it can be argued that ICT integrated digital economy has brought new horizons for economic development.

Nevertheless, countries that would be able to re-model their business dynamics to take advantage of the new technology would have competitive edge over countries that would be slow in adapting to the new environment. The multinational companies that have the advantage of their multicultural knowledge and operate in more than one country are pioneer to use new technology. These companies are open to new ideas and are under pressure for re-modeling their workflow to take advantage of ICT. The companies are increasingly outsourcing their non-core services to countries where these functions could be performed at a much lower price. It is no longer necessary for a product to complete its manufacturing process in a single country. A product may start its journey from a region where the raw material is cheaper, then shift to a country with lower labour rates or other manufacturing overheads, then again move to another country for high class finishing processes and finally end its journey in a country that will pay the best price.

Another emerging trend is of "tele-working", that is offering services over telecommunication channels. This model has a good prospect for growth in low income countries; with educated human capital. A critical success factor for this business model is quality of services provided by the "tele-worker". It is sometimes argued that ICT has replaced old occupations through automation, and thus created unemployment, but it should be seen that it has also created new jobs like Web designers or call-center workers and software developers, which did not exist earlier.

ICT is redefining trade routes and establishing new maps based on "digital integration" of trading partners. A map that is not traced on physical proximity of the regional trading countries, but one that is dependent on digital integration of the trading partners and business process synchronization between them. There is a great potential for growth in developing countries, but new jobs and business opportunities will require higher ICT skills. Developing countries can harvest benefits from new opportunities through better ICT business integration, but before that they must plan to develop the right amalgamation of human capital, ICT infrastructure, and business environment.

ROLE OF ICT IN NATIONAL DEVELOPMENT

The role of ICT in national development is presently a hot issue for debates, and a number of authors have expressed their views on the subject (Madon, 1994; Avgeron, 1998; Wilson & Heeks, 2000; Sein & Ahmed, 2001). The views expressed are both against and in favour of using ICT for accelerated development. The governments mostly appear to favour ICT promotion and are with the view that it will ultimately cause higher economic growth.

The development debate can be categorized into three perspectives. These are modernization, dependency, and human-centeredness (Sein & Harindranath, 2004). The modernization perspective promotes that underdeveloped countries do not have the ability to breakout from the outdate mode of production, because these countries lack the necessary knowledge and resources. The developed countries have capital, technology, skilled work force, and entrepreneurial class to achieve the growth. However, the modernists believe that ICT can act as a provider for all this. If properly used, ICT can help poorer countries to reach at a much higher economic growth and with the help of knowledge integration between the developing and the developed world, poorer countries could frog leap stages of development.

The followers of dependency perspective domain criticizes modernists accusing that it does not take into account culture and local context, which are of vital importance for sustainability. The dependency perspectives views that poverty is not accidental, but is rather caused by very process that made developed countries rich. The developed countries exploited the underdeveloped and developing countries through colonization.

Table 1. Human development index: The HDI measures achievements in terms of life expectancy, educational attainment and adjusted real income.

HDI Rank High	HDI Rank Medium		HDI Rank Low
1 Norway	58 Libyan Arab Jamahiriya	113 Bolivia	146 Madagascar
2 Iceland	59 Macedonia, TFYR	114 Mongolia	147 Swaziland
3 Australia	60 Antigua and Barbuda	115 Moldova, Rep. of	148 Cameroon
4 Luxembourg	61 Malaysia	116 Honduras	149 Lesotho
5 Canada	62 Russian Federation	117 Guatemala	150 Djibouti
6 Sweden	63 Brazil	118 Vanuatu	151 Yemen
7 Switzerland	64 Romania	119 Egypt	152 Mauritania
8 Ireland	65 Mauritius	120 South Africa	153 Haiti
9 Belgium	66 Grenada	121 Equatorial Guinea	154 Kenya
10 United States	67 Belarus	122 Tajikistan	155 Gambia
11 Japan	68 Bosnia and Herzegovina	123 Gabon	156 Guinea
12 Netherlands	69 Colombia	124 Morocco	157 Senegal
13 Finland	70 Dominica	125 Namibia	158 Nigeria
14 Denmark	71 Oman	126 São Tomé and Principe	159 Rwanda
15 United Kingdom	72 Albania	127 India	160 Angola
16 France	73 Thailand	128 Solomon Islands	161 Eritrea
17 Austria	74 Samoa (Western)	129 Myanmar	162 Benin
18 Italy	75 Venezuela	130 Cambodia	163 Côte d'Ivoire
19 New Zealand	76 Saint Lucia	131 Botswana	164 Tanzania, U. Rep. of
20 Germany	77 Saudi Arabia	132 Comoros	165 Malawi
21 Spain	78 Ukraine	133 Lao People's Dem. Rep.	166 Zambia
22 Hong Kong, China (SAR)	79 Peru	134 Bhutan	167 Congo, Dem. Rep. of the
23 Israel	80 Kazakhstan	135 Pakistan	168 Mozambique
24 Greece	81 Lebanon	136 Nepal	169 Burundi
25 Singapore	82 Ecuador	137 Papua New Guinea	170 Ethiopia
26 Slovenia	83 Armenia	138 Ghana	171 Central African Republic
27 Portugal	84 Philippines	139 Bangladesh	172 Guinea-Bissau
28 Korea, Rep. of	85 China	140 Timor-Leste	173 Chad
29 Cyprus	86 Suriname	141 Sudan	174 Mali
30 Barbados	87 Saint Vincent & the Grenadines	142 Congo	175 Burkina Faso
31 Czech Republic	88 Paraguay	143 Togo	176 Sierra Leone
32 Malta	89 Tunisia	144 Uganda	177 Niger
33 Brunei Darussalam	90 Jordan	145 Zimbabwe	
34 Argentina	91 Belize		
35 Hungary	92 Fiji		
36 Poland	93 Sri Lanka		
37 Chile	94 Turkey		
38 Estonia	95 Dominican Republic		
39 Lithuania	96 Maldives		
40 Qatar	97 Turkmenistan		
41 United Arab Emirates	98 Jamaica		
42 Slovakia	99 Iran, Islamic Rep. of		
43 Bahrain	100 Georgia		
44 Kuwait	101 Azerbaijan		
45 Croatia	102 Occupied Palestinian Territories		
46 Uruguay	103 Algeria		
47 Costa Rica	104 El Salvador		
48 Latvia	105 Cape Verde		
49 Saint Kitts and Nevis	106 Syrian Arab Republic		
50 Bahamas	107 Guyana		
51 Seychelles	108 Viet Nam		
52 Cuba	109 Kyrgyzstan		
53 Mexico	110 Indonesia		
54 Tonga	111 Uzbekistan		
55 Bulgaria	112 Nicaragua		
56 Panama			
57 Trinidad and Tobago			

The colonized countries were forced to use their resources for manufacturing products solely for the benefit of the richer countries. This view of the ICT strengthens the position of the developed countries, implying that the offshore computing and manufacturing of ICT commodities are only serving the purpose of the richer nations.

Aside from these two perspectives on ICT, the third perspective is the "human development" or "people-centered development". The focus in this perspective is on creating a society where individual potential can be realized (Sein & Ahmed, 2001) by enhancing the choices available to the masses and providing them better opportunities to compete on a level field, with improved systems. These basic systems to improve standard of living like education, health, development of democratic systems by introducing transparency, and participation at all possible levels. Human Development in a region can be measured by using human development index (HDI), gender development index (GDI), gender equity measure (GEM), or human poverty index (HPI) (see UNDP, 2001; UNDP, 2005). The human development perspective argues that the use of ICT in business and government will improve the systems in the region, thus providing masses an environment for better development.

Table 1 ranks countries on human development index as calculated by UNDP in its human development report 2005. The countries that rank low on the HDI should take advantage of ICT and achieve higher development goals by focusing their ICT strategy on human development perspectives. These countries should also consider viewing ICT directed at specific sectors like health, education, law, and order to provide better services to the masses. Countries ranking in the mid on HDI ranking may develop ICT strategy for facilitating general development activities. These activities could include development of business and economic linkages. But if a country has high education level and developed ICT infrastructure it may prefer to view ICT as a driver of economy.

In all cases, policy makers should set their priorities in the light of constraints and challenges that their specific countries face before deciding to follow a certain approach. It is quite likely that policy makers may like to follow an integrated approach.

Table 2. Development indicators for ICT (Note: Data on selected countries have been obtained from UNDP, Human Development Report, 2005)

	Telephone mainlines (per 1,000 people)		Cellular subscribers (per 1,000 people)		Internet users (per 1,000 people)	
Country	**1990**	**2003**	**1990**	**2003**	**1990**	**2003**
Norway	502	713	46	909	7	346
Iceland	510	660	39	966	0	675
Australia	456	542	11	719	6	567
Lesotho	7	16	0	47	0	14
Mauritania	3	14	0	127	0	4
Haiti	7	17	0	38	0	18
Kenya	8	10	0	50	0	..
Rwanda	2	..	0	16	0	..

CONCEPTUALIZING ICT USE AND IMPACT

It is critical for policy makers to have a clear conception on the use of ICT in social network in the region. The perspective in which ICT is viewed will have a direct impact on its use and ultimately effect the development of the region. Kling (2000) views ICT as a socio-technical network and argues that it is socially shaped. The investment figures or development indicators for ICT (see Table 2) usually quoted in the development documents does not provide the true picture as these figures are only the half of the picture. There is a genuine need to examine whether ICT was used appropriately and what was its impact on the development goals of the country. Sein and Harindranath (2004) give four conceptualizations (Vignette 1) of uses of ICT in national development.

One perspective is to view ICT as a commodity, and from this view ICT is considered as a commodity to earn foreign exchange through export of computer systems manufactured or assembled in the country, offshore software development, or back office processing. It is worth mentioning that ICT has been one of the fastest growing technologies in production and consumption in the last four decades, and the future prospects of this technology are promising (see Gurbaxain et al. 1990). The association of use and production of ICT are both found to be positively related with national development indicators, as investigated by OECD (1987) and Kraemer & Dedrick (1994). Some authors have however question rightly the potential of ICT as commodity by pointing out that every country can not become an ICT exporter.

The second perspective is to view ICT as a facilitator of general development activities and for better management, like providing reliable information to the right person and at the same time collect, organize, and process information for critical decision making. In this perspective, it can be said that ICT is used as a management tool to improve general management of a country, and as a result of this improvement in management, resources available are better utilized to create higher economic value for the system.

The third perspective for viewing ICT is as a driver of economy. This is perhaps the most active view of ICT, where it is conceptualized to have a direct impact at macro-level. In this view ICT is fully integrated with the business process, and provide the main driving force in the economy. ICT provides strategic advantage to the industry to compete globally. It is used to up-grade the industry and create economic linkages. This perspective views ICT as a critical infrastructure in the global knowledge driven economy (Hanna, 1994; Flamm, 1987) arguing that the investment in ICT is to simulate economic development and improve national productivity (OECD, 1988). Fourthly, ICT may be extensively utilized in a specific sector or development project to provide

Vignette 1. Conceptualizing ICT use in national strategy

1. ICT as a commodity
2. Facilitator of general development activities and for better management
3. As a Driver of Economy
4. ICT directed at specific development sectors or projects

immediate relief to the masses. This view focuses on immediate advantages of using ICT rather than a long term plan.

Assessing the impact of ICT on national and social development would again depend largely on its use as discussed earlier. It is, however, advantageous to have some conception on a framework to assess the impact of ICT at national level. A framework for the purpose is suggested by Malone and Rockart (1991) and later on adopted by Sein and Ahmed (2001) to measure the impact of ICT on national development. The framework suggests measuring the impact at three levels. The first-order or primary effect is noted when it is seen that ICT has simply replaced the old technology. This is evident when an increase in usage of computers in offices replacing typewriters, increasing numbers of telephones and Internet connections have been observed. The impact from this perspective would be dominated with statistics on increase in number of telephones, Internet connections, or new computers purchased. This primary view will, however, not be able to indicate the true development in the social setup of the region.

The second-order or secondary effect is the development in the communication process enabled by the technology. Probable indicators for this may be the number of research articles co-authored by the scholars in the developing countries with the developed countries or between the developing countries themselves. The references have been collected and used from Internet in the research conducted in the developing countries.

Finally the third–order or tertiary effect defined by the framework is the impact measured by the development of new technology related businesses and the social change ICT has brought to the region. A region could see development of new businesses such as services and training in the electronic communication, e-business portals, and electronic communication enabled businesses as virtual organizations. The tertiary effect can also be seen in the social setup as the society becomes more open in sharing knowledge.

Sein and Harindranath (2004) have proposed an "Integrated Framework" of ICT development by integrating the above; arguing that ICT perspective as viewed by policy makers and its use is strongly linked to the impact of ICT on national development. If policy makers are able to view ICT in its correct perspectives by keeping in mind the country's requirement, the policies developed by them will cater society's requirements and ultimately result in human development. Kling (2000) states that the "tool" view of ICT often underestimates the cost of complexities of computerization that would be associated with implementing the policy. He referred to Sein and Harindranath (2004), which advocate that despite the elevated talk on ICT impact on development, most of the developing economies have focused ICT as a "tool" and are developing policies to cultivate their ICT production facilities only. It should be kept in mind that the impact of ICT on human development can not be taken for granted. Public policies should be developed while maintaining the linkages between different sectors. Bhatnagar and Schware (2000) discussed specifically India's software industry noting that the ICT policies designed were related to "tools" view of ICT and as the industry developed it was realized that India could not fully exploit its potential without developing the linkages with rest of the sectors, like telecommunications sector, stating further the researchers add that in a number of projects introduced in the developing countries, it was later realized that social and cultural aspects are as important as the technology itself. India's computerized rural information system project (CRISP) is an example (Madon, 1994) where lack of strategic directions from Central Government for planning at the local level and issues relating to social structures obstructed the project. Summarizing the above discussion, it can be said that ICT should not be viewed as an object, but rather it should be viewed with all of its social and cultural interactions.

STRATEGY DEVELOPMENT AS A PROCESS

The risk of strategy failure can be minimized by following a strategic management framework. A policy developed without following a framework is most likely to fail. But before going into the details of framework, it would be appropriate to discuss the perspectives for approaching strategy development. The two popular approaches on strategy development are environment led "fit" and the resource led "stretch" approach. In the "fit" perspective the strategy is viewed as matching of the resources and activities of an organization or a country, whereas in the "stretch" perspective, strategy development is viewed as leverage of resources and competence to provide competitive advantage or create new opportunities. When deciding on adaptation of certain perspective for a country strategy, the policy makers should develop both perspectives and then later on implement the most productive one from available alternatives.

Strategy management is the process of ensuring that an organization benefits from the use of an appropriate strategy (Certo, 1997). The process itself can be explained as five sequential and continuing steps given in Table 3.

A feedback arrangement has to be established between all steps (from step 1 to step 5). The feedback information ensures that policy can be reviewed at all times and arrangements can be made even at later stages to keep the policy on desired track. The same steps can be followed to develop ICT policy of a country. In the first step, environment analysis has to be conducted to identify factors that can significantly influence a country's development. The analysis will help policy makers to understand complex dynamics; both inside and outside their country, and also ensure that the policy reflects country's inspiration. Environment analysis can be viewed at three levels. Firstly, the internal environment, constituting planning and controlling aspects; secondly, the operating environment which include components normally having specific and immediate implications like demand and supply, human resource component, competition, suppliers, and so forth (the list is not limited as additional components can be included to make analysis more comprehensive), and finally, the third level that is general environment. The general environment contains components that have long-term implications, which may be legal aspects, technology and social aspects, political implications, and most importantly the economic aspects. In general the policy makers must study meticulously the impact of World Trade Order (WTO) developments on their country's trade, regional trade agreements taking place between competitor countries and the status of their present ICT infrastructure. Environment analysis is of critical importance for developing any strategy,

Table 3. Sequential and continuing steps in strategy management

Step 1: Environment analysis
Step 2: Establishment of development objectives/ directions
Step 3: Strategy formulation
Step 4: Strategy implementation
Step 5: Strategy control

as it forms the foundation of the strategy. A multidisciplinary approach is preferred for such an analysis as it will balance the challenges posed by the increasing complexity in the environment.

The next step following environment analysis is determining development directions. The step determines the development path that the country is going to follow. Strategic plan is generally viewed as long-term planning that focuses on the nation as a whole. Therefore, before finalizing the orientation of the development policy, the policy makers should be clear about the mission, vision, goals, and objectives of the policy (Certo, 1997). The objectives of the ICT policy should be defined; as precisely as possible and even better, in quantitative terms. The progress and strategic control of the policy can be implemented only if there are quantifiable goals.

After deciding on the development direction for the country the policy makers must focus on the initiatives that are best suitable to achieve the vision. This starts the strategy formulation stage (see Table 3, step 3). It is a search for effective positioning of a country in relation to competitor countries and to achieve advantage within a changing environment. The information obtained from environment analysis, conducted in the earlier stage should be used by policy makers to determine available options and specific "pay off" for the country (using SWOT or critical question analysis tools, decision theory, etc.). This would provide them with a fair picture to decide on actions that would be needed to accomplish the vision set in the policy at the earlier stage. The strategy formulation stage materializes the actions need to achieve the vision.

The policy implementation stage which follows formulation stage is at least as important as strategy formulation. The policy makers often have to work back and forth between strategy formulation and implementation as every step of the policy can not be planned in detail and secondly the working environment keeps on changing. The implementation stage is action oriented and often requires change management approach. The policy makers should assess the change in the present setup that will be required to implement the strategy and the impact of that change on the rest of the society. They must take into account both formal and informal changes that will take place, when the ICT strategy is implemented.

In order to develop a successful strategy the policy makers should not only formulate ICT strategy but should also ensure that it is implemented correctly. Unfortunately policy makers are inclined to assume that effective implementation of the strategy will occur, but this is not the case in real life situations. Strategy implementation can be assured by establishing a strategy control mechanism. The strategy control focuses on monitoring and evaluation of the strategy being implemented. Policy makers developing the mechanism should understand the process of strategic audits, developing performance indicators, and the role that information systems can play in strategy control process. It is self-evident that any strategic management process would only be as good as the information on which it is based. Therefore the control process will have to be provided correct information from quantitative and qualitative data gathered by periodic surveys or by some other mechanism.

ELEMENTS OF STRATEGY MANAGEMENT

Strategy development is a complex issue because it has to deal with ambiguous situations most of the time, but even then the risk of pursuing wrong directions can be reduced, if strategic management model like the one outlined by Johnson and Scholes (2001) is followed. The model advocates three important components of successful strategic management process. If these fundamental components are mapped to ICT policy development, it can be said that, firstly the policy makers should have a comprehensive understanding of the strategic position of their country in global scenario. Secondly, the strategic choices that are available to the country for adopting the future line of action should be identified, and thirdly the policy makers should have a clear implementation plan for materializing the strategy. The three elements are viewed as integrated processes in strategic management framework (see Figure 1).

Furthermore, when analyzing the position of a country in order to develop a strategic policy the policy makers should focus on:

- Environment analysis
- Expectations and influence of stakeholders
- Resources and competences of the country

The environment of a country includes political forces influencing the region, plus economical, social, and technological status of the country with respect to its competitor countries. Environment analysis is one of the most difficult stages in developing a strategy and it is becoming even more complex with increasing globalization. The ambiguity in environment is balanced by introducing dynamism in the strategy. Dynamism in strategic policy means that the environment is constantly monitored and the strategy is managed accordingly to balance the changes in the environment, or in other words it can be said that the strategy has become a living document, which must be reviewed and adjusted from time to time accordingly.

Identification of stakeholders is of critical importance in development of the policy because leaving out a stakeholder could be fatal for the policy. Similarly attaching priority to the stakeholders is also critical as giving high priority to wrong stakeholder would naturally disorient the policy. An example of this is the clash of interest between computer vendors which are interested in selling a specific solution and public sector department that are focusing on public needs. The computer vendors would be more interested in selling their state of art products than solving a problem faced by a public sector organization.

Figure 1. A model of strategic management elements modified from Johnson and Scholes (2001)

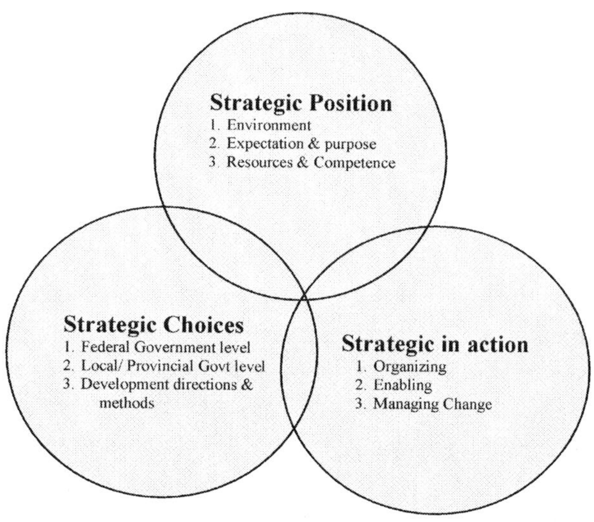

The resources and competencies are a constraint for achieving goals, a policy maker must be able to analyze the resources available to start out for setting the goals in the first place, and at the same time he/she should also have an understanding of the competency level, so that the path decided for development is achievable.

Comprehensive analysis of the environment will lead to identification of strategic choices that are available to the policy makers for directing the country's development path. The process basically involves understanding the bases of future strategy at country and local government (or provincial) levels. Factors which policy makers should consider carefully at the time of finalizing strategic choices are:

- The national development strategy
- Line department strategic policies
- Development directions and methods

ICT policy of a country is not an end in itself; its objective is basically to facilitate the country to achieve its development goals. The goals set for ICT strategy at country level could be to increase commerce to a certain level, improve transparency in government, promote democracy, and so forth. When identifying strategic choices it should be kept in mind that there should be maximum alignment between a countries development policy and its ICT strategy. The ICT policy should be designed to facilitate development goals target in the National Development Policy. It would be natural to see that policies developed by line departments would inherit goals from the National Development Policy and therefore many of the goals would be same, meaning that line departments should be working in synchronization to achieve national goals. It also means that National Development Policy should be available before the work on ICT is started or at the least, work on ICT and national development policy should start simultaneously.

A strategy is not just a good idea or a plan that is developed after a lot of efforts and then shelved to collect dust. The strategy is only meaningful when it is actually carried out. The policy makers should be able to foresee the supporting changes that would be needed to implement the new plan. Major issues that should be considered to ensure success of a strategy are:

- Organizing / or re-structuring needed
- Enabling success
- Managing change

As strategic decisions change the development path of a country, the operational decision making in the country's machinery (departments/ ministries) is also affected. New roles and processes need to be envisioned at the time of policy formation to ensure that the strategy developed will be implemental. This could also include restructuring of departments and legislating laws to support the emerging new structures. Government rules and regulations (processes), departmental boundaries, and interactions need to be updated for the smooth performance of the new system. The role of aligned departments would also need to be modified to "enable success" of the strategy. The changes in legal, educational, or infrastructural support provided by line departments bring in light another issue of extreme importance in strategy development which is "managing of change" as discussed earlier, ICT has a strong impact on social setup of countries. Societies and organizations which are slow to adapt to the changing environment feel the friction created from the disparity between the previous and the new system. Societies in the developing countries may particularly show resistance to the new technology especially when the new technology is going to have a direct impact on their way of life. The policy makers should adapt steps to manage the change in the society and promote new social setup in the larger interest of their countries. With reference to ICT policy for a country, the policy makers should consider a country as an interdependent network for creating value; keep in view interaction of policies in education, telecommunication, trade, and so forth.

A number of Performance and Strategic Management tools are available, which can be used to aid implementation of ICT strategy. Performance Prism (Cranfield University) or Balance Scorecard framework by Kaplan and Norton (1996) are examples of these new generation of management tools. These concepts are being used successfully to improve working of public sector organizations in developed countries. Implementing these tools does require a certain level of technical knowledge and experience, but the benefit gained from using these tools will certainly outweigh the expenditure on them.

It would be interesting here, to briefly review strategic framework for the Information Economy 2004–2006 developed by the Department of Communications, Information Technology, and the Arts Australia. The framework provides a good example of a well structured policy. The document begins with environment analysis and evaluation of opportunities and challenges facing Australia in the coming years. And then goes on to defining Australia's broad objectives for the information economy 2004-06 as:

- To promote social cohesion by ensuring that particular sectors, groups of Australians, and regions are not left behind
- To secure Australia's information economy against external and internal threats and to promote Australia's interests in the emerging global information economy
- To remove barriers to information economy development
- To make government an exemplar in the use of ICT to improve citizen engagement, efficiency, and effectiveness of service delivery

These objectives are translated into four strategic priorities (Table 4) and 16 associated strategies for the information economy. The objectives are aligned with national priorities. It can be seen that the priorities are interdisciplinary and broadly

Table 4. Four priorities for the information economy (Source: Department of Communications, IT and the Arts—Australia)

Strategic Framework for the Information Economy 2004–2006

1. Ensure that all Australians have the capabilities, networks, and tools to participate in the benefits of the information economy.
2. Ensure the security and interoperability of Australia's information infrastructure and support confidence in digital services.
3. Develop Australia's innovation system as a platform for productivity growth and industry transformation.
4. Raise Australian public sector productivity, collaboration and accessibility through the effective use of information, knowledge, and ICT.

defined. This facilitates synchronization among different government departments.

The policy initiatives or action plan is then developed in the light of the above. It may be noted that an integrated, rather than a narrow single facet approach is followed in developing the policy.

WHY STRATEGIC POLICIES FAIL

The reason for which strategies fail in the underdeveloped countries is not so difficult to identify. Partial and sometimes even complete failure of the policy in achieving its objectives is not uncommon in these countries. A general reason for this can be traced to the weak management practices. It is the third world countries, which have to face mammoth challenges with scanty resources and in the most difficult circumstances. Thus these countries require the best of management acumen but unfortunately, they have least developed management systems. The planning processes, implementation plans, performance management concepts are mostly unknown in these countries.

The style in which public services are provided to the population of these countries must be changed from controlling perspective to that of management. The concept of citizenship needs to be understood by the government machinery as well as the population. Asking for the above cultural change in developing countries may not be possible in near future. Focusing on down to earth management processes it is concluded that policy failures in developing countries can be traced to the causes given in the alert list (Table 5). The list is not made to follow a strict priority order. It is generally seen that the policy makers are usually blinded by their preconceived views on the subject. This bias can be removed by ensuring a participation of maximum number of stakeholders in policy development process. This involvement of stakeholders will also aid in comprehensive scanning and understanding environment reaction to the policy under development.

The involvement of government machinery that will be leading the implementation phase of the policy is equally important. It is quite normal, in developing countries, to find a government official heading a project, while the official himself/herself does not believe in the objectives of the project. The government official implements the project halfheartedly, just because he/she has to do it. These halfhearted attitudes are disastrous for projects. The projects are doomed as they lack leadership. It is this lack of leadership that later on transforms to emerge as low coordination,

Table 5. Alert list on why strategies fail?

- Inadequate environment scanning
- Inability to predict environmental reaction
- Over-estimation of resource competence
- Lack of coordination
- Lack of center / Federal Government commitment
- Failure to obtain departmental/ Local Government commitment
- Under-estimation of time requirement
- Failure to follow the plan
- Failure to manage change
- Lack of feedback mechanism

lack of commitment. Policies without leadership soon loose their directions and die an unknown death.

Absence or lack of feedback mechanism is another important reason for strategy failures. Change management requirements that should have been envisioned at strategy development stage are sometimes overlooked; resultantly environment reaction to the new setup is pessimistic. A new mindset with team building attitude is needed to overcome this challenge. Maximizing stakeholders' participation with continuous monitoring and modification of strategy implementation can increase strategy success rate. Fortunately, a number of strategic management tools have lately emerged to aid top management in implementing strategies; these tools should be integrated in the management information systems to guide the development strategy on correct path.

CONCLUSION

The chapter begins with a general discussion on popular perspectives relating to role of ICT in national development and how these perspective views held by policy makers effect ICT role in national development. It is seen that the role, which ICT can play in national development is dependent on the perspective with which policy makers view ICT. It is now released that ICT has much more to offer than just to facilitate present business models. ICT is not only re-organizing the way to do the business, but it is also redefining the world's geography by drawing an e-map for the digital economy. The countries, which have higher data transfer connections, are physically closer on e-map than the countries which have low data transfer capacity.

Secondly, the chapter presents a bird's eye view of strategy development as a process from strategic management perspective. The purpose is to understand policy development and build capacity for making sustainable policies. The details of the process are deliberately not discussed, but the issues highlighted should be enough to ring alarms in the minds of policy makers. Though nothing can guarantee success of a policy, as the best of the policies can fail, due to a number of factors which policy makers cannot predict at the time of policy formation, but policy makers can reduce failure risk by following a proper framework. Policies developed by confirming to an appropriate framework are most likely to be free from inherent flaws and thus more likely to succeed. Strategic policy is a live document, which must be constantly reviewed and appropriate changes must be made to the policy to guide it on the correct path. A new generation of strategic management tools can be integrated with the regular monitoring and evaluation information system at the implementation stage of the policy.

In a nutshell the chapter highlights that developing an ICT or any other policy should be considered from strategic management perspective. A policy should be considered complete only if it includes a comprehensive analysis of environment (strategic position), an understanding of strategic choices available for future, and most importantly a plan for turning strategy into action.

FURTHER RESEARCH DIRECTIONS

In summary, it has been argued that policy development should be viewed as a systematic process. Each stage in policy development should be well thought out, as it gives critical inputs to the subsequent stages and ultimately towards the success of the policy. The development process should not be viewed with a mechanistic approach; where a strategic decision maker envisions changes, allocates resources, and finally commands its implementation. Strategic policy development is rather a creative process involving complex social systems with variety and diversity. Therefore developing a successful policy requires "out of box" thinking. Cultural and organizational issue in the society should be considered in multi-dimensional perspectives, as these are the very factors which bring a society to life.

The research has raised a number of issues that warrant further investigation. In particular research is needed on developing a comprehensive analysis framework similar to PESTEL or Porter's five forces framework, which can provide policy makers with standard tools for environmental analysis, and aid them in determining a countries' strategic position and strategic choices that they can opt for.

Secondly research can be initiated in the domain of improving performance measurement systems. Frameworks for evaluating immediate as well as secondary and tertiary level impact of policies on the society will be a functional tool in the hands of the decision makers. Analyzing the impact of strategic policy on economic, cultural, and political systems would provide a strong feedback and strengthen policy control mechanisms. Balance scorecard approach can be extended and used to monitor a countries' progress towards its strategic and development goals.

It is observed that numerous new issues emerge as the new policy is implemented. This brings into picture, the research issues relating to change management and social network analysis (SNA). We are living in integrated and complex social systems. Therefore the success of a policy would greatly depend on culture and psychology of the beneficiary society. The policies developed should be rooted in, or as close as possible to, the present cultural parameters of the society and must have the capacity to change the social parameters to the desired values. Studies exploring solutions or suggesting a path for sustainable development initiative can draw strength from research in the field of SNA. Case studies exploring the social factors leading to success or failure of policies initiated previously or case studies conducted in other countries faced with similar situations will provide valuable insight to the practitioners.

REFERENCES

Anderson, D.L., & Lee H.L. (2000). The Internet-enabled supply chain: From the first click to the last mile. *Achieving supply chain excellence through technology* (pp. 15-22). Montgomery Research Inc.

Avgerou, C., & Walsham, G. (Eds.). (2000). *Information technology in context: Studies from the perspective of developing countries.* London, Ashgate (2000) ISBN 0 7546 1480. From: Sein, M. K., & Harindranath, G. (2004). Conceptualizing the ICT artifact: Towards understanding the role of ICT in national development. *The Information Society, 20,* 15-24.

Beer, S. (1994). *Beyond dispute: The invention of team syntegrity.* John Wiley & Sons.

Bhatnagar, S., & Schware, R. (2000). *Information and communication technology in rural development.* Case Studies from India. Washington, DC: World Bank Institute.

Certo, Samuel C., & Peter, J. P. (1995). *Strategic management; Concepts and applications.* Chicago: Austen Press/ Irwin.

Certo, S. C. (1997). *Modern management.* NJ: Prentice Hall International.

Cranfield University (2006). Performance prism measuring stakeholder value. Available at http://www.valuebasedmanagement.net/methods_performance_prism.html

Flamm, K. (1987). *Targeting the computer: Government support and international competition.* Washington, DC: Brookings Institution.

Ghosh, S. (1998). Making business sense of the Internet. *Harvard Business Review,* March-April, 126-35.

Gurbaxani, V., King, J. L., Kraemer, K.L., Mc Farlan, F.W.,Raman, K.S., & Yap, C.S. (1990). Institutions and international diffusion of information technology. In J.I DeGross, M. Alavi, & H. Oppelland (Eds.), *Proceedings of the Eleventh International Conference on Information Systems* (pp. 97-98). Copenhagen.

Hanna, N. (1994). *Exploiting information technology for development: A case study of India.* World Bank Discussion paper 264. Washington, DC: World Bank.

Johnson, G., & Scholes, K. (2001). *Exploring corporate strategy: Text and cases.* FT Prentice Hall.

Kalim, R., & Lodhi, S. (2004). Digital economy and its impact on employment in the developing countries. *The ICFAI Journal of Applied Economics, 3*(3), 61-73.

Kaplan, R. S., & Norton, D. P. (1996). *The balanced scorecard: Translating strategy into action.* Boston: Harvard Business School Press.

Kling, R. (2000). Learning about information technologies and social change: The contribution of social informatics. *The Information Society, 16*(3), 217-232.

Kraemer, K., & Dedrick, J. (1994). Payoffs from investments in information technology: Lessons from the Asia-Pacific region. *World Development, 22*(12), 1921-1931.

Learner, E., & Levinsohn, J. (1995). International trade theory: The evidence. In *Handbook of Internal Economics, Vol. 3.* Amsterdam: North Holland.

Madon, S. (1994). Designing information system for development planning. Alfred Waller, Oxon. (From : Sein, M. K. and Harindranath, G. (2004). Conceptualizing the ICT artifact: Towards understanding the role of ICT in national development. *The Information Society, 20*,15-24.

Malone, T. W., & Rockart, J. F. (1991). Computers, networks, and the corporation. *Scientific America, 265,* 128-136.

OECD (1988). *New technologies in the 1990s: A socio-economic strategy.* Paris: OECD.

OECD (1987). *Information technology and economic prospects. Information, Computer and Communications Policy Series.* ICCP No. 12. Paris: OECD.

Rappa, M. (2002). Business models on the Web. Managing the digital enterprise. Retrived from http://digitalenterprise.org/models/models.html

Sein, M. K., & Ahmad, I. (2001, June 4-6). A framework to study the impact of information and communication technologies on developing countries: The case of cellular phones in Bangladesh. In *Proceedings of the BIT World International Conference*, Cairo, Egypt.

Sein, M. K., & Harindranath, G. (2004). Conceptualizing the ICT artifact: Towards understanding the role of ICT in national development. *The Information Society, 20*, 15-24.

Turban, E., King, D., Lee, J., Warkentin, M., & Chung, H.M. (2002). Electronic commerce: A managerial perspective. Upper Saddle River, NJ: Prentice Hall.

United Nations Development Programme (2001). *Human Development Report 2001: Making New Technologies work for human development.* Oxford: University press.

United Nations Development Programme: Human Development Report (2005). Retrieved from http://hdr.undp.org

Wilson, G., & Heeks, R. (2000). Technology, poverty and development. In T. Allen & A. Thomson (Eds.), *Poverty and development: Into the 21st century* (pp. 403-424). Oxford University Press.

ADDITIONAL READINGS

Aaker, D.A., & Day, G.S. (1982). *Consumerism: Search for the consumer interest* (4th ed.). New York: The Free Press.

Ackoff, R. L.(1981). *Creating the corporate future.* New York: John Wiley & Sons.

Andrews, K. (1971). *The concept of strategy.* Homewood, IL: Dow Jones-Irwin, Inc.

Ansoff, I.H. (1965). *Corporate strategy.* New York: McGraw-Hill.

Arnold, D.R., Capella, L. M., & Smith, G. D. (1983). *Strategic retail management.* Reading, MA: Addison Wesley Publishing.

Bennis , W., & Nanus, B. (1985). *Leaders: Strategies for taking charge.* New York: Harper & Row.

Bhide, A. (1986). Hustle as strategy. *Harvard Business Review,* September-October, 59-65.

Byars, L. L. (1987). *Strategic management: Planning and implementation, concepts and cases.* New York: Harper & Row.

Chamberlain, N. W. (1982). *Social strategy and corporate structure.* New York: Macmillian.

Diffenbach, J. (1983) Corporate environmental analysis in large U.S corporation. *Long Range Planning, 16*(3), 107-116.

Donaldson, T., & Werhane, P. H. (1983). *Ethical issue in business* (2nd ed.). Englewood Cliffs, NJ: Prentice-Hall.

Drucker, P. F. (1980). *Managing in turbulent times.* New York: Harper & Row.

Fahey, L., & Narayanan, V. K. (1986). *Macro-environmental analysis for strategic management.* St. Paul: West Publishing Company.

Foote, S.B. (1984). Corporate responsibilities in a changing legal environment. *California Management Review,* Spring, 217-228.

Foster, L.W. (1985). From Darwin to now: The evolution of organizational strategies. *Journal of Business Strategy, 5*(4), 94-98.

Frederickson, J. W. (1986).The strategic decision process and organizational structure. *Academy of Management Review, 11*(2), 280-297.

Glueck, W.F., & Jauch, L. R. (1984). *Strategic management and business policy.* New York: McGraw-Hill.

Godiwalla, Y.M., Meinhart, W.A., & Warde, W.A. (1981). General management and corporate strategy. *Managerial Planning, 30*(2), 17-29.

Hofstede, G. (1980). *Culture's consequences: International differences in work related values.* Beverly Hills, CA: Sage.

Lorsch, J. W. (1986). Managing culture: The invisible barrier to strategic change. *California Management Review,* Winter, 95-110.

Mac Millan, I. C., & Jones, P. E. (1986). *Strategy formulation: Power and politics.* St. Paul: West Publishing Company.

Midas, M.T. Jr., & Werther, W.B. Jr. (1985). Productivity: The missing link in corporate strategy. *Management Review,* March, 44-47.

Mitroff, I. I. (1983). *Stakeholders of the organizational mind.* San Francisco: Jossey-Bass.

Porter, M. E. (1980). *Competitive strategy. Techniques for analyzing industries and competitors.* New York: Free Press.

Prescott, J.E. (1986). Environment as moderator of relationship between strategy and performance. *Academy of Management Journal, 29*(2), 329-346.

Steiner, G. A., & Steiner, J.F. (1985). *Business, government and society* (4th ed.). New York: Random House.

Stenier, G. A., Miner, J. B., & Gray, E. R. (1982). *Management policy and strategy.* New York: Macmillan.

Thompson, A.A., Jr., & Strickland, A. J. III. (1987). *Strategic management: Concept and cases plano.* Texas: Business Publication, Inc.

Yavitz, B., & Newman, W. H. (1982). *Strategy in action*. New York: The Free Press.

Chapter III
Developing Strategies for Competitive Advantage

Tamio Shimizu
University of Sao Paulo, Brazil

Marly Monteiro
University of Sao Paulo, Brazil

Fernando Jose Barbin Laurindo
University of Sao Paulo, Brazil

UNDERSTANDING THE COMPETITIVE ENVIRONMENT

As in any game, the success of competitive strategies depends on knowing the rules as well as the other players. Nevertheless, in the real competitive environment, the rules are not always clear and can change very quickly, demanding quick responses and making it increasingly difficult to map the moves of the main players.

Strong monitoring channels must be maintained to follow changes in the competitive environment and to discern new trends. Changes in the environment can result from emerging technologies or changes in society's behavior, and thus in clients' demands. They can also be the result of competitors′ moves in the market.

Thus developing a good competitive strategy depends on mapping the environment and constantly monitoring it to be able to capitalize quickly on opportunities or neutralize threats.

The focus of this chapter is on mapping the environment. It will show the main aspects of the model for industry analysis and explain the process of forming strategic groups and pressure maps.

INDUSTRY ANALYSIS

Porter (1979) developed a model to analyze the environment based on mapping five competitive forces in order to better understand the rules of the competitive game and to help companies to find a position in the industry with greater clarity. During the 1990s, it was often remarked that environmental analysis had little to contribute given the turbulent state of the new economy, but Porter (2001) was able to demonstrate how important environmental analysis is, especially when there are a large number of uncertainties.

A good map is most useful during a storm!

To develop a map of an industry, one needs to know how the main players behave and how the battle for greater profitability works. According to Porter (1979), there are five competitive forces that run through an industry: customers, suppliers, current competitors, new entrants and substitute products or services.

Understanding these five forces is vital to formulating a company's competitive strategy. The model for Industry Analysis is shown in Figure 1.

The pressures exercised by these forces are structural determinants both of competition and for an industry's profitability (Porter, 1979). In Figure 1, the battle for profitability in the supply chain, i.e., the bargaining power of customers and suppliers is laid out along the horizontal axis. The relationship among competitors, whether those already established, potential entrants or substitute products is shown along the vertical axis.

It is important to emphasize that industry, in this context, means an industrial sector and not a single company.

The structure of an industrial sector depends on how the competitive forces are configured. When they are stronger, there is less potential for profitability. It is easy to imagine that within an industry with fierce competition, the rate of return tends to be less. On the outside limit, an industrial structure with all five forces intensified, approaches the economic model of "perfect competition."

On the other hand, there are industries where the five competitive forces are weak, i.e. the rivalry among existing competitors is not intense, the entry barriers for new competitors are difficult to overcome and there are no substitute products. Allied to this favorable scenario, the sector is the driver of the supply chain, with high bargaining power with the others tiers in the chains, whether

Figure 1. The five competitive forces (Adapted from Porter, 1979)

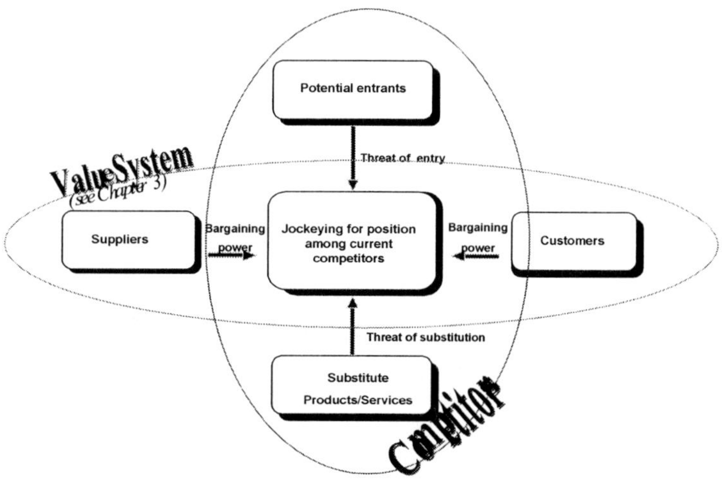

customers or suppliers. This type of industrial sector is called "five star" since is show a high potential for profitability.

Industry analysis allows us to identify which competitive forces most influence the industrial sector. It is common for one force in the competitive industrial environment to stand out and the competitive strategy should focus on it. It is then possible to influence the competitive environment in such a way as to assume a strategic position that can neutralize it, and thus to obtain a sustainable competition advantage.

It is worth emphasizing that the map of the sector structure is dynamic, since the industry can undergo changes and, depending on the magnitude of the competitive movement of each player, the shape of the five forces can be altered, having an impact on its attractiveness.

Mapping the Competitive Forces in the Supply Chain: Relationships with Customers and Suppliers

Bargaining power in the relationship with customers and suppliers has decisive impact on both an industry's profitability and attractiveness. Imagine the vehicle assembly plants and their auto parts suppliers seated around the bargaining table. It is east to predict that the scale will tilt toward the side of the assemblers.

When the customers are concentrated and direct large volumes, such as the case of the assembly plants, they manage to bargain for better prices, increasing demands for quality, flexibility in the product mix and the volume of production, as well as access to information via Integrated Management Systems – ERP. They thus diminish the industry's freedom to establish its own rules. This situation is aggravated when the products supplied by the industrial sector are standardized, are not critical to the quality of the customers' production, and can be substituted without generating costs for the change (Carvalho et al., 2003).

Table 1 shows some aspects, which determine the bargaining power of an industry, tilting the balance when they are present.

When the forces in the supply chain are intense, it is important to outline a strategy for buyers and suppliers, with criteria for choosing buyers and for selecting target customers and also for developing a buying strategy, paying attention to strategic aspects and not just operational ones.

This matrix proposed by Cox et al. (2001) is a useful tool for analyzing bargaining power between buyers and suppliers. Figure 2 shows the matrix, where the horizontal axis shows the dominance of the supplier and the vertical, the dominance of the buyer, and four quadrants: buyer dominance, interdependence, independence, and supplier dominance.

The quadrants buyer dominance and supplier dominance indicate who has power over the other. In the independence quadrant, buyers and suppliers have attributes of power, and both must be involved in order to carry out an agreed upon exchanges. In the independence quadrant, neither of the two have attributes of power, capable of demanding that the other fulfill their requirements and both have other opportunities for doing business.

Yet in the more complex supply chains, a more detailed analysis of the relationship among the links must be done. It is necessary to identify which link commands the chain, i.e. dominates the strategic activities which add value. Two aspects take on special importance in this analysis—governance of the chain and the value system.

Analyzing the Competitors: Current, Potential, and Substitutions

Competition in a given industry can be a factor that significantly limits profitability. In an industry with an intense rivalry, this can lead to predatory actions, eroding the attractiveness and profitability of the industry as a whole.

Table 1. Analysis of customer and supplier bargaining power

Customers	Suppliers
Customers are concentrated or make large volume acquisitions.	Supplier product is a critical input. It is unique or differentiated.
Purchases represent a significant part of their costs.	The group of suppliers is dominated by few companies and is more concentrated than the industry being analyzed.
The industry's products are standardized and/or are not important for the customers' product quality.	The industry analyzed is not an important customer for the group of suppliers.
The customers have all the information.	The suppliers products are differentiated and do not compete with substitute products.
The customers represent a concrete threat of backward integration	The suppliers are a concrete threat of forward integration

It is important to note that rivalry should not analyzed just in the context of existing competitors, but also in the extended rivalry which includes new entrants and substitute products and services, as illustrated in Figure 3. In industries that are threatened by substitute products or have a high potential for new entrants, competition tends to be sharper and often sets off price wars.

There are many aspects that have an impact on the rivalry among established competitors in an industry, in the battle for market share and better competitive positioning. Imagine for example a sector where growth is slow or stagnant! For a company to grow, it has to take market shares from the competitors, increasing rivalry among the established competitors, which is not the case when the industry is growing overall.

The main points to be verified in analyzing the forces of competition are:

- Numerous or balanced competitors
- Slow growth of the industry
- High fixed costs
- Absence of differentiation
- Increased capacity in large increments
- Competitors with divergent strategies
- High exit barriers

When rivalry is extensive, the threat of new entrants to an industry is strongly related to the effective entry barriers and the expectation of potential competitors regarding retaliation from already established players.

The main source of entry barriers for new competitors is the economies of scales, whether for production, research, marketing or services, which require that the entrant operate on a large scale or which place the entrant at a cost disadvantage. In addition to this barrier there are cost disadvantages, which are independent of scale, such as patent costs, access to raw material, location, official subsidies and learning curves. For a long time, Xerox took advantage of its patent for copying machine to remain all alone in the market, however, when the patent expired a lot of entrants such as Canon, HP and others came on the scene.

According to Heizer and Render (1999), learning curve models were first observed for the aeronautics industry. It has been calculated that for the period 1925-1957, the number of worker hours per unit was reduced by 80%.

Learning curves, when applied to the context of strategy, allow us to determine changes in the volume-cost trade off, and to evaluate performance for the business and the industry. The

Figure 2. Buyer X supplier power (Adapted from Cox et al., 2001)

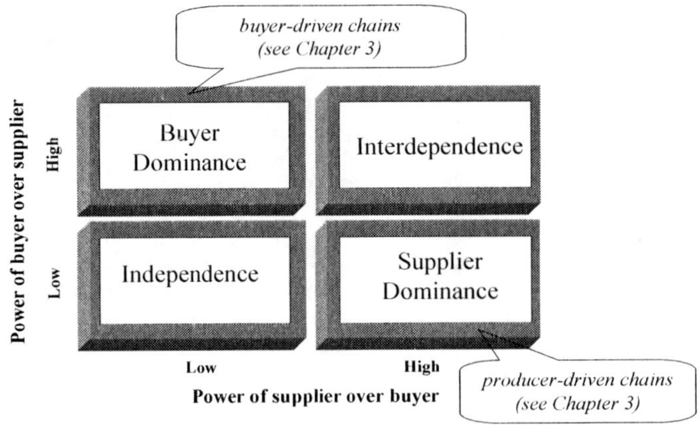

Figure 3. Forces of competition: Competitors, entrants, and substitutes

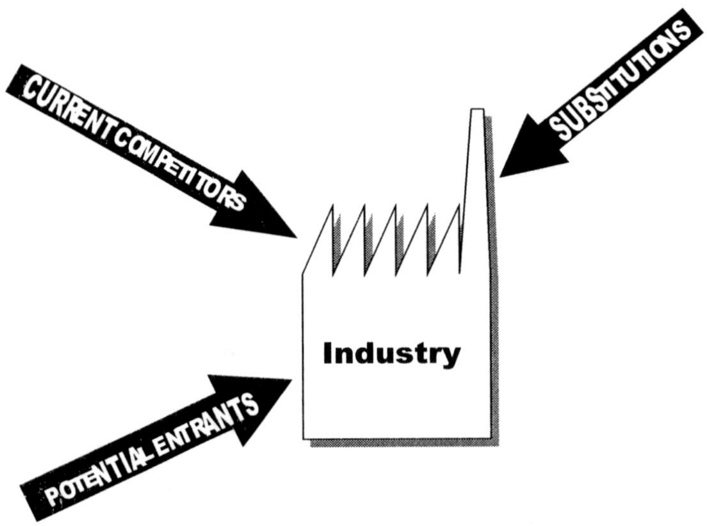

Figure 4. Factors that affecting learning curve (Adapted from Heizer and Render, 1999)

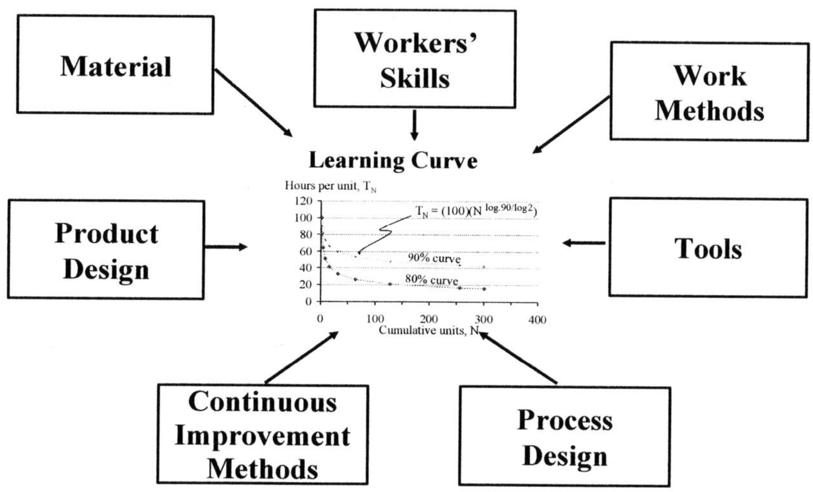

major restrictions to these models are related to the wide range of variables, which need to be analyzed, as illustrated in Figure 4.

Other important sources of barriers are: *product differentiation, capital requirements, change cost* and *access to distribution channels*.

The threat of new entrants can also effect expectations regarding retaliation from already established competitors, especially when excess capacity and/or capital are available to maintain and sustain a direct battle.

Finally, the last force of competition to be analyzed is the pressure of substitute products/services. Substitutes can be defined as those, which carry out the same function, even with another technology base (another industry). S substitutes, which are subject to price/performance, trade off with products from the industry and/or those produced by five star industries merit the most attention.

TAKING A LOOK INSIDE THE INDUSTRY

As we saw earlier, even though all the companies within an industry are subject to the same forces of competition, they adopt different strategic positions with different results. Therefore, the analysis of the overall industry can be deepened, extending to a structural analysis within the industry

The steps for doing a structural analysis inside the industry are:

- **First step:** Characterize the strategies of all the significant competitors in all their dimensions.
- **Second step:** Evaluate the dimensions and the composition of the mobility barriers, which protect each group.
- **Third step:** Evaluate the relative bargaining power of each strategic group in the industry *vis-a-vis* their buyers and suppliers.
- **Fourth step:** Evaluated the relative position of each strategic group with regard to substitute products.

- **Fifth step:** Evaluate the patterns of interdependence in the market among the strategic groups and their vulnerability to conflict initiated by other groups.

Dimensions of Competitive Strategy

The breadth of different strategies throughout an industry depends on the nature of the industry. The choice of strategy has repercussions in the strategic dimension, which demands trade-offs. According to Porter (1987) the following dimensions can be identified:

- **Specialization:** The degree to which the company concentrates effort on the breadth of its line, target-client segments and geographic markets.
- **Brand identification:** The degree to which a company seeks recognition of its trade name through marketing.
- **Channel policy:** The degree to which a company seeks to develop trade name recognition directly with the end consumer, or supports the distribution channel in selling its product.
- **Choice of channel:** The choice of distribution channels varying the company's channels, specialized or general.
- **Product quality:** The level of product quality in terms of raw material, specifications, tolerances and characteristics.
- **Technology leadership:** The degree to which the company seeks technology leadership instead of imitative behavior.
- **Vertical integration:** The amount of value added reflected in the level of integration forward and backward, involving aspects such as captive distribution channels and exclusive retail outlets, among others.
- **Cost position:** The degree to which the company seeks a position of lower cost in manufacture and distribution.
- **Service:** The degree to which auxiliary services are provided for its product line, such as technical assistance, credit, etc.
- **Price policy:** Relative price position on the market, related to cost and quality.
- **Leveraging:** The degree of financial and operational leveraging.
- **Relationship to headquarters:** Demands form headquarters; this can be a department of a diversified conglomerate, a link in the vertical chain, part of a group of businesses related to a general sector or a subsidiary. It influences goals, resources and the division of operations.
- **Relationship with governments of the country of origin and the host country:** Multinational companies have a relationship with home country governments and those of the countries in which they operate, in terms of supplying resources, regulations and other types of aid.

Strategic Groups

Strategic groups allow clustering the firms in groups of similar companies, consisting in an analytical tool that makes easier the industry's analysis. Strategic groups method assumes that similar firms react similarly in the some external disturbances (Feka et al., 1997).

Strategic groups cluster firms in groups of similar companies, forming an analytical tool that makes it easier to analyze the industry. The strategic group method assumes that similar firms react in a similar way to external disturbances. (Feka et al., 1997) While the company should stake out a singular position, it is possible to identify strategic groups that are following the same or a similar strategy within an industry. These strategic groups usually diverge in their approaches to the product or to marketing. Even though there could be cases of a single strategic group in an industry, these are rare.

An industry's historical development explains differences in strategy and the barriers to mobility, which follows the players' different resources and potentials as well as their differences in terms of goals and positioning with regard to risk. This leads to the formation of strategic groups within the industry. Yet, changes in the industry's structure can facilitate the formation of new strategic groups or make the groups homogenous.

Strategic groups affect the barriers to mobility, bargaining power up and down the supply chain, the threat of substitutions and rivalry within the industry. Global barriers to entry differ greatly depending on the specific strategic group the newcomer intends to join, as well as creating barriers to changes of position from one strategic group to another. These barriers to mobility end up discouraging changing strategic position. High barriers explain differences in profitability between companies.

Strategic groups can also affect bargaining power in the chain, since they confer different degrees of vulnerability on common buyers and suppliers. Strategic group permit one to focus on different parts of the product line, to serve different customers, to operate with different levels of quality and technological innovation, to have different cost positions, minimizing the level of exposure to competition from substitute products.

The presence of several strategic groups almost always affects the level of rivalry inside the industry, since there can be interdependence among the groups in the market or in their degree of overlap in targeting customers. The possibility of product differentiation achieved by the groups, the number and relative size of the strategic groups, strategic distancing between groups or the degree of divergence of strategy are other factors that have impact on the global rivalry of the industry.

A company's viability, as a function of its chosen strategic group, is affected by the structural analysis within the industry, as we have seen. Still, other factors affect the profitability of a company or group, such as the position of the company inside its group, the degree of competition within the group, the scale of the company in relation to others in the group, the cost of entry into the group and the company's ability to execute and implement its chosen strategy in operational terms.

It is worth stressing that it is not always true that a company with a larger market share is more viable. Larger companies will be more viable if they compete in strategic groups more protected by barriers to mobility and with a stronger position in the chain, and the smaller companies will be more viable if there are no large economies of scale and if they adopt differentiated strategies for service and technological innovation. This exception should not be taken as a rule, since that could lead to a commodity trap. The view that cost position is the sole sustainable factor on which a competitive strategy can be built, which can invade the territory of other areas of strategy, is not easily sustainable and, in global low cost positioning, is not always necessarily important.

SWOT Analysis and Strategic Groups

The analysis and choice of which strategic groups one should compete has implications in formulating strategy. Hence, these decisions have to take into consideration the comparison between a company's strengths and weaknesses, whether they are structural or related to implementing the selected strategy, and they must highlight distinctive competencies in order to maximize opportunities and minimize threats arising from the competitive environment. The SWOT analysis allows us to visualize the impact of strategic groups on strategy in a structured way as illustrated in Table 2.

Strategic Groups Map

The map of strategic groups presents a graphic demonstration of competition in an industry, identifying the relative position of the diverse adversaries within it.

The map of strategic groups is laid out along two axes, which contain the critical dimensions for industry analysis, usually dealt with at two levels, high and low, as illustrated in Figure 5. Hence, the first step in mapping is to select the strategic variables that will be used along the axes. In order to do so, one needs to decide which variables determine the most important barriers to mobility for the industry.

An analysis using maps of the strategic groups permits:

- Identification of barriers to mobility
- Identification of the marginal groups
- Graphic representation of the direction of strategic moves
- Trend analysis
- Forecast of reactions

Feka et al. (1997) state that total revenue can measure the size of each company, while its business focus can be seen in the percentage of sales in relation to the company's total revenue. The market share of the analyzed company's

Table 2. SWOT analysis and strategic groups

S Strengths Factors that build barriers to mobility Factors that reinforce bargaining power of the group with buyers and suppliers Factors that isolate the group from the rivalry of other companies Larger scale in relation to the strategic group Factors allowing for lower costs for entry into its strategic group Strong ability to implement strategy in relation to competitors Resources and abilities which allow it to overcome barriers to mobility and penetrate more interesting strategic groups	**W Weaknesses** Factors that weaken barriers to mobility Factors that weaken the bargaining power of its group with buyers and suppliers Factors that expose its group to rivalry of other companies Smaller scale in relation to its strategic group Factors provoking higher costs for entry into its strategic group Much lower capacity to implement its strategy in relation to competitors Lack of resources and ability that allow it to overcome barriers to mobility and penetrate more interesting strategic groups
O Opportunities Creating a new strategic group Move to a strategic group in a more favorable position Strengthening of the structural position of the existing group or the company's position in the group Move to a new group and strengthening of this strategic group's structural position	**T Threats** Other companies join its strategic group Factors that reduce barriers to mobility of the strategic group, diminishing its power in relation to buyers and suppliers, worsening its position in relation to substitute products or increasing its exposure to greater rivalry Investments that aim to improve the company's position increasing barriers to mobility Attempts to overcome barriers to mobility to join more interesting or entirely new strategic groups

Figure 5. Division of the industry in homogeneous groups (Adapted from Scaranello, 2004)

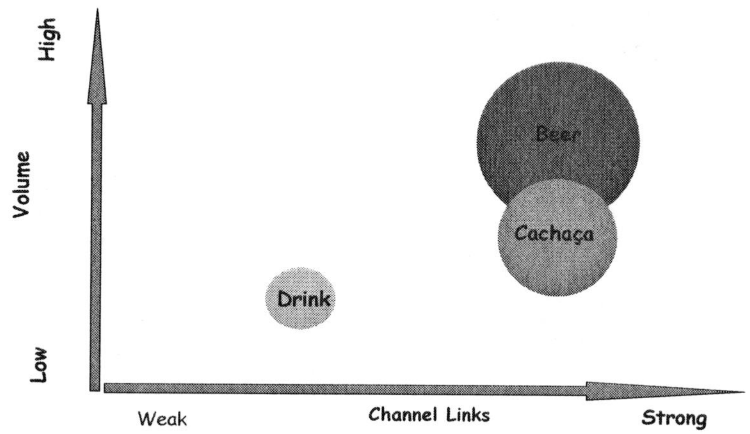

product in relation to its competitors determines its market positioning. However, it is important to stress that other measurements could be used to represent better the competitive environment in a specific industry.

Competitive Pressure Diagram

Another instrument to map the competitive environment is the competitive pressure diagram (D'Aveni, 2002). According to this author: "The more two firms' products or geographic markets overlap, the more pressure they exert on each other. The pressure is proportional to the importance of markets to each firm and their degree of penetration. This simple concept enables organizations to quantify the degree of pressure that one rival puts on another." As D'Aveni (2002) writes, pressure systems can never be frozen; the best that can be achieved is a kind of 'dynamic stability'.

D'Aveni (2002) states that the competitive pressure from a given incursion into a rival's geographical or product market can be measured by two factors: the importance of the market (I.M.) and the size of the incursion (S.I.). This pressure is given by the formula seen in Vignette 1.

Similarly, to the strategic groups, the pressures map is valid just for a short period. In this diagram, each circle represents one of the market players and arrows represent the pressure each player is exercising on the others. The diameters of the companies' representative circles mean the size of the respective market share at that moment. The width of the arrows in the diagram indicates the intensity of the pressures.

These pressure maps can be also useful to take decisions about, offensive or defensive moves, in conjunction with other decision making tools like the rivalry matrix, decision trees and Game Theory. However, Furrer and Thomas (2000) warn that the analyzed market should have few decision variables and a predictable environment.

In both kinds of competitive environment maps, data about the market should be obtained from secondary sources in order to construct these diagrams.

Vignette 1.

Competitive Pressure = I.M. x S.I. Where: ▪ I.M. - the percentage of the targeted firm's total revenue represented by the market; ▪ S.I. - the market share achieved by the invading firm.	(1)

INTERNAL ANALYSIS OF THE COMPANY TO ADEQUATE POSITIONING

Identifying the competencies a company needs in order to gain competitive advantage in a market, which is in constant mutation, is an arduous task. However, the capacity for knowing its potential and developing it in a consistent way can be decisive for a successful positioning on the competitive scene.

Self-knowledge propounded by Sun Tzu and other philosophers is a decisive weapon in the competitive game.

The focus of this chapter is to discuss how a company should position its competitive strategy in order to maximize its competencies, but without losing sight of the competitive state of the industry overall.

Core Competencies

The word competence comes from the Latin, *competere*. The derivation of the concept of competencies can be seen by taking apart the Latin word: ***com***, which means together and ***petere***, which means effort. Identifying the set of forces that will be able to lead the organization to build a sustainable competitive advantage is vital in the context of strategy. (Rabechini Jr & Carvalho, 2003).

Prahalad and Hamel (1990) relate the competitive success of a corporation to its ability to identify, cultivate and exploit its core competencies.

For these authors, competitiveness in the long run derives from the ability to build core competencies more rapidly and at a lower cost than the competition. To build them, it is necessary to meld a broad range of abilities in production and technology with competencies the enable individual deals and adapt rapidly to new opportunities, generating competitive advantage for the corporation.

In cultivating core competencies, the corporation manages to share costs, without needing to invest more than the competition in research and development.

Figure 6 represents the diversified corporation as a large tree. The trunk and larger branches are the core products, the smaller branches are the business units; the leaves, flowers and fruits are the final product. The roots are core competencies, whose function is to distribute nutrients and provide stability to the tree.

Each part of the tree determines a different plane of competition: core competencies, core products and final products. The battles for market partners are engaged at each of these levels, and leadership in final products does not imply leadership at the other levels. Given a dominant position in relation to core products allows a company to mold the evolution of application and finals markets and managing to achieve economies of scale and of scope, since the market share of core products can come from multiple industries.

Core products incorporate one or more core competencies, making the link between core competencies and final products. These products are the key components of various final products and contribute to generating value in the final product, materializing core competencies. According to Prahalad and Hamel (1990), Matsushita holds 45% of the market in the manufacturing of key components for VCRs (core product), while its brands participation in the final product (Panasonic, JVC) represents 20% of the market. Another of Matsushita's core products is the compressor, with 40% of the world market, while its participation in air conditioners and refrigerators is quite small.

The concept of core competencies is decisive both for services and manufacturing.

In contrast to final products that are consumed upon use, core competencies are enhanced when applied or shared, even though if they are not well constructed, they can be lost since knowledge also deteriorates.

Core competencies require a commitment to working on organizational frontiers, involving different business units, several levels of hierarchy and different functions. Consequently, the traditional concept of a corporation unnecessarily limits individuals' abilities by making functional cut offs that patchwork the corporation into autonomous business units, makes core competencies more difficult to achieve.

Identifying Core Competencies

According to Prahalad and Hamel (1990) it is hard to build world leadership when based in more than five or six core competencies. Consequently, if a corporation has not yet managed to achieve a lean list of competencies, it can apply at least three tests to identify them, as follows:

- **First test:** Provides access to a broad variety of markets
- **Second test:** Makes a significant contribution to the benefits perceived by consumers in the final product
- **Third test:** It is hard for competitors to imitate

To aid in this process, it is interesting to have recourse to the idea of the tree, extending the concept of core competencies to core products, and finally, to end products. To make this exten-

Figure 6. The tree: A diversified corporation (Adapted from Prahalad & Hamel, 1990)

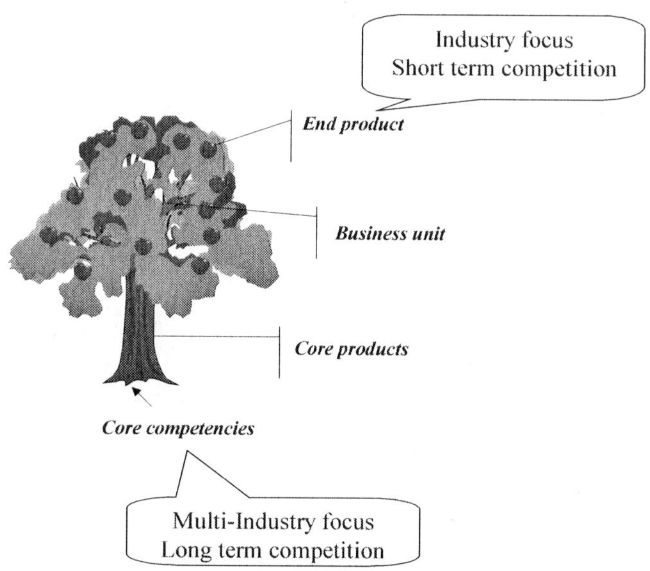

sion in a consistent way, it is suggested that the company reflect on the following questions:

- How long can competitiveness in business survive if the core competencies are not controlled?
- How central are these core competencies, in terms of generating value for the client?
- What future opportunities will be lost if these particular competencies are lost?

It is important to keep in mind that the complex harmonization of individual technologies and abilities needed to build core competencies cannot be simply acquired by a competitor. It might even acquire some of the technology, but would not be able to duplicate the pattern of internal coordination and learning.

Another important component for self-knowledge is to identify which generic strategies will permit a more adequate positioning in order to construct sustainable competitive advantages.

Generic Competitive Strategies and Positioning

As Porter (1979) suggests, the development of any company in a given industry can be broken down into two parts, the first deriving from the average performance of all its competitors, and the second to the company's performance relative to the sector.

The issue to be discussed in this chapter is the company's positioning in order to obtain sustainable competitive advantage. This advantage should guarantee the company greater income that the average of the sector where it is active.

According to Porter (1979), there are two basic kinds of competitive advantage: leadership in cost or differentiation. Besides, the scope of strategic action should be defined. The target can be a broader market as a target, with several or restricted, acting only in a specific segment.

Based in these two dimensions, there are four possible generic strategies, as illustrated in the

Figure 7. Generic strategies (Adapted from Porter, 1987)

quadrants of Figure 7. The generic strategies are: leadership in cost, differentiation, focus on cost and focus on differentiation.

Defining a generic competitive strategy aims at establishing a favorable, profitable, sustainable competitive position, against the forces that comprise the competition in an industry. Not doing clear positioning in one of these strategies can result in inferior performance, with a high risk that the company has an average performance because of the different trade off of these two kinds of strategies.

In addition, the company disposes itself to deliver two distinct types of value packages for the client (cost and differentiation) or to oscillate between them over time, and this can confuse its image with customers and cause it to lose credibility and its reputation.

The alignment of all an organization's activities around a generic strategy makes its implementation easier, as well as its communication with customers, employees and stockholders. Finally, when the strategic positioning is not clear and the trade offs are not made explicit, it is hard to build the needed resources, creating problems with internal coordination Porter (1996). In this manner, the most critical position in terms of strategic positioning is the central region, as highlighted in Figure 7.

Nonetheless Deschamps and Nayak (1995) warn that this or-or type proposition can lead to errors of interpretation called the Commodity Trap. The Commodity Trap comes from an error of interpretation that associates strategies of differentiation to small, specialized companies active in segments or niches of the market, leaving it to the large companies to adopt strategies of low cost.

Before defining the generic strategy, it is important to verify the positioning of the competitors, by making a detailed structural analysis inside the industry, highlighting the major strategic groups and their dimensions. Remember that the essence of good strategic positioning is selecting activities that distinguish the company from its competitors.

We stress that, in principle, the strategy of cost leadership is not intrinsically better than that of differentiation or vice versa. Everything depends on the external context and the unique conditions of each company!

Cost Leadership

The company that seeks leadership in cost has to focus the entire organization to achieve this objective, creating a culture of low costs, working constantly to reduce waste and remaining constantly attentive to the cost positions of its competitors.

In order to obtain this advantage, the company needs to make trade-offs, such as letting go of greater variety and flexibility in favor of lower costs. On the other hand, this does not mean that the company that adopts this strategy will have poor quality products.

The corporation that competes in this strategy has to create a gap in cost performance in relation to other competitors in the industry, which allows it to increase margins of profit, as illustrated in Figure 8. If the company manages to establish a level for costs which is much lower than the competition, but at the same time sets off a price war, this will reduce the margins and consequently its profitability. The goal is to work with costs well below average for the industry but to set prices only slightly lower than the competition.

Companies such as the airline GOL, are always searching in different areas, involving quick in flight service, that doesn't require preparation and facilitates the later cleaning of the airplane, quick boarding and debarking procedures, making equipment available more rapidly for new flights, as well as investing in technology and maintenance focused on operational efficiency.

The bases for competition in this strategy are shaken up when competitors are able to quickly overcome the gap, eroding the bases for cost lead-

Figure 8. Generic strategies: Cost leadership

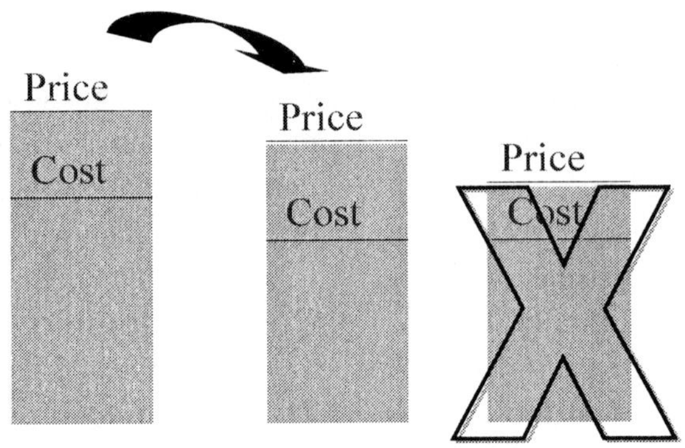

ership, or when technology changes, eliminating advantages of scale, for example.

Differentiation

The rationale for a generic strategy of differentiation is also to increase the company margin, but starting from the price variable not the cost variable.

The goal of the companies that compete for differentiation is to identify and satisfy customers' needs, but those needs for which the client is inclined to pay a premium price. Therefore, the greater viability obtained in this business profile derives form the price premium. This does not mean, however, that companies that compete with this generic strategy can disregard the price variable. Once the target for differentiation has been defined, only costs associated to differentiation can be absorbed. In other aspects, the company should also exercise careful daily cost management, if it doesn't, it will be eliminating the margin obtained from the premium price with excessive costs as illustrated in Figure 9.

Companies such as the Brazilian airline TAM seek out added value by offering differentiated services to the client which allows it to charge more, and it clearly communicates the difference to the client—"TAM's way of flying." To add value, it offers a more customized service in the air as well as on land. In opting for this approach, the trade offs with the lower cost positioning are explicit. .

Results of competitive advantage in differentiation usually lead to investments in innovation and technology, always with the focus on the specific needs of the customers which add value, as well as the use of more specialized labor who in counterpart obtain higher remuneration.

The bases for competition in this strategy are shaken when the competitors are able to quickly manage to imitate the target of differentiation offering very similar packages of value or the differentiated object loses importance due to changes in the environment. For this reason, these companies have to be always in motion, improving performance and innovating more rapidly than their competitors.

This strategy implies higher risks, since the target of differentiation chosen by the company will not be able to obtain a premium price. In this case, the company will have costs higher than the industry average, without being able to charge more for its extra costs, since customers

Figure 9. Generic strategies: Differentiation

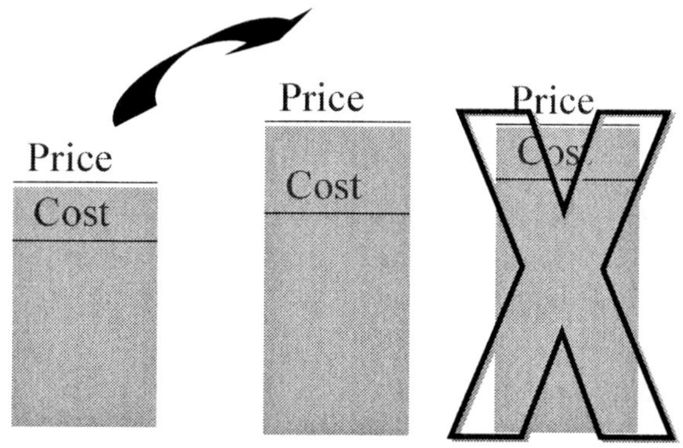

don't see value in the package offered. This happened with several retail companies that decided to differentiate their products through selling on the Internet, but didn't manage to add value, since the average consumer hasn't even got a computer, and they suffered losses with B2C (business to consumer).

Focus

The generic strategy of focus, on cost as well as differentiation has many things in common with the strategies of cost leadership and differentiation, respectively. Basically, they differ in competitive scope.

The key to this strategy is to find the lucrative, sustainable target segment to exercise the strategy, whether of cost or differentiation. This focus on a segment allows the company to develop performance superior to its competitors, who have multiple targets. This segmentation can be a function of a geographic cut, or of scale, or an activity that reaches a specific type of consumer.

Once the target segment has been selected, the company has to devote all its efforts consistently in this direction, clearly communicating its option to the market. To have greater success in selecting the segment to implement the strategy of focus, segments that are more needy than others must be prioritized.

As with the strategy of differentiation, here the risks are also high since in addition to the specific trade offs of strategy of cost or differentiation, there are trade offs related to the target segment, for example, abandoning distribution channels. In case the trade offs made do not result in value for the customers, these choices end up generating performance inferior to the industry average.

Defining generic strategy has a decisive impact on manufacturing strategy and the definition of competitive criteria that orient it, as we will see shortly.

CASE STUDY: STRATEGY AND DECISIONS IN PERSPECTIVE

Case: Competition in the Automobile Industry

Auto assembly plants engaged in much more respectful competition in the past than now.

Partnership such as Autolatina in Brazil, shared between Volkswagen and Ford, from 1987 to 1995, were viable in the closed market, but ceased to be so with the opening of commerce. The number of competitors with plants in the country multiplied, from those already installed before the opening, GM, Fiat and Toyota, were joined by the French – Renault, Peugeot and Citroën; and the Japanese, Honda. Presently the majority of models produced in Brazil are of European origin.

To aggravate the situation, the automotive industry began to experience slow growth during that period, which led to even sharper competition.

The assemblers depended more on the supply of components produced and supplied by the first tier suppliers, with higher added value, which related to with second and third level tiers in the supply chain, who manufactured sub-sets and individual components.

Case: Analysis of Competition in Belem

The absence of large Brazilian retail chains in the city of Belém, in the Amazon State, is due to difficulties in meeting the specifics of the local market and the complexity of supplying it.

In the 1980s, the Pão de Açúcar Group, the largest supermarket chain in Brazil, was the largest major supermarket chain present in Belém, but ended up beating a retreat during the group's restructuring. The Bom-Preço channel also gave up fighting over Belem's consumers, closing ten stores in 1993.

Local groups were very aggressive in defending their territory, balancing modernizing of their stores with their regional roots. Peculiarities of local demand are very marked. Regional customs making the residents of Belem very particular consumers of things such as fruits and vegetable range from fruit and vegetables typical of the region, such as jambu, bacuri, cupuaçu, teperebá or uxi, to the substitution of fresh milk with powdered milk. Imagine the effort needed to supply customized products for department stores or supermarkets in this city!

Due to their good knowledge of the characteristics of the local market, the local retail groups, such as Yamada, Leader and Visão have managed to hold their own. Among the major retail chains in the country, only C&A, Arapuã, Riachuelo and Lojas Americanas dispute the local market.

Source: Adapted from Revista Exame (2000)

Case: Bologna Competitive Market

The bologna market in metropolitan São Paulo, Brazil, is highly competitive. The main players are six high competitive companies with strong local brands: Cardeal, Ceratti, Marba, Rezende, Sadia and Perdigão.

Two pressure maps were drawn and each diagram is a picture of the São Paulo's bologna market for each period analyzed. For the studied industry, it was assumed that the importance of the market (I.M.) was the percentage of bologna in the company's total revenue and the size of the incursion (S.I.) was the market share achieved reached by each analyzed player.

The strongest pressure ordinarily gets the quickest or the most effective response. Figure 10 shows a competitive pressures map, which represents the beginning of a great offensive movement of Ceratti over its stronger competitor, Marba, with significant reflexes on the others. It seems that Marba did not make any effort to avoid the imminent aggression from Ceratti.

Figure 11 shows a significant expressive retaliation by Marba over Ceratti for the purpose of recovering its market share. Cardeal also made sales efforts to gain market shares from Ceratti, Sadia and Perdigão. Apparently, Perdigão had directed its efforts to other regions in Brazil, and maintained its margins in São Paulo.

Source: Adapted from Pereira, Carvalho and Laurindo (2004)

Figure 10. Competitive pressures map in moment 2 (2001)

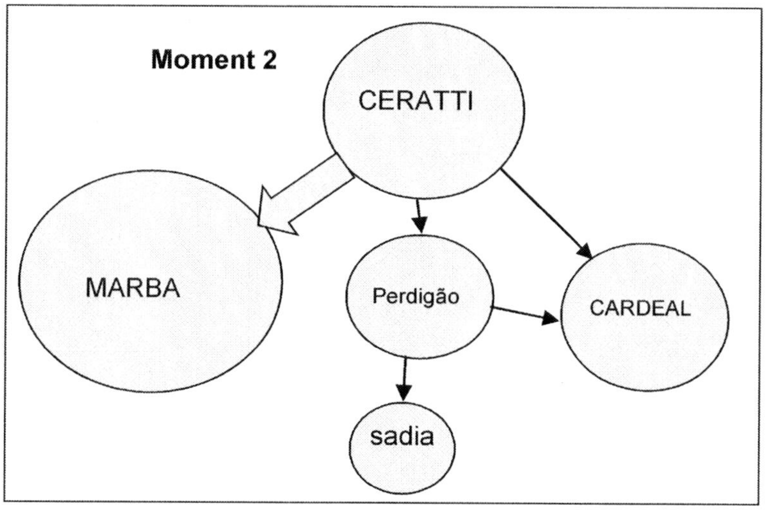

Figure 11. Competitive pressure map in moment 3 (2002)

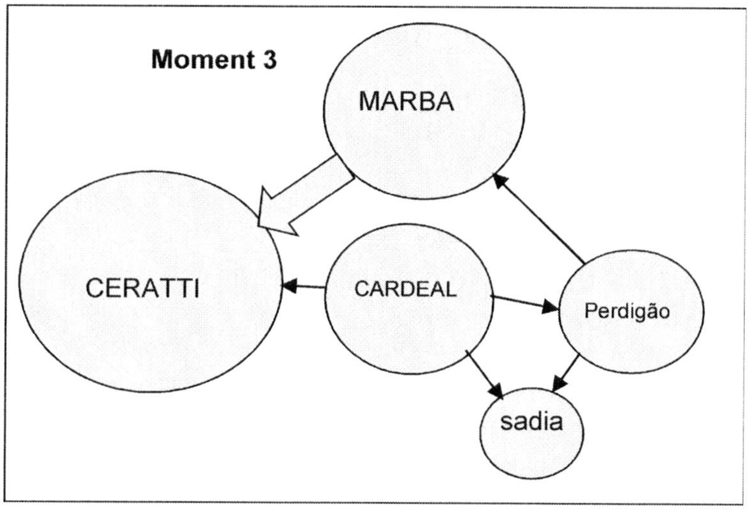

Case: Building Competencies at NEC

At the end of the 1980s, NEC emerged as a leader in the area of semiconductors and as one of the major competitors in producing telecommunications and computers. In addition to consolidating its position in mainframes for computers and acting in the areas of public commuters and transmitters, it began to produce products for everyday use such as cell phones, faxes, laptops—linking telecommunications and office automation.

NEC is the only company among the five major revenues in telecommunications, semiconductors and mainframes industries.

Retuning to the beginning of the 1970s, NEC articulated a strategy to explore the convergence of computers with communications, which was called "C&C". The success of this management strategy resulted in the acquisition of competencies, especially in semiconductors. NEC formed the "C&C Committee" composed of high-level executives to supervise the development of its core products and core competencies alongside its individual business interests. In accordance with its strategic structure, NEC transferred enormous resources to fortify its position in components and central processors. By employing collaborative arrangements to multiply its internal resources, NEC was ready to accumulate a vast range of core competencies.

NEC carefully identified three channels for technological evolution and interrelated markets, involving things from its large mainframes to distribution processes, components from simple ICs to VSLI and communications from mechanical exchange bars to complex digital systems that we now call ISDN. Finally, NEC determined that semiconductors would be its most important product.

To build up its competencies rapidly and at low cost, NEC got involved in numerous numbers of strategic alliances, which numbered over one hundred just in 1987. In mainframes, the best-known alliance was with Honeywell and Bull. Almost all the arrangements in the area of semiconductors and components were oriented toward access to technology. The research director at NEC summed up this acquisition of competencies during the 1970s and 1980s thusly "From the point of view of investment, it was much cheaper and quicker to use outside technology. There was no need for us to develop new ideas."

Source: Adapted from Prahalad and Hamel (1990)

Case: In Search of a Generic Strategy at Bunge

The Bunge group, the third largest processor of soy in the world, with billing of ten billion dollars per year, is active in Brazil, Argentine and the United States.

Brazil is responsible for about half of world sales, and it is only here that the group has a consumer goods division, Santista Foods, with 20 factories and over 100 brand names that run from breads to margarines.

About three years ago, the Bunge stockholders resolved to focus on the agricultural commodities market to compete with the Americans, Cargill and ADM, the two largest corporations in the sector.

Based on this decision, several business units were sold, but Santista, which had been put up for sale in 1998, remained. According to a former Santista executive: "the consumer area always represent the least part of Bunge's billing and 90% of its headaches," constituting a kind of foreign body in the Bunge organism, since the integration of this company into the global strategy has required major efforts by its executives.

Santista has been devalued by about ten million dollars since its acquisition by the Bunge group, with significant liquid losses in 2000, causing Bunge's entire Brazil operation to close in the red.

The total profit on other operations—pulled by the fertilizer area—was about 60 million dollars Despite the slips in the consumer areas, the group has been very successful with its strategy in the area of commodities and fertilizers, say Luciana Massaad, investment analyst for the Itaú Bank.

The solution to problems caused by maintaining a division foreign to Bunge's focus has to do with the challenge of valorizing Santista Foods and Sears, separating out the businesses that are of interest to the group, integrating them into the global strategy and selling off the rest, preferably for a good price.

Many of Bunge's difficulties with the consumer area come from the group's traditional vocation in the area of agricultural commodities. This has been the group's area of competency since 1818, when it founded the Koninklijke Bunge in Amsterdam to import grain from the Dutch colonies. Looking for wheat, Bunge arrived in Argentina in 1874. In 1906 it landed in Brazil, acquiring Moinho Santista. During periods of prosperity for wheat, the food group began to develop around these mills, which became accustomed to large volume business. The executives trained at Bunge were never close to the consumer market as were the teams from companies such as Unilever, one of the principals of Santista Alimentos in the area of margarine and mayonnaise. Until a short while ago, sales personnel spoke of tons of bread sold and not in units, as this is normally referred to in other consumer goods companies. "We will never be like a Unilever or Nestlé in terms of brand name management," says Weisser.

The strategy of the team led by Weisser, at least up until now, has been to eliminate the less profitable brands from Santista Alimentos. Some products have also bee redirected to large institutional clients, such as bakeries. "Publicity efforts are not needed in order to serve this market," says analyst/Luciana, from Itaú. "What is important is volume. And Bunge understands volume."

The idea is to integrate Santista more and more into Bunge Foods, Bunge's Food Company in the United States that does not have consumer brand names and is totally dedicated to the sales segment for bakeries, industries and to the area of its own brand names in supermarket chains. The group is seeking to create synergy between the two operations.

Source: Adapted from Mano (2001).

QUESTIONS FOR DISCUSSION AND REFLECTION

1. What is the impact of B2B (*business to business) relationships on bargaining power in the links of the chain?*
2. The merger of Brahma and Antarctica, which created AmBev, aimed to adapt the two national capital companies to the commercial blocks of the Americas, in order to confront the commercial opening. Give the asymmetry of the companies, the market was concerned Brahma would swallow Antarctica with the merger. What has been the impact of the merger that created AmBev on the shape of the beer industry? Which force was most affected?
3. Name a war of substitute industries that has been broadcast in the media.
4. What are generic strategies? Cite an example of each type of strategy.
5. What are core competencies? How can they be identified?
6. Considering the case in item 2.5.1, answer the following questions:
   - Analyze the five forces of competition in this industry
   - Did the changes in the patterns of growth in this industry affect the intensity of the forces of competition? Which ones?
7. Considering the case in item 2.5.2, answer the following questions:

- Which strategic dimensions would represent well an analysis inside the industry in this case?
- Which would be more adequate for an analysis of strategic groups?
- Make an analysis of competition in this context.

8. Considering the case in item 2.5.3, answer the following questions:
   - Compare the strategic groups map and competitive pressures map.
   - Make an analysis of the competition in this context.
9. Considering the case in item 2.5.4, answer the following questions:
   - What are NEC's core competencies and products?
   - What are its final products? Comment on the range of markets reached by each of these products.
   - How was the concept of competence implemented at NEC?
10. Considering the case in item 2.5.5, answer the following questions:
    - Using Figure 2.7 as a reference, locate Bunge Foods and Santista in one of the quadrants.
    - Bunge Food has to harmonize its companies with disparate generic strategies, what are the main points that need to be balanced? What generic strategy should be adopt?

REFERENCES

Carvalho, M. M., Laurindo, F. J. B., Pessôa, M. S. De P. (2003). Information technology project management to achieve efficiency in Brazilian companies. In S. Kamel, *Managing globally with information technology* (pp. 260-271). Hershey, PA: IGI Global.

Cox, A., Sanderson, J. E., & Watson, G. (2001) Supply chains and power regimes: Toward and analytic framework for managing extended networks of buyer and supplier relationships. *The Journal of Supply Chain Management, 37*(2), 28-35.

D'Aveni, R. A. (2002). Competitive pressure systems: Mapping and managing multimarket contact. *Harvard Business Review*, Fall.

Deschamps, N. (1995). *Product juggernauts.* Harvard Business School.

Feka, V., Xouris, D., & Tsiotras, G. (1997). Mapping strategic groups: An international example. *Journal of Business & Industrial Marketing, 12*(1), 66-75.

Furrer, O., & Thomas, H. (2000). The rivarly matrix: Understanding rivarly and competitive dynamics. *European Management Journal, 18*(6).

Heizer, R. (1999). *Operations management.* Prentice-Hall.

Mano, C. (2001). Não sei se caso ou compro uma bicicleta. *Revista Exame,* i. 735.

Pereira, P.L.S., Carvalho, M. M., & Laurindo, F.J.B. (2004). Competitive mapping in a Brazilian food industry. In *EUROMA2004 - European Operations Management Association International Conference, 2004, Fontainebleau. Proceedings of EurOMA2004* (Vol. 1, pp. 247-256).

Porter, M.E. (1979). How competitive forces shape strategy. *Harvard Business Review*, 137-145.

Porter, M.E. (1987). From competitive advantage to corporate strategy. *Harvard Business Review,* May-June.

Porter, M.E. (1996). What is strategy? *Harvard Business Review*, 61-78.

Porter, M.E. (2001). Strategy and the Internet. *Harvard Business Review*, 63-78.

Prahalad, C.K., & Hamel, G. (1990). The core competence of the corporation. *Harvard Business Review,* 79-91.

Rabechini Jr, R., & Carvalho, M.M. (2003) O perfil das competências em equipes de projeto. Revista de Administração de Empresas—RAE Eletrônica—FGV.

Revista Exame (2000). Na terra do açaí. São Paulo Nov.

Scaranello, B.M. (2004). Análise da Concorrência: Um Estudo de Caso do Setor de Bebidas. Trabalho de Formatura. Escola Politécnica da USP. Orientadora: Marly Monteiro de Carvalho. São Paulo.

Chapter IV
The Internet Challenge to Public Service Broadcasters

Juan Luis Manfredi Sánchez
SEK University, Spain

ABSTRACT

The expansion of the social influence of the Internet has led both general interest and national public broadcasters to a structural challenge. The new online paradigm demands a full version of the public enterprise, which include all the available tools to achieve its mission. The Internet would be the ideal complement to the public service. Before this, public broadcasters should define its core business to attend the public demand and to collaborate in the creation of the public sphere. Finally, the clear definition of the public activities and the separation of public and commercial finance will be a main point. To confirm the benefits of the intervention of the public broadcasters the cases of Spain and United Kingdom will be studied.

INTRODUCTION

The social influence of the Internet, the popularization of both asymmetric digital subscriber line (ADSL) networks and wireless, in addition to the expansion of the public sites that offer free access (universities and cyber cafes), have led both general interest and national public broadcasting stations to a structural challenge similar to that occurred in the eighties. It was during that decade when the liberalizing wave entered the sector. In 1980 in Western Europe there were 36 public television operators and only three private companies. Other similar countries as Japan, Canada, and Australia showed a similar scene with the hegemony of the NHK, the CBC, and the ABC respectively. At the end of the nineties, the figures change. Private companies represent the 56% of the whole market, moving from 3 to 59 operators (McQuail & Siune, 1998). The number of users and the time they spend surfing the Internet grow exponentially while time devoted to watch television is reduced.

Since the appearance of the Internet, several decisions have been taken to face the declining audience ratings, some examples are the popularization of the contents, the participation increase (through SMS and telephone calls), or the implementation of the programme contract. These measures have been proved to be inadequate, as the most of the public operators have not developed a global strategy for both broadcasting and Internet. Furthermore, this time the challenge is much more important than that taken up in the eighties since the rationality of a public service television company is much more based on analogic technologies, limited spectrum, and advertising (McGonagle, 2001, p.6).

Nowadays, public service television companies are much more organized than ever before, as it exists a real danger of both vertical and horizontal oligopoly in the market and because after 20 years of liberalization, political and social functions of communication have been scarcely attended by commercial operators. The media's hypercommercial growth, the concentration of the property, and the pressure to fuse news and entertainment are examples of how the "kind" media industry is really far from the real citizens' necessities. Croteau and Hoynes (2001) explain that "the result is a reduced public sphere—the discursive space which is a central component in the democratic society—and a weakened democracy" (p. 208).

Internet and functions derived from its use can improve and complement the public service broadcasting mission statement, since it is difficult that the private operators assume the economically unprofitable obligations. Karol Jakubowicz believes that "audiovisual public services welcome the digital technology as they are finally offered an opportunity to make their work properly" (Jakubowicz, 2002). The audiovisual reconfiguration represents an only opportunity to widen the range of services, renew and innovate, and make public television become the engine of the public policy of communications. The digital technology is introduced as the opportunity to establish a strategic leadership (benchmarking) and a communication policy for the whole of the society.

These ideas carry out a central function in the national media ecosystem. It appears in the Report on the Reform of the State-Owned Public Media, which is inspired by the new Spanish legal framework after 25 years of stagnation, and it also appears in the audiovisual public service review in United Kingdom with the British Broadcasting Corporation's (BBC) leadership. The Spanish Report goes deep in this matter and reuses the idea of a public service as the fuel to migrate to the digital age. It argues that the objectives of the public service are valid if they fit the socioeconomical and technological context in order to relate the action and the public interest to the knowledge driven society. Radiotelevisión Española (RTVE) will benchmark in the digital age, if it widens the benefits of the digital services (Report on the Reform of the State-Owned Public Media, 2005).

The BBC's reform to face the digital age is justified as it defends the social interest, because it supports the communities' development and connects civic interests. Secondly, it defends the interest of democracy, because it strengthens the public information that allows the citizens' free choice. And thirdly it defends the cultural interest, because it maintains both national and independent production (Thompson, 2003). The board of governors has accepted the digital challenge and now consider the Corporation as an organization that consists of radio, television, and new media that enrich the services offered by the two traditional media. The interactive services (BBCi) and the web site (www.bbc.co.uk) are the natural extensions of service and show singled out objectives. The inversion on digital services during the financial year 2003/2004 grew to 451,52 million euros, 29 million more than during the previous one (BBC, 2004a).

In both cases, the Internet has become a central point to carry out the entrusted mission, conceiving public communication and cultural policies in a wider way. This redefinition of the mission demands a comprehensive management of the public broadcasting enterprise, a leading unit and a specific budget appropriation.

BACKGROUND

The Internet will not eliminate television from everyday life, because both of them play different roles. The television set requires some distance between the screen and the user, who uses a remote control as a tool for connection. Time devoted to consumption is longer than 200 minutes. On the other hand, personal computers place the screen at a distance of thirty centimetres. The keyboard is the instrument to participate and to navigate in the personal computer. Time devoted to entertainment does not exceed that devoted to television and, above all, its conception, design, and way of navigation favour an individual consumption.

In the field of mass media, the growth of the number of connected users has not damaged the global consumption of the population in general. But the change in common net users is being easily guessed, and they do reduce the time they devote to television in an 18% in order to surf the Internet (AIMC, 2006). As we can see in Figure 1, Internet users dedicate less and less time to television consumption because they are connected to the Net. The structural change will come from the generational relief that means the gradual incorporation of new public services. If public service does not fit its mission to the social changes, it will lose its meaning.

Internet will be a substitutive (to search for information and news as well as for entertainment), complementary (it will add information to that seen on television or read in press), and alternative media (it will make easier to establish new ways for personal relationships and socializing). The objective of the public television is to work as a guide in a knowledge driven society as an information catalyst, redistributing contents, and creating values in a virtual world full of facts and information. The guide function turns facts into knowledge.

It is a question of adding together all the possibilities offered by technology, as practice has shown that the increase in number of channels does

Figure 1. Comparison between average population TV consumption and Internet users

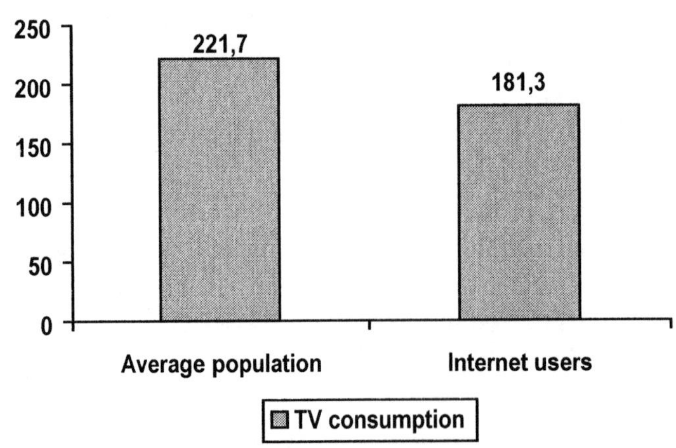

not complement the public service statement. On the contrary, an excess of offers has been generated, because the audience ratings have become stagnant and do not grow proportionally to the channel offers. This situation has provoked the atomization of the audience ratings and changed in quality of the model of the television business. The audience ratings will not keep on growing in the European context and, paradoxically, "in spite of the increase in the invested capital and the multiplication of the television sets, the audience could choose a specific and limited kind of programme or less-quality information" (Lamizet & Weymouth, 1996, p.214). In Spain, where the maximum time of television consumption is about 210 minutes per person and day (Anuario SGAE, 2006, p.127), it is considered that "the consumer's demand is less than fifteen channels. This number of channels would be enough to satisfy the general and specific contents required by the consumer" (García Mestanza & Fernández, 2002).

From a technological point of view, a wide consensus about the maintenance of public services tasks in the Internet exists as lots of the original television model characteristics coincide. The nucleus is still invariable: Public use of waves or transmitting wires; media intrusion in people's houses; increase in the use of the media for communication, in terms of contents and access; need for informative, educational, and entertaining contents; the nation's cultural identity; childish protection; attention to diversity; competition; and privacy (Grainger, 1999).

Therefore, if consumption is directed towards the Internet, if the mission is not complemented with the creation of more channels depending on the use and if technology shares the original nucleus with the public services, it will be necessary to conceive a communication public policy which clearly distinguishes public actions from those considered commercial.

THE NEW MISSION STATEMENT

The Internet challenge consists of the necessity for defining clearly which is the public television mission, establishing criteria and indicators to carry out public interest obligations, having long term financing methods and long term reviewing policies at disposal to know the reach of the audiovisual public action. This definition is also demanded by the collective heritage, that is pressing to define the specific tasks of each mission with both qualitative and quantitative indicators to make the public service clear and avoid it to distort the business activity (Bustamante, 2003). RTVE has not started its review and it is already immersed in the European television's new stage, while the BBC is ready for the change.

The ideal archetype includes the audiovisual public action in the public sphere model which considers the individuals as citizens with communication requirements and not only as consumers (Dahlgren, 2002). The financial resources are allocated to the entrusted mission fulfilment, and mission includes the active citizen promotion. In the production field, a public sphere service fosters both innovation and diversity through the contraction of contents from all kind of producers. This model would compensate the commercialization of the private companies (Croteau & Hoynes, 2001, p. 35).

RTVE and the BBC, European public companies, have assumed the commitment to lead the migration and widen their obligations to a digital world as a core of the national media ecosystem with different success in each case. In both cases, the Internet is the quantum leap. In Spain, this matter had been kept practically in silence, in spite of what is shown in the sixteenth additional disposal in the Law 24/2001 about fiscal, administrative, and social order measures that urges the active promotion of the information society development through the participation in technical development, the use of all the ways and means of distribution and diffusion, the new programme production and diffusion techniques, and the audiovisual communication services, the development of new services, including digital and online ones, and the approach between citizens and the services and actions of the public service. The Report on the Reform of State Owned Public Media goes deep in this matter rescuing the idea of a public service as the fuel to make the migration to digital age possible. This report argues that "the public service objectives currently in force need, however, to be adjusted to the new circumstances appeared from the evolution of the socioeconomical and technological scene. Making public services' activities in a digital world—activities which are led by the general interest, social profitability, and the principles inherent in its public service tradition—be adapted to the final purpose of an information and Knowledge society that serves the whole of the citizenship" (Report on the Reform of the State-Owned Public Media, 2005, p. 79). To carry out its new mission, RTVE must play an important role in the strategic leadership in the digital age, establishing associations with other public and private channels to act as the engine of the development of hertzian waves digitalization in Spain and, finally, as an accelerator to the popularization of the information society (Report on the Reform of the State-Owned Public Media, 2005, p.120). Nowadays, the digital mission has not been developed. The main Web is underutilized in comparison to other public media in Europe. The statistics are clear: the Web www.rtve.es/ receives less than 105,000 visitors a month behind all the reference media in Spain, including print, audiovisual, or multimedia press (AIMC, 2006). The public operator does not give the data, so it has been verified by private entities.

On the other hand, the Department for Culture, Media and Sport published in 2000 the white paper, 'A New Future for Communications' that signed the audiovisual public service objectives and explored the industry tendency for the next decade. The document defended keeping the audiovisual public service as the nucleus of the Digital Terrestrial Television and pleaded for completing the digital migration before 2010. The method proposed by the labour government was based on fostering a strongly competitive market, guaranteeing the universal access and the capacity of election among a wide range of quality services, as well as protecting the individual rights through the creation of the Office of Communications. Beside these proposals coming from the government, the BBC has published a wide number of documents that defend the digital television, interactivity, and the Web as natural extensions of the audiovisual public service (Graham & Davies, 1997).

The revolution caused by the digital world implies a reconsideration of the financing methods. With a fragmented audience, the fee has a large lack of meaning, because the digital television, theoretically, satisfies all the tastes and requirements the audience may have. In order to counteract this school of thought, the BBC has strategically defined its position as a source of the audiovisual diversity and as the defender of the democratic values (BBC, 2004) while it increases the value of the money it uses to pay the fee. The objective number five in the annual revision of the Corporation is based on the philosophy of value for money, which ensures that the public money is allocated to activities related to its mission, the close relationship between the audience, and the operator and the optimization of business management processes (BBC, 2005, p.18).

The BBC has developed a public policy for the benefit of all the audiences. The Web can be visited in different ways. First at all, the visitor can choose the access through the preferred operating system (Windows 95, 98, 2000, Me, and XP versions, Apple Mac OS 8, 9, and OS X versions, and Linux), as well as the different versions for disabled people. The creation of accessible sites is a main objective in order to include all the citizens, for instance blind people, subtitled contents for deaf citizens, and contents for children with special education needs. These options give to the citizenry a broad sense of community. It does not mean your condition, because you will have a choice in the BBC Web (the motto is "my Web, my way").

The definition of the way continues with the selection of the version. The "UK version" gives prominence to the British news, sports, and weather along with the BBC radio and television, meanwhile the "international version" rewards the BBC international radio and TV services, also including news, sports, and weather. It is the preferred version for people living outside the Great Britain. Finally, the user can also choose the text only version, ideal type for no band with connections, and contents specially designed for Internet mobile (mobile phones and PDAs).

The expansion of communications public policies does not mean a never-ending list of obligations or actions. The first step for policy makers is to establish the mission, which will define the financial issues and the audiovisual contents, as have been demonstrated. The case of the British Corporation confirms this statement.

THE MATTER OF CONTENTS

In the domain of the online markets, public operators act as facilitators for some of the main demands of contents. As the engine of the local economy, public operators have to commit themselves in buying and distributing others' contents to be used by radio, television, and multimedia. The external production fosters the creativity and dynamism in the audiovisual industry, even more when public televisions do not have the monopoly on production, programme planning, and diffusion. The healthy outsourcing goes through the order diversification, in such a way that avoids common analogical vices ("sunflower producers" or movement of managers from public companies to private ones) and fosters innovation and creativity. Without the appropriate stimuli from the public operators, the market has tended, for the last 20 years, to take advantage from its privileged position, fragmenting audiences, concentrating the property, and increasing of the digital divide (BBC, 2003, pp. 9-15). The passive contribution of public operators includes the legal commitment for the acquisition of local products produced by national companies. The constant demand favours the expansion of the audiovisual production because it protects the private investment and foster independent production. The commercial operation of buying rights must not be taken as a subvention on permanent loan, on the contrary, it must stimulate the industry through some published criteria available for anyone interested in them.

As a bidder of audiovisual contents, a new mean of distribution is open on a worldwide scale. It is necessary to develop a content selling area for all the radio and television channels: news, exclusive stories, documentation, library pictures, and biographies of relevant personnel, among other products that are constantly demanding. Direct selling saves costs for the company and increase incomes coming from its distribution channel. On a citizen scale, technology allows to pass from pc to television, so, the download of contents (films, documentaries, cartoons, but also a weekly news programme) will favour the citizen's freedom to choose at what time of the day they want to consume those products. The final objective of the different actions must be the creation and consolidation of an owned industry that guarantees the audiovisual and economical autonomy. As a result of the new strategic method, public companies would favour the content creation hub for television, computing, and telecommunication industries. These new products, close to the demanded television, ask for a precise study for the necessities of the users, who do not demand this service specifically. The function of the public television, in any case, deals with the coordination and standardization of technological applications and contents. This is a function that shakes the market.

The impact can be negative for private operators when they are not in the level playing fields. The disappearance of a private company or the lack of investment is damaging for both the market and the citizen because some possibilities of choice are eliminated. On the contrary, the lack of investment on public operators can cause that they fail in their fulfilling of the public service. In United Kingdom, the Corporation defends the commercial activity since it reflects the BBC's values for public service and because it does not abuse its condition of public investment to ensure itself any special advantages in the market (BBC, 2004b). The experience of the digital terrestrial television is significant. The BBC participation in Freeview Platform with two complete multiplex (eight television channels and ten radio channels) and BBCi's interactive services has become a real success and has provoked the increase in demand of services offered by the digital television. By the end of 2005 the number of subscribers reached to five million and a second private platform, namely Top Up TV appeared.

When using a specific software for viewing, compressing archives, and multimedia archives some kind of technologies are more common than others. This action damages market freedom, so it is recommended the use of processes and tools based on free software and on open code instruments. These tools are compatible with those used by private entities. This decision favours in the same way the public enterprise, which will know beforehand which is the strategy in new media and will be able to decide in which direction it wants to invests its money. Furthermore, the use of open code software gives back to the citizens the possibility to choose against the imposition of a specific group of computer applications in their computers. Freedom of speech, the access to the diversity of sources and plural opinions, the access to public information, and audiovisual diversity also deal with the choice ability and with redistributing the media power. Public operators play an important role as counterweight of media corporations. Public operators can establish a unchanging market of contents favouring the demand increases. It is essential, therefore, to pay attention to all the users' taste, to promote the independent producers and to protect the citizenship's capacity of election.

CRITICISM FROM PRIVATE SECTOR

These functions are original and outside the traditional television public service point of view and they have the opposition from a large number of private actors who consider that public operators must not act in the cyberspace. In Spain, the low development level in RTVE's commercial activity, the scarce demand for digital services, and the indifference of journalistic groups justify the absence of opposing groups to the public action for the Spanish state television and radio. A typical case was the management of the show "Operación Triunfo" (Star Academy). This TV show started in October of 2001 and it was a huge success for its audience ratings and its social impact. Its weekly shows led the audience ratings and modified other channels weekly schedule. The turnover exceeded, before the summer tour, fifty million euros, but only ten million went to RTVE funds, which had made the quiz possible investing six million euros in its production. The project had been pushed away by other two general interest channels so, under those circumstances, TVE was in an excellent position to negotiate about margins without taking risks with its own capital. Vale Music and Manga Films paid TVE for the royalties, and TVE divided this money up to pay Gestmusic in a proportion of 60/40.

The profits coming from advertising both in daily resumes and weekly shows were exclusively for TVE. This success is even more remarkable if it is compared with the final show ratings for the most seen programmes in Spanish television history in terms of total impact. The final show was seen by 12873,000 spectators, one of the three most seen programmes in history. In spite of the demand, TVE did not create its own channel in the Internet as the producer, Gestmusic (Endemol group), did in portalmix.com. The official Web became a site of reference among the millions of spectators who downloaded pictures, videos, and interviews of the contestants. The programme encouraged spectators to visit the Web, because the new stars were obligated to answer e-mails and to take part in chats. Like this programme, many others have a loyal audience that demands wide on line contents, according to the nature of the programme. "Redes" (first level scientific spreading) or "Memorias de España" (historical spreading) could be completed with contents, educative games, forum for exchange of documents, literary criticism, and so forth.

In United kingdom, the lobby British Internet Publishers' Alliance (BIPA), which defends private enterprise with commercial activity online interests, like Capital Radio, Chrysalis, Guardian Unlimited, Independent Digital, or Telegraph Group Ltd. argues that the government should redefine the extension of BBC activities on line in order not to damage the market expansion, explain clearly which is the budget allocated to the new media limiting self promoting and crossed marketing activities, and close down those services that do not follow these guidelines. The definition of a public service in the Internet must be clearly justified and differentiated from commercial offer, mentioning exclusively sports and pop music, so as not to offer products existing in other Web sites. The BIPA invited the BBC to develop innovative contents which did not affect the private sector commercial interests, because it could accuse the company before the European commission for disloyal competence and double financing.

In the same line, the lobby asked the government for valuing in which way these actions were distorting—definitely or potentially—the commercial base of the Internet's services. With this objective, it proposed the establishment of some qualitative and quantitative criteria similar to those already developed in the Royal Charter for television and radio (British Internet Publishers' Alliance, 1999).

The lobby's fears have shown to be unfounded. Six years after the government's authorization for the creation of the online section of the Corporation, the KPMG consulting agency published in June of 2003 a revealing report on the activity of the digital public services (BBC, 2003b). In first place, the BBC does not compete in markets of Internet supplier services or achieves electronic marketing transactions, which represent more than the 96% of the sector's incomes. Web sites bound to the BBC have not got incomes derived from the online advertising market. In addition to it, free contents could damage private enterprises, which offer paying products. But the truth is that it is only one more actor from the wide range that can be found in the Net, which proliferate as paying versions of other successful media. In the scene of the incomes coming from advertising, the corporation raised about 2.2% of the whole figure which does not distort the commercial activity of the competence. In any case, the BBC has committed for the systemic review of its actions in the digital world in agreement with the employers' organization and the European Commission's argument in terms of public financing. In 2004, the Department for Culture, Media and Sport asked Philip Graf to write the first report, whose guidelines redefined the online activities (BBC, 2004c). Together with the general corporation's objectives, the Web pays special attention to the democratic and educative values through the supply of an innovative content, distinguished and available for all the audience. Likewise, it is a guide for new media use increases the fee's added value and reinforces transparency. It has transformed the business model and almost 10% of the budget allocated to the online contents has been readdressed to other projects more related to the mission of the public service. It has also committed that the 25% of the contents is from other's owned production, so the market will be stimulated in terms of demand (BBC, 2005, pp. 39-40).

The facts, in the United Kingdom's case, confirm that the intervention of the public operator in the cyberspace is positive for private interest. The annual revision of the public service and the adequation of the philosophy "value for money" increase the number and the quality of the actions related to public service, as well as it stimulates the demand of digital contents (BBC, 2005, p.18). For instance, the digital television experience confirms that the intervention of the public service, far from commercial activities, re-launched Freeview and the interactive services.

STRATEGIC PLANIFICATION AND ALLIANCES FOR PUBLIC SERVICE BROADCASTING COMPANIES

Political functions: The Informed Society

Information is a social good, whose objective is to get people to develop their role as citizens. The social responsibility for television and the new media resides in the obligations constructed towards the society and the public interest. Information is an internal good, because it complements the citizen's character, and an external good, because it is a preconceived condition for contemporary democracy. Communication of public policies has a catalytic function as they try to generate a better society. This can intensify the citizenship's experience related to people's identity and different notions of community and membership of the group. The public service is conceived as a right for all the members of society, that is, citizens; and here is the core of the relationship between television, citizenship, and democracy. Equal opportunities and social stability support this triangle and its continuity is an essential supply from contemporary democracy. Public companies can support the equal opportunities of the citizens to participate in the public sphere and ensure that all of them have the opportunity to get and use knowledge in a globalized world. With the appropriate tools, citizenry can choose how and when to receive the necessary information to cope in the environment around them.

Media's political and democratic function and cyber democracy's dream cannot falsify the base of democratic deliberation, but they can help citizens to participate actively in the progress of society. However, the strategic planning must have:

- Creation of its own searching engine
- Usability criteria
- Blog, e-mail, chat, and forum services
- Syndication of contents (Really Simple Syndication or similar procedures)
- Creation of virtual communities
- Generation of a search and help bar
- Informative programmes open to citizenship's participation

Social Functions

Digital public policies guarantee that all the citizens have an information space for the public debate about local, national, European, and international matters. In this line, the media's public space must be open to all the cultures, tastes, and citizen's demands. The Internet multiplicities the emission sources, and provides an opportunity for segmented attention to ways of life in collaboration with public and private institutions.

The following actions may be initiated to increase the audiovisual diversity for the next generation:

- The elaboration of a programme planning that attends the necessities and demands of the different communities
- Programme planning for different kind of target groups

- Promotion of Independent products
- Promotion of the different communities' representation
- Extent and variety of genres, minor genres, and formats
- Publishing investment criteria compared among genres
- Availability of contents in prime time
- Spaces open to artistic and creative expressions
- Definition of a digital curriculum
- Interconnection online between the educative system and the learning about how media work
- Alliance among museums, libraries, and schools to change into information and contents suppliers
- Privileged relationship with the open university
- Educative programme
- Promotion of values like health, environment respect, childish, and disadvantaged protection
- Technologic literacy
- Audiovisual learning and comprehension of language
- Establishment of a programme planning devoted to immigrants: language, news from their native country, their own productions
- Development of cyber voluntary work

Economic and Industrial Functions

The public company demands an integrated development as reference node for audiovisual, informatics, and telecommunications industries. Dealing with public financing, television, and Internet public policy objectives must be cleared. Public financing objective is to attend those demands that are not usually satisfied by the market. The expectation is that the content financing causes the public service objectives to banish all doubts about possible unfair competence and set the conditions for the acquisition of audiovisual rights. Public television can participate in processes of creation of value, adjusting its contents and increasing the quality level and, in consequence, widening the bottleneck, understood as the spectator (addressee, user, customer, or audience).

Furthermore, the public company can lead this jump through the addition of services as a one stop offer (bundling services) and establishing prices of reference (price bundling), which would stimulate the audiovisual market in the Internet. All the actors would have profits, because it homogenizes the service, establishes the minimum quality accepted, and guarantees a space of representation and identification in the public sphere for the minorities.

The main activities would be:

- Elaboration of a catalogue of criteria on the audiovisual public services quality assessment
- Qualitative investigation of the audience ratings
- Creation of a net of local and regional audiovisual producers
- Promotion of creators and experiments independent from big multimedia groups
- Support to training and education
- Elaboration of programme contracts to foster the production and covering of social minorities

The final objective of the actions must be the creation and consolidation of an owned industry that guarantees the audiovisual and economical autonomy through the establishment of the content creation hub for the television, informatics, and telecommunications industry.

CONCLUSION

RTVE has not adapted itself to the digital environment and is tiring out its model in the analogic world, in spite of being a company with leadership ability (30% of the audience with two channels) in the analogic world. The lack of strategic planning and the absence of a previous definition of public service broadcasting have delayed the group action as the promoter of the digital migration. In her last appearance before the House of Commons, the general manager Carmen Caffarel presented a five line work project that includes the promotion of presence and group's contents online and the reinforcement of the role played by TVE in the DTT development. It is urgent that the measures are held by qualitative and quantitative indicators and long term reviews to check the degree of fulfilling the public audiovisual mission in the digital environment. The Web site does not offer general interest contents for the most of the audience or a news service. Successful programmes do not have a digital version where spectators can share their common interests (social matters, local information, hobbies). Only the marketing department is developed because it allows the on line hiring of advertising spaces and all this activity can be carried out online.

United Kingdom has chosen a mixed solution. The BBC has been pioneering in political changes applied both to its public service function and its business model. The Board of Governors has accepted the digital challenge and considers the corporation as an organization of radio, television, and new media that enrich the services offered by the two traditional media. Interactive services (BBCi) and the Web site (www.bbc.co.uk) are the natural service extensions and show different objectives, but always keeping in mind the fulfilling of the BBC's mission.

The investment on the digital curriculum has lightly increased for the last 4 years. In 2003, 451,52 million euros were spent, 29 million euros more than in the last financial year. The figure was kept in 2004 was 451,80 million euros and experimented a qualitative increase in 2005 when it reached 512,87 million euros (BBC 2002, 2003, 2004, 2005). These investments are justified by the constant investment on infrastructures, contents, and different services. It is directed to the "public value" philosophy, whose objectives are the increase in services and defence of democratic values in a globalized world, and the "money for value" that tries to optimize the fee's value (BBC, 2004b).

The objectives of Web services are to complement the audiovisual offer by increasing the depth, interactivity and the easy access to information, training, and entertainment. The Internet services work in a very acceptable way and they monthly receive more than ten million users during 2005, with an increase in 1,7 million since the previous year (BBC, 2005). As Lycos50.com statistics show, BBC News is the second source of world information after the CNN. It has RSS technology, download for mobile phones and PDA, and is the British site with more registered contents that is more than two million pages, specialized sites, virtual communities, associated blogs, and so forth. The public mission has been even more adjusted after private sector's complains and the publishing of the Graf report (BBC, 2004c), which included the review of the corporation's functions in the digital world. Lots of services have been cancelled, and the same has happened to those offers that distorted the competence. In all, the 10% of the budget allocated to the fulfilling of public service has been relocated, and at the same time, has accepted that the 25% of the contents is other's owned production. In 2006 it is being considered to increase the access possibilities for disabled people. Success has made easier the online commercial expansion through two actions: The BBC Creative Archive, that offers since September of 2003 more than 2,000 videos and audiovisual archives (more than 100 hours of contents), and the Interactive Media Player (IMP) system, that allows the download of emitted radio and television programmes. They are available freely for those who pay the fee and in paying version for the rest of the users.

The BBC's Web site has been consolidated as a strategic asset with more than two million Web sites with contents and a decided commitment to informative services. During 2004 Christmas time, the tsunami occurred in Southeast Asia, visits were doubled and the informative dossier was downloaded 2.5 million times. The audience ratings reached 1,875,000 printings, 813,400 of news and sports. One of the most important successes in 2005 was the reappearance of BBC Radio Player, that allows the download of a wide range of contents during the week after its emission. In the last term of 2004/2005, 34 million hours from Radio1 and Radio 4 were downloaded. Nowadays, MP3 format download is being developed for the podcasting. The investment on services of the Web site reached 112,87 million euros in 2005, 10,13 million more than 2004. The increase is due to the increase in demand and the investment allocated to cover the Olympic Games in Athens (BBC, 2005, pp.13, 41, and 144). The allocation of the investments coincides with the mission committed in annual reports and creates a new financial income line based on high quality products.

Developing successful ICT strategies in public service broadcasting, as those we have seen the BBC has implemented, not only serves to address the problem of coping with the new markets and new types of audiovisual and multimedia consumption, it also serves the purpose of educating the population in the era of the media citizenry and of giving them the feeling that their tax money is well spent.

Future Research Directions

The article explores the challenges the Internet pose to the traditional public service broadcasters before the expansion and consolidation of the Web 2.0. There is not a clear definition on what this concept include, but it is sure that the Net is more "under construction" than never. The user wants to participate, to create. At the time, there are few chances for citizens to participate in the creation of specific contents in most of the public broadcasters. How can people be part of the public service project? Is there any chance to create an authentic network of citizen journalist to inform about public issues?

As study of case, it would be very interesting to analyse the recent agreement between the British Broadcasting Corporation and the portal YouTube, because it confirms the new ways the audience is consuming television. But will the agreement increase or reduce the audience? Will people have in advance the future contents (e.g., soap operas.)? Does the BBC benefit a private company against others? Shall the BBC produce a share-video competitor? Besides, the agreement impacts the property rights. Will be BBC pay the authors due this broadcast? And, what about the contracts of producers, artists, actors, or musicians?

Likewise the growth of the bandwidth have made very easy for users to watch public service television in real time over the Internet. There are no more geographical boundaries for audience: people can watch all over the world without paying any fee. Will the public corporations attend these new audiences? Will they produce specific contents for an international audience? What about the migrated people? Is there any obligation to those who emigrated?

Finally, it could be important to analyse if there is any European political decision to develop an integrated strategy for public service broadcasters. Is the European Commission (EC) worried about the decline of audience? Can the EC create a network of European contents? Is it good or bad for the European common market? Can a European strategy contribute to the creation of the European public sphere and an European citizenry?

REFERENCES

AIMC (2006). *Estudios General de Medios.* Madrid: AIMC.

British Broadcasting Corporation (2002). *Annual Report and Accounts 2001/2002*. London: BBC.

British Broadcasting Corporation (2003a). *Annual Report and Accounts 2002/2003*. London: BBC.

British Broadcasting Corporation (2003b). *Market Impact Assessment of BBC's Online Service*. London: BBC.

British Broadcasting Corporation (2004a). *Annual Report and Accounts 2003/2004*. London: BBC.

British Broadcasting Corporation (2004b). *Building Public Value. Renewing the BBC for a Digital World*. London: BBC.

British Broadcasting Corporation (2004c). *Report of the Independent Review of BBC Online*. London: BBC.

British Broadcasting Corporation (2005). *Annual Report and Accounts 2004/2005*. London: BBC.

British Internet Publishers' Alliance (1999). *BIPA Submission to the Davies Review Panel*. London: BIPA.

Bustamante, E. (2003). *Comunicación y cultura en la era digital. Industrias, mercados y diversidad en España*. Barcelona: Gedisa.

Council of the State-Owned Media Reform (2005). *Report on the Reform of the State-Owned Public Media*. Madrid: Secretaría de Estado de Comunicación.

Croteau, D., & Hoynes, W. (2001). *The Business of Media. Corporate Media and the Public Interest*. Thousand Oaks: Pine Forge Press.

Dahlgren, P. (2002). Reconfiguring Civic Culture in the Evolving Media Milieu. In J. Corner & D. Pels (Eds.), *Media and political style: Essays on representation and civic culture*. London. Sage.

García Mestanza, J., & Fernández, V. (2002). El valor estratégico de los contenidos en el mercado de la televisión de pago. Una perspectiva de recursos y capacidades. Boletín económico de ICE, 2.719.

Grainger, G. (1999). *Broadcasting Coregulation and Public Good*. The 1999 Spry Memorial Lecture. Retrieved July 8, 2006, from www.comumontreal.ca/spry/spry-gg-lec-html

Jakubowicz, K.(2002). *We know it when we see it? Public Service Broadcasting: Definitions, Descriptions and Policy Dilemmas*. Changing Media-Changing Europe Working Papers, 2. Retrieved July 5, 2006, from http:/info.lut.ac.uk/research/changing.media/K.J.-Paper.html

Lamizet, B., & Weymouth, T. (Eds.). (1996). *Markets and Myths: Forces for change in the European media*. London: Longman.

McGonagle, T. (2001). *Changing Aspects of Broadcasting*: *New Territories and New Challenges*. IRIS Plus, Issue 2001-10.

McQuail, D. & Siune, K. (Eds.). (1998). *Media Policy. Convergente, Concentration and Commerce*. Euromedia Research Group. London: Sage.

SGAE: *Anuario SGAE de las Artes Escénicas, Musicales y Audiovisuales*. Madrid: SGAE.

Thompson, C. (2003). *The Mission for Public Service Broadcasting in a competitive digital environment*. Retrieved July 6, 2006, from www.bbc.co.uk/pressoffice/speeches/stories/thomson-caroline_oxfordmediaconv.shtml

Wiio, J. (2004). *From the Information Society to a Society of Information: Public Service Digital Television as a Player in the Knowledge Society of the Future*, RIPE@2004. Paper presented at Mission, Market and Management. Public Service Broadcasting and the Cultural Commons, Copenhagen and Arhus, Denmark.

ADDITIONAL READINGS

Anderson, C. (2006). *The long tail. Why future of business is selling less of more.* New York: Hyperion.

Arampatzis, A. (2004). Online business models in Greece and the United Kingdom: A case of specialist versus generic and public versus privately owned online news media. *International Journal on Media Management, 6*(1&2), 88-101.

Bakker, P. (2006). *Public broadcasters and the Internet. Webcasting in a commercial environment.* RIPE@2006. Paper presented at Public Service Broadcasting in a Multichannel Environment: Programmes and Platforms.

Barnett, S. (2006). *Public service broadcasting: A manifesto for survival in the multimedia age (a case study of the BBC's new charter).* RIPE@2006. Paper presented at Public Service Broadcasting in a Multichannel Environment: Programmes and Platforms.

Cave, M., Collins, R., & Crowther, P. (2004). Regulating the BBC. *Telecommunications Policy, 28*(3-4), 249-272.

Centre for Public Integrity (2004, October 28). *Networks of influence: The political power of the communications industry.* Retrieved February 2007, from http://www.publicintegrity.org/telecom/report.aspx?aid=405

Dahlgren, P. (1995). *Television and the public sphere: Citizenship, democracy and the media.* London: Sage.

Departament for Culture, Media and Sport (2006). *A public service for all: The BBC in the digital age*, Cm 6763. London: HMSO.

European Commission (2007). *Media pluralism in the member states of the European Union.* Commission Staff Working Document SEC (2007) 32.

Freedman, D. (2006). Dynamics of power in contemporary media policy-making. *Media, Culture and Society, 28*, 907.

Harrison, J., & Wessels, B. (2005). A new public service communication environment? Public service broadcasting values in the reconfiguring media. *New media & Society, 7*(6), 834-853.

Hesmondhalg, D. (2005). Media and cultural policy as public policy: The case of the British Labour Government. *International Journal of Cultural Policy, 11*(1), 95-109.

Holtz-Bacha, C. (2006). *Medienpolitik für Europa.* Wiesbaden: VS Verlag für Sozialwissenschaften.

Iosifidis, P. (2006). Digital Switchover in Europe. *Gazzette. The International Journal for Communication Studies, 68*(3), 249-267.

Iosifidis, P. (2006). *DTT and the Role of PSBs in Driving Digital Switchover.* RIPE@2006. Paper presented at Public Service Broadcasting in a Multichannel Environment: Programmes and Platforms.

Iosifidis, P., Steemers, J., & Wheeler, M. (2005). *European television industries.* London: British Film Institute.

Lloyd, J., & Seaton, J. (2006). *What can be done? Making the media and politic better.* Oxford: Blackwell.

Manfredi, J.L. (2006). El reto digital de las televisions públicas en Europa. Las estrategias de la BBC y de RTVE. *Telos*, 68. Retrieved February 2007, from http://www.campusred.net/telos/articuloperspectiva.asp?idarticulo=3&rev=68

Manfredi, J.L. (2006). Servicio público en el entorno digital: teoría y práctica. *Quaderns del CAC*, 23-24, p. 179-186. Retrieved February 2007, from http://www.audiovisualcat.net/publicaciones/q23-24casmanfredi.pdf

McChesney, R. (2004). *The problem of the media*. New York: Monthly Review Press.

McFayden, S., Hoskins, C.G., & Finn, A. (2004). *Media economics: Applying economics to new and traditional media*. London: Sage.

Meier, H. (2003). Beyond convergence. Understanding programming strategies of public broadcasters in competitive environments. *European Journal of Communication, 18*(3), 337-365.

Picard, R.G. (2005). *Corporate governance of media companies*. Jönköping Sweden: Jönköping International Business School.

Picard, R. G. (2002). *The economics of financing media companies*. New York: Fordham University Press.

Syvertsen, T. (2003). Challenges to public television in the era of convergence and commercialization. *Television and New Media, 4*(2), 155-175.

Thomas, B. (2006). *Public service broadcasting beyond national boundaries: Its performance for European Public Spheres*. RIPE@2006. Paper presented at Public Service Broadcasting in a Multichannel Environment: Programmes and Platforms.

Wiedeman, V. (2004). Public service broadcasting, state aid and the Internet. Emerging EU Law. *Diffusion Online, 47*.

ZEMOS 98 (2006): *La television no lo filma*. Retrieved February 2007, from http://www.zemos98.org/

Chapter V
Civil Liability of ISPs, the Internet, and Law

Onyeka K. Osuji
University of Manchester, UK

Emmanuel Eze
University of Hull, UK

ABSTRACT

This chapter examines the Internet and law relationship. Using rules and decisions on ISP liability for civil defamation to illustrate, the chapter attempts to identify and proffer solutions to the challenges posed by the Internet to law. It suggests legal recognition of the distinctive nature of the Internet by policy makers and administrators as the key to addressing the legal issues continually raised by the Internet with the result that appropriate legal strategies would be applied to the Internet. The chapter also advocates international legislative action on the Internet issues. Law, including its substantive and procedural contents, interpretation and administration, is an important factor for encouraging or hindering continued Internet development.

INTRODUCTION

Law is an important factor in any field of human activity. Depending on the contents of the relevant legal regime, law is capable of encouraging or hindering the development of such activity. The Internet is not an exception. Talking about the Internet and/or its development, expansion, or prospects without discussing the law's role, amounts to leaving out an indispensable aspect of the structure. The Internet has been playing increasingly critical role in the modern world. The relationship between the law and the Internet, therefore, requires close scrutiny. Examination of the past and present relationship is as important as looking at the future. A look at the past and the present may reveal difficulties, and may also offer the benefit of hindsight in suggesting models of the Internet law for the future. This would assist the policy initiators in assimilating appropriate ingredients for developing successful strategies for Internet deployment.

The successful evolution of the Internet has resulted in the use of the Internet for a whole range of operations different from its original purpose, which was basically to facilitate communication and data sharing among researchers. Recent increase in processing power and cheap availability of network bandwidth has brought many businesses and millions of users on the Internet. This exponential growth is expected to continue in the coming years. An English judge once described the Internet as "a new phenomenon" (John Bunt v. David Tilley [2006] EWHC 407, para.9 (Eady J.)). But like most other fields of human or commercial activity, "the Internet is not and never has been a law-free zone" (Cummins, 2003, p. 10). The geographic reach of the Internet is an important issue. It has been rightly observed that the Internet has the "ability to effortlessly reach people" (Cummins, 2003, p. 1) across jurisdictional boundaries. It is also clear that the Internet "undermines the feasibility and legitimacy of laws based on geographic boundaries" (Bettelheim, 2006, p. 24; Kohl, 2004) because "the Internet has no territorial boundaries" (Digital Equip. Corp. v. Altavista Tech. Inc., 1997, p. 462). It is also unclear what portion of the Internet a country may legitimately treat as within its jurisdiction (Rundle, 2006, p. 12).

The Internet has continued to raise several (and sometimes) novel issues for the law in areas such as intellectual property rights, defamation, breach of confidence, and right of privacy. The use of the Internet for activities relating to these areas of law, arguably, exposes users to liability. The scope of civil liability is not closed; the dynamic nature of law and the realities of modern life determine that new areas of liability may be added by legislation or judicial decision while old rules on liability may be modified as the need arises. For example in Campbell v. MGN Ltd. (2004), a fashion model sued a newspaper for covertly taking and publishing pictures of her receiving therapy from a self-help group. A split UK House of Lords reversed the lower court and held that the action constituted a breach of confidence and invasion of privacy. In Douglas v. Hello! Ltd. (No.3) (2005), the English Court of Appeal upheld the lower court's judgment in favour of the claimant film actors. The defendant magazine's action in taking and publishing photos was held as invasion of privacy and breach of confidence. These two cases established that breach of privacy is now an area of civil liability in English law. It is unlikely that the two decisions would have been different if the Internet medium had been used for publishing the offending pictures.

This chapter addresses the question of the liability of Internet service providers (ISPs) for defamatory materials created by third parties. The discussion is restricted to civil defamation only and does not extend to criminal defamation. The objective of this chapter is to reveal the difficulties with the current regime on Internet defamation. The concern is with the appropriate relationship between legal rules and the Internet. Has the law responded positively and actively to the challenges posed by the Internet? Are the present legal, regulatory, and institutional mechanisms sufficient or adequate safeguards for the ISPs? Are there difficulties with the application of old or traditional rules at both national and transnational levels? Is unilateral, bilateral, regional, or global solution to the Internet-related issues to be preferred? Some of the decided cases reveal difficulties, which indicate the law's critical role in the continued development of the Internet. The first step in overcoming these difficulties is the awareness of their existence by legal policy makers, administrative agencies, and judicial authorities. Recognition of the Internet as being of a distinctive class and having a peculiar character is the key to effective Internet legal policy, administration, and interpretation.

Although the focus is on civil defamation, the overall aim is to illustrate the difficulties with the apparent absence of separate treatment of Internet claims by legislative and judicial authorities. This chapter also champions a new legal regime for the Internet where international action is applied in addressing pertinent issues and difficulties.

This chapter is divided into five sections. The next section presents a background study of ISP and the Internet, and their relationship with law. The third section examines the issues associated with the Internet and law and also proposes solutions and recommendations. The fourth section discuses the future and emerging trends; while the last section concludes the chapter.

BACKGROUND

A definition of an Internet service provider (ISP) is not without difficulties. An ISP is a company that offers users access to the Internet and related services for a fee. About a decade ago, most ISPs were run by the telephone companies but now things are different. ISPs have grown exponentially and can now be run by just about anyone. They provide services such as domain name registration and hosting, dial-up, DSL, or wireless Internet access, leased line access, Internet transit, collocation, and so forth. The list of services provided by ISPs is not closed; the list seems to be dynamic and extending with time. ISPs transmit third party content from their computers to others without any human intervention from the ISP.

Regulation 2(1) of the UK Electronic Commerce (EC Directive) Regulations 2000 incorporates Article 2(a) of the Directive on Electronic Commerce 2000 (Directive 2000/311EC of the European Parliament and Council), which in turn borrows the definition in Directive 98/34/EC. In the legislative instrument an "information service" is defined as "any service normally provided for remuneration, at a distance, by electronic means and at the individual request of a recipient of services". Mathew Collins has argued that the definition covers "commercial Internet intermediaries such as ISPs, bulletin board operators, and Web hosting services" while public libraries, universities, and schools would fall outside the definition since access to such services would not be provided for remuneration, or at a distance (Collins, 2005, paras.15.38, 15.43). But Milmo and Rogers (2004, p. 168, para. 6.27) have argued that in the case of a university or a fee-paying school the Internet service is provided for remuneration from the fees "even though no separately identifiable charge is made for the service". Milmo & Rogers (2004, p. 168, para. 6.27) have also pointed out that that the definition excludes "any business organisation operating an internal network."

However, in this chapter the term ISP has been used, to cover both commercial and not-for-profit Internet intermediaries. The term is not restricted to its technical meaning or the meaning it might have been assigned in any applicable legal instrument. The realisation that Internet service or operation is highly dynamic requires the broadest definition of Internet intermediaries for the purposes of this chapter.

The Internet has been described as promising to be the "supreme mechanism" (Fried, 2003, p. 3) for defamation. Defamation is the common name for the torts of libel and slander, which protect reputation (Milmo & Rogers, 2004, p. 3, para. 1.1). Generally, a statement in a permanent form or in the nature of a broadcast or a theatrical performance is libel while slander is in transient form (Milmo & Rogers, 2004, p. 3, para.1.1, p. 28, para. 2.1, p. 141, para. 6.1). The cause of action arises when there is publication "to a third person words or matter containing an untrue imputation against the reputation of the claimant" (Milmo & Rogers, 2004, p. 7, para. 1.3). Publication is an essential part of a defamation claim (Milmo & Rogers, 2004, p. 7, para. 1.4). Publication is used here in the sense of making information known to the public, or more correctly, to a third party. In Steinberg v. Pritchard Englefield (A Firm) (2005), the English Court of Appeal upheld the judge's decision that publishing on the Internet assumed widespread publication.

The liability of an ISP for third-party Internet defamation and other cases of infringement on civil rights may likely be contributory. Contributory liability for any infringement imposes liability for assisting, directing, or aiding and abetting the primary infringer (Reese, 2005, p. 7). A contributory infringer "induces, cause or materially contributes to the infringing conduct of another" (Gershwin Publishing Corp. v. Columbia Artists Management Inc.,1971, p. 1162). One may be liable as a contributory infringer by distributing devices or products that can be used for infringing purposes (MGM v. Grokster, 2005; Brown, Armon, Ploeger, & Traynor, 2005, pp. 21-22). In Godfrey v. Demon Internet Ltd. (2001, pp.208H-209A) the judge held that "whenever [the ISPs] transmit and whenever there is transmitted from the storage of their news server a defamatory posting, published that to any subscriber …who accesses the newsgroup containing that posting". These cases point to the clear possibility of ISPs being held liable for Internet defamation.

LAW AND THE INTERNET

This section examines the critical issues arsing from the apparent absence of separate treatment of Internet claims by legislative and judicial authorities. Although the focus is on civil defamation, this section illustrates the difficulties that are generally encountered in the relationship between law and the Internet. The first subsection explores the associated issues, while the next subsection proffers possible solutions and recommendations to tackling to issues raised.

Issues, Controversies, and Problems

Some of the rules on issues relating to the Internet were developed by the courts, or enacted by the legislatures, at periods that were not as technologically advanced as the present world. For instance, in defamation the generally applicable rule is that each separate publication of a defamatory statement constitutes a separate cause of action. In the common law world, courts place reliance on the 1849 decision of the English Court of Queens Bench in Duke of Brunswick v. Harmer (1849). In that case an 1830 newspaper contained an article defamatory of the claimant. Seventeen years later, an agent of the claimant bought a back copy of that newspaper. Another copy was also obtained from the British Museum. The applicable limitation period for libel actions was then 6 years. The defendant argued that the suit was time-barred. This contention was rejected by the court, which held that the delivery of the back copy to the claimant's agent constituted a separate publication that the suit could be based on. Is there any need for another look at this old rule in view of the modern Internet reality? Is there any need for the applicable law to reflect national, local, or global Internet perspectives?

One major difficulty with the current regime on Internet defamation is the absence of a clearly defined role of the ISPs in the publication process. In Estelle Smet-Hallyday v. Valentin Lacambre, the Tribunal de Grande Instance (District Court) Paris, stated that:

The provider of an Internet service has an obligation to ensure the proper moral conduct of those whom he accommodates, their respect for ethical codes governing the Web, and their respect for laws and regulations and the rights of third parties. The provider of the service, like any Internet user, has the opportunity to verify the content of the site which he manages and therefore, if occasion arises to take any action which may be necessary to put an end to distress which could be caused to a third party.

This is, indeed, a wide statement. It has left several questions unanswered. It is unclear what are the "codes governing the Web" or which laws and regulations that an ISP should recognise. Who are the third parties and what are the rights of such parties that must be respected? In Church of Spiritual Technology, Religious Technology Centre, New Era Publications International APS v. Dataweb B.V. & 21 Others (2004), the defendants in a copyright infringement suit included a number of ISPs. The Court of Appeal of the Hague held that the ISPs were not liable since they provide only technical facilities for others to present information.

In Lynda Lacoste (née Heineman) v. Multimania Production SA (2001), the court made a distinction between a supplier of hosting services and other ISPs. The court viewed the ISPs role as "limited to ensuring the transfer of data instantaneously and without the chance of monitoring their content." According to the court, the host provides a continuing service of storage of information that goes beyond the "mere provision of technical services." Such a host has a "general obligation of prudence and diligence...to avoid a violation of the rights of third parties" by alerting "creators of a site on a number of occasions to the need to respect the rights of third parties". But the court also conceded that a host is "not obliged to exercise a minute and profound surveillance of the content of the sites it accommodates". However, the court's decision and observation do not provide much guidance. For instance, what is "general obligation of prudence and diligence"? How does one distinguish this duty from "minute and profound surveillance"?

A pertinent question is at what point the liability of the ISP starts or stops. This is a controversial issue. In Antonia-Belika Koubareli v. Evangelos Volotas and Forthnet S.A (2003), the claimant sued the recipient of Internet services and the ISP for infringing her economic and moral rights. Upon being notified of the suit, the ISP ceased providing Internet services to the recipient who then contracted with another ISP based out of the court's jurisdiction. The court granted injunctive relief against the recipient only and declined to make an order against the ISP on the ground that it had no complicity in the recipient's activities and had taken all reasonable steps to abate them.

Dow Jones & Co. v. Gutnick (2002) was a libel case decided by the High Court of Australia. In that case, the defendant magazine publisher's server was in New Jersey, but there were about 1,700 paid subscribers of the online publication in Australia. The publisher contended that the publication was brought to Australia by the claimant's unilateral act. The contention was rejected by the court which held that the publication of the online article in question occurred at the time it was downloaded by a third party. The court also stated that the publisher accepted the application of the libel laws of Australia by accepting Australia resident credit card subscribers.

Re the Law to Promote Confidence in the Digital Economy (2005) concerned an application challenging a French law which imposed civil and criminal liability on ISPs who had actual knowledge of the illegal nature of the information stored on behalf of the recipients of their services or who were aware of the facts and circumstances which made the illegal nature of the information apparent. The basis for challenging that law was incompatibility with the constitutionally guaranteed right to respect for private life, right to freedom of the press and of communication, and right to a fair hearing. The French Constitutional Court rejected the objections to the law on the ground that liability is attracted only if the information is of manifestly illegal character, or if there is a judicial order for the removal of the information.

In Euromarket Designs Incorporated v. Peters, the suit was for breach of trademark law over an advertisement in a magazine circulating in both Ireland and the UK. The court stated that the function of the advertisement should be considered before a conclusion is made as to whether there was an infringement. The court also stated that there was no such infringement if the defendant in that case had no trade in the UK and none of their customers bought their goods for use in the UK. The court further stated that the fact that the Internet was available throughout the world did not necessarily imply infringement in every part of the world.

An important issue in Internet defamation is the jurisdiction on cases arising from it. In Lynda Lacoste (née Heineman) v. Multimania Production SA (2001), the Tribunal de Grande Instance (District Court) Nanterre, France held that in Internet cases, normal tort jurisdiction rules apply and damage occurs "at any place where the data available on the server are [down] loaded by any person desiring to connect to the Internet". In effect, the decision suggests the possibility of multiple or concurrent jurisdictions wherever data on the Internet could be downloaded.

In Shevill v. Presse Alliance SA (1996), an English woman sued a French newspaper for an article published in that paper in French language. The newspaper had a circulation of about 200,000 copies in France, but only 250 copies in England and Wales. The European Court of Justice held that Article 5(3) of the (European) Brussels Convention on Jurisdiction and Recognition and Enforcements of Judgments applied with the result that the claimant could sue in any country where the publication was distributed or where the claimant claimed to have suffered damage. (The Convention is now superseded by the Council Regulation (EC) No. 44/2001 of December 22, 2000 on jurisdiction and the recognition and enforcement of judgments in civil and commercial matters.) At the UK House of Lords, it was confirmed that the presumption that a defamatory statement is harmful to a claimant without specific proof of damage is sufficient for the application of that Article 5(3) (Shevill v. Presse Alliance SA, 1996, p. 983, Lord Jauncy of Tullichettle).

In Berezvousky v. Michaels (2000), two Russians who were frequent visitors to England sued a U.S.-based magazine for libel. The circulation of the magazine was about 785,000 in the U.S. and Canada, 1,900 in England and Wales, and 13 in Russia. The UK House of Lords held that the publication in England of an internationally disseminated libel constituted a separate tort on which a suit could be based in England. The court also held that England is an appropriate forum where a claimant has significant connections with, and reputations to protect, in England. Although the decision does not relate to an Internet claim, it is of relevance to cases of "internationally disseminated libel" including the Internet. The case shows judicial acceptance of the availability of a global medium for claimants.

In André and Jean-Pierre Vasarhelyi v.Guiness World Records Ltd. (2003), the French court held that a French court had jurisdiction over a suit where the offer for sale of a book alleged to have infringed copyright was directed at French consumers through Web sites that were accessible in France. The implication of the decision is that the French jurisdiction is always available whenever any website could be accessed in France.

Forum shopping is a distinct possibility in Internet defamation cases even in cases where the parties are resident or based in the same jurisdiction. In Jameel (Yousef) v. Dow Jones & Co. Inc. (2005), a U.S. citizen sued a U.S.-based publisher over an alleged defamatory article on the magazine's Web site. Although the Web site was available to subscribers in England, the publisher claimed that only five subscribers in England had accessed the Internet article and that consequently, the claimant had suffered no or only minimal damage to his reputation there. The claimant disputed the number of subscribers who actually accessed the article, but conceded that there had been minimal publication in England.

The English Court of Appeal upheld the judge's decision to strike out the part of the publisher's defence that sought to prevent the claimant from relying on the legal presumption of damage without proof. The court pointed out that the correct method for the publisher was to challenge the claimant's resort to English jurisdiction, make an application for striking out the case as an abuse of process, or apply for costs sanctions against the claimant. However the court reversed the trial judge and struck out the suit as an abuse of process since the publication within jurisdiction was minimal and did not amount to a real and substantial tort with the result that the damage to the claimant's reputation was insignificant.

Where there is an apparently unlimited jurisdictional choice the consequence is forum or rule shopping. Even where the rules appear settled in a jurisdiction, the risk of exposure to rules of foreign jurisdictions is real (Ryan, Nader, and Lihan, 2003). In the U.S., the Digital Millenium Copyright Act (DMCA) was enacted to provide "relief for ISPs" (Gosnell, 1999, p. 28). For instance, section 512 of the DMCA entitled the Online Copyright Infringement Liability Limitation Act of 1998 limits the scope of the liability of ISPs for copyright infringement (Ciske & Webb, 2005). However in King v. Lewis (2004), a U.S. resident boxing promoter sued a British boxer, the British boxer's U.S.-based promotion company and a New York attorney who acted for the boxer and the promotion company. The suit was for alleged libellous statements posted on two California-based Web sites. The defendants appealed against the order granting permission to the claimant to serve the claim out of jurisdiction. The parties agreed that under English law, libel was committed wherever publication took place. They also agreed that each publication generated a separate cause of action and that a text on the Internet was published at the place it was downloaded.

The English Court of Appeal dismissed the appeal. The court held, among others, that the starting point for consideration of the appropriate forum for libel is the place where the libel was committed. The court stated that the more tenuous the claimant's connection with English jurisdiction and the more substantial the publication abroad would make the consideration weaker. The court confirmed that in Internet cases, each publication constituted a separate tort, with the result that a defendant publisher on the Web is, at least theoretically, vulnerable to multiple actions in different jurisdictions. The court observed that the fact that a posting on the Internet is made available to "all and sundry" without any geographic restriction is obscured when there is a focus on the breadth of the reach of the particular forms of communication. According to the court, communicating on the Internet is a choice of a global medium, which targets every jurisdiction in which the text might be downloaded. However, the court pointed out that the claimants in such cases would not be given a free hand in forum shopping, although the court's discretion would tend to be more open textured because of the choice of the Internet medium.

There may be procedural or other juridical advantages in choosing a particular forum. For example in Berezousky v. Michaels, Lord Hoffman admitted that the claimants were "forum shoppers in the most literal sense...[and] have weighed up the advantages to them of the various jurisdictions that might be available...". In the same case, Lord Steyn confirmed that English defamation law presumes harm from defamatory statement without the requirement of proof (p. 1012). The presumption of damage was also accepted in Shevill v. Presse Alliance SA.

One pertinent issue is whether a posting on the Internet should be regarded as continuous publication until that posting is removed. In Loutchansky v. The Times Newspapers Ltd. (2001), the English Court of Appeal rejected the defendants' suggestion that the accrual of cause of action in defamation should be limited to the date of the initial publication. In the criminal case of Jean-Louis Costes v. Ministere Public, the French court held that documents on the Internet are subject of continuous publication until they are actually removed from the Web site in question.

In international plane, the absence of uniformity of rules and practice is a big difficulty in view of the global nature of the Internet. It also appears that uniformity of practice may even be absent in the same jurisdiction. In Canada, it has been observed that "cases demonstrate that the legislation in place has left the Canadian courts ill-equipped to render convincing and certain decisions regarding online defamation" (Handa, 2004, p. 20). Rosenthal (2003) has a carried a survey of a number of Internet-related trademark suits decided by the U.S. courts and has found that there is a lack of uniformity in the application of the rules. Ambiguous or inconsistent judicial decisions "supported by weak or little rationale" offer Internet participants "very little notice and the potential for unpleasant jurisdictional surprises" (Bettelheim, 2006, p. 22).

SOLUTIONS AND RECOMMENDATIONS

Denegar has argued that a statutory regime for Internet defamation is unnecessary (Denegar, 2001). According to Denegar, the common law has dealt "effectively [with Internet defamation and] has kept pace with technological advancement and can keep pace with such technology" (Denegar, 2001, p. 18). This is a very optimistic conclusion that is not supported by the facts. The cases in the previous section show the difficulty the courts in various jurisdictions are facing in applying traditional legal rules on defamation to the Internet. A legislative intervention in this area is called for, to provide certainty, predictability, and comprehensiveness. Justice Gaudron in Dow Jones & Co. v. Gutnick (2002, p. 6) pointed out legal rules on civil liability for Internet activities need "national legislative attention and... require international discussion in a forum as global as the Internet itself."

There is need to recognise that the Internet is a different medium from the other media of communication. Internet communication is not limited by space—it transcends national boundaries. Cheap availability of bandwidth and hardware like storage attracts more users to the Internet and also provides limitless space for content provision and dissemination. Unlike other traditional media, the Internet does not impose any time restriction or communicator roles. It is almost like a "playground" for just about anybody to communicate by authoring and accessing contents as they wish. Contents can be authored and published at any time and can take any form—multimedia or plain text. Coverage can happen in real-time, without any delay for production and distribution, and in some cases, without even any form of quality control.

In human communication, in particular, in journalism, "gatekeeping" which is the process through which ideas and information are filtered for publication, is employed. Gatekeeping occurs at all levels of the media structure—from a reporter deciding which sources are chosen to include in a story to editors deciding which stories are printed, or even covered. Gatekeeping is inevitable and in some circumstances it can be useful. Gatekeeping can also be dangerous, since it can lead to an abuse of power by deciding what information to discard and what to let through. Nevertheless, gatekeeping is often a routine, guided by some set of standard provisions. However, the Internet treats censorship as a noise and is designed to work around it. The Internet is packet switched. Even though data packets can be traceable to source and destination computers, the identities of senders or authors are often very difficult to trace especially if done from a public network. This technical fact affects the kind of legislation that can be imposed on the Internet and also raises the issue of what to trust on the Internet.

There appears to be recognition of the different nature of the Internet in one case. S.A. Groupe Progres v. Syndicat National Des Journalistes (2001) concerned a suit for breach of copyright. The French court held that the exploitation rights granted to the defendant employer by the employees were exhausted after the first printing in newsprint, and that the employer was in breach of copyright by republishing the copyright material on the Internet. The court distinguished electronic publication and storage on a server from newsprint distribution on the ground that "the reduction to typographic form and the presentation of an article in a publication corresponding with the conceptions in the mind of the author at the conclusion of the contract no longer exist, the readership is enlarged and the duration of the distribution is different."

The role of the ISP in the publication should, therefore, be clearly defined. In Lunney v. Prodigy Services Co., the New York Court of Appeals drew an analogy between an ISP and a telephone company "which one neither wants nor expects to superintend the content of subscriber's conversations." In Anderson v. New York Telephone Co (1974), the court observed that "the telephone company's role is merely passive". Collins has suggested that ISPs should be regarded as "mere facilitators of Internet publications passing through their computer systems" and that ISPs should not be responsible for publishing them in the same way telephone carriers are "mere facilitators of telephone calls" (Collins, 2005, paras.15.38, 15.43) and not liable for defamatory telephone calls.

Defining the role of the ISPs in the publication process requires clear delimitation of the different activities carried out by the Internet intermediaries. For example, an Internet hosting service allows organizations and individuals to serve content to the Internet by running Internet servers. There are various levels of service and various kinds of services offered. These include: Web hosting, DNS hosting, e-mail hosting, and so forth. The Internet servers have good Internet connection and allow clients to run just about anything they want including Web servers, DNS servers, and any other server they may want. Proxy caches, also referred to as forward proxy caches, are usually deployed by ISPs, especially for schools and corporations to save bandwidth. A Web cache typically stores copies of documents passing through it; subsequent requests may be satisfied from the cache if certain conditions are met.

There is need to re-examine the separate publication rule. On the face of it the separate publication rule appears to be clear. But when one asks of the definition of publication, difficulties arise. Prosser (1953, p. 961) has described the rule as possibly more appropriate to "small communities and limited circulation" and "potentially disastrous today." However, the rule has been applied to the Internet without qualification by courts in places such as England and Australia. Legal rules in any field should be sufficiently dynamic to meet the challenges of "political, social and economic changes" and offer fresh definitions of the "exact nature and extent" of such legal rules (Davidson & Bryant, 2001, p. 8; Warren & Brandeis, 1890, p. 5).

Other traditional rules relating to the Internet should also be re-examined in the light of the challenges and opportunities provided by the Internet. The Internet requires separate rules and standards (Bettelheim, 2006, p. 28). The major fear with cases like Dow Jones v. Gutnick is the possibility of "floodgates for a wave of libel actions" (Cummins, 2003, p. 1) in multiple jurisdictions. At present the degree of exposure to national or global jurisdiction for Internet activities may cause "unpleasant surprise" for businesses (Bettelheim, 2006, p. 22). As the court in Young v. New Haven Advocate (2002, p. 263), stated liability for Internet activities or conduct should be based on principles going beyond "posting and accessibility." Clear understanding of the issues is important. For instance, in Internet defamation cases with trans-national elements, prevention of multiplicity of actions, and the choice of applicable law are related but separately important issues that are often confused (Dow Jones v. Gutnick, p. 34; Cummins, 2003, p. 9).

The unlimited choice of forum available to claimants in several jurisdictions should be addressed. In Dow Jones v. Gutnick, the court stated that notwithstanding the breadth of the means of communication, "those who make information accessible by a particular method do so knowing their information may have" (p. 39). This point has been described as an "overreaching principle" (Cummins, 2003, p. 10). The use of the Internet is now a business method that does not necessarily involve an acceptance that the information made available on the Internet would be used to determine jurisdictional questions.

There is need to examine the current jurisdictional rules on Internet defamation in the light of the generally internationally recognised right to freedom of speech. The current rules may "serve to create a crippling effect" (Handa, 2004, p. 20) on the right to freedom of speech if publication on the Internet amounts to unlimited exposure to different jurisdictions with access to the publication on the Internet. This fear was also observed by a Canadian court in Braintech v. Kostiuk (1999). However fashioning an international agreement on jurisdiction over Internet defamation or an international code on freedom of speech on the Internet is problematic because of the "disparate free speech rights afforded citizens around the world" (Cummins, 2003, p. 1).

The Internet "blurs national boundaries" and "creates an international channel of communication" (Rustda & Koenig, 2006, p. 3). The question is at what level should the problem of civil liability of ISPs be addressed? This issue is an important one because of the global reach of the Internet. The question is whether a solution should emanate from a national, regional, or international authority. Solutions to Internet issues may be designed or emanate from national, regional, or international authorities. The Irish Single Publication Act is an example of a domestic approach to issues relating to the Internet. Blumenfeld, Carone & Jain (1999) has rightly pointed out that "the response of legal systems at the international, national, and local levels to [the] challenges [presented by the Internet] has been very uneven" (p. 20).

Effective legal regime on the Internet requires international solutions and cooperation (Sableman, 1999, p. 12). Handa (2004, p. 20) has suggested clear and "fresh policy analysis with a view to adopting or adapting to international norms, coupled with some legislative amendment" for online defamation. Cummins (2003, p. 9) has advocated legislative intervention instead of court decisions for resolving jurisdictional questions relating to the Internet.

Getting an international agreement on the issue is actually the least practicable, although it is the most desirable. Inherent in the process of obtaining agreement is the possibility that parties may disagree. This possibility of disagreement increases with the rise in the level of negotiation. The process of international agreement is particularly difficult. Obtaining the ultimate decision is often difficult. A major reason is the conflict of national systems and interests in international economic relations. Political compromises are commonplace in international agreement making. Where there is such compromise, the quality and effectiveness of the rules may be undermined by gaps in the rules introduced by such compromises.

Two states may agree on the resolution of an issue that cuts across their borders. Bilateral agreements may, without doubt, be a solution to "regulatory diversity between states" (Muchlinski, 1999, pp. 110-111). But is doubtful whether a bilateral action would be an appropriate response to the issues relating to the Internet. The global reach of Internet indicates that such bilateral actions may not be very effective. Being agreements between two states only, bilateral treaties are "limited solution" since "they represent specific regimes applicable only to [those two] signatory states" (Muchlinski, 1999, p. 111). Although involving more than two countries, a regional treaty suffers from the same weakness of "jurisdictional limitation" (Muchlinski, 1999, p. 111) like a bilateral agreement. The geographical scope of Internet is clearly beyond the competence of any regional regulatory institution.

The consequences of substantial differences in national rules and practices are uncertainty and unpredictability. In any field, knowledge of the level of risk or other exposure is very useful for the ISPs, businesses, and other participants on the Internet. International uniformity of rules has the advantage of saving regulators, decision-makers, and other stakeholders the cost of developing, maintaining, and learning a multiplicity of jurisdictional and other rules relating to Internet activities. The legal environment of business of any kind ought to be "an atmosphere of knowledge" devoid of "legal ambush, legal surprise, [and] legal incomprehension" (Wood, 1995, p. 22). Voltaire's comments over 200 years about the problems of multiplicity of rules in pre-revolution France are quite relevant to today's Internet-connected global community. In Oevres de Voltaire VIII (1838) Dialogues (p. 5) quoted in Wood (1995, p. 22) Voltaire had asked:

Is it not absurd and terrible thing that that which is true in one village is false in another? ... When you travel in this kingdom you change legal systems as often as you change horses.

How "local," "national," "regional," or "global" is the Internet generally, or the issue of liability of ISPs in particular? How does one explain the increasing trans-national litigation over the operations of the Internet? The fact is that the Internet-based liability is an international issue. The Internet operates across the limits of any national legal jurisdiction. Tying the Internet to a particular country is a very difficult task. The applicable European Union instrument recognises the inherent weakness of any regime that addresses international problems nationally. The objective of the EU Directive on Electronic Commerce (Directive 2000/311EC) was stated in its recitals to be the removal of "existing and emerging disparities in member States legislative and case law concerning liability of service providers acting as intermediaries."

Although it is possible to (or attempt to) legislate or enforce extra-territorial rules but extra-territorial prescriptive rules and enforcement powers are usually frowned upon by other jurisdictions as an affront to their sovereignty. In Re Enforcement of an English Injunction (1997), a court of another European country strongly criticised an anti-suit injunction from an English court, and refused to enforce the injunction. Extraterritorial application and enforcement of law can have "serious political effects" (Muchlinski, 1999, p. 109). In 2005, the French subsidiaries of MacDonald and Eaide Technologies requested permission to implement some aspects of a domestic U.S. legislation. The requests were rejected by the French regulatory authorities (Eecke & Truyens, 2006, p. 22).

There may be risk of "trouble with international legal principles" (Popov, 2005, p. 19) when a domestic legislation is given extraterritorial application. In such cases counter measures by other jurisdictions may result in the abundance of often conflicting rules and other requirements (Seidl-Hohenveldern,1987, pp. 17-22). In U.S. v. Aluminium Co. of America (Alcoa) (1945), U.S. regulatory authorities applied a domestic U.S. anti-trust legislation to a Canadian company's participation in an international aluminium cartel outside the territorial boundaries of the U.S. This regulatory decision was "heavily protested" (Popov, 2005, p. 25) against by the international community (Born, 1992, pp. 32-33).

In Yahoo Inc. v. La Lique Contre le Racisme et L'Antisemitismea, French court ordered the ISP, Yahoo! to block French citizens' access to Nazi materials displayed or offered for sale on Yahoo! U.S. Web site. But a U.S. District Court in California declared that the French court's order was unenforceable in the light of the U.S. constitutionally guaranteed right to free speech. This decision was criticised by French citizens (Cummins, 2003, p. 1). The nature of the Internet ensures that most national legislative or judicial measures on its operations have international impact, although such measures may have no international counterparts.

One of the major obstacles to legal uniformity is even the existence of different legal rules, concepts, and traditions. A survey of the substantive defamation laws of Canada, the UK, and the U.S. shows substantial differences in content (Fried, 2006, p. 3). For instance, section 1(3) of the UK Defamation Act 1996 has been criticised for failing to recognise that an ISP may not monitor or edit online content (Fried, 2006, p. 5). In contrast, the U.S. laws offer relatively stronger protection for ISPs, but as Sableman (1999) has observed, the U.S.-based ISPs "may be vulnerable to content-based claims under the laws of other nations" (p. 11). Sableman (1999) illustrates that some foreign advertising and promotion laws may differ significantly from U.S. laws on the subject (pp. 11-12). This obstacle to legal uniformity is not peculiar to any legal field; it is an ever-present danger to international harmonisation in any area of law. But international cooperation and harmonisation have continued in different fields despite this problem. The existence of different national rules, concepts, and traditions may even be an advantage in a search for international legal uniformity. The "awareness of the range of possible solutions" (Wood, 1995, p. 22) in existing national systems is clearly a positive factor.

Authors suggest international legislative solutions to the problems and issues relating to the operation of the Internet, and go further to suggest a permanent international institution on law and the Internet. The dynamic nature of the Internet requires the existence of a global intergovernmental body to advise the international community on legal issues relating to the Internet. The proposed international Internet Law Commission may be charged with the responsibility of identifying relevant issues and problems, preparing draft treaties and model laws on Internet issues, and recommending changes to existing treaties and legal rules. This suggestion is not completely without precedence.

The Internet Corporation for Assigned Names and Numbers (ICANN) is an internationally organized, non-profit corporation that has responsibility for Internet protocol (IP) address space allocation, protocol identifier assignment, generic (gTLD) and country code (ccTLD) top-level domain name system management, and root server system management functions. ICANN is responsible for coordinating the management of the technical elements of the DNS to ensure universal resolvability so that all users of the Internet can find all valid addresses. It does this by overseeing the distribution of unique technical identifiers used in the Internet's operations, and delegation of top-level domain names (such as .com, .info, etc.). Other issues of concern to Internet users, such as the rules for financial transactions, Internet content control, unsolicited commercial e-mail (spam), and data protection are outside the range of ICANN's mission of technical coordination. It is instructive that there have been calls by some governments for ICANN's functions to be transferred to a multilateral body (Rundle, 2006, p. 14).

FUTURE TRENDS

The Internet is certainly the most valuable information resource there is, with over 800 million people form all walks of life using it everyday. Many reputable mass media companies have an online version of their paper-based publications. People are seeing a lot of online TV and radio stations. Online shopping is on the increase. Most major Internet players like Yahoo, MSN, and Google are innovating Internet-based advertising and are already making huge turnovers from it (Briggs, 2006). The global community will continue to see more and more users who would prefer to access online contents and do most things on the Internet due to the convenience it offers and other numerous strengths. In the coming decade, every household will potentially have Internet access and the Internet will be a social necessity just like television, radio, and telephone. It is clear that the Internet is here to stay.

The Internet trend appears to be the increasing globalisation of the world economy (Drury, 1998; Rammelo, 2001, pp. 1-2). The recognition of this fact may encourage greater global cooperation in this area. For instance, the creation of the .eu top level domain, (.eu TLD) was an initiative of the European Commission through the following instruments: Regulation (EC) No733/2002 of 22 April 2002, of the European Parliament and of the Council (OJ L 113, 30.04.2002, pp. 1-5); and Commission Regulation (EC) No874/2004 of 28 April 2004 (OJ L 162 30.04.2004, pp. 40-50). Those legislative instruments were followed by a contract signed between the European Commission and the selected European Registry for Internet domains (EURid) in 2004. In 2005, EURid and ICANN signed an agreement "recognising EURid as the official body to manage the .eu TLD for the next 5 years" (Van Eecke & Rombouts, 2005, p. 26).

It appears that international cooperation on the regulation of the Internet will be issue specific. More issues requiring global action and resolution may be identified with the passage of time. For instance, there is now an international Convention on Cybercrime. This agreement was signed in 2001 by 45 European countries, the United States, Canada, Japan, and South Africa with a commitment "to suing common definitions of what constitutes cybercrime and common procedures for criminal enforcement" (Rundle, 2006, p. 15). The eConsumer.gov is an international consumer ombudsman established to "monitor the development of consumer problems connected with electronic commerce and work together with officials in other countries to develop common solutions" (Rustad & Koening, 2006, p. 7). Non-governmental organisations will be playing active roles on issues relating to the Internet. There are such organisations already in existence (Rundle, 2006, p. 14).

Unless there is legislative intervention in this area, it appears that the courts in different jurisdiction would create rules limiting the global extent of the choice of jurisdiction that seem to be available to claimants. For instance, in Berezouvsky v. Michaels (2000, p. 1032) Lord Hope pointed out that in internationally disseminated defamation cases, the claimant "must limit his claim to the effects of publication in England".

CONCLUSION

ISPs are key players in the online world. As they have control over the content, one key issue is to clarify if and to what extent they should be made responsible for third party content. Generally, the process of legal change and development usually "lingers behind society" and "delays to translate newly formed social relations into enforceable rights and obligations until (in many cases) long after they have been fully formed" (Davis, 1961, p. 209). This is the case with the relationship between the Internet and the law. The fact is that, despite years of its availability and use by the public globally, "Internet law is still in its early stages of development" (Popov, 2005, p. 27).

Internet defamation like other issues relating to the Internet "is inherently trans-national in nature" (Fried, 2006, p. 3). It is surely "impracticable for each country to apply its own legal regime to the whole of" (Rundle, 2006, pp. 12-13) the Internet. It is "inevitable" (Rustad & Koenig, 2006, p. 3) for domestic legislation on Internet activities from different jurisdictions to conflict. The global scope of the Internet requires that an international level of response is a very efficient solution and the most appropriate response in the circumstance. As a consequence of its nature and scope, internationally accepted rules are required for establishing effective rules for the operation of the Internet. The pertinent issues transcend national boundaries. An international commission on the Internet law is the most effective way to address the question of the relationship between the Internet and the law.

Policies initiated in the recent past by WSIS 2003 and 2005, ITU, ICANN, and other UN bodies (UN ICT TF) related to information and content, and very recent approaches taken by IGF and GAID would presumably come up with more applicable rules, procedures, and regulations, compromising Internet's universal access for all.

FURTHER RESEARCH DIRECTIONS

Experience has shown that the Internet is in constant flux. The development of the Internet is continuing in expansion and diversity. Consequently there will continue to be the need for the law to adapt to the challenges posed by, and attempt to address the issues raised by, the Internet in its different or emerging forms. As a result, the relationship of the various aspects or components of the Internet and the law offers exciting and wide-ranging research opportunities. This point may be illustrated with some examples.

One of the issues that will continue to require exploration is the governance or control of the Internet. Various national authorities have continued to regulate Internet activities and contents. Some of the questions that may arise include the relationship between national regulatory measures and international law and practice, and the effectiveness of national measures in particular areas or aspects of the Internet. It may also be necessary to examine particular national rules. The role of inter-governmental organisations and non-governmental organisations in the governance of the Internet may also raise issues to be addressed.

As the scope of the Internet widens, human rights-related issues will continue to pose challenges. Particular or emerging aspects of the Internet may raise issues of protection of privacy, freedom of speech, freedom of association, and other rights, for example. Legal rights are subject to re-definitions and clarifications in the light of developments in the society. There may be a continual need for researching appropriate definitions of legal rights if the Internet continues to develop. For example, the use of the Internet for healthcare delivery may raise issues for research.

Another area that may be of research concern is corporate disclosure regulation. Various regulatory authorities now require web-based disclosure forms. There may be need to inquire into the effectiveness of these disclosure requirements in general. Research may also be required in particular disclosure rules.

A huge research area is the Internet and intellectual property rights. Much has been written on different aspects of this area. However, the area is as dynamic as the Internet itself. Intellectual property issues continue to arise from novel ideas, inventions, or materials that use the Internet as medium. The law will have to cope with the increasing number and diversity of such ideas or materials. There may also arise need to study particular intellectual property issues affecting Internet-based creators, users, intermediaries, or hosts. For instance, the status and liability of ISPs and other Internet intermediaries are not settled.

Research possibilities on the Internet and the law are limitless. The Internet and the law relationship is an area that will continue to require further empirical, historical, analytical, or experimental research into its diverse and emerging forms.

REFERENCES

Legislation

Convention on Cybercrime (2001)

(EU) European Brussels Convention on Jurisdiction and Recognition and Enforcements of Judgments (1968).

(EU) Council Regulation (EC) No 44/2001 of 22 December 2000 on jurisdiction and the recognition and enforcement of judgments in civil and commercial matters

(EU) Directive on Electronic Commerce 2000 (Directive 2000/311EC of the European Parliament and Council)

(EU) Directive 98/34/EC

(EU) Regulation (EC) No733/2002 of 22 April 2002, of the European Parliament and of the Council (OJ L 113, 30.04.2002)

(EU) Commission Regulation (EC) No874/2004 of 28 April 2004 (OJ L 162 30.04.2004)

(UK) Defamation Act 1996

(UK) Electronic Commerce (EC Directive) Regulations 2000

(US) Digital Millenium Copyright Act (DMCA), 1998

(US) Online Copyright Infringement Liability Limitation Act of 1998

Cases

Anderson v. New York Telephone Co., 35 NY 2 d 746 (NY,1974).

André and Jean-Pierre Vasarhelyi v. Guiness World Records Ltd., No. 01/19480 Tribunal de Grande Instance de Paris [2003] ECDR 14.

Antonia-Belika Koubareli v. Evangelos Volotas and Forthnet S.A. No. 1639/2001, Athens Court of First Instance ECDR 19 (2003).

Berezousky v. Michaels, 1 WLR 1004, 1024 (2000).

Berezousky v. Michaels, 1 WLR 1004,1032 (Lord Hope, 2000).

Berezousky v. Michaels, 1 WLR 1004 (2000).

Braintech v. Kostiuk (1999) 171 D.L.R. (4th) 46 (B.C.C.A).

Campbell v. MGN Ltd. [2004] UKHL 22; [2004] 2 WLR 1232.

Case of the Barcelona Traction Company (Belgium v Spain), ICJ Rep. 3, 47 (1970).

Church of Spiritual Technology, Religious Technology Centre, New Era Publications International APS v. Dataweb B.V. & 21 Others, No. 99/1040 (Court Appeal of The Hague) ECDR 25 (2004).

Digital Equip. Corp. v. Altavista Tech. Inc., 960 F. Supp. 456 (D. Mass. 1997).

Douglas v. Hello! Ltd., No.3, EWCA Civ. 595 (2005).

Duke of Brunswick v. Harmer, 14 QB 185 (1849).

Estelle Smet-Hallyday v. Valentin Lacambre, ECC 439 (1999).

Euromarket Designs Incorporated v. Peters, ETMR 1025 (Jacob J., 2000).

Gershwin Publishing Corp. v. Columbia Artists Management Inc., 443 F. 2d 1159, 1162 (2d Cir. 1971).

Godfrey v. Demon Internet Ltd. QB 201 (Morland J., 2001).

Jameel (Yousef) v. Dow Jones & Co. Inc. EWCA Civ. 75 (2005).

Jean-Louis Costes v. Ministere Public,) ECC 451 (Cour D'Appeal, Paris, 2000).

King v. Lewis [2004] EWCA Civ. 1329; [2004] I.L.Pr. 31.

Loutchansky v. The Times Newspapers Ltd., EWCA Civ. 1805 (2001).

Lunney v. Prodigy Services Co., 250 AD 2d 230 (1998).

Lynda Lacoste (née Heineman) v. Multimania Production SA, E.C.C. 22 (2001).

MGM v. Grokster ,125 S.Ct. 2764 (2005).

Re Enforcement of an English Injunction, ILPr 320 (1997).

Re the Law to Promote Confidence in the Digital Economy, 1 CMLR 14 (French Conseil Constitutionnel Constitutional Court, 2005).

S.A. Groupe Progres v. Syndicat National Des Journalistes, ECC 8 (Cour D'Appeal, Lyon France 2001).

Shevill v. Presse Alliance SA,AC 959 (1996).

Steinberg v. Pritchard Englefield (A Firm), EWCA Civ. 824 (2005).

US v. Aluminium Co. of America,148 F. 2d 416 (2d Cir. 1945).

Yahoo Inc. v. La Lique Contre le Racisme et L'Antisemitisme.169 F. Supp. 2d 1181 CND (Cal. 2001).

Young v. New Haven Advocate,315 F. 3d 256 (4th Cir. 2002).

Books and Journals

Bettelheim, N. (January 2006). Personal jurisdiction and the Internet: Cyber differences shed new light on existing conflicts. *Journal of Internet Law, 22.*

Blumenfeld, E., Carone, P.J., & Jain, S. (January 1999). *Journal of Internet Law,* 20.

Born, G. B. (1992). Reappraisal of the extraterritorial reach of U.S. Law. *Law and Policy in International Business, 24,* 1.

Briggs, J. (February 2006). *64 billion impressions per month for Google and Yahoo PPC adverts.* Retrieved July 18, 2006, from http://othermedia.com/industry/64-billion-impressions-per-month-for-google-and-yahoo-ppc-adverts,158,TA.html

Brown, M.D., Armon, O., Ploeger, L., & Traynor, M. (2005, Dec). Secondary liability for inducing copyright infringement, MGM v. Grokster Infringement-Prevention and Product Design. *Journal of Internet Law, 21.*

Ciske, M.M., & Webb, M.C. (2005, September). Suing the downloading doe: Post Verizon ISP liability and infringemnet prosecution under the DMCA. *Journal of Internet Law, 3.*

Collins, M. (2005). *The law of defamation and the Internet* (2nd ed.). Oxford: University Press.

Cummins, G.R. (April 2003). Global liability risks emerge in cyberspace. *Journal of Internet Law, 1.*

Davidson, S.J., & Bryant, D.M. (2001, November). The right of privacy: Historical development, international discord, and the interface with intellectual property law. *Journal of Internet Law, 8.*

Davis, J.P. (1961). *Corporations: A study of origin and development of great business combinations and of their relation to the authority of the state.* New York: Capricorn Books.

Denegar, A.P. (December 2001). ISP liability for third-party defamation in Canada: Adopting a US-style statutory regime. *Journal of Internet Law, 18.*

Drury, R. R. (1998). The regulation and recognition of foreign corporations: Responses to the Delaware Syndrome. *Cambridge Law Journal, 165.*

Fried, E. (June 2003). Issues and challenges presented by defamation on the Internet. *Journal of Internet Law, 3.*

Gosnell, G. R. (September 1999). Internet-related intellectual property issues. *Journal of Internet Law, 23.*

Handa, S. (February 2004). Online defamation: Broadcasting and republication in Canada. *Journal of Internet Law, 18.*

Kohl, U. (2004). The rule of law, jurisdiction and the Internet. *International Journal of Law and Information Technology, 365,* 12.

Milmo, P., & Rogers, W.V.H. (Ed.). (2004). *Gatley on Libel and Slander 10th edition*. London: Sweet & Maxwell. *Dow Jones & Co. Inc v Gutnick* [2002] HCA 56.

Muchlinski, P. (1999). *Multinationals and the law* (2nd ed.). Oxford: Blackwell Publishing.

Popov, A. R. (May 2005). The extraterritorial reach of the Lanham Act over trademarks on the Internet: Adopting the Lanham Act to global e-commerce or an unjustified extension of US laws? *Journal of Internet Law, 19.*

Prosser, D. (1953). Interstate publication. *Michigan Law Review, 959*, 51.

Rammelo, S. (2001). *Corporations in private international law: A European perspective*. Oxford: University Press.

Reese, R. A. (2005, December). The problems of judging young technologies. *Journal of Internet Law, 1.*

Rosenthal, R. E. (April 2003). From ZIPPO to ALS: Jurisdiction over IP violations on the Internet. *Journal of Internet Law, 12.*

Rundle, M. (April 2006). Beyond Internet governance: The emerging international framework for governing the networked world. *Journal of Internet Law, 1.*

Rustad, M.L., & Koenig, T. H. (May 2006). Harmonizing Internet law: Lessons from Europe. *Journal of Internet Law, 3.*

Ryan, Nader, J.R., & Lihan, T.P. (2003, Sept). Internet-based exercise of personal jurisdiction. *Journal of Internet Law, 12.*

Sableman, M. (June 1999). Business liabilities on the Internet. *Journal of Internet Law, 1.*

Seidl-Hohenveldern, I. (1987). *Corporations in and under international law*. Cambridge: Grotius Publications Ltd.

Van Eecke, P., & Rombouts, D. (October 2005). The EU domain: The sun is rising. *Journal of Internet Law, 26.*

Van Eecke, P., & Truyens, M. (2006, March). Recent events in privacy legislation. *Journal of Internet Law, 22.*

Warren, S.D., & Brandeis, L.D. (1890). The right to privacy. *Harvard Law Review, 4*, 5.

Wood, P.R. (1995). *Comparative financial law*. London: Sweet & Maxwell.

ADDITIONAL READING

Afuah, A., & Tucci, C.L. (2003). *Internet business models and strategies: Texts and cases* (2nd ed.). London: McGraw-Hill.

Arkadiusz, L. (2006). Security in the Grid: Key legal aspects. *Computer and Telecommunications Law Review, 12*(8), 272.

Barnes, F., Florez-Duncan, F., & Niels, G. (2006). Voice over broadband. *Competition Law Insight, 5*(3), 8.

Campbell, D. (Ed.). (2005). *E-commerce and the law of digital signatures*. Dobbs Ferry, NY: Oceana Publications.

Castells, M. (2001). *Internet galaxy: Reflections on the Internet, business and society*. Oxford: Oxford University Press.

Charlesworth, A. (2006). SCL forum: Decentring Internet regulation. *Computers & Law, 17*(3), 5.

Crowcroft, J., Handley M., & Wakeman, I. (1999). *Internetworking multimedia*. London: Taylor and Francis.

Deane-Johns, S. (2006). Counter-regulation. *Computers & Law, 17*(1), 20.

Flint, D. (2007). Computers and Internet. *Business Law Review, 28*(2),34.

Freestone, O.M. (2001). *Internet-related consumer aberrant behaviour: A cross-cultural study.* Manchester: UMIST.

Gamble, J. (2006). From virtual reality... 228 *Childright,* 15.

Gkoutzinis, A.A. (2006). *Internet banking and the law in Europe: Regulation, financial integration and electronic commerce.* Cambridge: Cambridge University Press.

Goldsmith, J.L., & Wu, T. (2006). *Who controls the Internet?: Illusions of a borderless world.* Oxford: Oxford University Press.

Gower, I. (Ed.). (2005). *Internet usage in business* (8th ed.). Hampton: Key Note.

Hardy, S., & Base, M. (2001). *Law and lawyers on the Internet: An essential guide and resource for legal practitioners.* Plymouth: Internet Handbooks.

Hays, T. (2007). Secondary liability for copyright-protected works: Part 2. *European Intellectual Property Review, 29*(1), 15.

Henry, J. (2007). Music: Campaign for new internet provider legislation. *Entertainment Law Review, 18*(1), 6.

Hornle, J. (2007). Germany- Internet Forum- Liability for illegal postings. *Electronic Business Law, 8*(2), 13.

Howitt, S. (Ed.). (1999). *The Internet usage in business* (3rd ed.). Hampton: Key Note Ltd.

Jacobs, D.C., & Yudken, J.S. (2003). *The Internet, organisational change, and labor: The challenge of virtualization.* London: Routeledge.

Jagdish, N.S., Eshghi, A., & Krishnan, B.C. (2000). *Internet marketing.* London: Harcourt Brace College.

Jones, J.C. (2000). *The Internet and everyone.* London: Ellipsis.

MacLean, D. (2005). *Internet governance: A grand collaboration.* New York: United Nations.

Oates, S., Owen, D., & Gibson, R.K. (2006). *The Internet and politics: Citizens, voters and activists.* London: Routledge.

Paré, D.J. (2003). *Internet governance in transition: Who is the master of this domain.* Lanham, MD, Oxford: Rowman and Littlefield.

Pinedo, A.T., & Tanenbaum, J.R. (2007). The dangers of blogging. *International Financial Law Review, 26*(1), 38.

Rasdale, M. (2006). Legislating for robots and zombies. *Computer Law and Security Report, 22*(3), 222.

Reed, C. (2000). *Internet law: Text and materials.* London: Butterworths.

Smith, G. (2007). Here, there or everywhere? Cross-border liability on the Internet. *Computers and Telecommunications Law Review, 13*(2), 41.

Smith, G.J.H. (Ed.). (2005). *Internet law and regulation* (4th ed.). London: Sweet & Maxwell.

Stefik, M.J. (1999). *The Internet edge: Social, technical and legal challenges for a networked world.* Cambridge, MA; London: MIT Press.

Tadmore, N. (2007). Source taxation of intellectual supplies: Concepts, history and evolution into the digital age. *Bulletin for International Taxation, 61,* 2.

Thomas, G. (2006). Offshore Internet gambling. *Chambers Client Report, 17,* 72.

Whine, M. (2006). Cyberhate, anti-semitism and counter legislation. *Communications Law, 11*(4), 124.

Section II
Education and Research

Chapter VI
Disintermediated Education

Helena Mitev
University of Tasmania, Australia

David Crowther
De Montfort University, UK

ABSTRACT

In this chapter, it is argued that one of the benefits of IT is its role in enabling the disintermediation of education to provide an environment in which all can participate. In the context of this chapter however, the focus is on the role of the Internet as a disintermediating technology. There are many ramifications to this but the focus is upon the implications as far as education is concerned. The controlling mechanisms of education have been cost and geographical access. The Internet has changed both of these through its disintermediating function, potentially allowing access to all. The purpose of this chapter is to theorise this in considering the prognosis for development in various parts of the world.

INTRODUCTION

Traditional models of learning focus upon the needs of students in terms of preferred learning styles and learning needs with teaching strategy dependant upon those factors. All such models consider the learning process as an interaction between teacher and learner (Crowther & Carter, 2002), with technology providing the subordinate role of supporting the learning and teaching strategies. It has been argued however (Davies & Crowther, 1995; Crowther & Davies 1996) that there is a largely unquestioning acceptance discourse that the subject matters being taught are appropriate for the needs of students. Consequently the only topic for debate within the discourse is the mechanism for transferring the knowledge contained within these subjects from the holders of that knowledge (the academics in higher education) to the persons desirous of receiving that knowledge (the students). The problem with this approach to learning is that it is predicated in the traditional mode of students attending at a location in order to receive wisdom from the teacher. While this has been established for centuries and works fine in developed countries it is problematic in less developed countries where geographical distance and cost make prohibitive barriers. In this chapter therefore, we argue that one of the benefits of IT is its role in enabling the disintermediation of education to provide an environment in which all can participate. And education is of course essential to change developing countries into developed countries.

The Internet is often portrayed as a panacea for the enabling and empowering of individuals and grass-roots activity: a 'distributed' technology capable of providing a counter-balance or even an alternative, to the omnipresent and omnipotent organisations and institutions of the twenty-first century. Indeed, there exist some powerful examples of successful e-collectivism and online action (Holmes & Grieco, 2000; Steward et al., 2000; Hogan & Greene, 2002). However, despite such successes the virtual world that is emerging and unravelling before us continues to be contested space; on the one hand, the Internet is an enabling technology that allows for the empowerment of grass-root networks, through the liberation of their communication paths and practices, and the potentialities for network building and boundary-spanning. On the other hand, in the struggle for the retention of 'old economy' positions, as well as the development of 'new economy' power-bases, organisations and institutions are strategising and manoeuvring in order to constrain and shape such networks and communications. This containment can occur, for example, through technological means (e.g., surveillance and 'walled gardens') and the structuring of the virtual world (e.g., portals).

Much of the discourse surrounding the opportunities available through the Internet is concerned at a societal level with the effects of internet technology upon society, and only by implication, upon individuals within society. It is however only at the level of the individual that these changes can take place. Indeed, access to the Internet and the ability to communicate via this technology to other individuals, without regard to time and place, can be considered to be a revolutionary redistribution of power (Russell, 1975). Moreover, the disciplinary practices of society (Foucault, 1977) breakdown when the Internet is used because of the lack of spatial contiguity between communicants[1] and because of the effective anonymity of the communication which prevents the normalising surveillance mechanisms of society (Clegg, 1989) to intercede in that communication.

Meanwhile, over the last quarter of a century higher education in the most Western countries has been transformed. In the UK, for example, this has meant a change from an elite activity, in which only 5% of school leavers participated, to a mass activity in which over 40% of school leavers participate. The government has plans to expand this further over the next few years to ensure that 50% of school leavers participate. This trend has been paralleled throughout the Western world, while in other parts of the world there has been a similar trend towards a greater participation in higher education, albeit starting from a lower base. For many would-be participants in higher education from many countries around the world the opportunity to study at a university is desirable but the opportunity to study at a university in the Western world is even more desirable and the limiting factor to the achievement of this desire is the cost involved. Nevertheless, a significant and increasing number of young people do manage to achieve this desire. This inevitably means of course that wealth is one factor which determines a person's access to higher education. Those lacking this wealth tend to study at universities in their own country—provided of course that they have the necessary wealth to even do this. For those without the financial resources to study at a university in this manner, there is recourse to other options. Thus around 30 years ago the UK established its Open University to enable those otherwise unable to attend university to participate in higher education. This trend has been followed in many countries around the world and the option to study by means of distance learning while working full or part time is now firmly established as a route to achieving a degree level qualification. In recent years, the range of options has been extended further with the advent of electronic learning (over the Internet) and virtual universities—the subject of this chapter.

Access to university education therefore has expanded dramatically and access has been given to many who previously could never aspire to this form of education and the ensuing benefits which can be obtained there from. It is widely promulgated in the UK that a university degree leads to a 40% increase in income over the working life of the average graduate with an unquestioning acceptance that this is both a desirable objective and a legitimation of the expansion of the higher education system by whatever means are considered necessary. In the UK and much of the Western world, this expansion means the opening up of the higher education system to market forces and the increased marketisation of education as a corollary to the massification of such education (Crowther, 2003). Alongside this marketisation have come the nostrums of the market—strategic alliances (between universities around the world), centres of excellence, branding, efficiency, and competition.

Teaching Strategy and Learning Theory

Teaching strategies adopted within the higher education and management development areas are based upon the learning theories which are predominant at any particular point in time. In a review of the existing theories of learning, Burgoyne & Stuart (1977) identified eight schools of thought and the teaching strategies appropriate to each school. We argue however that such schools of thought may have been relevant to the pre-computer environment but that current practice has become much more student centred. Thus we argue that basis of all the current schools of thought and all teaching can be considered within the context of Bloom's (1956) taxonomy of the six aims of teaching and Kolb's (1984) analysis of the four stages in the learning cycle. Recent work has concentrated upon the preferred learning styles of students and the elements essential to the teaching process to facilitate learning. Thus Gibbs (1992) states that for effective learning the material being studied needs to provide motivation to the student and incorporate material requiring activity from the learner and interaction with others, within the context of a well structured knowledge base. Houldsworth (1994) however, has argued that students have responsibility for their own learning but that the learning experience must meet both the experiential and cognitive needs of the student in order for effective learning to take place. She contrasts this with the enjoyability derived from the novelty factor of the use of multimedia as a learning vehicle and argues that this novelty factor does not in itself provide a vehicle for learning.

Learning theory therefore is concerned with addressing the differences in the way people learn and with meeting their needs within the teaching strategy adopted, in order to maximise the educational value of that teaching. This learning theory is grounded in cognitive psychology and Kahney (1993) considers the problem solving aspect of learning and identifies that novices and experts differ in the way in which they structure problems to be solved and in the strategies adopted for solving those problems. This suggests that different teaching/learning strategies are needed for novices and experts. This therefore implies that the teaching of managers, who can be regarded as experts because of their possession of a more sophisticated knowledge base, needs to be different from that of undergraduates, who can be regarded as novices in the subject being taught. Rumelhart & Norman (1983) however, consider the way in which knowledge is represented in memory and suggest that schemas are created which enable new knowledge to be fitted into meaningful contexts within the existing knowledge framework. Cognitive psychology however, is concerned with the acquisition of knowledge and does not separate the knowledge itself from the process by which it is acquired.

Models of learning, generally, are premised upon the fact that it is the process of teaching which facilitates knowledge acquisition and that the process is equally appropriate whatever the subject matter being taught and whatever the knowledge base from which the student starts. It is argued by Crowther and Carter (2002) however, that this is not the case and that the learning process must inevitably differ according to the subject matter being taught and according to the nature of the students going through the learning process. It is argued specifically, within the context of the use of information technology and the Internet as a facilitator of the learning process, that these factors are highly pertinent and due consideration of them leads inevitably to the need for the construction of a new paradigm for teaching when using multimedia as a vehicle for delivery.

The different needs of different groups of students is recognised by Grupe and Connolly (1995) who argue for the different needs of adult learners and by Filipczak (1995) who argues, within an organisational learning context, that training must match the preferred learning style of the people involved in that learning process in order for that training to be effective. All modern learning theory and the teaching strategies developed from that theory, is founded in the student centred learning paradigm initiated by Rogers (1951, p. 389) who states, "We cannot teach another person directly: we can only facilitate his learning."

Rogers also considers that learning takes place best in the context of a relevant situation and occurs most readily within the context of problem solving. He states, "Significant learning occurs more readily in relation to situations perceived as problems" (1961, p. 286).

He bases his arguments about effective teaching upon his experience as a therapist, and relates learning to his experiences specifically when he states (1961, p. 288):

In therapy the resources for learning one's self lie within. There is very little data which the therapist can supply which will be of help since the data to be dealt with exist within the person. In education this is not true. There are many resources of knowledge, of techniques, of theory, which constitute raw material for use. It seems to me that what I have said about therapy suggests that these materials, these resources, be made available to the students, not forced upon them. Here a wide range of ingenuity and sensitivity is an asset.

The advent of information technology into education as an integral part of the teaching process can be considered to match ideally the student centred paradigm of learning initiated by Rogers (Laurillard, Swift, & Darby, 1993). This is particularly true of the development of the Internet which has been considered to be a means of putting learning entirely in the hands of the student—effectively disintermediating education—and giving him/her complete responsibility for his/her learning (Tooley, 2000). Indeed this facility has been used as a justification for the seizing upon such technology as the vehicle to use for the enhancement of the education process (Dillenbourg, 1999). At the same time this technology aids the achievement of the educationalist agenda of improving efficiency in the teaching process.

THE NEW LEARNING ENVIRONMENT

It has been argued (Crowther & Davies, 1996) that in order to maximise the benefit from using information technology in the education process it is necessary to cease considering it as merely a technological extension to the current set of teaching aids and to start to consider it as an opportunity to create a new learning environment. In order to do so it is necessary to create courseware that is appropriate to the multimedia medium of delivery. This requires a redesign of the material to be delivered through this medium and not merely a transfer of existing course material into a new medium. The educational objectives of the course need to be redefined more precisely to be appropriate for the method of delivery. In this context it is important to remember that the Internet is a technology which lends itself to exploration and analysis in learning rather than merely knowledge transfer. This might be taken to imply that the Internet as a teaching method cannot totally replace other methods of teaching and that the role of textbooks and lectures is not replaced by this technology—or at least not in total. Using this argument, therefore, the Internet is not appropriate when the educational objective of teaching is student-lecturer interaction, the development of presentational skills, or collaborative interaction. When used with knowledge transfer objectives, the key benefit of the Internet is its ability to set that knowledge in an appropriate context to enhance the learning process through the perceived relevance of the knowledge to a real world situation with which the student is familiar. This changes knowledge transfer from an abstract process to a relevant contextual process. The Internet therefore needs to be considered as more than another method of delivery appropriate to some of the educational objectives of higher education but not others because of the way that it can be used not just to enhance the learning process of existing students but also to provide education to those who would not otherwise have such access. See for example the case of Ansted University, Malaysia. According to its prospectus, one of the goals of this university is to:

...foster the unity of our world community through education. ... It aims to produce men and women who ... have the breadth of vision, a sense of civic and moral responsibility, as well as a devotion to the fundamental values of human life. ...Our aim is that of changing the global village into a global community.

The subject areas which are most appropriate for the use of Internet technology require careful consideration, as does the type of student for which this method as a means of learning is most appropriate[2]. Beishuizen (1992) has suggested that a complex knowledge domain is best engaged with not by a linear exploration of that domain but by navigating around the domain using different routes. This type of knowledge domain therefore seems more appropriate for the multimedia internet technology than does a simpler or more linear knowledge domain. It can be argued therefore that more complex subjects are the ones which can best be taught more effectively using the internet as the teaching vehicle and it is argued that most business and management[3] subjects fall into this category. This is particularly true when it is considered that in most cases the solutions to problems in management are not arrived at linearly but depends upon an interaction between various factors and the necessity of undertaking often complex analysis in order to arrive at possible solutions to a problem.

Baker, Simon, and Bazeli (1986) have demonstrated that business managers, graduate students, and accounting undergraduate students differ from other undergraduate students in their learning style preference. These students prefer active experimentation and abstract conceptualisation in the learning process rather than more straightforward knowledge transfer mechanisms. It can be argued therefore that this is the student population that will gain maximum benefit from using the internet in the learning process, and will also have the maximum motivation for engagement with material presented using this medium. This reinforces the previous argument that the Internet is not a universal panacea appropriate in all instances and without regard for subject matter and type of student. Given that management students are people who are likely to be exploring complex knowledge domains, requiring the synthesis of learning from different disciplines with their actual experience, this reinforces the argument for the use of multimedia in the teaching of management subjects, and as part of the management development activities of a higher education institution, rather than considering it to be an appropriate method of teaching in all subject areas. To be effective however, even in this subject area the design and content of the courseware needs careful consideration. This is particularly true when the simulation of "real world" situations is attempted through the use of the Internet, and the social implications of attempting this, given that it cannot be fully successful, need to be considered.

It has been argued therefore that the role of the Internet within the learning process is different from that of other technology, even other computer based teaching technology, and it is not merely an extension of other forms of CBT[4]. Its use needs to be considered within the context of learning theories, motivation, and educational objectives. It is clear however that there is a role for the Internet which is central to the function of higher education and which does not subjugate the technology to the teacher-learner interaction. This therefore requires the construction of a new paradigm but furthermore requires a consideration of the subject areas and student populations which are most appropriate for the introduction of this new paradigm. To date however, developments in Internet education, and even the agenda of the discourse in this area, do not recognise the potential of the technology to create a new learning environment (Crowther & Davies, 1998; Crowther, Davies, & Barnett, 1997). Hence current developments are unlikely to gain maximal benefit from the technology in terms of effectiveness, and hence efficiency, in the learning process. Examples of the use of Internet based teaching for distance learning education abound. For example, in this research we consulted some of the most important distance learning programs in business administration in North America, Latin America, and the UK. As such, some of the Web sites consulted are: Athabasca University (Canada, http://www.mba.athabascau.ca/Titan/aucimwebsite.nsf/frmHome?OpenForm), TéléUniversité (Canada, http://www.teluq.uquebec.ca/), Open Business School (UK, http://www.open.ac.uk/oubs/), Harvard Business School-Executive Education (USA, http://www.exed.hbs.edu/products/tel/index.html), Penn State (USA, http://www.worldcampus.psu.edu/iMBA.shtml), and *Universidad Virtual del Tecnológico de Monterrey* (Mexico, http://www.ruv.itesm.mx/portal/principal/oe/).

The existing paradigm of teaching is grounded in the person centred subset of the humanistic psychology school. Consequently, technology is considered to be a vehicle for assisting in this student centred approach and this argument applies particularly to multimedia. The benefits of this technology in the learning process are however subject to dispute. Thus Cole (1994) is able to claim that evidence exists which suggests that computer based teaching is 70% more effective at transferring knowledge than traditional forms of teaching while knowledge retention is also higher. On the other hand, Maul & Spotts (1993a) claim that statistical evidence shows that there is no significant difference in learning between computer based teaching and more traditional methods but that less instructional time is involved (i.e., greater efficiency). It is argued that this debate concerning the effectiveness of IT in the education process is based within the current teaching paradigm. This paradigm considers that the learning process consists of an interaction between teacher and learner and that technology is merely a tool to help the teacher in this process. Thus IT is considered to be merely the next step in the linear development of teaching aids from blackboards through overhead projectors to computers. Multimedia itself is thus just a new development within this sequence. The current paradigm can therefore be modelled as follows:

While this paradigm continues to be accepted the debate about effectiveness and efficiency will continue because no clear definition of these terms within the context of CBT in general and multimedia in particular can emerge. The discourse will consist of this debate and consideration of the technologies involved (e.g., Gayeski, 1993) and how to design better courseware (e.g. Maul & Spotts, 1993b). At the same time Cochrane is able to criticise the lack of change in teaching methods in responding to technological advance. He states:

With all this IT you might have expected that teaching methods would have changed radically—not so! In many respects we have seen a regression. (Cochrane, 1995, p. 17)

He continues to argue that changes in society require changes in the education process, stating:

It is important to recognise that the acceleration of society through technological change is creating a new paradigm that education has to respond to. (Cochrane, 1995, p. 17)

It is argued here that this technological change itself necessitates a change in the teaching paradigm due to the advent of multimedia into the educational agenda. It is further argued that the discourse needs to recognise this changed paradigm and concentrate the agenda of the discourse upon how to gain maximum benefit from this new learning environment created by the advent of this new technology.

It has been argued elsewhere (Crowther, Davies, & Barnett 1997; Crowther & Davies, 1998; Crowther, 2003) that to maximise the benefit from using technology in the education process, it is necessary to cease considering it as merely a technological extension to the current set of teaching aids and to start to consider it as an opportunity to create a new learning environment. In order to do so it is necessary to create courseware which is appropriate to the computer based medium of delivery. This requires a redesign of the material to be delivered through this medium and not merely a transfer of existing course material into a new medium. The educational objectives of the course need to be redefined more precisely

Figure 1. Traditional paradigm for teaching with IT

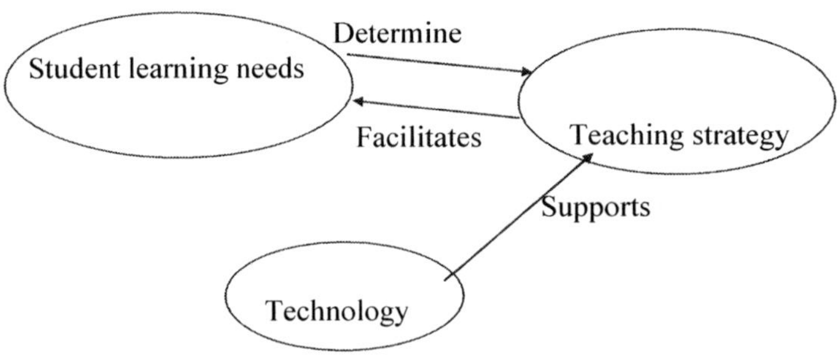

to be appropriate for the method of delivery. In this context it is important to remember that the Internet itself is a technology which lends itself to exploration and analysis in learning rather than merely knowledge transfer. This implies that computer based teaching as a teaching method cannot totally replace other methods of teaching and that the role of textbooks and lectures is not replaced by this technology—or at least not in total. Thus this technology is not appropriate when the educational objective of teaching is student–lecturer interaction, the development of presentational skills, or collaborative interaction. When used with knowledge transfer objectives, the key benefit of multimedia is its ability to set that knowledge in an appropriate context to enhance the learning process through the perceived relevance of the knowledge to a real world situation with which the student is familiar. This changes knowledge transfer from an abstract process to a relevant contextual process. These information and communication technologies therefore needs to be considered as another method of delivery appropriate to some of the educational objectives of higher education but not others. The new paradigm proposed for the teaching environment recognises this and can be modelled in Figure 2.

The Aston University Approach: The Witherspoon Ltd. Case

In order to give an example of how the technology can be applied to enhance the learning process we have chosen to give an example of the material developed at Aston Universtiy, UK, by one of the authors. Thus the authors describe their experience of developing a management accounting multimedia case study for use with business studies undergraduate students taking an advanced managerial accounting option.

The case study is based on a pottery company, Witherspoon Ltd., which produces a limited range of ornamental pottery products. The company is a subsidiary of a multinational group and is coming under increasing pressure from group level to improve its profitability. Two alternative investment proposals have been generated and each has been evaluated according to return on capital employed and payback criteria. The first alternative involves a major investment in a new computer-controlled kiln with increased firing capacity. The second option involves automating the production process and will require a significant number of production employees to be made redundant.

Figure 2. Paradigm for the technology based learning environment

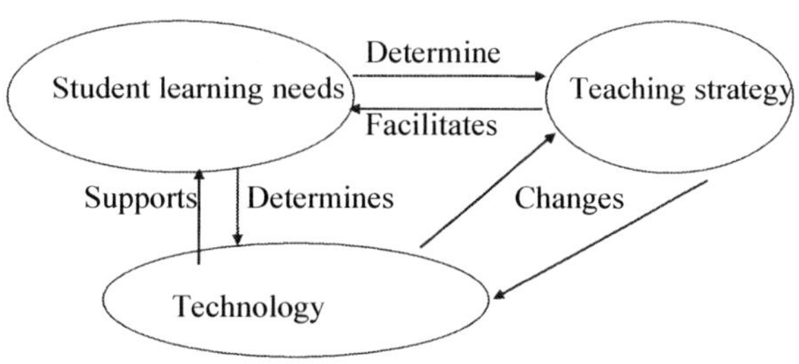

The main objective of the case is to provide a business setting which allows students to apply their knowledge of management accounting concepts and techniques, enabling them to gain a deeper understanding of the subject, whist at the same time enabling them to develop some key transferable skills. There is no single right answer to the case. Instead, the key requirement is for the student to build a set of recommendations. This in turn requires the ability to deal with an ambiguously defined practical problem through careful analysis of the case details, the presentation of a set of clear and logical arguments, supported with appropriate evidence and analysis and the recognition of the importance of assumptions, and any extra information required.

The authors believe that it is particularly important that students consider the interrelationships between management accounting and other subjects which they study as well as between different topics within management accounting itself. The case, therefore, provides scope for students to draw upon their studies of subjects such as strategic management, marketing, quantitative methods, and organisational behaviour, and consider the ways in which these subjects interact with management accounting. In addition, a number of interrelated management accounting topics are included in the case: traditional standard absorption costing, activity-based-costing, contribution analysis, product mix decisions, pricing decisions, investment appraisal techniques, performance measurement, and behavioural considerations.

It is also considered important that students are aware of the practical uses and limitations of concepts and techniques. Conventional textbook and lecture treatment of management accounting can give the impression that business problems can be easily and precisely formulated and then solved using the appropriate technique. It is considered extremely important that students are aware of the difficulties encountered when attempting to apply techniques and concepts in practice. The case, therefore, contains incomplete and partially conflicting information, some facts, and many opinions.

The program runs as a form of simulation exercise, with the student acting as a consultant to the company, called in to help decide which investment alternative should be chosen. On entering the program the student first receives an introduction to the company from the managing director which is followed by a factory tour. The factory tour involves actual footage of the Royal Worcester Pottery company, with a voice-over explaining the production process. Once the student has viewed the factory tour, they are then able to view an animated version of the production process which helps to clarify the process involved and also highlights how product costs accumulate within the process.

These two parts of the program form part of the background material, enabling the student to become familiar with the company and its operations. The main part of the program, however, centres around the 'consultant's office', from which the student is able to access various pieces of potentially useful information, which may be viewed, printed, or saved onto floppy disc. Pre-recorded interviews with key company officials can also be accessed and Excel spreadsheet software is fully integrated with the program, as is a word processing facility.

The student can access case information, interviews, tutorials, and spreadsheets via various 'hot-spots' in the consultant's office. The office filing cabinet contains three drawers, all of which contain potentially useful information: copies of internal memorandums, company documents (costing manual, marketing plan, for example), and various tutorials (case study analysis, report-writing, activity-based-costing, for example). The computer on the consultant's desk provides the route into the spreadsheet software in which pre-prepared worksheets can be viewed, manipulated, saved, and printed. A number of pre-recorded interviews with the three functional directors of the company can also be viewed. The student specifies a query or an area of interest and the program searches for key words which link to the relevant video clip. There are three or four video clips per director.

One other feature has been included to enhance the realism of the case and also to add a little fun to proceedings. After the student has reached a certain point in the program, the telephone on the consultant's desk will ring and the student will get a message that the company's shop steward would like a meeting to discuss his concerns (one of the investment alternatives involves a significant number of redundancies for shop floor employees). The student may decide not to see the shop steward at that moment, but the program has been written so that the student can only be ignored twice. The third time the telephone rings, if the student again decides not to meet the shop steward, he bursts into the office and makes his views known in no uncertain terms!

Integral to the case is the inclusion of a set of tutorials on various topics pertinent to the case; the student can access these as required from within the simulation. This again emphasises one of the differential advantages, linking knowledge transfer material with the simulation interactively to enhance the learning process in a way which is not possible with either a traditional paper-based case or a textbook. The rationale of this software development was based on several objectives: to provide a way of enabling students to direct their own learning by the way they choose to navigate around the case study; to integrate the knowledge acquired from a number of discretely taught modules, and to provide an experience of the real world as ambiguous and uncertain in a way that is not possible by other means.

DEVELOPING TRANSFERABLE SKILLS

Transferable or generic skills are not specific to a particular subject, course, or job. They include communication, analytical, interpersonal, quantitative, problem-solving, decision-making and learning skills, as well as computer literacy. The importance of acquiring these as part of the learning process is evidenced by the Enterprise in Higher Education Initiative of the Department of Employment[5] which commenced in 1987 with the aim of helping Higher Education Institutions to equip students for the "world of work," through the development of transferable skills which will be required in their future working lives.

Further evidence of the importance of transferable skills is provided by the Accounting Education Change Commission's (AECC) Position Statement Number 1 (1990), which outlined the committees' views on the objectives of education for accountants. The AECC stated that an accounting degree programme should prepare students to become professional accountants, not to be professional accountants. The AECC also developed a taxonomy of objectives consisting of three components: skills, knowledge, and professional orientation. Greater emphasis on communication, intellectual, and interpersonal skills was advocated, along with an appreciation of ethical issues. Indeed, with regard to the knowledge requirement, accounting knowledge was seen as only one important aspect along with general knowledge, and organisational and business knowledge. Also, the focus of accounting knowledge should be on developing analytical and conceptual thinking, not on memorising professional standards. Such an approach might be likened to focusing on the higher objectives in Bloom's taxonomy referred to earlier.

It is clear, therefore, that there is a growing recognition of the importance of what students can do, as well as what they know. This has been accompanied by a shift away from the traditional teaching culture of higher education, to that of a learning culture, in which students are encouraged to take responsibility for their own learning. It is argued that the use of case studies has a significant role to play in promoting the development of transferable skills within this more student-centred learning environment.

Many authors have advocated the use of realistic scenarios and case studies—for which the internet is ideally suited—in business education. For example, Argyris (1980) describes the following benefits:

- Hearing others views
- Confronting differences
- Making decisions
- Becoming aware of the complexity of reality
- Realising that there are rarely right or wrong answers since cases are incomplete, and so are real-life situations.

It can be seen that Argyris' list of benefits refers mainly to transferable skills as opposed to subject-specific objectives.

Although Argyris' work relates to the use of cases in business education, Libby's work (1991) provides specific evidence that cases are useful in an accounting context. Libby lists the following advantages of using case studies for accounting students:

- **Affective benefits to students:**
    - Student motivation
    - Student interest in material
    - Development of confidence
- **Skill development of students:**
    - Oral communication skills
    - Written communication skills
    - Group interaction skills
- **Cognitive benefits to students:**
    - Development of problem-solving skills
    - Development of judgment skills
    - Development of ability to solve problems addressing multiple issues
    - Development of the ability to deal with ambiguity
    - Development of an understanding of the real world
    - Comprehension of the material by students

Again, most of these benefits relate to the development of transferable skills.

Drawing upon the above evidence which is generally consistent with their own experiences, the authors believe that such scenarios and case studies are an extremely effective teaching tool (Caliyurt & Crowther, 2006), both for facilitating the achievement of subject-specific learning outcomes and facilitating the development of transferable skills and qualities. Perhaps most importantly, students seem to enjoy the case study approach and respond positively to the challenges posed.

DISINTERMEDIATING EDUCATION

One of the most significant—but least emphasised—roles of the Internet is its ability to disintermediate interactions through its virtual presence—eliminating the need for either geographical or temporal proximity (Barnett & Crowther, 1998). Thus it is no longer necessary to have intermediaries for many transactions. This has a significant impact upon the economies of most developing countries, particularly for farmers and independent manufacturers and traders. In this chapter, however, the concern is with education which can be disintermediated—because teachers are little more than intermediaries between students and knowledge—just as effectively as economic exchange. More significantly it means that the use of, for example, the Witherspoon case study described earlier is not restricted to on campus teaching but can be—and has been—used by distance learning students wherever in the world they are geographically located.

Distance learning is an increasingly utilized way of delivering a service that traditionally has been offered requiring a physical presence. Its occurrence dates primarily from the British and

Australian experiences based upon the concept of an "Open University" at the beginning of the 1900's, when in order to bring education to places geographically distant from university centres (Gury-Rosenblit, 1999), they implanted the system of distance learning through paper courier between students and professors, and to a lesser extent by telephone—in an attempt to reproduce a synchronous interaction. In those times, interaction was not treated by mediated technologies. Even nowadays, with the IT boom, most of universities that use IT tools are located in developed countries. On the contrary, most universities in developing countries that had opened a distance learning division still use the traditional writing mail and phone interaction. However, that is not always the case, and examples of IT based distance learning exist in such countries as Mexico and Malaysia (Crowther & Davila Gomez, 2006). Nonetheless, the traditional way of distance learning is not exclusive to developing countries, because many universities, particularly in developed countries offer at the same time, distance learning in both ways: online and written documentation dispatch.

Although there is a discrepancy between the need for IT development and its acquisition, due to high costs, the subject's sense and meaning can continue to be studied, because this discrepancy represents only one component of the meaning: that of the social axiology imposed (or accepted at the highest political level). The interest also lies in other axiological aspects, those of the cultural orientation (sometimes, those of the familial nucleus surrounding a particular personal experience) and those of the ontological individual principles. Hence, the reflection continues with focus on subjects that have had contact and have dealt with IT tools in distance learning.

In this perspective, it can be identified that many traditional institutions that want to incorporate this new way of doing, must confront not only a re-engineering process, but in many cases, conserve both process and structures because the two ways of delivering the service continue simultaneously (the former: in presence; the latter: the distance). This new subjects of the production must gain the professor's confidence and at the same time, the professor must explain its abilities to others in order to transform the same product (service of teaching a course) to a different level (before in presence, now in distance with IT support). Those characteristics are somehow not the most privileged when an IT solution is introduced in whatever organization. It has been said before that productivity is the primary goal pursued with IT, because it is possible to diminish costs (less personnel and time processes) and increase income (more units to sell in less time). However, in the distance learning case, it is noticed that service excellence could involve more expenses when IT is introduced (more people in the "production department": technical, pedagogical). Thus it is certain that not always does IT produces immediate savings (or the planned mid-term investment return). Depending on the institutional axiological sense, IT can be introduced to accomplish community and social goals such as the democratisation of education.

There are of course, a number of problems with IT, such as lack of availability to everyone, the cost involved, and the lack of IT skills of potential students. Although this is a broad subject which is outside the scope of this chapter, it can be stated as an example, that from the research it is possible to identify examples such as the public university which recognises that its social objective has been a factor in the decision to introduce an IT approach to distance learning, although that can represent a more costly approach. In addition, it can be compared to the axiological explicit goal imposed by owners and directors (democratisation of education) with those of subject's (professors, students, and the new subjects: technological and pedagogical subjects). Furthermore, the distance factor demands a harder effort on human contact between clients and front-end organization. Hence, for the specific case of distance learning service, it is argued that additional to new experts roles—technical and pedagogical—there are others whose function is to filter, answer, and handle the logistics between client, professor, and institution. Those new roles are the coordinators and the tutors. This highlights the importance to rescue the management process of coordination claimed by Fayol (1841-1925) as imperative, which is nowadays forgotten and included into that of direction. This rescue obeys the need for the coordinator to obtain other people's collaboration without having hierarchical authority, which is a "given" for the direction processes. Given the imminence of those entire new subjects, their ontological sense must be examined, and they must be contrasted with the axiological (the imposed and the natural). For instance, even though an IT introduction is not necessarily a risk taking situation, for the previous owners of the process (the professor), experiences fear of not developing technical skills to deliver the same course content in a new way (distance learning), or experiences defiance in sharing his secrets with others (the new subjects). Thus, a professor who could experience fear of sharing his knowledge at the beginning of the IT installation was able to tell us 2 years later how he has learned and developed new technical skills and a network of academic colleagues. It is here when the relevance of introducing into the framework of sense-making, the experience signification is introduced (which is called the continuous construction of meaning). In synthesis, the reflections of this point allow to be examined, in a more detailed way, what are the different meanings of subjects involved in IT solutions.

The argument here is that IT provides opportunities and threats, no matter what the situation of the person concerned. In the context of this chapter however, the focus is on the role of the Internet as a disintermediating technology (Schwarz & Watson, 2005). There are many ramifications to this but the focus is upon the implications as far as education is concerned. It is indisputable that one of the prime factors in social and economic development of any country (and indeed any person) is education. But education has been the preserve of the affluent with the ensuing privilege being passed down the generations seemingly as of right. The controlling mechanisms have been cost and geographical access. The Internet has changed both of these through its disintermediating function, potentially allowing access to all. The purpose of this chapter is to theorise this in considering the prognosis for development in various parts of the world.

CONCLUSION: SOCIAL RESPONSIBILITY AND HIGHER EDUCATION

The marketisation of higher education is premised in the myth of the free market as being beneficial to all (Crowther, 2002a). It is widely accepted—almost unquestioningly—that free markets will lead to greater growth and that we will all benefit from this growth. In opposition to the myth of the market is the concept of social responsibility which stems from stakeholder theory (Freeman, 1984). The acceptance of a stakeholder approach to organisations and the existence of multiple perspectives upon the objectives of an organisation inevitably imply a rejection of a monistic view of those organisations (Crowther, 2002b). Thus there can be no one single view of the objectives of the organisation and consequently no one single evaluation of the performance of that organisation. It is therefore inevitable within this paradigm that if monism is rejected as a view of organisations then this implies that the economic rationality view of organisations and their behaviour must also be rejected. For a university however, the objectives of equipping students for their future life and thereby benefiting society must not be forgotten in the drive for marketisation. Growth in student numbers and/or turnover are not objectives in their own right for a university. Indeed this sort of objective, and concomitant performance measure, are not wholly subscribed to in the untrammelled free market of the commercial organisation (Cooper, Crowther, Davies, & Davis, 2001; Crowther, 2002c) and so certainly cannot be justified for a university with claims to providing tertiary education (Maskell & Robinson, 2001). Thus by treating higher education as a commodity (Crowther, 2003) which can best be managed by the operations of the market, and rewarding the decision makers accordingly, the objectives of such education have been betrayed and lost within a drive for growth—on the unquestioned assumption that more equates to better. Thus the socially responsible ethos of higher education has been subordinated to the forces of marketisation and both society and the participants in such education are the poorer for it.

The central argument of this chapter however, is that Internet technology has the opposite effect and is liberating (Crowther, 2000) in its possibilities. Of particular significance in the context of this chapter is its disintermediating effect for education. And education is of course the most significant factor in the development of a country.

Future Directions for Research

When we consider the way in which an organisation operates we normally tend to think in terms of the formal structure of the organisation—in other words the formal hierarchical structure of the organisation. We tend to think that information and decisions flow upwards and downwards through this hierarchy. In this way the flow of information and decisions is related to the formal power structure of the organisation. But power exists throughout the organisation and we all potentially have the capacity to exert power and influence over what happens within the organisation. One of the disintermediating aspects of the Internet is change this and empower people—a fruitful area for research.

In recent years there has been much hype in relation to the emergence and importance of virtual teams, organisations, and communities. Yet the question that still remains to be answered, despite this hype, is whether or not the spatial and temporal dislocation that goes hand-in-hand with virtual interaction, qualitatively, and quantitatively impacts upon the social structure and 'effectiveness' of that social structure. In other words, whether it is possible for networks or communities to be built and maintained virtually. This question can be explored by studying the phenomenon itself, and indeed there exists literature on virtual teams, organisations, and communities. An alternative approach is to look at examples of virtual communities that existed prior to cyberspace in order to identify the distinguishing characteristics which have made such communities possible. A wide range of such virtual communities have existed throughout history, and indeed still continue to exist. Examples include such things as the eco-protest network, the anti-globalisation network, and the fanzine counter-culture in the UK. Each of these 'virtual communities' was able to develop coherence and a strong sense of identity despite separation by time and space. However, what is evident among these cases is the existence of a strong sociometric proximity of those within these communities or networks. This leads to a consideration of whether this is a prerequisite for the sustainability of 'strong' virtual networks. This in turn leads to a consideration of whether the potential for sociometric diversity within Cyberspace will lead to highly dynamic and unstable virtual communities and networks.

The increasing availability of access to the Internet has instigated a discourse that considers the present and likely future impact of this means of communication upon the construction of society and upon the lives of individual members of that society. Much of this discourse is based upon an expectation that the Internet and the World Wide Web will have a significant impact upon the way in which society operates. Thus some argue that this technology will be more liberating, participatory, and interactive than previous cultural forms while others argue that it will lead to increasing globalisation of politics, culture, and social systems. Much of this discourse is concerned at a societal level with the effects of Internet technology upon society, and only by implication, upon individuals within society. It is however only at the level of the individual that these changes can take place. Indeed access to the Internet, and the ability to communicate via this technology to other individuals, without regard to time and place, can be considered to be a revolutionary redistribution of power. The Internet provides the ideal environment for a community or network to exist. The Web structured organisation of the Internet means that no separate geographical existence is necessary for a web-based community, and the nature of the Internet is such that people are only part of an Internet community when they actually choose to participate in the activities of that community. Thus an Internet community is a truly post modern virtual community. The information architecture of the Internet means that it is relatively easy for such a community to establish its existence and therefore examples of such communities abound. It is relevant to observe however, that such communities have existed for a considerable period of time in such areas as academic life and social life. One essential feature of post modernity however, is the changing informational architecture of society and this both makes virtual communities more prevalent and also provides one possible infrastructure for determining community identity. Thus both the territorial and temporal constituents of community disappear, or at least assume diminished significance, in a post modern environment. At the same time this argument has been extended to the proposition that any community need not have either any discrete geographical existence or any continuous temporal existence but may exist as a virtual community having sporadic temporal existence and no territorial existence.

There are a lot of exciting things happening concerning the disintermediating role of the Internet in facilitating education and many avenues of further research to explore. The issues raised here are just a small number of the possibilities.

REFERENCES

Accounting Education Change Commission (1990). *Objectives of education for accountants (Position Statement Number 1)*. Bainbridge Island: AECC.

Argyris, C. (1980). Some limitations of the case method: Experiences in a management development programme. *Academy of Management Review, 5*(2), 21-34.

Baker R.E., Simon J.R., & Bazeli, F.P. (1986). An assessment of learning style preferences of accounting managers. *Issues in Accounting Education, 1*(1), 1-12.

Barnett, N. J., & Crowther, D. (1998). Community identity in the 21st century: A postmodernist evaluation of local government structure. *International Journal of Public Sector Management, 11*(6/7), 425-439.

Beishuizen, J. J. (1992). Studying a complex knowledge domain by exploration or explanation. *Journal of Computer Assisted Learning, 8*(2), 104-117.

Bloom, B. S. (1956). *Taxonomy of educational objectives: Cognitive domain*. London: Longmans.

Burgoyne, J., & Stuart, R. (1977). Implicit learning theories as determinants of the effect of management development programmes. *Personnel Review, 6*(2), 5-14.

Caliyurt, K. T., & Crowther, D. (2006). The necessity of fraud education for accounting students: A research study from Turkey. *Social Responsibility Journal, 2*(3/4), 322-328.

Carter, C., & Grieco, M. (2000). New deals, no wheels: Social exclusion. tele-options and electronic ontology. *Urban Studies, 37*(10), 1735-1748.

Clegg, S.R. (1989). *Frameworks of power.* London: Sage.

Cochrane, P. (1995). The virtual university; *Business of Education, 5,* 17-18.

Cole, G. (1994). Learning with computers. *Accountancy* (UK), *113,* 1209.

Cooper, S., Crowther, D., Davies. M., & Davis, E.W. (2001). *Shareholder or stakeholder value.* London: CIMA.

Crowther, D. (2000). Corporate reporting, stakeholders and the Internet: Mapping the new corporate landscape. *Urban Studies, 37*(10), 1837-1848.

Crowther, D. (2002a). *The importance of corporate social responsibility.* Escola Superior de Tecnologia e Gestao da Guarda. Working paper no 05/02.

Crowther, D. (2002b). *A social critique of corporate reporting.* Aldershot: Ashgate.

Crowther, D. (2002c). *Creating shareholder value.* London: Spiro.

Crowther, D. (2003). Social responsibility and the marketisation of higher education. *Staff and Educational Development International, 7*(1), 23-40.

Crowther, D., & Carter, C. (2002). Legitimitating irrelevance: Management education in higher education institutions. *International Journal of Education Management, 16*(6), 268-278.

Crowther, D.E.A., & Davies, M. L. (1996). Multimedia, mesmerism and teaching strategy; *CALECO 1995 Conference Proceedings.* University of Bristol, 11-17.

Crowther, D., Davies, M., & Barnett, N. (1997). Evaluating computer based courseware: A stakeholder approach. In *Proceedings of the 8th CTI-AFM Conference* (pp. 32-39).

Crowther, D., & Davies, M. (1998). Integrating Multimedia technology into a management accounting curriculum: an evaluation of effectiveness in the learning process. In *Proceedings of the 9th CTI-AFM Conference* (pp. 85-93).

Crowther, D., & Davila Gomez, A. M. (2006). Stress in the back office. In *Proceedings of India-the processing Office to the World,* Kochi, India (pp. 27-38).

Davies, M. L., & Crowther, D.E.A. (1995). The benefits of using multimedia in higher education: Myths and realities. *Active Learning, 3,* 3-6.

Dillenbourg, P. (1999). Introduction: What do you mean by 'collaborative learning'? In P. Dillenbourg (Ed.), *Collaborative learning-cognitive and computational approaches* (pp. 1-19). Oxford: Pergamon.

Fayol, H. (1970 edition). Administration industrielle et générale: prévoyance, organisation, commandement, coordination et contrôle, extrait du *Bulletin de la société de l'industrie minérale.* Paris: Dunod.

Filipczak, B. (1995). Different strokes: Learning styles in the classroom. *Training, 32*(3), 43-48.

Foucault, M. (1977). *Discipline and punish.* trans A Sheridan. London: Penguin.

Freeman, R. E. (1984). *Strategic management: A stakeholder approach.* Boston: Pitman.

Gayeski, D. (1993, April). De-mystifying multimedia. *IABC Communication World,* 27-32.

Gibbs, G. (1992). *Improving the quality of student learning.* Bristol: Technical & Educational Services.

Grupe, F. H., & Connolly, F .W. (1995). Grownups are different: Computer training for adult learners. *Journal of Systems Management, 46*(1), 58-64.

Gury-Rosenblit, S. (1999). *Distance and campus universities: Tensions and interactions. A comparative study of five countries.* Unesco-Pergamon.

Holmes, L., & Grieco, M. (1999, June). *The power of transparency: The Internet, e-mail and the Malaysian political crisis.* Paper presented to Asian Management in Crisis Conference, Association of South East Asian Studies, UNL.

Hogan, J., & Greene, A.M. (2002). E-collectivism and e-commerce: On-line action and on-line mobilisation. In L. Holmes, D. M. Hosking, & M. Grieco (Eds.), *Organising in the information age.* Aldershot: Ashgate.

Houldsworth, E. (1994). Multimedia in management education: Theoretical framework and methodological choices. *Henley Management College working paper* no. HWP9420.

Kahney, H. (1993). *Problem solving.* Buckingham: Open University Press.

Kolb, D. A. (1984). *Experiential learning: Experience as the source of learning and development.* NJ: Englewood Cliffs.

Laurillard, D., Swift,B., & Darby, J. (1993). Academics' use of courseware materials: A survey. *Association for Learning Technology Journal, 1*(1), 4-12.

Maskell, D., & Robinson, I. (2001). *The new idea of a university.* London: Haven Books.

Maul, G.P., & Spotts, D.S. (1993a). Developing computer-based instructional courses. *Industrial Management, 35*(6), 9-11.

Maul, G. P., & Spotts, D. S. (1993b). A comparison of computer-based training and classroom instruction. *Industrial Engineering, 25*(2), 25-27.

Rogers, C.R. (1951). *Client centred therapy.* London: Constable & Co.

Rogers, C.R. (1961). *On becoming a person.* London: Constable & Co.

Rumelhart, D.E., & Norman, D.A. (1983). Representation in memory. In R.C. Atkinson, R. J. Herrnstein, G. Lindsey, & R. D. Luce (Eds.), *Handbook of experimental psychology.* London: Wiley & Sons.

Russell, B. (1975). *Power.* London: Routledge.

Schwarz, G., & Watson, B. M. (2005). The influence of perceptions of social identity on information technology-enabled change. *Group & Organization Management, 30*(3), 289-318.

Steward, F., Conway, S., Mahay, V., Yearley, S., Bailey, P., Garcia, C., Joly, P., Lemarie, J., & Hansen, A. (2000). *Consumer and sectoral networks in the societal management of food technology.* Final report to the EC Environment and Climate Research Programme, Contract ENV4-CT97-0695.

Tooley, J. (2000). *Reclaiming education.* London: Cassell.

ADDITIONAL READINGS

Alam, K.F. (1999). Ethics and accounting education. *Teaching Business Ethics, 2*, 261-272.

Albrecht, W.S., & Sack, R.J. (2000). *Accounting education. Charting the course through a perilous future.* Sarasota, FL: American Accounting Association.

Ammons, D.N. (1994). The role of professional associations in establishing and promoting performance standards for local government. *Public productivity & Management Review, 17*(3), 281-298.

Armstrong, M.B. (1993). Ethics and professionalism in accounting education: A sample course. *Journal of Accounting Education*, (11), 78-92.

Armstrong, M.B., Ketz, E.J., & Owsen, D. (2003). Ethics education in motivation and ethical behavior. *Journal of Accounting Education, 21*, 1-16.

Ball, C. (1995). IT: An impending revolution. *Business of Education, 7*, 7.

Bampton, R., & Cowton, C. (2002). Pioneering in ethics teaching: The case of management accounting in universities in British Isles. *Teaching Business Ethics, 6*(3), 279-295.

Bottoms, D.T. (1994). Multimedia delivers the message; *Industry Week, 243*(7), 70-72.

Crowther, D., & Green, M. (2004). *Organisation theory*. London: CIPD.

Guha, A. S. (2005). Enhancing teaching and learning: Integrated approach to pedagogy. In D. Crowther & R. Jatana (Eds.), *Social responsibility in India* (pp. 32-38). Leicester: Social Responsibility Research Network.

Guha, A. S., & Crowther, D. (2005). Auditing and academic quality in India and the UK. In D. Crowther & R. Jatana (Eds.), *Social responsibility in India* (pp. 79-87). Leicester: Social Responsibility Research Network.

Holley D., Sinfield, S., & Burns, T. (2006). It was horrid, very very horrid: A student perspective on coming to an inner-city university in the UK. *Social Responsibility Journal, 2*(1), 36-41.

King, T. (1994). Authoring for distance learning. *The CTISS File, 17*, 30-32.

Laurillard, D., Swift, B., & Darby, J. (1993). Academics' use of courseware materials: Asurvey. *Association for Learning Technology Journal, 1*(1), 4-12.

Libby, P.A. (1991). Barriers to using cases in accounting education. *Issues in Accounting Education, 2.*

Lilley, P. (1995). Taking advantage of CD-R. *Management Accounting* (UK), July/Aug, p. 20.

Marx, W. (1995). The new high tech training. *Management Review, 84*(2), 57-60.

Maskell, D., & Robinson, I. (2001). The new idea of a university. London: Haven Books.

Noble, D. (2002). *Digital diploma mills: The automation of higher education.* New York: Monthly Review Press.

Satterthwaite, J., Atkinson, A., & Martin, W. (Eds.). (2004) *The disciplining of education: New languages of power and resistance.* Stoke on Trent. Trentham Books.

Sinfield, S,, Burns, T., & Holley, D. (2004). *Outsiders looking in or insiders looking out?* Widening Participation in a post 1992 University IN.

Sinfield, S., Burns, T., Wallace, M., Holley, D., Dobson, C., & Albon, D. (2005, April). *Friend or foe? The role of an online tutorial in an institution experiencing rapid change.* Paper presented at Discourse, Power & Resistance.

Smith, J. (2005). From flowers to palms: 40 years of policy for online learning ALT-J. *Research in learning Technology, 13*(2), 93-108.

Tilak Jandhgala, B.G. (1995). Privatisation of higher education in India: The capitation fee colleges. *Higher Education In India: In Search Of Quality, Association of India Universities.* New Delhi.

Trainor, R. (1992). Computers, arts based teaching and rising student numbers. *The CTISS File, 13*, 3-6.

Ward, M. (1994). Hypermedia learning. *Information Today, 11*(8), 15-16.

White, S. D. H. (2002). In S. M. S. Ketteridge & H. Fry (Eds.), *The effective academic: A handbook for enhanced academics practice.* Kogan Page.

Williams, B.C., & Nicholson, A.H.S. (1994).Values and value for money in courseware design. *Account*, *6*(1), 2-3.

ENDNOTES

[1] See Carter and Grieco (2000) regarding the emerging electronic ontologies and Barnett and Crowther (1998) regarding organising without contiguity.

[2] It is of course recognized that when cost or geographical constraints are a problem then this method of education is vastly better than nothing; it is merely argued that some benefit more than others.

[3] The authors themselves—separately—have designed and delivered courses in the area of accounting making use of this technology.

[4] Computer based training

[5] This is a part of the UK government mechanism.

Chapter VII
Towards an ICT4D Geometry of Empowerment:
Using Actor-Network Theory to Understand and Improve ICT4D

Donna D. Rubinoff
University of Colorado – Boulder, USA

ABSTRACT

This chapter introduces actor-network theory (ANT) as an approach to the analysis and improvement of the use of ICT for development (ICT4D). It argues that ANT helps to conceptualize ICT beyond the technological systems of the conventional "digital divide." ANT supports thinking about the sociotechnical networks that incorporate humans, hardware, institutions, texts, and policies, and so forth, into ICT networks. It also supports the inclusion of marginal actors, helps to address development problems from the perspective of those populations, and traces the networks of power that supports their participation or exclusion. The author hopes that the ideas in this chapter will promote further discussion on the topic and the refinement of an analytical framework for ICT4D.

INTRODUCTION

The term "digital divide" was originally coined to connote differential access to ICT along class lines between global north and south. Yet this "divide" persists even in the developing world, where ICT access tends to be privileged within government bureaucracies and technical/university settings. In rural spheres where development organizations and telecenters provide the main access to ICT, there still exists differential access to technology by professional status, education, gender, age, disability, and so forth. The digital divide often persists even in progressive development organizations, where access to digital tools and resources can be barred by institutional and system gates and obscured by invisible and often unintended social and organizational barriers. Ultimately, the terminology of the "digital divide" sets up preconditions for incorrect analysis and solutions to these problems. The term suggests a binary opposition in which people either have or do not have direct access to communications technology. If they do, they win, if the don't, they lose. Solutions tend to be centered on providing direct access, training people to use computer hardware and other direct solutions. Yet this epistemological framework simplifies the nature of peoples' engagement with ICTs, trivializes the wide array of background conditions that contribute to both development success and failure, and in the end masks the systemic nature of the whole process of using ICT for development (ICT4D). It also limits the possibility for analytical frameworks that can demystify these systems and promote enhanced ICT4D for a wider range of constituents.

The need for attention to this problem is not a new topic. In a recent World Bank seminar on the WSIS Gender Caucus (World Bank 2005, September 14), Eva Rathgeber spoke on the need for "new conceptual frameworks and methodologies to better understand relationships between ICTs and gender." She showed that when the WSIS Gender Caucus issued a request for proposals for research grants on three topics including: 1) documentation, analysis, and evaluation of efforts to mainstream gender into ICT Policy; 2) applications and content; and 3) theories and methodology; there were no responses in the third category.[2]

This is not to say that academic researchers have not addressed the issue of theoretical and methodological frameworks with which to analyze structures and impacts of ICT4D. A survey of recent research shows that researchers in the ICT4D arena explicitly positioned their work in an array of theories. Noteworthy among these are conference publications on "Women's ICT Based Enterprise for Development" held at the Institute for Development Policy and Management, Manchester, UK, in June, 2006. The research team of Morgan, Heeks, & Arun explored theoretical/analytical approaches to ICT4D in their three publications from 2004, including an enterprise approach, a livelihoods approach, and a feminist theoretical perspective (Heeks, Arun, Morgan, 2004; Arun, Heeks & Morgan, 2004; Morgan, Heeks, & Arun, 2004). They further developed this analysis in Heeks, Arun, Morgan (2005). Morgan (2006) has also addressed this topic with her discussion of new institutionalism as an analytical/theoretical framework; and Litho's recent (2006) analysis stems from social construction of technology and African feminist standpoint theories. All this work is important in terms of evaluating theory for the best approaches to understanding how ICTs support various development objectives. It calls for a deeper analysis of the institutions of ICT4D, and Morgan's work on the new institutionalism is especially useful in this regard. It highlights the regulative, normative, and cognitive power relations that shape information flow and management across the structures of formal and informal institutions. In addition, this approach allows for agency and transformation in the construction of networks. Morgan's work is an important move towards unpacking the "micro social" relations and structures of power that regulate flows of information and power across ICT4D networks.

All of the aforementioned works on theoretical frameworks move the analytical framework project ahead, but towards a specific focus on the analysis of ICT4D (as opposed to more theoretical frameworks), the landscape is sparser. In the most recent commentary on this problem along the context of gender analysis in ICT4D, Hafkin & Huyer (2006, pp. 4-5) point out that "ICTs have been used to promote social development only during the past few years, and little large-scale analysis of the results is yet available, concerning either gender equality or the benefits and results of these projects." They go on to argue, however, that "there are some promising signs of progress," including examples in their publication discussing the most recent use of ICT4D for women's empowerment. Working towards the goal of more precise, refined, and useable analytical frameworks, several chapters in Hafkin & Huyer are relevant. Hafkin's (2006) chapter on the issue of disaggregated ICT4D data collection, statistics, and indicators by government agencies is a crucial move towards the quantification of existing conditions and impacts: without data, disaggregated analysis is impossible. Huyer's (2006) discussion of the "enabling conditions" that support ICT4D and women's empowerment lays important conceptual groundwork for the analysis of ICT4D as it: 1) outlines the links between ICTs and empowerment and 2) shows research outcomes of dozens of projects on this topic, showing impacts on economic, sociocultural, psychological, education, political, and legal empowerment. However, although Huyer (p. 35) rightly argues that there is a need for more precision and depth in the analysis of "relations and environments," "analysis of existing models of ICT use," "facilitating conditions," and other aspects of ICT4D, she does not discuss how this might be accomplished. Most importantly, she does not raise the question of how this precise analysis in different domains or aspects of enabling conditions might be linked together, so that it is possible to understand how they work iteratively to produce outcomes and impacts, and how they are connected systematically.

This issue is broached in Nath's (2006) discussion in Hafkin & Huyer (2006) entitled: "Empowerment of Women through ICT-enabled Networks: Toward the Optimum ICT-impact Model." Nath (2006) makes important strides towards the creation of a systematic ICT4D analytical tool in two ways: 1) he envisions a "geometry of information flows" (p. 199) resulting from policies, projects, and programs; and 2) he implicitly picks up Morgan's new intuitionalist approach to show how missing segments of an institutional geometry could be constructed to facilitate information flows. Here, Nath has moved further towards the construction of an analytical tool or model in ICT4D and it is a welcome step in that direction.

Nath's work and the others mentioned above should be seen as the first foray into a concerted and coordinated dialogue on the "conceptual framework and methodology problem;" and all of this work together should be incorporated into a systematic effort to address this problem. Ultimately, a useful systematic analysis of ICT4D should be constructed that incorporates both quantitative and qualitative data; that addresses institutions, enabling conditions, and barriers to ICT4D; that shows flows of information, communication, and empowerment; and which shows how all these factors link together systematically. A key requirement would be that a systematic analysis would provide for the incorporation of spheres and factors not considered by western researchers. This would be similar to the best ethnography, which centers its meanings and its study on the experience of the subject; and would result in what Escobar (2000) calls an "ethnography of networks."

In this chapter, the author wishes to call for an ongoing and focused dialogue around the problem of ICT4D analysis, and made an additional contribution to the effort, which she believes has the potential to provide coherence and organization to the disparate efforts noted above. In the discussion below, she introduces actor-network theory (ANT) and explores its value as a systematic analytical tool to evaluate, refine, and expand success outcomes of ICT4D projects and programs; and to conduct ethnographies of existing, nonexistent, or proposed ICT4D networks. The author suggests that by using ANT analysis (alone or perhaps in conjunction with aforementioned approaches), it becomes possible to both "draw" networks and to "color" them; to trace quantitative and qualitative flows of power and empowerment; to discover gaps and opportunities for improvement, and ultimately to develop ICT4D programs that work better from the outset. At the same time, it is recognized that ANT has its critics, and possibly its limitations for this endeavor. Importantly, this chapter presents only a first step in developing a "hands-on" practical procedure for the analysis of ICT4D: it is not a guidebook but an introduction of ANT to the ICT4D world. Therefore, it is expected that this chapter will provoke response from both theoretically and practically oriented ICT4D advocates, and that this will be the beginning of a fruitful dialogue that explores and develops the potential of ANT for ICT4D.

Below, ANT has been introduced and historicized as it has evolved within the sociology of science and technology, how the concept relates to ICT4D has been demonstrated, and finally, a few examples have been briefly presented to show how this approach might work to trace or evaluate impacts and outcomes of ICT4D in the course of research on rural women in Central America. The emphasis of the chapter will be to show how ANT can be used to move ICT4D analysis beyond simple binary ideas of "access" to technology, towards showing how it can be used to describe the sociotechnical networks that propel marginal peoples from being (in the terminology of Cornwall & Gaventa, 2001) "users and choosers" of technology to "makers and shapers" of their development agendas. However, this chapter will not apply ANT to specific cases. The author calls on other researchers to move this agenda ahead by: 1) extending the thinking in this chapter to develop the analytical framework and 2) to test it with specific cases.

CONTEXT: HISTORICAL EVOLUTION OF ICT4D

During the last decade the vision and practice of ICT4D has moved from the simple technological silver bullet (depicted by the term "access" and represented by the ubiquitous image of the person sitting in front of a computer screen), to the slightly more socially just but equally ubiquitous image of a village tele-center in which people collectively accessed the technology. In both cases the silver bullet was direct access to computers, and the digital divide was broached when one or two people sat next to one computer. However, both of these visions have evolved over time into more complex ICT4D programs that address contextual factors such as skills, economics, literacy, and content, and which sophisticate the simplified vision of direct access. Furthermore, the conceptual flow of information has morphed from its early unidirectional form in which information was downloaded from "on high," into multidirectional pathways of knowledge production. Yet, analytical frameworks have not kept up with the ongoing sophistication of projects and case studies.

A tracing of the evolution of ICT4D projects looks back to the first simplistic and techno deterministic solutions for bridging the digital divide. These depicted a computer terminal/portal out of which poured the information necessary to receive knowledge from others. Information diffusion, in the form of market conditions, health care, disaster prediction, and the like, were the first wave of benefits to be promulgated through ICT. Next, the computer was seen as a tool for entrepreneurialism, Internet sales, and disintermediation of markets. Flower growers in Costa Rica could skip the 'middle man' in San Jose and sell directly to their markets in North America and Europe, as long as they were online (Rubinoff, 2003). But markets were joined, they were not made in the vision of the new participant, and this phenomenon was critiqued by those who saw its neoliberal and imperialistic agendas (Harcourt, 1999; Cornwall & Gaventa, 2001; MacLean et al., 2002).

Soon, though, opportunities arose to shift this paradigm in the arenas of national e-policy development and knowledge production. ICT4D advocates raised a vision of information, influence, and power that flowed not only in a north/south or rich/poor direction. When it appeared that e-governance (the digitizing of government systems) would change the political landscape, a rush to influence national e-policies emerged, especially during the World Summit on the Information Society (WSIS) conferences in 2003 and 2005. Perhaps the most refined programs for socially just and locally driven ICT4D policy come from the WSIS Grassroots Caucus and WSIS Gender Caucus recommendations, which highlight the need for a broad range of interventions in the creation of e-governance that span knowledge production, access, management, and use (Grassroots Caucus Recommendations to the WSIS Plan of Action, 2005 and Seoul-Gyeonggi Declaration on the Equal Participation of Women in the Information Society, 2005). Both of these initiatives showed the culmination of a surge of ICT4D advocacy and net activism. These policy advocates built their platforms on a growing body of research that addressed gender and ICT as well as broader civil society concerns in this domain.[3]

Although the issue of control over content has been articulated since the beginning of discussions about the digital divide, substantial local production of knowledge and content has only recently emerged in practice. The shift in the direction of knowledge flow from "top-down" to "bottom-up" and sideways represented an important change in the conceptualization of what access meant, a rise of political and social empowerment, and one of the most encouraging trends in ICT4D. Although discussion and chat groups had been operating for some time, the PRCA First Mile project initiated the focus on social and political connectivity at the grassroots (Anyaegbunum, et al. 1998). More recently, innovative and truly content producing learning groups are emerging in full force, for example, in ICT4D projects run by the international fund for agricultural development (IFAD) in Tanzania, the CELAC project in Uganda, and UNESCO. In the IFAD project, farmers groups in Tanzania are linked with other local farmers through the "Linking Local Learners" program and together, they co-produce knowledge and "do" ledge. They are also being trained to gather and compile information by writing and using other media techniques (Bringing Markets Closer in the United Republic of Tanzania, 2006; and LLL-an Internet Learning Support Service). In Uganda, the CELAC project (Collecting and Exchanging Local Agricultural Content) also follows this approach, using mixed media such as SMS and radio, as well as music, dance, and drama to convey information in appropriate and relevant ways (CELAC, 2006). UNESCO's Ethnographic Participatory Research project aims to teach people to conduct research on their community ICT related projects, with the goal to improve them and contribute to the reciprocal evolution of these projects to suit people's needs (Tacchi, Slater, & Hearn, 2003). Along these same lines, the telecenter movement has developed in a more systematic way, including a wide range of relevant elements for their socially just use and promotion, as noted in *From the Ground Up: Evolution of the Telecentre Movement* (Carvin & Sermin, 2006).

This history of ICT4D shows a remarkable evolution in complexity along its trajectory. Yet, even with the strides made in ICT4D projects, programs, and case studies noted above, persistent problems remain, especially for marginal groups based on gender, class, and other categories of difference. In India, Uganda, and Jamaica, studies show the continued and in some cases exacerbated, exclusions from ICT, as well as increased violence that exist for women and other marginalized groups under some ICT4D initiatives (Patel & Parmentier, 2005; Litho, 2006; Nadamato, 2006). These negative outcomes insist on the injection of a cautionary and urgent note into the call for improved ICT4D analysis. They demonstrate that the stakes in this agenda are not just to move from neutral to improved impacts of ICT4D, but to avoid negative or deteriorating social impacts. In the next section, the actor-network theory has been introduced and its relevance to the problem at hand has been discussed.

ACTOR-NETWORK THEORY: A HISTORY AND PRIMER ON ITS APPLICATION TO ICT4D[4]

It is the contention of this chapter that actor-network theory (ANT) is a useful analytical framework for the improvement and expansion of successful ICT4D because it allows the analysis of technological/social networks as complex systems, not just as simple binary points of access to technology. In general, ANT presumes that it is possible to trace out sociotechnical networks comprised of multiple entities across multidimensional spatialities and times. ANT allows for the analysis of ICT4D systems in which networks are "drawn" (using descriptive and quantitative patterns) and "colored" (using analysis of power and influence). ANT has the capacity to view sociotechnical networks from the margins; and also across multidisciplinary or trans-historical spheres, such as state/market, group/individual, economic/cultural, political/spiritual, or any categories that might seem useful to the improvement of the network.

Actor-network theory originated as a technique with which to explore and analyze the production of large technical and scientific systems.[5] ANT originated within the sociology of science and technology studies (STS). For example, in one of the original histories of technology, Hughes (1983, cited in Law 2000b) studied the spread of electrification by following Thomas Edison into his journey through political, economic, legal, and architectural spheres as he coalesced the necessary components of electrification beyond his simple invention of the light bulb. In another seminal work of ANT, Latour (1988) featured Louis Pasteur as the driving force behind the broad and complex institutionalization of pasteurization, but he showed how the system was much more extensive than just the man or the technology. These original ANT theorists, in turn, drew from Serres' "sociology of translation" which traced the "punctualization" (in materials, technologies, patterns of behavior, bureaucracies, and texts, for example) of the networks that comprise them. These terms and ideas are developed below.

A central aspect of ANT was its incorporation of technological, non-human entities together with humans into an analysis of the "mechanics of power" (Law, 1992, p. 1). For example, human connections with technology are treated the same as human/human relations; and technology is given equal weight with other actors, including discourses, texts, humans, buildings, and so forth. For the purposes of this chapter, the key value of ANT is that it explores the assemblages of multiple entities across boundaries of human, technical, legal, institutional, material, semiotic, and so forth, to better understand how a technical or scientific entity or system is produced and/or holds together. In the context of ICT, Graham (1988, p. 179) explains, "each application has associated with it whole multiplicities of human actors and institutions, who must continually struggle to enroll and maintain the communications technologies along with other technologies, money and texts, into producing some form of functioning social order."

The subtle intersections between humans and technology extend from the obvious to the invisible and unexpected. For example, Law (1992) suggests that all human interaction is mediated by technology, even invisible technology, such as that which weaves the material for our clothing. "If human beings form a social network it is not because they interact with other human beings. It is because they interact with human beings and endless other materials too" (Law, 1992, p. 2). This notion can be extended to imply that long chains of interaction, not just direct access to computers, are also important for understanding the connections between humans and communication technology, and therefore it is important to draw out these chains to begin to understand the impacts of ICT and opportunities presented by it. Consequently, regardless of whether actors have direct access to ICT, or whether they are conscious of ICT related processes, they can be implicated in and impacted by ICTs. For example, within the ICT4D sphere, powerful actors, such as development institution staff, utilize ICTs to shape the development process for so-called "beneficiaries." At the global and state level, ICT4D policymaking influences the development arena in ways that are invisible to many ICT4D actors.

Beyond simply tracing the patterns of networks and making them visible, though, ANT enthusiasts such as John Law and others were concerned with how these assemblages were held in place, how they came to be identified as an entity or even an actor in and of themselves. As Law (1992) points out:

This then, is the core of the actor-network approach: a concern with how actors and organizations mobilize, juxtapose and hold together the bits and pieces out of which they are composed; how they are sometimes able to prevent those bits and pieces from following their own inclinations and making off; and how they manage, as a result.... to turn a network from a heterogeneous set of bits and pieces each with its own inclinations, into something that passes as a punctualised actor. (p. 6)

Answers to these questions involved ANT theorists' explorations of how structures came to exist through the examination of micro social interactions and the power relations through which they are *repeatedly performed* into the patterns of the macro social. The exploration and analysis of actor-networks require the use of three key concepts: *enrollment, translation, and punctualization*. Actors are randomly dispersed and they are brought into the network by "enrollment," which is essentially a power struggle, aiming for engagement into the network. But this power struggle operates as much through a process of relation building as through domination or coercion. Enrollment occurs during the process of translation, which Star (1991) says involves "translating the images and concerns of one world into that of another, and then disciplining or maintaining that translation in order to stabilize a powerful network" (p. 32).

According to Law (1992, p. 8) the strategies of translation can be determined by asking the following questions:

What are the kinds of heterogeneous bits and pieces created or mobilised and juxtaposed to generate organisational effects? How are they juxtaposed? How are resistances overcome? How it is (if at all) that the material durability and transportability necessary to the organisational patterning of social relations is achieved? What are the strategies being performed throughout the networks of the social as a part of this? How far do they spread? How widely are they performed? How do they interact? How it is (if at all) that organisational calculation is attempted? How (if at all) are the results of that calculation translated into action? How is it (if at all) that the heterogeneous bits and pieces that make up organisation generate an asymmetrical relationship between periphery and centre? How is it, in other words, that a centre may come to speak for and profit from, the efforts of what has been turned into a periphery? How is it that a manager manages?

The study of translation is central to conducting an actor-network analysis; and it is this study that can uncover the power relations that perform and stabilize the structure, as well as the struggles that destabilize it. At the same time, it is also important to note that ANT sees agency as being dispersed throughout the network; and the dispersal of agency in micro social interactions illuminates important opportunities and constraints to the participation of people at the margins in ICT4D.

The final step in the production of actor-networks must include the periodic sedimentation of power relations into something solid. John Law (1992) suggests that networks are "translated" into solid "punctualizations" which embody and simplify underlying networks of power and relations:

Thoughts are cheap but they don't last long, and speech lasts very little longer. But when we start to perform relations—and in particular when we embody them in inanimate materials such as texts and buildings—they may last longer. Thus a good ordering strategy is to embody a set of relations in durable materials. (p. 6)

Despite some form of permanence though, actor-networks are always an ongoing achievement based on the constant unpredictable re-performance of the entities that comprise them. Graham (1988) reminds the reader that actor-networks are "always contingent, always constructed, never spatially universal, and always embedded in the micro social worlds of individuals, groups and institutions, that they 'always represent geographies of enablement and constraint'." (Law & Bijker, 1992, p. 301, cited in Graham, 1988, p. 179). Thus, as argued in Rubinoff (2005, p. 63) "in cybernetworks, every point on the constellation of nodes and flows that comprise them becomes a point of struggle, a switch that can be opened or closed." But the more they are performed to open, the more solid the open switch becomes. Conversely, it is possible to explore not only the circumstances under which Law's "bits and pieces" are held together, but also the circumstances under which they fail to do so; how those "bits and pieces" follow their own or someone else's inclinations and "make off" on their own (Law, 1992, p. 2). In this respect, ANT has been useful in studying technological failure (such as a crashed airplane), but it can also be useful in finding out why something did not come to fruition in the way that it might have, such as a development cybernetwork.

Making an accurate analysis of socially inclusive and relevant ICT4D project requires a capacity for incorporating the view from the margins, but it wasn't until the 1990's that ANT writers began to address this issue. Early ANT focused on enrollment of bits and pieces by the centers of power, the "heterogeneous engineers" at the center of a network. In one of the original ANT analyses, Latour featured Louis Pasteur as the driving force of the institutionalization of pasteurization (1988). In a pivotal critique that also drew on feminist theories of difference, Star (1991) argued that ANT was too managerialistic: that the analysis of pasteurization should include the viewpoints and experiences of not only the senior scientists, but the lab assistant and the janitor as well, must also understand the construction of social networks from the perspective of the enrolled. This critique opened the door for the inclusion of what Latour called the 'missing masses' into ANT analysis (1992, cited in Hetherington & Law, 2000).

Another major feminist analysis that contributed to the development of ANT came from Haraway (1993), whose metaphorical cyborg is the classic example of an actor that spans the boundary of human and technological spheres for actor-network theorists.[6] Among other things, Haraway was concerned about "relationality," the various forms of interaction that connect actors across thresholds and boundaries, that dissolve boundaries, as well as the barriers that enhance them and thwart these moves. Haraway's attention to relationality is a valuable contribution to the study of dynamics of techno social structures of power and brings several advantages to the study of ICT4D. As a feminist approach, it obviously supports attention to inclusion of women and other marginal communities and thus extends the networks to include invisible and marginal participants in network building. Secondly, though, as a feminist epistemology, it makes a fundamental shift away from western subjectivities, which privilege the rational aspect of knowing, and which isolate, bound, and attribute wholeness to individual subjects. In a feminist "way of knowing," knowledge comes from a holistic array of sources and thus establishes a more complex context for knowledge and content paths within ICT4D. Furthermore, within feminist epistemology identities are fluid and defined in response to various situations, being produced and changing, depending upon the context and the relation to another person, process, or place. This notion reinforces the fluid aspects of networks and necessitates attention to how this phenomenon would influence their production and maintenance. Finally, a feminist analysis suggests that identities are located not only in minds, but also in bodies situated in multi-scalar contexts, and which are multiply aspected (gender, age, race, ethnicity, sexuality). Bodies operate through the "performativity" of speech, writing, and action to solidify social processes, relations, and identities. A feminist approach to ANT, as suggested by Haraway, would complicate and refine the analysis of relations and segments through which actor-networks are produced and analyzed. It would facilitate an investigation of the micro social relations and flows of power through network segments, nodes, and paths.

After the early explication of ANT in the 1980's and early 1990's, an internal revolution occurred within the body of ANT theorists. According to one of its major proponents (Latour, 1999), ANT was "recalled" like a defective automobile, and which led to the revision of several aspects of the theory, which some theorists called "after-ANT." In this internal critique, Lee and Brown (1994) took Star's assessment one step further, arguing that ANT had no place for the "other," not only the marginalized, but that which was outside and not containable by the network. As Law argued, "ANT is involved in an intellectual and political refusal to try to squint beyond the possible. To find, to make, the undiscovered continent and discover the shady and heterotopic places, the places of Otherness that lie beyond the limits of the current conditions of possibility" (Law, 2000a).

Taking this line of thinking one step further, some after-ANT theorists have voiced concern for the incorporation of the "Other," which is outside of the network in a way that is not contained within the logic or the discourse of the network. In this vein, Law (2000a) argues that it is necessary to think about not only networks, but also the fluid space outside those networks; and Hetherington & Lee (2000) have argued for the importance of blank space as an under evaluated and under-appreciated structural factor within networks. This move opens the door for a variety of cultural differences and understandings of the world in its visible and "invisible" forms. This concept also allows the analyst to concretely incorporate gaps, silences, or vacancies in an analysis as if they were finite objects, not just simply problems or nothingness. Spaces become entities that deserve attention, or that can be seen to function as connectors between other equally important parts of the chain. In principle, this notion makes it possible to identify spaces and gaps, and perhaps reconsider them as opportunities.

Other after-ANT theorists see the network as a useful, but limited conceptualization for understanding the spatiality of social life. These include, most notably, Haraway's (1994) image of a cat's cradle, in which parts of the network are convoluted, variable, broken, or misshapen; Escobar's (2000) ideas on meshworks; and metaphors of fluidity that allow actors to shift identities and meanings across space and scale (Mol & Law, 1994; Law, 1999; Hetherington & Law, 2000). These discussions expand the spatiality of networks beyond the "lines" or "threads" that connect actors; and they open the possibility of contemplating spatiality of networks from non-modern or Euro-centric perspectives.

Finally, anthropologist Strathern (1996) has addressed the concept of "cutting the network," in which she argues that networks can conceivably go on forever, but that they are arbitrarily cut in ways that determine power relations. Ultimately, struggles over the sites of cutting and extending enable different flows of power and resources and enable different visions and outcomes. For example, patents imply a "cutting of the network" at a particular point at which the benefits of research flow to the owner of the patent. Indeed, property ownership in general implies a cutting of the network. But, conceptual and discursive frameworks can also imply a cutting of networks, and it is important that networks are not cut arbitrarily before important flows of power and resources are considered. At the same time, it is important to acknowledge that networks could conceivably go on forever, and so focusing on them becomes worthless if the analysis does not ever end. In this sense, "cutting the network" becomes a necessary but power laden activity.

Despite significant questions raised by these debates, the ongoing evolution of ANT, (especially subsequent to Star), demonstrates an increasing sophistication that suggests its consideration as a potential analytical framework for the evaluation of ICT4D. ANT has also refined existing elements of its technique that increasingly align with a feminist approach to research and analysis, making it even more appropriate for a study of rural women and other marginalized actors. For all these reasons, ANT should be part of the ongoing debate about ICT4D frameworks and methodologies.

APPLYING ANT TO ICT4D ANALYSIS

The development of a practical, "hands-on" ANT analysis will be a collective achievement that is beyond the scope of this chapter, as detailed in the conclusion. However, three brief examples are being introduced on how this approach might be applied, drawing from research conducted in 2000-2002 with the international fund for agricultural development (IFAD) and its local projects in Central America.

As described in greater detail in Rubinoff (2003 and 2005), IFAD has been on the forefront of using ICT for its development activities and also mainstreaming gender into its projects. As noted in the introduction, it has embraced a broad and complex implementation of participatory and empowering ICT4D, which is the outcome of its long history of engagement with ICT4D and lessons learned over the last decade. Nevertheless, analysis of IFAD's early engagement with ICT4D presents important lessons for the improvement of similar programs and the development of analytical frameworks and methodologies.

One of IFAD's first ventures into cyber networking was the creation of FIDAMERICA, a Santiago, Chile based network designed for the development staff of IFAD projects in Latin America. It was followed by similar IFAD cyber networks, ENRAP in Southeast Asia and FIDAFRIQUE in Africa. FIDAMERICA trained IFAD staff to use computers and the Internet and created several cyber network groups that engaged staff to practice using ICTs, addressing specific themes such as agricultural techniques, project management, and rural financial mechanisms. In the mid 1990's, IFAD engaged strongly with gender and development, and as part of this agenda, Fund directors wanted to "sensitize" project staff to gender mainstreaming. They designed a project that would simultaneously accomplish gender objectives and expand the use of ICT4D strategies. They had FIDAMERICA conduct a life history gathering project, whose objective was to identify local women leaders associated with the IFAD projects. Set up as a "*Concurso*" or contest, and conducted within the structure of an international electronic conference, the project literature states that it was intended to:

stimulate the telling of life histories by women leaders, so that other peasant and indigenous women in other countries could learn lessons from their experiences. Across this dialogue, we wish to strengthen communication and exchange between peasant organizations across Latin America. Also, we hope that this project will permit rural development organizations at all levels to learn from these autobiographies, and that they will go on in their work to support the development of new women leaders (Concurso Autobiografía de Mujeres, 1997).[7]

Submissions were encouraged by monetary prizes ranging from $500-$1000 for the winners and their organizations, and FIDAMERICA received life histories for over 100 local women leaders, most of whom were beneficiaries of local IFAD projects. Subsequently, the life histories were published on the FIDAMERICA Web site, and after the "judging," IFAD published a small book with the stories of the winners. But the results of this experiment for the women participants were never documented, for IFAD or for the women themselves. Indeed, except for the winners, no one knew about the outcomes and none of the women knew that their life histories were online. Contrary to the intended objectives of the *Concurso* objectives noted in the quote above, its benefits accrued primarily to staff and observers, not to the participants themselves.

The contradictions in this situation set up a research agenda under which the production of IFAD's ICT4D networks was explored with their impacts on rural women (Rubinoff, 2003, 2005). Central to the study were questions of the connections (or gaps) between rural women and ICT and the potential opportunities for improvement. Research was conducted with *Concurso* entrants and relevant IFAD projects in Honduras, El Salvador, and Costa Rica. A fuller discussion of the IFAD ICT4D system has been developed elsewhere (Rubinoff, 2003), where the use of ANT was recommended and initiated. In the following three examples, the author drew on that study and another paper to briefly show how an ANT approach can be used to better understand or improve rural women's connections to ICT. In all cases, the analysis was supported by principles of ANT, drawn from the discussion in the previous section.

In the first example, Rubinoff (2005) discussed the use of an ANT approach to conduct an in-depth analysis of the life history gathering aspects of the *Concurso* process and the networks of engagement produced for rural women, and to critique the positive and negative outcomes for them. For the analysis, typologies of engagement were created that would trace the production of life histories online, trace rural women's invisible connections to the Internet, and help to understand how life history writing could support electronic networking. The analysis emphasized not only a tracing of sociotechnical networks, but also analyzed the social relations that they comprised. In this respect, the analysis strove to use the most highly developed ANT approach, which started at the margins and also detailed social relations from a feminist theoretical perspective. A key conclusion of this analysis was that direct access to ICTs is not necessary for rural women to be able to access ICT resources and benefits. The author concluded that drawn-out networks of engagement through development institutions, family, and other contacts could be as useful for rural women as direct contact with a computer. She also showed how the production of texts that would be placed on line can create a powerful emotional linkage to the technology and showed how this could be a useful strategy for helping to fully engage people with ICT4D.

The second two examples focus on bringing background or otherwise invisible factors (such as women's own experiences and understanding of networks) into the network analysis, in order to promote intensified and meaningful connections to ICT4D. Both of these analyses conceptualize women's experiences and understandings as networked entities themselves, which flow through the individual and which can be attached or linked to existing or planned ICT4D projects/networks. In this way, by connecting these internally originating networks, it becomes possible to link all kinds of experience and knowledge that will enhance rural women's connection to and benefits from ICT.

For the second and third analyses, cognitive mapping techniques were used to assess rural women's understanding of the Internet. As the first example shows, direct access to computers only highlighted women's familiarity with technological hardware, but it didn't correlate with an understanding of how cybernetworking might work. For example, the cognitive map (Figure 1) was drawn by Alma Galanos,[8] a Salvadoran woman who ran a chicken wholesaling business. Although she seems to have the screen clearly in mind, her comments contradict her knowledge:

I want you to know something, that we live here in a world at the end of the world, where we live almost alone, I have heard people speak about computers and the Internet and via television I know that you can do business and augment your capital and your accounts but how you do this, I don't know. Right? So, it's like I don't know the potential that this has, and I don't have access to these possibilities, so later if I have the opportunity, great. But right now, here, if you talk to me about the Internet, I don't know what the Internet is, what a field it will be, I don't know. (Interview, Alma Galanos, August 2001)

In the second example (Figures 2 and 3), Lydia Gutierrez in Talgua, Honduras, evokes a vague sense of opportunity, but in her case, success is thwarted by a sense of entanglement in the crisis in her country (Interview, Lydia Gutierrez, August 2001). Lydia's people, whose heads are big because they are full of unrealized ideas, look to the sky, which contains an airplane, with a sense of wishing that they could get out of the place of poverty they inhabit: "we feel like we are in a noose." At the same time, Lydia comments that she knows that ideas are like lifesavers, and Lydia's drawings (Figure 2 and 3) also contains an envelope (conventional mail), a radio, and a computer terminal with a keyboard, but they are on different sides of the morass, as if she knows that they could be linked but doesn't see how they will get through to the other side. Lydia says:

We see some alternatives, we see them there at the bank when we visit, not to cash a check, nor to take out a loan.... but to sell tamales and nothing else. Then we see people who communicate, right, by way of a computer. At the same time that we see this big tangle that ensnares us, and that

Figure 1. Alma Galanos' cognitive map of the Internet

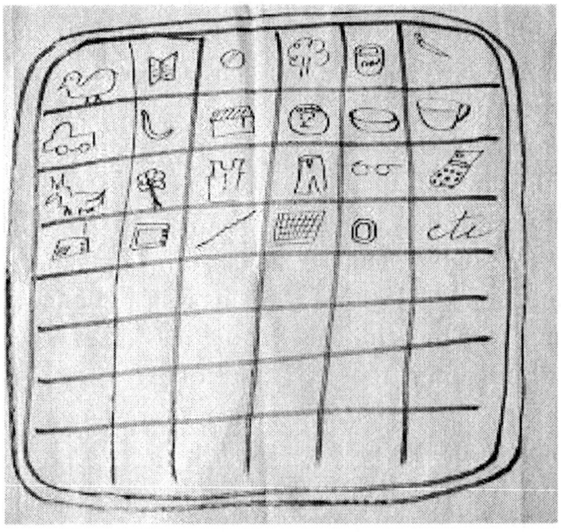

it is not just money but ideas [that we need in order] to leave the past behind. (Interview, Lydia Gutiérrez, August 2001)

In the next example, Marta Esmerelda Cartes, the leader of a flower farming women's co-op, who actually had an (unused) computer in her house, expressed a sense of confusion and even a sense of ambivalence about the cultural dynamics of the Internet as she explains:

OK, this is the most complicated and odd and crazy [story], but this signifies, for me, the universe, this is the enormously large antenna that is above collecting all the connections that are directed to a satellite. I think that here where the entire world comes and goes, and the huge antenna, this is the universe, this is the world, this is everything, but here it is accessing everything. Look, one person has heir hair on end [pelo parado] because he does not know what the Internet is and he thinks it's a monster, right? It's a vision. (Interview, Marta Esmeralda Cartes, May 2001)

Marta Esmeralda goes on to describe her concern that the Internet will control future generations, (highlighted in Figure 5):

Also others are just watching. And here is the mother with a baby, but the baby is already on the Internet, but it doesn't even have eyes yet. Some people are entering [the Internet] and some are running away.[9]

In these cognitive mappings of the Internet, both the understandings and concerns or rural women become apparent in their drawings. These become rich sources of insight as to the opportunities and constraints for linking them to ICT4D and for shaping the ICT projects to meet their needs, capacities, and concerns.

The final analysis was named "networking consciousness" and it evaluated women's previous experience with organizational networking on local, regional, national, and international scales and connected this to their ability to grasp Internet principles. Many of the women in this study had been involved in extensive networking activities, some as far away as the US or Europe, and their experiences in this area set an important backdrop for their ability to understand and envision the potential for ICT4D. Indeed "networking consciousness" appeared to be even more important than access to computers in helping people to imagine ICT potential.

Figures 2 and 3. Lydia Gutierrez cognitive map of the Internet

Figure 4. Marta Esmeralda Cartes' cognitive map of the Internet

Figure 5. Close up Marta Esmeralda Cartes' cognitive map of the Internet

In perhaps the most concrete example of this point, Lorena Sanchos, from western Honduras, exhibits a relatively clear, (if simplistic) idea of the technology of the keyboard and wiring systems that interconnect people across borders; but a very unambiguous understanding of the transnational linkages that the Internet provides, as well as an explanation of what people might do with these connections.

Describing her cognitive map (Figure 6), Lorena says:

I drew this motion [line] because we understand that I am going to tell about my experience, you are going to learn from me, and you are going to tell about yours, and then I am going to learn from you; so that we all will learn and share. Because we will go on living together. We are going to organize all the [innovative producers]. (Interview, Lorena Sanchos, August 2001)

Lorena alludes here to the value of collective organizing within the model that she knows at the local/regional/international level, based on her own networking and organizational experience; and which she has here applied to the use of the Internet across transnational borders. This is clear to her, despite the fact that she has no electricity in her home and at the time there were no telecenters in her village.

Together, these examples represent possibilities of the kinds of analysis that stems from an ANT approach. The creative options for applying ANT are endless, and they will depend on the objectives of the analyst. In most cases, a good place to start is with the individuals being targeted and supported in development agendas. The potential networks can be derived from their issues, concerns, and lives, with their guidance and eventually in analyses that they conduct themselves. In the end, the key is to utilize the metaphor of the actor-network, as it has been outlined in this chapter to begin the process and to use the creativity of the collective team involved, to both draw and color the network.

CONCLUSION

In this chapter, the actor-network theory approach was introduced as a valuable analytic tool for the study and improvement of ICT4D. The advantages of ANT for this work stem from its ability to link technological and human processes in one clean line, because it sees all technological processes as mediated by social relations, and

Figure 6. Lorena Sanchos' cognitive map of the Internet

all human behavior as mediated by technology to some degree. But beyond tracing of networks (which is extremely valuable in and of itself), ANT is useful because it considers not only the visible elements of networks, but also the gaps and spaces in networks. In this respect ANT allows for a consideration of the broader picture and the inclusion of perhaps unlinked segments into the analysis. Finally, ANT supports a focus on the power relations that shape development networks and processes, how these relations hold networks together (or allow them to dissolve). It is a useful technique with which to examine the ultimate ability of marginalized groups such as rural women and others to participate in and influence development processes and to show how empowerment occurs. In this way, also, it is possible to explore failed or unresolved networks. Ultimately, the goal is to be able to shape and control the production of ICT4D networks that are useful to those who need them most to support their development goals.

FUTURE RESEARCH DIRECTIONS

As was argued in the introduction to this chapter, this discussion is seen as another voice in an ongoing conversation about how refine, deepen, and make more precise the analysis and potential of ICT4D. There are several possibilities for further research and development of the approach and for its testing. For example, it would be useful to work with seasoned ANT analysts to develop a protocol that could be used as a standard model. The development of this protocol could possibly take place as an academic exercise, as work needs to be done on how to fit together various elements of a cybernetwork and how to trace power flows. Individuals could work on resolving questions around subsets of the problem, or they could experiment with its application to an ICT4D problem. These might include (but are in no way limited to) research on questions such as:

- How to model the systems approach (are economic, mathematical or algorithms useful here?)
- How to trace flows of power
- How to link development theory categories such as social capital (social networks and social movements) with human capital (capacity building), policy, formal institutions, technological access and others
- How to link (or translate) between identity related elements such as self-esteem, religion and spirituality, emotional life, leadership, or others with institutions that would support their emergence
- How to solicit and invite the participation of ICT4D beneficiaries into this project so that it becomes meaningful

The research context for these questions could center on a wide range of development contexts including: development institutions, donors, and NGOs; government ministries and programs; development projects; or transnational networks operating through telecenters and Internet cafes. Mixed and integrated ICT categories should also be investigated with an ANT approach. Some interesting contexts for research could apply an actor-network analysis to cases that have been analyzed previously with different methodologies or to new development strategies as they are implemented. Especially in cases where an explicit ICT4D strategy has been identified by special government ministry or policy agenda, it would be useful to trace and evaluate success outcomes. Of course, comparative analysis of post WSIS digital governance strategies would also be useful research agendas. Comparative analysis across an array of similar projects would help to test and evaluate the analytical approach.

Ultimately, it would be useful to have workshop or conference in which interested parties convened to share their thinking in presentations, and then spend time collectively work shopping the problem, to address this and other ICT4D related analytical approaches. A conference could support not only movement towards the consolidation of an ICT4D analytical framework, but it could also move the process ahead to the practical analysis of the burgeoning number of projects and programs.

The intent of this chapter has been to pique the interest of practitioners, academics, and others involved in the ICT4D arena. The next step is to build on previous effort and to move ahead a process of focused attention on the research, integration, application, testing, and evaluation of this and other analytical frameworks for ICT4D.

REFERENCES

Anand. A., & Uppal, M. (2002). *Engendering management and regulation of ICTs*. Prepared for the UN/INSTRAW Virtual Seminar Series on Gender and ICTs Seminar Two: Women and ICTs Enabling and Disabling Environments. July 15-26, 2002. Retrieved July 26, 2002, from http://www.uninstraw.org/en/research/gender_and_ict/virtual_seminars.html

Anyaegbunum, Mefalopulus, & Moetsabi(1998). Participatory Rural Communication Appraisal Methodology (PRCA): The application of participatory approaches to communication programme design and implementation for sustainable human development. In D. Richardson & L. Paisley (Eds.), *The first mile of connectivity*. Rome: FAO.

Arun, S., Heeks, R., & Morgan, S. (2004). *Researching ICT-based enterprise for women in developing countries: A livelihoods perspective*. Women's ICT-Based Enterprise for Development project. Manchester, UK: IDPM, University of Manchester. Retrieved June 2006, from http://www.womenictenterprise.org

Balit, S. (1999). *Voices for change: Rural women and communication*. Rome: FAO.

Bringing Markets Closer in the United Republic of Tanzania. (2006, February). *IFAD Update*. 2006: 2. Retrieved May 15, 2006, from http://www.ifad.org/newsletter/update/2/6.htm

Carvin, M., & Sermin A. (2006). *From the Ground Up: Evolution of the Telecentre Movement*. Ottawa, CN: IDRC. Retrieved July 2006, from http://www.telecenter.org

CELAC, *Collecting and Exchanging of Local Agricultural Content Program in Uganda*. Retrieved July 15, 2006, from www.celac.or.ug/

Concurso Autobiografía de Mujeres.(1997). Electronic Conference. Santiago: RIMISP. Retrieved June 1999, from http://www.fidamerica.cl/seccion.php?seccion=156

Cornwall, A. (2001). *Making a difference? Gender and participatory development*. Working Paper 378. Brighton, UK: IDS. Retrieved March 2002, from http://server.ntd.co.uk/ids/bookshop/details.asp?id=597

Cornwall, A., & Gaventa, J. (2001). *From users and choosers to makers and shapers*. Working Paper 127. Brighton, UK: IDS. Retrieved March 2002, from http://server.ntd.co.uk/ids/bookshop/details.asp?id=618

Cummings, S., et al. (Eds.). (1999). *Women's information services and networks*. Amsterdam: KIT Press and Oxfam GB.

Escobar, A. (2000). Notes on networks and anti-globalization social movements. Available at http://www.unc.edu/depts/anthro/faculty/fac_pages /escobarpaper.html

Forestier, E., Grace, J., & Kenny, C. (2002). Can information and communications technologies be pro-poor? *Telecommunications Policy, 26*(11), 623-646.

Garcia, C. (1999). Weaving webs of unity: The experience of Asia and the Pacific. In S. Cummings, et al. (Eds.), *Women's information ser-*

vices and networks. Amsterdam: KIT Press and Oxfam GB.

Gittler, A. (1999). Mapping women's global communications and networking. In W. Harcourt (Ed.), *Women@Internet.* London: Zed.

Grace, J., Kenny, C., & Qiang, C. with J. Liu and T. Reynolds. (2001). *Information and communication technologies and broad-based development: A partial review of the evidence* (Draft). Washington, DC: World Bank.

Graham, S. (1988). The end of geography or the explosion of place? Conceptualizing space, place and information technology. *Progress in Human Geography, 22*(2), 165-185.

Grassroots Caucus Recommendations to the WSIS Plan of Action. (2005). Retrieved November 2005, from http://www.grassroots-caucus.org

Hafkin, N. (2002). *Gender issues in ICT policy in development countries.* An Overview. Prepared for the UNDAW Expert Group Meeting on "Information and communication technologies and their impact on and use as an instrument for the advancement and empowerment of women. Seoul, Republic of Korea. Retrieved from http://www.un.org/womenwatch/daw/egm/ict2002/documents.html

Hafkin, N. (2006). Women, gender and ICT statistics and indicators. In N. Hafkin & S. Huyer (Eds.), *Cinderella or cyberella? Empowering women in the knowledge society.* Bloomfield, CT: Kumarian Press.

Hafkin, N., & Huyer, S. (2006). *Cinderella or cyberella? Empowering women in the knowledge society.* Bloomfield, CT: Kumarian Press.

Haraway, D. (1994). A game of cat's cradle: Science studies, feminist theory, cultural studies. *Configurations, 2*(1), 59-71.

Haraway, D. (1990) [1985]. A manifesto for cyborgs: Science, technology, and socialist feminism in the 1980's. In L. Nicholson (Ed.), *Feminism/postmodernism.* New York, NY: Routledge.

Harcourt, W. (Ed.). (1999). *Women@Internet.* London: Zed Books.

Heeks, R., Arun, S., & Morgan, S. (2005). *Researching women's ICT-based enterprise for development: Methods, tools and lessons from fieldwork.* Women's ICT-Based Enterprise for Development project. Manchester, UK: IDPM, University of Manchester. Retrieved June 2006, from http://www.womenictenterprise.org/

Heeks, R., Arun, S., & Morgan, S. (2004). *Researching ICT-based enterprise for women in developing countries: An enterprise perspective.* Women's ICT-Based Enterprise for Development project. Manchester, UK: IDPM, University of Manchester. Retrieved June 2006, from http://www.womenictenterprise.org

Hetherington, K., & Law, J. (2000). After networks. *Environment and Planning D: Society and Space, 18*(2), 127-132.

Hetherington, K., & Lee, N. (2000). Social order and the blank figure. *Environment and Planning D: Society and Space, 18*(2), 169-184.

Huyer, S. (2006). Understanding gender equality and women's empowerment. In N. Hafkin & S. Huyer (Eds.), *Cinderella or cyberella? Empowering women in the knowledge society.* Bloomfield, CT: Kumarian Press.

Huyer, S. (1997). *Supporting women's use of information technologies for sustainable development.* Manuscript prepared for the Gender and Sustainable Development Unit.

Latour, B. (1988). *The pasteurization of France.* Cambridge, MA: Harvard University Press.

Latour, B. (1999). On recalling ANT. In J. Law & J. Hassard (Eds.), *Actor network and after.* Oxford: Blackwell Publishers.

Law, J. (2000a). *Objects, spaces, others, draft manuscript.* Centre for Science Studies and Department of Sociology, Lancaster University, Lancaster. Retrieved August 28, 2001, from http://tina.lancs.ac.uk/sociology/soc027jl.html

Law, J. (2000b). *Networks, relations, cyborgs: On the social study of technology,* (Draft). Centre for Science Studies and the Department of Sociology, Lancaster University. Retrieved August 28, 2001, from http://www.comp.lancs.ac.uk/sociology/soc042jl.htm

Law, J. (1999). Introduction: Monsters, machines and sociotechnical relations. In Law & J. Hassard (Eds.), *Actor network theory and after.* Oxford: Blackwell Publishers.

Law, J. (1992). *Notes on the theory of the actor network: Ordering, strategy and heterogeneity.* Centre for Science Studies and the Department of Sociology. Lancaster, UK: Lancaster University. Retrieved August 28, 2001, from http://www.lancs.ac.uk/fss/sociology/papers/law-notes-on-ant.pdf

Law, J., & Hassard, J. (1999). *Actor network theory and after.* Oxford: Blackwell Publishers.

Lee, N.M., & Brown, S. D. (1994). Otherness and the actor-network: The undiscovered continent. *American Behavioral Scientist, 37*(6), 772-790.

LLL: An Internet Learning Support Service for Linking Local Learners. (n.d.) Retrieved July 15, 2006, from http://www.linkinglearners.net/

Litho, P. (2006, June). *ICTs, 'empowerment' and women living in rural Uganda.* Paper presented at Women, ICTs and Development Workshop. Manchester, UK: IDPM, University of Manchester.

Lucas, F.B. (1999). *A radio broadcasting model for rural women and farm households.* Prepared for FAO, Regional Office for Asia and the Pacific.

MacLean, D., Souter, D., Deane, J., & Lilley, S. (2002). *Louder voices: Strengthening developing country participation in international ICT decision-making.* London: Commonwealth Telecommunications Organization and Panos Institute.

Massey, D. (1994). *Space, place, and gender.* Minneapolis: University of Minnesota Press.

Mol, A., & Law, J. (1994). Regions, networks and fluids: Anaemia and social topology. *Social Studies of Science, 24*(4), 641-671.

Morgan, S. (2006, June). *Studying gender, ICTs and development using new institutionalism.* Paper presented at Women, ICTs and Development Workshop. Manchester, UK: IDPM, University of Manchester.

Morgan, S., Heeks R., & Arun, S. (2004). *Researching ICT-based enterprise for women in developing countries: A gender perspective.* Manchester, UK: IDPM, University of Manchester. Retrieved July 2006, from http://www.womenictenterprise.org

Nadamato, S. (2006, June). *Gender and ICTs: Empowerment of women in telecentres in Jamaica.* Paper presented at Women, ICTs and Development Workshop. Manchester, UK.

Nath, V. (2006). Empowerment of women through ICT-enabled networks. In N. Hafkin & S. Huyer (Eds.), *Cinderella or cyberella? Empowering women in the knowledge society.* Bloomfield, CT: Kumarian Press.

Patel, R., & Parmentier, M.J. (2005). The persistance of traditional gender roles in the information technology sector: A study of female engineers in India. *Information Technologies and International Development, 2*(3), 29-46.

Radloff, J. (1999). Celebrating the power, diversity and strength of African women's networks. In S. Cummings, et al.(Eds.), *Women's information services and networks.* Amsterdam: KIT Press and Oxfam GB.

Reilly, K. (2002). *Government ICT use in Central America...a view from civil society.* Unpublished Manuscript. Retrieved March 2002, from http://katherine.reilly.net/publications.html

Research Program: Engendering ICT Policy—Call for Grant Proposals, WSIS Gender Caucus. Retrieved March 2005, from http://wsis.ecommons.ca/node/view/502

Rubinoff, D. (2003). *Campesina leaders, life histories and ICT for development: Cybernetworking and rural women in Central America.* Unpublished Doctoral Dissertation, University of Colorado-Boulder.

Rubinoff, D. (2005). Life histories in cyberspace: Life writing as a development tool for rural women. *Geoforum, 36*(1), 59-75.

Seoul-Gyeonggi Declaration on the Equal Participation of Women in the Information Society (2005). Retrieved October 2005, from http://women.or.kr/wsisforum/

Star, S. L. (1991). Power, technologies and the phenomenology of conventions: On being allergic to onions. In J. Law (Ed.), *A sociology of monsters? Essays on power, technology and domination, Sociological Review Monograph.* London, Routledge.

Strathern, M. (1996). Cutting the network. *Journal of the Royal Anthropological Institute, 2*(3), 517-535.

Tacchi, J., Slater, D., & Hearn, G. (2003). *Ethnographic participation research.* New Delhi: UNESCO Regional Bureau for Communication and Information. Retrieved July 15, 2006, from http://portal.unesco.org/ci/en/ev.php-URL_ID=16619&URL_DO=DO_TOPIC&URL_SECTION=201.html

Women and ICTs: Creating Global Transformation. (2005, June). Conference Proceedings. Retrieved July 2005, from http://portal.acm.org/toc.cfm?id=1117417&coll=portal&dl=ACM&type=proceeding

World Bank. (2005, September 14). *Gender and the World Summit on the Information Society.* [Conference Streaming Video]. Washington, DC: World Bank. Retrieved March 2006, from http://web.worldbank.org/WBSITE/EXTERNAL/TOPICS/EXTGENDER/

0,,contentMDK:20620521~menuPK:489311~pagePK:148956~piPK:216618~theSitePK:336868,00.html and http://web.worldbank.org/WBSITE/EXTERNAL/TOPICS/EXTGENDER/

0,,menuPK:336874~pagePK:149018~piPK:149093~theSitePK:336868,00.html

ADDITIONAL READING

Association for Progressive Communications (1997). *Global networking for change. Experience from the APC Women's Programme.* London: APC.

Callon, M. (2005). Why virtualism paves the way to political impotence. *Economic Sociology - The European Electronic Newsletter, 6*(2), 3-20. Retrieved March 10, 2007, from http://econsoc.mpifg.de/archive/esfeb05.pdf

Callon, M. (1991). Techno-economic networks and irreversibility. In J. Law (Ed.), *A sociology of monsters: Essays on power, technology and domination.* London: Routledge Publishers.

Callon, M. (Ed.). (1998). *The laws of the markets.* London: Blackwell Publishers.

Callon, M. (1987). Society in the making: The study of technology as a tool for sociological analysis. In W. Bijker (Ed.), *The social construction of technical systems: New directions in the sociology and history of technology.* London: MIT Press.

Callon, M. (1986). Some elements of a sociology of translation: Domestication of the scallops and the fishermen of St Brieuc Bay. In J. Law (Ed.),

Power, action and belief: A new sociology of knowledge. London: Routledge & Kegan Paul.

Callon, M. (1980). Struggles and negotiations to define what is problematic and what is not: The socio-logic of translation. In K. Knorr (Ed.), *The social process of scientific investigation*. Dordrecht: Reidel Publishing.

Castells, M. (1996). *The rise of the network society.* Oxford: Blackwell Publishers.

De Laet M., & Mol, A. (2000). The Zimbabwe Bush pump. *Social Studies of Science, 30*(2), 225-263.

Dicken, P., et al. (2001). Chains and networks, territories and scales: Towards a relational framework for analyzing the global economy. *Global Networks, 1*(2), 89-112.

Escobar, A. (1999). Gender, place and networks: A political ecology of cyberculture. In W. Harcourt (Ed.), *Women@Internet.* London: Zed Books.

Farwell, E., et al. (1999). Global networking for change: Experiences from the APC Women's Programme. In W. Harcourt (Ed.), *Women@Internet.* London: Zed Books.

Food and Agriculture Organization of the United Nations (FAO). (2001). *Discovering the "Magic Box": Local appropriation of information and communication technologies (ICTs).* Prepared by the Communication for Development Group, June 2001. Rome: FAO. Retrieved August 19, 2001, from www.fao.org/sd/2001/KN0602a_en.htm

Food and Agriculture Organization of the United Nations (FAO). (1999). *Proceedings of the High-Level consultation on Rural Women and Information.* Rome. FAO.

Food and Agriculture Organization of the United Nations (FAO). (1999). *Participation and information: the key to gender-responsive agricultural policy.* Prepared by the Women and Population Division, Sustainable Development Department. FAO, Rome.

International Food Policy Research Institute (IFPRI) (2001). *Empowering women to achieve food security: Overview.* Retrieved Aug. 21, 2001, from http://www.ifpri.org/2020/focus/focus06.htm

International Food Policy Research Institute (IFPRI). (2000). *Bridging the digital divide.* Retrieved October 14, 2000, from www.ifpri.org/2020/newslet/nv_0900a.htm

Latour, B. (2005). *Reassembling the social: An introduction to actor-network-theory.* Oxford, UK: Oxford University Press.

Latour, B. (2004). *Politics of nature: How to bring the sciences into democracy* (translated by C. Porter). Cambridge, MA: Harvard University Press.

Latour, B. (1999). *Pandora's hope: Essays on the reality of science studies.* Cambridge, MA: Harvard University Press.

Latour, B. (1996). *Aramis, or the love of technology.* Cambridge, MA: Harvard University Press.

Latour, B. (1993). *We have never been modern.* Cambridge, MA: Harvard University Press.

Latour, B. (1992). Where are the missing masses? The sociology of a few mundane artifacts. In W. Bijker & J. Law (Eds.), *Shaping technology/building society: Studies in sociotechnical change.* Boston: MIT Press.

Latour, B. (1987). *Science in action: How to follow scientists and engineers through society.* Cambridge, MA: Harvard University Press.

Latour, B., & Woolgar, S. (1979). *Laboratory life: The social construction of scientific facts.* Los Angeles, CA: Sage.

Law, J. (2004). *After method: Mess in social science research.* London: Routledge.

Law, J. (2002). *Aircraft stories: Decentering the object in technoscience.* Durham, NC: Duke University Press.

Law, J., & Mol, A. (Eds.). (2002). *Complexities: Social studies of knowledge practices*. Durham: Duke University Press.

Massey, D. (1993). Power-geometry and a progressive sense of place. In J. Bird, et.al. (Eds.), *Mapping the futures: Local cultures, global change*. London: Routledge.

Metcalfe, A. (2001). Nothing: The hole, the holy, the whole. *Ecumene, 8*(3), 247-263.

Paris, T., Feldstein, H., & Duron, G. (2001). *Empowering women to achieve food security: Technology*. In International Food Policy Research Institute. Empowering Women to Achieve Food Security: Overview. Retrieved August 21, 2001, from http://www.ifpri.org/2020/focus/focus06.htm

Ribiero, G. L. (1998). Cybercultural politics: Political activism at a distance in a transnational world. In S. Alvarez, E. Dagnino, & A. Escobar (Eds.), *Cultures of politics/politics of cultures: Re-visioning Latin American social movements*. Boulder: Westview Press.

ENDNOTES

1. ICT for Development
2. For the original RFP for this program, see Research Program: Engendering ICT Policy—Call for Grant Proposals, WSIS Gender Caucus. http://wsis.ecommons.ca/node/view/502
3. For gender and ICT, see Anand and Uppal, 2002; Balit, 1999; Cornwall, 2001; Cummings et al., 1999; Garcia, 1999; Gittler, 1999; Hafkin, 2002; Harcourt, 1999; Huyer, 1997; Lucas, 1999; Radloff, 1999. For civil society and ICT, see: Reilly, 2002; Forestier et al., 2002; Grace et al., 2001; MacLean et al., 2002. Also see: Women and ICTs—Creating Global Transformation. (2005, June). Conference Proceedings.
4. For a good introduction to actor-network theory, as well as an extensive bibliography, see John Law's Web site, http://tina.lancs.ac.uk/sociology/jlaw.html, especially Law, 1992. Also see Law and Hassard, 1999.
5. The term ANT suggests that it is a theory of network formation, however, many critics argue that it is more an analytical framework than a theory, and this chapter does not argue with this critique.
6. Haraway's work is situated in a large body of feminist work on theories of difference and relationality. For more discussion of this, see Rubinoff, 2003, Chapter 2.
7. Although the *Concurso* literature bills it as an "autobiography" contest, I am using the term "life history," because in most cases, the texts were collected as oral histories, sometimes reworked and almost always typed by intermediaries. Furthermore, as I will show below, these words are politically loaded, and I have chosen the one that seems to align most clearly with the authors' intentions, although FIDAMERICA and IFAD staff and authors used the words autobiography, life history and testimonial interchangeably.
8. All names are pseudonyms.
9. Author red emphasis on drawing to emphasize baby being connected to Internet

Chapter VIII
Beyond Localization:
Effective Learning Strategies for Cross-Cultural E-Learning

Patrick Dunn
Networked Learning Design, Ltd., UK

Alessandra Marinetti
Networked Learning Design, Ltd., UK

"Instructional design cannot, and does not, exist outside of considerations of culture."
--Henderson, 1996, p.85

ABSTRACT

Instructional systems are products of the cultures in which they are developed. Culture, which we define here as "the collective programming of the mind which distinguishes the members of one group or category of people from another" (Hofstede, 2001), has a pervasive influence on instructional systems, regardless of whether these systems involve self-paced e-learning, synchronous or asynchronous computer-based learning activities, or online communities of learners. The issue of culture's impact on instructional systems becomes most relevant and challenging where such systems are transferred across cultural boundaries, or developed for multiple cultures. This is currently happening in many large, globally dispersed organizations that use e-learning technologies to support the learning of their staff around the world. Theories of learning and of cultural dimensions suggest that the effectiveness of certain types of learning systems will be affected where they are used in culturally diverse environments. The aim of this paper is to highlight the issues that designers of a wide range of e-learning experiences face when designing e-learning for culturally diverse learner groups. We provide some models to support learning practitioners, focusing in particular on the importance of a conscious, culturally-informed selection of instructional strategies as the most critical part of the design and development process.

INTRODUCTION

It is clear to anyone who has taken on the task of designing learning products or services for different national or regional cultures, that such products or services—we will call them "instructional systems[1]"—are to some extent shaped by the cultures in which, and for which, they are developed. Culture, which for the purposes of this chapter is defined as "the collective programming of the mind which distinguishes the members of one group or category of people from another" (Hofstede, 2001, p.9), appears to have a pervasive influence on instructional systems, whether these systems involve classrooms, teachers, computers, videos, hand-held computers, cell-phones or informal communities of learners.

While the issue of cultural influences on instructional systems has for some time been of interest to academics and a relatively small cohort of globe-trotting trainers from large organizations, it is gradually becoming one of the more important challenges faced by developers of e-learning products. Theories of learning and of cultural difference suggest that the effectiveness of instructional systems may be reduced where such systems are transferred into cultures for which they were not designed, or into culturally diverse environments. Yet much of the rationale for e-learning, particularly in large organizations, rests on its ability to provide effective learning experiences, cost-effectively, to large, widely distributed audiences. As an increasing number of large, globally dispersed organizations use e-learning programs to support the learning of their people around the world, there is a growing need to support the designers of these programs in considering cultural factors.

Our particular interest is in forms of e-learning "...in which information and communications technology is used to promote connections: between one learner and other learners, between learners and tutors; between a learning community and its learning resources" (Goodyear, 2001, p.9). This is a view of e-learning in which a learning network is regarded as a means of connecting people, as much as a means of distributing learning resources. Indeed, we believe that cultural diversity is just one of the pressures that will erode the use of less connected e-learning approaches where learners interact solely with content.

Our aim in this chapter is to highlight the issues that designers of instructional systems are likely to face when designing e-learning for a culturally diverse learning environment, and to go some way in providing the basis for supporting them. In particular, our focus is on the importance of a conscious, culturally-informed selection of learning strategy as the most critical part of the design process. It is our contention that selection of learning strategy is most likely to be effective if it is:

- A conscious decision in the design process; "gut feel" decisions, based on "common sense" may lead to problems in a cross-cultural environment, as they will be based on the designer's own culturally induced worldview
- Informed by an understanding of the specific cultures involved
- Based on appropriate, researched models of cultural values
- Founded on the principle that networked learning technologies should aim to connect people, rather than merely distribute information

A theme running through this chapter is our view that what is currently called "localization" generally involves too shallow a process, as it does not usually include consideration of learning strategy.

The perspective of the authors of this chapter is primarily one that arises from the context of training and staff development in large organizations whose primary purpose is not educational, such as governmental and internal agencies and corporations.

LEARNING WITH TECHNOLOGY: CULTURE'S EFFECTS

One problem facing learning designers is that the effects of culture are experienced at various levels. Both Trompenaars, (1997) and Hofstede (2001) use the metaphor of an onion to show how the various layers of a culture relate to each other, and are experienced differently. Hofstede's view is that the outer layers of a culture are more visible, superficial and potentially changeable, whereas inner layers involve elements that are less visible and change more slowly. Hofstede's model is shown in Figure 1.

In the context of interactions within a culturally diverse environment, it is of course important to consider the outer layers of the onion. So, for example, when giving corrective feedback to learners in some east-Asian cultures, use of red highlights would be inappropriate, because this is the color of good luck, unlike in the west, where red tends to denote problems or errors. Similarly, where designing a relaxing entry page to a Japanese online module on effective employee relationships, it would be unwise to use shades of white because in Japan white is the color of mourning. Care also needs to be taken where using a language shared between many cultures, such as English or Spanish. For example, in American English, "tabling a motion" means the exact opposite of what it means in UK English.

However, an implication of Hofstede's model is that in attempting to understand the influences that a culture exerts or in analyzing a culture's response to a changing situation (such as the introduction of new learning technologies), it is important to consider all levels of culture, not just the more obvious, superficial ones. All software systems—of which instructional systems using technology are a subset—are visible embodiments of underlying cultural values that may be invisible, or at best undeclared. So understanding detailed but superficial variations at the level of symbols is a necessary condition, but is not sufficient. Henderson (1996) states: "When instructional design translates the topic into a tangible object, such as IMM software, it becomes an artifact of the culture in which it is embedded" (p.1). Watson, Ho and Raman (1994, cited in Collis, 1999, p.12) note that the customs of the culture that develops a particular piece of collaborative software are often reflected in it, resulting in modifications of both technical and social aspects, when it is used outside the original cultural context.

Hofstede's (2001) widely used model of value preferences is helpful in providing a basis for learning designers who need to understand how values vary in different cultures. The model is summarized in Table 1, with the LTO dimension adjusted slightly to reflect Trompenaars similar dimension[2].

Figure 1. Hofstede's "onion" model

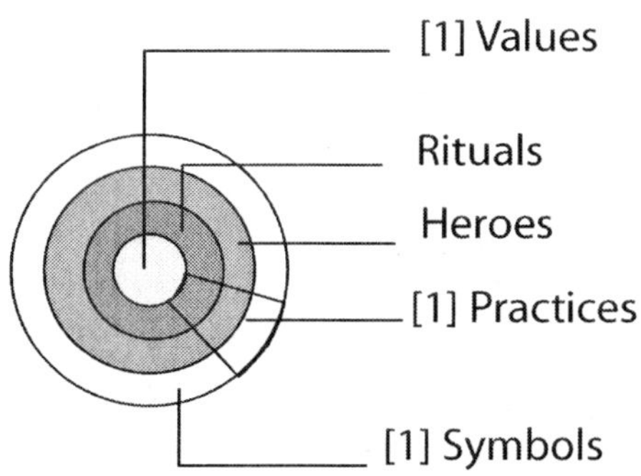

Table 1. Hofstede's five dimensions of cultural difference

In **High Power Distance (High PDI)** cultures, the less powerful members of society accept that power is not equally distributed.	In **Low Power Distance (Low PDI)** cultures, there is a general acceptance that power should be equally distributed.
High Individualism (High IDV) cultures show few ties beyond those of the nuclear family	In **High Collectivism (Low IDV)** cultures, people belong to strong, cohesive in-groups.
In **Masculine societies (High MAS)**, men are assertive, tough and concerned with winning material success, whereas women are more modest and interested in quality of life.	In **Feminine societies (Low MAS)**, both men and women are equally concerned with quality of life.
In **High Uncertainty Avoidance cultures (High UAI)**, people feel more threatened by uncertain or unknown situations, leading to a need for formality and clear rules.	**Low Uncertainty Avoidance cultures (Low UAI)**, are more comfortable in managing uncertainty, and are generally informal.
Long-term Orientation (High LTO) cultures value perseverance and thrift, and generally regard time as overlapping (synchronous)	**Short-term Orientation (Low LTO)** cultures think more short-term and generally regard time as linear and sequential.

There is an increasing number of studies that appear to show how cultural values influence the nature of instructional systems, and it is noteworthy that these cases are generally consistent with existing models of culture.

For example, Morse (2003) evaluated an undergraduate distance learning program delivered to two groups of learners from cultures described by Hall (1977) as "high and low context"[3]. He found that the high context cultures, which included students from China, Singapore, Sri Lanka and Pakistan used the online resources in different ways from the low context group, which included students from the US, UK, New Zealand and Australia. The high context group had different reasons for using the technology, emphasizing "the ability to say what is appropriate", in contrast to the low context group, which valued "personal convenience and the ability to study anytime, anywhere". Their values were reflected in their priorities as learners.

Selinger (2003) undertook a formative evaluation of the Cisco Networking Academy, a program involving more than 300,000 students in 149 countries. She discovered widely differing uses of the same online materials in differing cultures. For example, students in Denmark and Sweden were encouraged to take greater responsibility for their own learning than those in France. The Scandinavians had greater autonomy, collaborated more and relied less on the tutor than those in France, where there was little use of group work or peer support. These findings are broadly consistent with Hofstede's model, the Scandinavians being, in Hofstede's terms, more feminine (low MAS), lower power distance (low PDI) and lower in uncertainty avoidance (low UAI) than the French.

In one of the very few quantitative studies in this area, Edmundson (2004) showed that learners in the Indian office of a global organization had significantly different preferences and perceptions in a specific educational context (a simple technical e-learning course), particularly in relation to forms of motivation, degree of support for cooperative learning, learner control, teacher roles and value of errors. Other studies, such as those by Zakaria & Gould (2001), and Haulmark (2002) also indicate that learners in different cultures use learning technologies in significantly different ways, and imply that instructional systems that ignore cultural values may be less effective in producing the required learning outcomes.

But at what stage in a design process should a learning designer consider cultural values? In the next section we will argue that the most critical part of the design process is the point at which the learning designer makes decisions about what we call the "learning strategy".

LEARNING STRATEGY AND VALUES

Because not every instructional approach is effective in every learning context, it is necessary to determine when a particular approach is the best possible match for the learner's needs, the instructor's teaching style, the learning environment, and the instructional goals. It is also important to determine how an instructional approach should be used in a given context (Nelson, 1999).

The way that learning experiences are designed varies greatly in different situations. A school teacher, a corporate trainer and a personal development coach might not recognize each others' learning design processes. But at some point in any process, a decision is made—implicitly or explicitly—about what may be called a "Learning Strategy." And it is our contention that, for learning experiences to be effective in multi-cultural environments, this decision must take into account the cultural values of the learning designer and of the learners.

The term "strategy" in the context of learning design is, to use used Goodyear's (1999) description "a broad-brush depiction of plans - of what should be done to achieve certain objectives" (p.3). It is what film producers might call a "treatment", or an "approach"; what architects (such as Lawson, 1997, p.91) might call "the integrative idea". Traditional writers on instructional design, such as Gagne, Briggs and Wager (1992, p.28), explain the purpose of a strategy (they call it "instructional strategy") as to outline "...how instructional activities will relate to the accomplishment of the objectives". So, for example, a leadership learning program for senior school teachers may be based on a strategy of offering personally challenging experiential learning in a face-to-face setting, preceded by scheduled online discussions and supported by complex networked online simulations. Using current terminology, this is a form of "blended" strategy.

According to traditional instructional systems design (ISD) methodologies, selection of strategy should be the output of a sequence of planned stages such as needs analysis, objective setting and so on. Thus strategy selection is both planned and explicit. Certainly, such an approach is likely to avoid situations such as where a teacher walks into their classroom and asks "what shall I do today?" or the e-learning designer who strings together a loose sequence of multiple-choice quizzes or drag and drop exercises with no larger picture of the learning challenge in mind. But there is also a case for arguing that a highly experienced learning designer will be able to make strategy decisions intuitively and implicitly and still be successful. They can rely on their experience – their "common sense". Yet it is precisely by relying on common sense approaches that many problems arise in multi-cultural situations. Much of our "common sense" relies on our deeper values, which are an underpinning of our cultural conditioning, and are largely invisible to us. Don Norman (1996) in discussing product design makes this point strongly: "Cultural issues are perhaps the hardest to identify and deal with. Once people are acculturated, their thoughts, beliefs, and actions are biased, without their conscious awareness. Of all the problems...cultural issues are probably the most insidious" (p.236).

Various authors, such as Reigeluth (1999) emphasize the importance of values as the foundations of particular learning strategies, and this appears to lie at the heart of the problem facing the learning designer. In a multi-cultural situation, the learning designer will have to bring to the surface not only their own values and those of the learners, but also those of the particular learning strategies they may be considering.

Let us say, for example, that a designer working in a multi-cultural environment is considering, as a basis for their learning strategy, a collaborative problem solving approach. In order to provide a robust basis for this approach, they are reviewing Nelson's theory of Collaborative Problem Solving (In Reigeluth, 1999, p.242). Helpfully, Reigeluth's notes introducing Nelson's theory provide an explanation of the values on which it is based. Table 2 shows how some of the values underlying the theory suit some cultures more than others.

Given the example above, it appears that culturally determined values may influence both the selection of learning strategy, and the methods used within that strategy. Our experience suggests that this can be disturbing for learning designers who are accustomed (one might even say acculturated) to tackling learning problems in environments that are generally culturally homogenous, and of which they themselves are a part. Although there has been much written (such as in Hofstede's original source on the subject—2001) about the general characteristics of how people from different cultures may learn differently, there are few established tools to support a learning designer in grappling with selection of specific learning strategies based on an understanding of cultural values.

One basis for a decision support tool would be a structure, such as Reigeluth and Moore's framework for Comparing Instructional Strategies (1999, p.56), which categorizes learning strategies or theories. If a chosen model of culture identified the learning-related norms and preferences of specific cultures within the components of the framework, then mapped these against the positions of specific learning strategies and theories on the same components, such a framework might support decision-making as to which approaches (theories) are most appropriate for a culture.

Figure 2 illustrates how this approach might work, using four of the instructional theories from Reigeluth (1999): Mayer, Jonassen, Nelson and Schank, Berman and Macpherson. Although

Table 2. Cultural implications of Nelson's theory

Values from Collaborative Problem Solving – Nelson	Cultural implications of these values
Learning to use naturally effective collaborative processes	Different cultures value and use collaboration differently. The methods used to introduce and support collaborative learning will differ between cultures.
Learning environments that are learner-centered...	Learner-centered environments will tend to challenge the authority of tutors/teachers in high power-distance cultures. These environments will have to be introduced very carefully.
Cultivating supportive, respectful relationships among learners, as well as between learners and the instructor	Supportive relationships between learners may suit more feminine, and more communitarian cultures. The nature of "respect" in high power-distance cultures will be quite different from that shown in low power-distance cultures, with instructors in the former preferring a clear delineation of status. Methods and strategies to support this will have to vary between cultures.
Developing a desire for life-long learning...	Cultures with long-term orientation have always tended to regard life-long learning as normal, (for example through apprenticeship, guild and mentoring approaches), unlike those with shorter-term orientation. Introducing and supporting life-long learning in short-term oriented cultures will be more challenging.

there may be debate about the exact location of the theories on the particular components of the framework, that is not the main point of the exercise. We are attempting to show that a structure of this type may be useful in mapping theories in a way that is useful in highlighting cultural issues.

Figure 2 positions four example instructional theories on Reigeluth and Moore's framework, and provides example notes relating to how models of cultures may map onto the framework. The cultural positioning notes are samples only and are not intended to be comprehensive.

HUMAN MEDIATION AND LEARNING STRATEGY

The authors suggested in the introduction to this chapter that the networked learning technologies are regarded as a means of connecting people, as much as a means of distributing learning resources. However, in a multi-cultural networked environment, strategies that connect people together can present particular problems—and opportunities—to learning designers, and the role of human mediation needs to be carefully considered.

Figure 2. Mapping Reigeluth and Moore's framework against models of culture

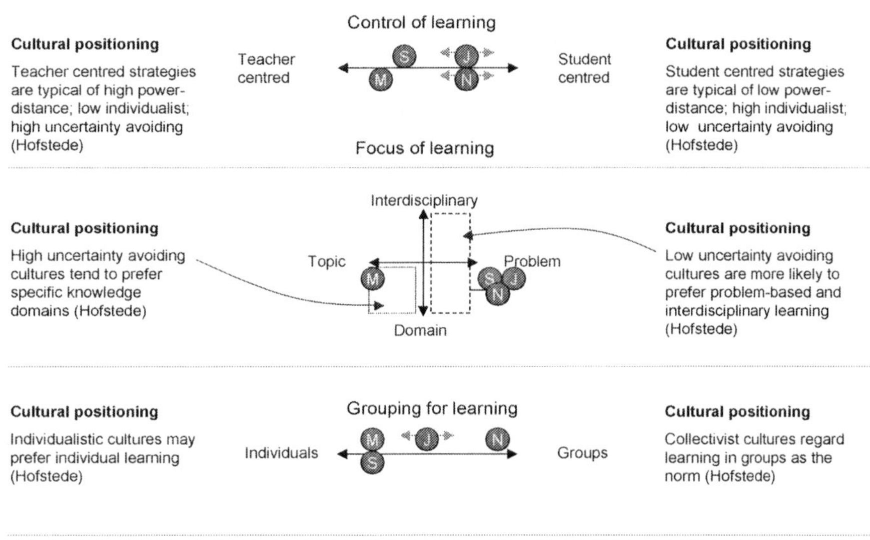

Instructional theory key:

- (S) Schank, Berman, Macpherson. - "Learning by doing" (1999, p161-182)
- (M) Mayer – "Designing Instruction for Constructivist Learning" (1999, p141-160)
- (J) Jonassen – "Designing Constructivist Learning Environments (1999, p215-240)
- (N) Nelson – "Collaborative Problem Solving" (1999, p241-268)

The nature and degree of human mediation in a networked learning environment varies somewhat. An increasingly common model is one in which a facilitator, moderator or tutor acts as a means of adapting learning resources to the local needs of culturally diverse learners. Perhaps this could be termed "human localization". For example, the Cisco Networking Academy (Selinger, 2003) uses networks to distribute learning materials to a large number of countries, while allowing tutors in those countries to customize the learning experience for learners. Local adaptation of materials by tutors means that culturally-based expectations on the part of learners are more likely to be met. The Cisco Networking Academy illustrates that point that, in many cases, the ability of an appropriately experienced person to identify particular requirements and respond sensitively to them will be greater than any current technology provides. Various models are available to assist tutors in adapting learning resources. For example, Henderson (1996, p.95-96) proposes a set of 14 "pedagogic dimensions of interactive learning", that provide a basis for considering the various ways in which a system may have to respond to varying cultural needs.

A different, more sophisticated form of human mediation is required where the strategy adopted is based on the principles of a community of practice (Wenger, 1999). Such communities are most often associated with forms of learning that involve the cultivation of understanding in learners, or "meaningful learning"; what various authors (such as Kember, 2000, p.104) refer to as "deep" learning. In the context of a community of practice, according to Wenger's account, learners "negotiate" meaning through discussion, and over time develop a sense of trust and mutual respect. As these conditions develop, so does a safe, supportive learning environment where learners are free to express themselves. Community members develop a sense of identity through their mutual engagement with one another, progressively making sense of their domain. Understanding develops through this process of negotiation of meaning.

Where a learner participates in a community that draws from a range of values that are very different from their own—as is often the case in multi-cultural environments—the negotiation process towards meaning-making may take a different shape, and make particular demands on community members. A sense of trust and mutual respect is only likely to develop where individuals initially heighten their awareness of their own cultural preferences and conditions, then gradually integrate the various cultural perspectives into their own. In a process similar to what Bennett (1993) describes as the three stages of ethnorelativism (acceptance, adaptation, integration), they are likely to have to take a mental journey towards a point where the differences within the community become part of their own identity. The skilled and sensitive facilitator can play a key role in progressing this journey by helping to surface cultural differences, using tools such as Hofstede's dimensions. These can provide a framework for interpretation and understanding of the behavior of diverse community members.

CONCLUSION: BEYOND LOCALIZATION

"The term "localization" is an impossibly light rendering of the real task. A truer model of knowledge transfer would be less like moving a piece of stuff and maybe sculpting it a bit to local needs, and much more like a local construction of knowledge at the source of need" (Patrick Lambe, 2002 p.16).

So far in this chapter we have largely avoided using the term "localization". So why introduce it now—apart from because it is in the title? We do so to highlight two key assumptions about how e-learning should be developed for multi-cultural environments that we believe are incorrect, so as to summarize our position.

Firstly, localization, as currently applied to e-learning products and services, assumes that a piece of e-learning originates in one culture, is based on that culture's values, and is then exported to, and adapted for, other cultures. Our research suggests that unless the values of the various cultures for which the learning is intended are taken into account to start with, the quality of the learning experience may be compromised. Secondly, current e-learning localization practice assumes "content" as a starting point. The surface features of this content are then modified somewhat to accommodate the surface feature preferences of other cultures. The authors' view is that e-learning's starting point should be an effective learning strategy, developed with cultural diversity in mind, and one intended to use learning networks to put people in contact with people.

REFERENCES

Bennett, M.J. (n.d.). Towards ethnorelativism: A developmental model of intercultural sensitivity. In R. M. Paige (Ed.), *Cross-cultural orientation: New conceptualizations and applications* (pp. 27-70). New York: University Press of America.

Collis, B. (1999). *Designing for differences: Cultural issues in the design of WWW-based course-support sites*. BJET.

De Jong, T., & Ferguson-Hessler, M. (1996). Types and qualities of knowledge. *Educational Psychologist, 31*(2).

Edmundson, A. (2004). *The cross-cultural dimensions of e-learning.* PhD Dissertation, Walden University

Gagne, R., Briggs, L., & Wager, W. (1992). Designing instructional systems. In *Principles of instructional design*, Wadsworth.

Goodyear, P. (1999). Pedagogical frameworks and action research in open and distance learning. *European Journal of Open and Distance Learning.*

Hall, E. (1977). *Beyond culture.* Anchor Press/Doubleday.

Hannafin, M. J., & Land, S. M. (1997). The foundations and assumptions of technology-enhanced learning environments. *International Journal of Learning and Cognition, 25*(3).

Haulmark, M. (2002). *Accommodating cultural differences in a web-based distance education course: A case study.* In the 9th Annual International Distance Education Conference, Texas A&M University

Henderson, L. (1996). *Instructional design of interactive multimedia: A cultural critique.* Educational Technology Research & Development.

Hofstede, G. (2001). Culture's consequences. *Comparing values, behaviors, institutions and organizations across nations* (2nd ed.). Sage.

Kember, D. (2000). *Misconceptions about the learning approaches, motivation and study practices of Asian students.* Higher Education, Kluwer Academic Publishers.

Lawson, B. (2002). *How designers think: The design process demystified.* Architectural Press.

Lambe, P. (2002). *The autism of knowledge management.* Retrieved from www.straitsknowledge.com

Morse, K. (2003). *Does one size fit all? Exploring asynchronous learning in a multicultural environment.* JALN.

Nelson, L. (1999). Collaborative problem solving In *Instructional: Design Theories and Models.*

Norman, D. (1996). Design as practiced. In *Winograd—Bringing design to software.* Addison Wesley.

Reigeluth, C. (1999). What is instructional-design theory and how is it changing? In *Instructional—Design Theories and Models.*

Reigeluth, C., & Moore, J. (1999). Cognitive education and the cognitive domain. In *Instructional- Design Theories and Models.*

Selinger, M. (2004). Cultural and pedagogical implications of a global e-learning programme. *Cambridge Journal of Education.*

Trompenaars, F. (1997). *Riding the waves of culture: Understanding cultural diversity in business.* Nicholas Brealey Publishing Ltd.

Wenger, E. (1999). *Communities of practice. Learning, meaning, and identity.* CUP.

Zakaria, N., & Gould, E. (2001). *Applying cross-cultural theory to instructional design.* Geneseo.

ENDNOTES

1. The term "instructional system" is the first of many used in this chapter that are, themselves steeped in cultural assumptions. It is a broadly North American term, and one that other English-speaking cultures, such as the UK, may find unfamiliar or imprecise. Based on Hannafin and Land's model of instructional systems (1997), it encompasses the technological, psychological, pedagogical, cultural and pragmatic components of systems.
2. We have found that Trompenaars' views on time-orientation are generally more readily understood by learning designers, so tend to work a blend of Hofstede's and Trompenaars' versions.
3. Which tends to correlate highly with Hofstede's IDV dimension (individualist/communitarian)

This work was previously published in Globalization E-Learning Cultural Challenges, edited by A. Edmundson, pp. 255-266, copyright 2007 by Information Science Publishing (an imprint of IGI Global).

Chapter IX
Strategies for Virtual Learning and E-Entrepreneurship in Higher Education

Juha Kettunen
Turku Polytechnic, Finland

Mauri Kantola
Turku Polytechnic, Finland

ABSTRACT

This chapter presents the strategies of higher education institutions and how they can be described using the balanced scorecard approach. The pedagogical ICT strategy describes the virtual learning and e-entrepreneurship in higher education. Strategic themes are presented to describe what management believes must be done to succeed and achieve the desired outcomes in virtual learning and e-entrepreneurship. Strategy maps are used to describe the strategy in a graphical representation. In addition, the study presents an example of the cooperation between a higher education institution and a spin-off company. This chapter helps the educational administrators to better describe and implement strategies for virtual learning and e-entrepreneurship.

INTRODUCTION

Higher education institutions (HEI) usually try to adapt their strategies to the local community, to society as a whole and to educational policy. HEIs are trying to ensure competitiveness and employability for their students. To a large extent the competitiveness is based on good skills in information and communication technology (ICT). They are linked to entrepreneurship, which is seen by the local stakeholders and educational authorities to create economic growth and welfare.

The primary purpose of this study is to explore the strategies for virtual learning and e-entrepreneurship in higher education. The pedagogical ICT strategy is a specific functional strategy, which describes the strategic outlines for virtual learning and e-entrepreneurship. The aim is also to explore the methods to communicate and implement the strategy in an understandable and efficient manner.

Strategic management is a matter of developing the organisation and its present activities to achieve the desired objectives in the future (Fidler, 2002; Davies & Ellison, 2003). The new strategies of HEIs typically reflect the existing strategies, which are tailored to meet the needs of the organisation and its stakeholders. The strategies are typically fairly stable, but they reflect the changes in society, economic development and educational policy.

The strategies typically focus the activities on specific fields of education according to the needs of the local community or society. Another typical strategy is the operations excellence theme. HEIs usually try to improve their quality, achieve more and reduce costs. These strategies can also be found in the business literature, where Porter (1990) has presented the strategies of focus and overall cost efficiency.

The balanced scorecard (BSC) approach developed by Kaplan and Norton (1992, 1993) is used in this study to describe a pedagogical ICT strategy. The strategy must be understood before it can be implemented. The balanced scorecard creates a shared understanding of the selected strategies, because it translates the strategy into tangible objectives and balances them into four different objectives: customer and regional development, financing, internal processes and structures, and learning and growth. The significance of the present study is to show how the competitive strategies and the balanced scorecard can be applied in HEIs.

A qualitative study is made based on the concepts of strategic planning and the balanced scorecard approach. The focus on interpretation in how the participants make sense of these rather than numerical exactness is the strength of qualitative research. When a qualitative study is carried out, qualitative data are required to clarify and illustrate the meaning of findings. Typically a relatively small number of cases is studied preserving the individuality of each of these in the analyses.

The overall strategy of Turku Polytechnic emphasises high quality learning. The pedagogical strategy is a functional specific strategy, which provides outlines for the development of education to promote the overall strategy. The pedagogical ICT strategy focuses on the pedagogical strategy to the virtual learning and e-entrepreneurship. The strategies of the educational departments are aligned with the functional strategies. The balanced scorecard can be used to communicate and implement both the overall and specific strategies.

The establishment of new businesses is an effective way to transfer new knowledge from higher education to the local environment. The spin-off activities result from the transfer of people and know-how from the HEI. The transfer of the skills and tacit knowledge embodied in the human capital differentiate this technology and knowledge transfer mechanism from technology sale, licensing, joint ventures and alliances (Davenport et al., 2002). The spin-off activities and e-entrepreneurship are means to implement the strategic plan of HEIs.

This study is organised as follows. Section 2 presents the overall strategic outlines of the HEI and presents how balanced scorecard can be used to communicate and implement the strategy. Section 3 presents the pedagogical ICT strategy including virtual learning and e-entrepreneurship. Strategic themes and strategy maps are used to describe the strategy. An example of a spin-off company and e-entrepreneurship are presented in Section 4. Finally, the results of the study are summarised and discussed in the concluding section.

STRATEGIC PLANNING IN HIGHER EDUCATION

Strategic Outlines for Educational Institutions

Strategic planning is needed to move an organisation from its present position to a desirable but uncertain future position. The strategic plan is a description of the route described by the mission to a desirable future position described by the vision (Wheale, 1991; West-Burnham, 1994). The strategic plan is a holistic description of how an organisation adapts to its environment and develops its activities for a better future.

The competitive strategies by Porter (1990) provide the general strategic framework for the planning of the strategy in different kinds of organisations. These strategies have also been applied successfully in educational institutions (Treacy & Wiersma, 1995; Kettunen, 2002). Turku Polytechnic has selected the focus strategy, which concentrates on the most promising clusters of Southwest Finland. The focus strategy is combined with the strategy of cost-efficiency, which enables resources for high-quality learning and research and development.

The focus strategy means that the organisation selects a market segment and builds bonds with the most important partners within the selected segment. Finnish polytechnics are higher education institutions, which aim to serve their geographical regions particularly well. Each polytechnic has selected its occupational groups, where they educate experts. The focus strategy enables the polytechnics to enhance their knowledge in their specific market segments.

The most promising growth clusters in Southwest Finland are ICT, biotechnology and the metal and maritime industries. ICT is especially suitable for entrepreneurship training and incubator activities, because the ICT industry is based to a large extent on the knowledge of graduates obtained in education. ICT is not a capital-intensive industry with high entry barriers for most graduates.

When an organisation selects a cost-efficiency strategy, it delivers a combination of price and quality that is recognised by customers and stakeholders. It is a natural choice for educational institutions, which typically have predetermined unit-priced funding and annual budgets. The strategy aims to remove the overlapping activities to achieve cost reductions. Even though the cost-efficiency strategy does not primarily aim to focus the activities it achieves also focusing on specific activities. The focus and cost-efficiency strategies complete each other.

Balanced Scorecard Describes the Strategy

The balanced scorecard is used to translate the strategy and vision into tangible objectives and measures, which can be communicated and translated to the staff and external stakeholders. The balanced scorecard approach can also be used to plan and evaluate strategies (Kettunen, 2004). The balanced scorecard has a balanced mix of objectives in the different perspectives to indicate the strategy. The perspectives can be defined according to the needs of the organisation.

The perspectives can be defined in higher education as follows:

1. **Regional development and customer:** The perspective includes the desired objectives of regional development. It also describes the value created for students and employers in the internal processes.
2. **Finance:** The financial perspective describes the public funding and external outcome. The funding is aligned with the internal processes and structures in the budget of an organisation.
3. **Internal processes and structures:** The internal processes and structures perspective describes the internal sequential processes and structures of organisational units. These processes create value for customers.
4. **Learning and growth:** The learning and growth perspective describes the drivers for future performance and what learning and capabilities are required in the internal processes.

These perspectives have been found to be necessary and sufficient across a wide variety of organisations in the private and the public sectors (Kaplan & Norton, 1996, 2001). The balanced scorecard was originally developed for business companies. The desired objectives of the private sector are typically in the financial perspective, but in the public sector organisations the financial result is typically not the primary objective. Therefore it is reasonable to place the customer or recipient of the services at the top of the hierarchy.

The customer-oriented process begins by defining the objectives of the regional development and customer perspective asking, "What value do we create for our customers in the internal processes?" Then the process continues by asking, "What are the objectives in the financial perspective to enable the internal processes?" and "What is the cost-efficient way of producing the services?" Finally the process asks, "What capabilities and learning are required to achieve the objectives in the internal processes?"

THE PEDAGOGICAL ICT STRATEGY

The pedagogical ICT strategy of Turku Polytechnic is based on the development plans of the Finnish Ministry of Education (1999, 2003a). The purpose is to increase the pedagogical research and develop virtual learning in order to increase the competitiveness. Education should provide skills for applying, managing and evaluating the information flow of the modern e-networks and knowledge society. The plans emphasise the importance of local networks, which combine ICT, modern cooperation methods, independent learning, public and private partnerships, regional development and entrepreneurship.

The pedagogical ICT strategy is part of the pedagogical strategy of Turku Polytechnic. The pedagogical strategy is a functional strategy, which provides general outlines for the development of education. The pedagogical ICT strategy emphasises the development of virtual learning and e-entrepreneurship. The pedagogical strategy is aligned with the other functional and department strategies.

E-business is defined in the European E-business Report as automated business processes (both intra-firm and inter-firm) over computer mediated networks (European Commission, 2004). It has been done in full accordance with the definition used by the OECD (2004). E-entrepreneurs can be generally characterised as persons who take advantage on the Internet. They are not necessarily expert in ICT or Internet technologies, but there may be many who are interested in including an Internet component in a more traditional business (cf. Benjamin & Wigand, 1995; Bakos, 1998; Coates, 2004). Virtual learning and e-entrepreneurship are tied to each other in education.

The pedagogical development includes new working and learning environments, new content production methods, content production, communication and interaction in the e-networks, guidance and evaluation, information security and copyright. The pedagogical development produces a wide range of teaching methods, contents and cooperation with working life. The purpose is to provide skills to meet the needs of the local, national and international partners in the public and private sector.

New forms of communication and virtual communities emerge in *e-networks*, when virtual learning is developed. The Finnish Virtual Polytechnic, which is a joint development network of all the 31 polytechnics in Finland, has adopted a widely used approach of developing virtual education. Guidance and support are important at the beginning. These are followed by independent learning and support. Finally, teachers and students can use various flexible methods of virtual teaching and learning (cf. Collis & Moonen, 2001). The approach of developing virtual learning can be described as follows:

1. **Guided and supported virtual learning:** Guided and supported virtual learning is based on teaching based on collaboration in groups where the teacher and students interact with each other using different kinds of digital software and equipment. The education may include pair and group tasks, which may entail discussions and negotiations on the e-networks.
2. **Independent learning in e-networks:** The students study and solve problems independently using the virtual material in e-networks supported by the instructions. Independent learning does not include interactive support given by the instructor or interactions with other students.
3. **Various methods of learning:** Implementation includes several kinds of flexible learning methods. The teaching has long-distance and contact sessions. The study takes several forms. The students may study independently, in pairs and larger teams or they may take lectures. The students use e-networks and the learning may take place on campus, at the working place or at home. The presence of the students is required both in the live meetings and virtual environments.

Strategic Themes

Strategic themes describe what management believes must be done to succeed and achieve the outcomes in the different perspectives. They are in line with the objectives and describe the causal relationships between them. Each organisation has a unique set of strategic themes, which are specific to the organisation's potential to create value in its environment. The strategic themes are also linked to the organisation's internal processes, because the essence of the strategy is in the activities (Porter, 1996).

The strategic themes of the pedagogical ICT strategy have their roots in educational policy, the needs of local community and the practices of the development work in virtual learning and e-commerce. The specification of strategic themes caused extensive thinking at Turku Polytechnic. The general strategic statement and strategic themes can be written as follows:

- Virtual learning skills and e-entrepreneurship for working life
- Various pedagogical methods are used in virtual learning.
- Virtual learning is included in the curriculum.
- Teacher teams are used to plan and implement the study modules.
- The learning material is used in a broader context.
- The change of virtual learning contents between the HEIs is increased.
- The entrepreneurship and intrapreneurship culture of the HEI is promoted.

The pedagogical methods include especially problem-based and virtual learning with the emphasis on the interactive skills and self reflectivity of students. Virtual learning requires a shared understanding and team work of teachers. The virtual learning material is mobile and it is used in other degree programmes, the Open Polytechnic and other polytechnics through the National Virtual Polytechnic. There are cooperative production teams for virtual learning material. The material is bought and sold by the partners of the production teams among the different institutions. This supports the entrepreneurship of teachers.

Strategy Map

The strategy map developed by Kaplan and Norton (2001, 2004) is a graphical representation of the functioning of the organisation. The strategy map helps the employees and stakeholders to understand why the objectives of the organisation have been set and how the desired objectives can be achieved. It is like a road map, which describes only the essential characteristics of the strategy in a simple way.

The description and communication of strategy requires an understandable "organisational theory" of value creation. Strategy maps can be used to translate the strategic themes into objectives located in the different perspectives. They also provide tools to describe the causal linkages between the objectives. The measurement system indicates the strategy through a sequence of relationships between performance drivers (leading indicators) and outcome measures (lagging indicators).

The strategy map clearly communicates the objectives of an organisation and describes why they have been set. The strategy must be understood before it can be turned into action. The objectives, corresponding measures and performance targets are derived from the organisation's strategy and vision and balanced into four different perspectives. The performance of organisational units and workers can be directly linked to the strategy.

Figure 1 describes the strategy map of the pedagogical ICT strategy of Turku Polytechnic. The regional development and customer perspective includes lagging indicators that report on the desired outcomes of an organisation. The financial perspective is always linked with the internal processes and structures. The internal processes illustrate the value chain describing the sequential internal processes and organisational units cooperating with each other. The learning and growth perspective includes the capabilities and learning of employees, which are the driving forces of future performance.

Regional Development and Customer Perspective

The first objective of this perspective is to equip all the students with good virtual learning skills so that they can use them in working life and adult education. According to the pedagogical strategy Turku Polytechnic is moving from a closed learning environment towards an autonomous and open expert organisation and network facilitating continuous learning. The polytechnic develops and provides learning and working environments, which enable the experiments and study of new activities and procedures together with the companies and other working life organisations.

Another objective is to provide entrepreneurship skills. Various learning methods include projects, practical training and theses, which are part of the entrepreneurship training and shared with spin-off companies and other customer organisations. Research and development produce methods of modern entrepreneurship training and incubator activities. In addition, continuing education includes virtual learning and e-entrepreneurship training.

The objective is also to achieve customer satisfaction with e-networking. Customer satisfaction is what virtually every organisation is trying to achieve. The purpose of the Polytechnic is to support customers' virtual learning skills, working life skills and e-entrepreneurship skills. The purpose is to create long-lasting customer relationships, which enable life-long learning and continuous cooperation between the educational institution and customer organisations.

The Financial Perspective

The first financial objective is to obtain sufficient funding from the Ministry of Education. The degree programmes and longer educational programmes in continuing education are funded by the Ministry. It also funds some part of the research and development and certain development projects of major national or regional importance.

Another financial objective is to obtain sufficient funding from the City of Turku. The funding from the Ministry of Education is paid to the City of Turku, which is the owner of Turku Polytechnic. The City of Turku contributes a minor share of the funding. The city supports some software, which is used at Turku Polytechnic. It finances and develops, for example, shared Intranet and Extranet solutions of the city organisations.

The objective is also to increase the external funding. External funding is obtained mainly for continuing education, research and development and services provided by the Polytechnic. The funding bodies include the European Social Fund (ESF), other funds, companies, public sector organisations and the Finnish Virtual Polytechnic. The external funding is used to create new contents and improve the quality of education.

Internal Processes and Structures Perspective

Research and development (R&D) is used to create content and methods of virtual learning. Content is also required in cooperation and networking with other educational institutions. The content of virtual learning is typically developed at the same time as the curriculum development. Research and development also supports the development of infrastructure, library and information services.

The Pedagogical Support Unit of Turku Polytechnic promotes educational development in many different ways. The support activities of virtual learning include the development and implementation of new methods of virtual learning. Educational development also includes many other development approaches such as problem-based, work-based and research-based learning. These approaches can be combined with virtual learning.

Quality assurance (QA) is based on documented procedures. The development of quality assurance has been agreed in the Bologna Process by the European Ministers (Berlin Communiqué, 2003). Quality assurance is at the heart of the setting up of the European Higher Education Area (EHEA) by 2010. Each HEI has the responsibility for quality assurance representing their academic and organisational autonomy. This provides the basis for accountability within the national quality framework. The purpose of the EHEA is to increase student and staff mobility in Europe.

The ICT infrastructure is a combination of data networks, systems, equipment and software. They include wireless networks, modern audiovisual solutions and personal mobile learning equipment. The ICT Unit of the Polytechnic is developing an electronic learning portal constructed on the existing information services, digital information sources, e-business instruments, virtual learning environments and existing network services of the departments. Physical and virtual learning environments are developed interactively with each other taking care of information security.

The library provides the main electronic sources and databases for students and staff. The library helps them in finding the relevant pieces of information and provides equipment for information processing. The librarians teach on the degree programmes and personnel training and participate actively in virtual learning projects. Cooperation between the support staff, teachers, students, researchers and the members of the working life organisations is necessary. Electronic journals and books reduce the proliferation of printed material.

Virtual learning is expanding at Finnish HEIs. The Government has set a target to provide at least 30 credits of virtual education for each bachelor student (Ministry of Education, 2003b). Virtual learning provides alternative ways of studying, especially for those who work during their studies. The replacement of contact learning with virtual learning helps the working students to avoid dropout. The virtual learning platform can also be used to carry out other activities such as tutoring during practical training, study counselling, international projects and continuing education.

Cooperation with other HEIs is an important characteristic of the internal processes and structures. Turku Polytechnic participates in many production teams of the Finnish Virtual Polytechnic. Typically the production team of several institutions makes a written contract, agreeing to produce virtual learning material. The contract also defines the financial transactions between the institutions. Cooperative and entrepreneurial teacher teams are usually needed within an institution to participate in the production teams.

Networking is a way of producing large study archives from small virtual learning materials. The network itself is used as the instrument of guidance, where the instructors and learners can select the information according their specific needs (Silander & Koli, 2003). The purpose is to use the learning material several times in different learning combinations or at different educational levels. Often the smaller modules correspond best to the various needs in working life situations of adult education.

Learning and Growth Perspective

Strategic awareness is created in the strategy process, where the expert team for virtual learning prepares the strategic plan with the experts and management team of the polytechnic. The existing strategy and the changes in the environment and technology are evaluated in order to update the strategy. The strategy is approved by the Board of Turku Polytechnic. Finally, the strategy is communicated and implemented using the balanced scorecard.

Personnel training is an important way of introducing new methods and software used in virtual learning and e-entrepreneurship. The departments

Figure 1. The strategy map of the pedagogical ICT strategy

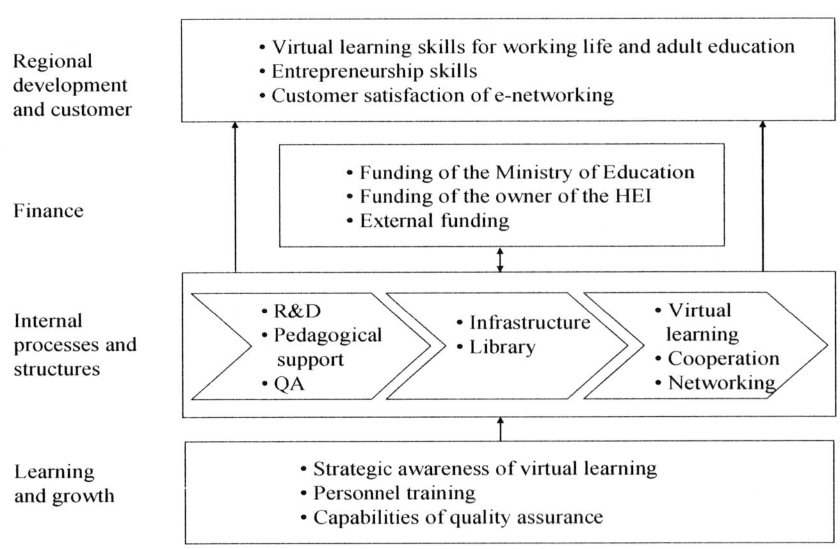

of the Polytechnic are responsible for the short-term education and working life periods of the personnel. The Personnel Development Unit of the Polytechnic arranges longer personnel training and creates preconditions for the implementation of the pedagogical ICT strategy.

The capabilities of quality assurance have been developed constantly while the quality system of Turku Polytechnic was developed. The procedures of quality assurance have been documented and they are also used in virtual learning. The procedures and documents include the evaluations of Finnish Higher Education Evaluation Council, the quality manual of the institution, internal audits, internal target discussions and feedback from students and employers.

E-ENTREPRENEURSHIP OF A SPIN-OFF COMPANY

A Spin-Off Company as a Partner

The spin-off companies result from the transfer of people and intellectual property from educational institutions. The continuous transfer of skills and tacit knowledge embodied in human capital differentiates the mechanism of technology transfer from technology sale, licensing, joint ventures and alliances (Davenport et al., 2002). The supporting of start-up companies is an effective way to transfer the expertise of HEIs to working life and make it commercial.

The case of Mansoft Tietotekiikka Ltd. is used as an example to illustrate how the pedagogical ICT strategy of Turku Polytechnic is implemented and how technology transfer takes place. Mansoft Tietotekniikka Ltd. is a spin-off company in software business and application development. The company was established by a senior lecturer at the Polytechnic. He is still the managing director of the company.

The purpose of Mansoft Tietotekniikka Ltd. is to develop the expertise to achieve customer satisfaction. In order to achieve this aim the products are tailored to meet the needs of the customers. The strategy of Mansoft Tietotekniikka Ltd. also includes the cooperation with Turku Polytechnic as described by Adamsson and Puukka (2004). The company has planned solutions for specific needs of knowledge-based organisations. It also carries out consulting, maintenance services and several boxed products with a fixed service. As a financially independent and customer-orientated company it aims to be a productive and safe partner in long customer relationships.

The company has been developed in a sustainable way. The business profit has always been ploughed back into the company and product development. All the eight young permanent employees of the company are graduates of Turku Polytechnic. One of the main principles of company policy has been to offer permanent positions to young experts starting their careers in the ICT business.

The managing director of Mansoft Tietotekniikka Ltd. is a member of the Advisory Board of Turku Polytechnic. The advisory boards of Finnish polytechnics include members from working life and help the polytechnics to develop the curriculum to meet the needs of companies and other organisations. The advisory board is a network of experts who helps the polytechnic and the participating companies to adapt to the changes of the environment and conceive new development ideas.

The recruitment of the company is focused on the final year students of Turku Polytechnic. Turku Polytechnic has arranged the education so that students can participate in the projects of the company. The arrangement of the ICT fair is another mode of cooperation between Turku Polytechnic and Mansoft Tietotekniikka Ltd. The ICT Fair is organised in Loimaa, where Turku Polytechnic operates. The Fair helps local small companies to recruit new staff, market their services and participate in regional development.

According to the company owner some customers have sometimes expressed their opinion that the company should have older experts to achieve credibility, but the managing director has felt that the young staff brings more flexibility and fresh ideas and fewer predetermined attitudes regarding the business culture. The company is represented by the managing director, who takes care of the business relationships and project management. The software and system development are left to the younger colleagues. The long experience of the manager and the fresh ideas of the young colleagues complement each other in the company. This is an important characteristic of the social capital and competitive advantage of the company.

Cooperation in Software Development

Turku Polytechnic has designed project management software for itself and other knowledge-intensive organisations. The software Projektori was created in association with Mansoft Tietotekniikka Ltd. Originally Projektori was created to manage development projects on the Intranet. Then the R&D unit of Turku Polytechnic expanded the use of the software into other project related activities. It can be used, for example, to manage project plans and disseminate project information on the Internet. These features make the software a useful tool in many publicly funded projects.

The software was designed to be used in data networks so that cooperative projects can use different kinds of Intranet and Extranet options. There is a general understanding about the roles between the partners. Turku Polytechnic is responsible for content planning and production while Mansoft Tietotekniikka Ltd. takes care of the database planning and implementation.

The guidelines for the cooperation can be found in the strategic plans and quality assurance manual of the Polytechnic. The cooperation with the spin-off company is regional development, which is an important characteristic of Finnish polytechnics. The quality manual contains guidelines for project management. Many of the basic principles of cooperation and software solutions can be traced form the commonly observed needs of databases and action models of large organisations (Groth, 1999).

The Projektori software is not only a technical instrument to arrange documents; the use of the software changes organisational behaviour into high-quality project management. The software guides the individuals and units to the process and team oriented organisational culture. It also provides an interactive connection and a virtual meeting place for public sector institutions, private sector companies and other working life organisations.

Turku Polytechnic has also sold the Projektori software to some other large knowledge-intensive organisations in Finland with the cooperation of Mansoft Tietotekniikka Ltd. The user rights of the software can be purchased by installing the application to the customer's server. It also can be rented as an application service provider (ASP) solution by locating the application and its database on the servers of Mansoft Tietotekniikka Oy. The rent includes both the application and its daily database back-up.

The Projektori software can be connected with other software such as the Puplikaattori software which is software designed for e-publishing. This software is the result of cooperation between the Publication Unit of Turku Polytechnic and Mansoft Tietotekniikka Ltd. Both software can be purchased and used as independent solutions. The software helps to increase the web-based resources, which are increasing in educational practice (Collis & van der Wende, 2002).

CONCLUSION

Information and communication skills are needed in the modern knowledge society. These skills are among to the basic professional skills in most industries. The ideal is an individual who is aware of the information sources, has the information reading skills, is capable of acquiring and communicating information efficiently, evaluating the information acquired and using the information in an ethically and legally correct way.

High quality and flexible learning is the overall strategic statement of Turku Polytechnic. The strategy focuses on the needs of the growing clusters in Southwest Finland. The ICT cluster is one of the most prominent clusters including software development and mobile telephones and TV production. Nokia Mobile Phones Ltd. is one of the best-known companies in this region. The strategy also focuses on entrepreneurship training, which includes e-commerce, incubator activities and other forms of entrepreneurship.

The pedagogical ICT strategy is a specific functional strategy aligned with the overall strategy, other functional strategies and department strategies at Turku Polytechnic. The pedagogical ICT strategy describes the strategic plan of virtual learning and e-entrepreneurship. The strategic themes of the pedagogical ICT strategy include various pedagogical methods, the inclusion of virtual learning in the curriculum, the role of teacher teams, the use of learning material in a broader context, transactions between other educational institutions and entrepreneurship culture.

The balanced scorecard approach is useful in accomplishing the strategic themes, objectives and measures for virtual learning and e-entrepreneurship. The strategy map was used to translate the strategy into a graphical representation which clearly describes the objectives in the different perspectives of the balanced scorecard. It was used to describe the main characteristics of the strategy to achieve better expertise in virtual learning and e-entrepreneurship.

An example was presented of the implementation of the pedagogical ICT strategy. The example illustrates a case of how a spin-off company was established and how it cooperates with the educational institution. Project-based education is followed by recruiting graduates. The competitive advantage of the spin-off company is based on continuous cooperation with the HEI and the innovative and fresh ideas of the graduates employed.

REFERENCES

Adamsson, V., & Puukka, J. (2004). Ihme, Yrittäjätarinoita Loimaan seudulta. Turun ammattikorkeakoulun oppimateriaaleja 13. Turku.

Bakos, Y. (1998). Towards friction-free markets: The emerging role of electronic marketplaces on the Internet. *Communications of the ACM. 41*(8), 35-42.

Benjamin, R., & Wigand, R. (1995). Electronic markets and virtual value chain on the information superhighway. *MIT Sloan Management Review, 36*(2), 62-72.

Berlin Communiqué (2003). Bologna Process Berlin 2003, Realising the European Higher Education Area. Communiqué of the Conference of Ministers Responsible for Higher Education in Berlin on 19 September.

Coates, B. E. (2004). E-entrepreneurship in a disadvantaged community: Project EdNet in California. *International Journal on E-Learning, 3*(3), 25-31.

Collis, B., & Moonen, J. (2001). *Flexible learning in a digital world: Experiences and expectations.* London: Kogan Page.

Collis, B., & van der Wende, M. (2002). *Models of technology and change in higher education: An international comparative survey on the current and future use of ICT in higher education.* Cheps, Center for Higher Education Policy Studies.

Davenport, S., Carr, A., & Bibby, D. (2002). Leveraging talent: Spin-off strategy at industrial research. *R&D Management, 32*(3), 241-254.

Davies, B., & Ellison, L. (2003). *The new strategic direction and development of the school.* London: Routledge Falmer.

European Commission (2004). The European e-Business Report. *A Portrait of e-Business in 10 Sectors of the EU Economy.* European Commission, Enterprise Directorate General, Enterprise Publications.

Fidler, B. (2002). *Strategic management for school development.* London: Paul Chapman Publishing.

Groth, L. (1999). *Future organizational design: The scope for the IT-based enter*prise. Chichester: John Wiley & Sons.

Kaplan, R., & Norton, D. (1992). The balanced scorecard: Measures that drive performance, *Harvard Business Review*, January-February, 71-79.

Kaplan, R., & Norton, D. (1993). Putting the balanced scorecard to work. *Harvard Business Review.* September-October, 134-147.

Kaplan, R., & Norton, D. (1996). *The balanced scorecard.* Boston: Harvard Business School Press.

Kaplan, R., & Norton, D. (2001). *The strategy-focused organisation.* Boston: Harvard Business School Press.

Kaplan, R., & Norton, D. (2004). *Strategy maps.* Boston: Harvard Business School Press.

Kettunen, J. (2002). Competitive strategies in higher education. *Journal of Institutional Research, 11*(2), 38-47.

Kettunen, J. (2004). The strategic evaluation of regional development. *Assessment and Evaluation in Higher Education, 29*(3), 357-368.

Ministry of Education (1999). Koulutuksen ja tutkimuksen tietostrategia sekä toimeenpano-suunnitelma 2000-2004. Helsinki: Ministry of Education.

Ministry of Education (2003a). Koulutuksen ja tutkimuksen tietoyhteiskuntaohjelma 2004-2006. Helsinki: Ministry of Education.

Ministry of Education (2003b). Koulutus ja tutkimus vuosina 2003-2008, Kehittämissuunnitelma. Helsinki: Ministry of Education.

OECD (2004). *The economic impact of ICT: measurement, evidence and implications.* Paris: OECD Publications Service.

Porter, M. (1990). *The competitive advantage of nations.* London: Macmillan.

Porter, M. (1996). What is Strategy? *Harvard Business Review,* November-December, 61-78.

Silander, P., & Koli, H. (2003). *Oppimisaihioista oppimisprosessiin.* Helsinki: Finn Lectura Oy.

Treacy, M., & Wiersma, F. (1995). *The discipline of market leaders: Choose your customers, narrow your focus. Dominate your market.* Reading, MA: Addison-Wesley.

West-Burnham, J. (1994). Strategy, policy and planning. In T. Bush & J. West-Burnham (Ed.), *The principles of educational management.* Harlow: Longman.

Wheale, J. (1991). *Generating income for educational institutions: A business planning approach.* London: Kogan Page.

This work was previously published in Entrepreneurship and Innovations in E-Business: An Integrative Perspective, edited by F. Zhao, pp. 107-123, copyright 2006 by IGI Publishing, formerly known as Idea Group Publishing (an imprint of IGI Global).

Chapter X
Computer-Based Assessment for Academic Evaluation

Devan Govender
Institute for Curriculum and Learning Development, South Africa

ABSTRACT

As student numbers and lecturer workloads increase, traditional methods of assessment make it rather difficult to undertake effective assessment and provide students with in-depth, relevant, specific, and speedy feedback. It is thus incumbent on tertiary academic institutions, more especially distance education providers, to consider alternative ways, means, and strategies towards the incorporation of computer-based assessment (CBA) methods. This automation can be easily realised by the ever-growing field of computer-aided assessment (CAA) software that is currently permeating the educational landscape of the world. While some individuals may argue that CBA software has been proven in only limited situations and that it is very costly to develop, it nevertheless remains a viable alternative technology that is beginning to challenge many traditional and out-dated modes of assessment. In this chapter the rationale for developing CBA has been discussed and thereafter pertinent issues around the types of CBA, advantages of CBA, proprietary, and non proprietary software have been analysed. The chapter concludes by firstly, highlighting strategic planning issues for consideration in engaging computer based assessment and secondly it provides modalities for future research directions within the arena of computer based assessment. These two components are key in designing ICT strategies for continuous academic evaluation.

INTRODUCTION

The growing call for the massification of higher education has led to commensurate increases in student numbers, class sizes, and lecturer workloads. In this regard, traditional methods make it progressively more difficult to undertake effective assessments and provide students with detailed, speedy, and effective feedback. Thus it appears only logical that academic institutions proceed to consider ways of automating their assessment methods, either fully or at least in part. The automation can be easily realised by adopting a well-structured and balanced computer aided assessment strategy using various proprietary and nonproprietary open source "computer automated assessment" (CAA) software that is readily and freely available. Though CAA software has been proven in some limited situations, it remains a technology with vast potential for refocusing and overhauling the current trends and practices of student assessment especially in an era dominated by globalisation and the information revolution. In this chapter, the pedagogical rationale has been outlined and thereafter an in-depth discourse has been provided as to the guiding principles for the incorporation of computer automated assessment into university teaching and learning.

Although computer automated assessment (CAA) is often touted as an antidote to the rising lecturer workload, there are various other reasons for the use of CAA within teaching and learning at higher education institutions. Whilst the key application of CAA is often argued as being in formative assessment, CAA often lends itself to various other forms of assessment such as summative evaluation, diagnostic testing, adaptive testing, portfolio assessment, group-based assessment, project assessment, and so on.

WHAT IS COMPUTER-BASED ASSESSMENT?

There are many terms, acronyms, concepts, and labels that are constantly being bundled around within the framework of computer based assessment. Acronyms such as computer assisted assessment (CAA); computer automated assessment (CAA); e-assessment (EA), online assessment (OLA), and computer based assessment (CBA) are often used within the vocabulary of individuals describing CAA. Although all of these terms refer in some way or another to the use of computers in assessment, Hogg (1997) underscores the need to distinguish if not differentiate between these concepts as each of them have fundamental differences. Hogg (1997) thus concludes that these terms cannot be used interchangeably as they will lead to inconsistency when referring to CAA. King (1994) also supports Hogg (1997) by stating that there must be a clear understanding of the various terminologies that are used interchangeably within the field of CAA as there are philosophical differences that are inherent in each of them. For example, King (1994) feels that there needs to be a clear distinction between CAA and CBA. King (1994, p. 213) argues that "CAA usually covers the use of computers in marking and administering optical mark reading cards while CBA on the other hand refers to the use of computers for the entire assessment process including delivery of the assessment, administration, management of the assessment, and the provision of feedback."

There is overwhelming consensus amongst educationists that computers can help with assessment in many ways; however, this article is limited to the consideration of "online assessment" of students. Online assessment specifically refers to the interaction between the learner and the computer within the bounds of assessment. Furthermore, this chapter given an in-depth looks into adaptation of 'online assessment' tools for improving educational quality and from this strategy not only academic institutes will be benefited, but also national education curricula implementation will be improved.

THE RATIONALE FOR CBA IN HIGHER EDUCATION

There are many reasons why students are being assessed and it is vitally important to consider these in the design of particular assessment strategies. Too often academics are drawn into repeating assessment patterns which they themselves have experienced or which they see in common use and assume that they must, therefore, nurture them. The syndrome of "one size fits all" is a common fallacy amongst academics evaluating and assessing student learning. However, CBA offers a very innovative, practical and fresh approach to this type of assessment.

Technology-enhanced strategies could be employed at various levels of assessment, ranging from the management of the assessment information to a fully automated assessment system. Using technology for the management of assessment information can enable information to be presented in different ways to meet the needs of different audiences (such as lecturers, students, course developers/organisers, and external examiners). The quality of presentation of assessment reports in terms of their range, content, frequency, and scope can be improved to a large extent by utilising technology for this purpose. At the other extreme, in a fully automated assessment system all aspects of the system ranging from the assessment task which the student completes to the processing and administration of the marks, including the overall management of the assessment information, can be completed with the utmost accuracy. Through the information management systems, CAA allows easy access and networking of institutions in terms of assessment practices.

Assessment strategies should be considered early, especially during the curse/module/ programme design phase. In this regard, Mansen (1996) states that it would be most useful to consider the following questions:

- What kind of outcomes do we want our students to achieve?
- What kinds of learning opportunities will be provided?
- What types of assessment task will be set?
- What methods/strategies of assessment will be used?

According to Taylor (1995), analysing the types of learning which is required to take place, has significant implications for the instructional design, assessment strategies, and methods. Sly and Rennie (1999, p. 21) in support of Taylor (1995), argue that since assessment involves "measuring and evaluation", it gives rise to problems such as in:

- Choosing a valid assessment instrument
- Finding a suitable unit of measurement
- Ensuring the test is measuring what it is supposed to measure
- Scorer reliability, especially if more than one marker is involved
- Using valid statistical methods and drawing valid inferences from measures
- Why is there a need to change the current practice of assessment in higher education?

Leddo (1996) maintains that there are various reasons for the need to change assessment strategies within most higher education institutions. He believes that globalisation coupled with the knowledge society is beginning to demand a new type of learner, that is a learner who will be able to develop competencies through-out his/her life and become an active citizen on the world stage.

Although not an exhaustive list, the following reasons could serve as a summative explanation for the need to restructure assessment types within the higher education framework:

- Drive and find effective new and improved assessment methods
- Focus on quality assurance
- Move towards student-centred learning instead of content-driven educator centred learning
- The changing aims of higher education (mix of academic, vocational, and general transferable skills)
- Emphasis on deep learning rather than shallow
- Interest in using technology-enhanced teaching and learning methods
- Need for greater cost-effectiveness to save staff time

THE ADVANTAGES OF USING CBA FOR FORMATIVE AND SUMMATIVE ASSESSMENT

Formative Assessment

With regard to formative assessment, CBA provides:

- Students to get immediate feedback which inherently keeps the activity and the result closely connected. This contrasts with typical feedback on paper and pencil essays and reports which always occurs much later than the writing itself and is thus divorced from it and seen perhaps as a separate activity by students (who inevitably focus primarily on the mark obtained).
- An immediate guide to student effort whereby lessons learned can be immediately put into action in the next piece of work/task.
- The diagnosis of problems in learning immediately. Staff can observe these problems and adapt teaching strategies to address them.
- Reflective student encouragement—students like to know how they are doing as fast as possible.

Summative Assessment

CBA has great potential in this direction in that many assessments can be run concurrently without any additional effort if they have already been developed for a dual formative-summative role. Deciding which of these functions is most important in your assessment is critical in designing effective and efficient assessment practices. This applies to all assessment but will strongly influence the way one decides on adopting CBA and how it will operate. It is obviously important that an individual element of assessment such as CBA should fit into the overall assessment strategy of the entire programme.

THE PRACTICAL RATIONALE FOR CBA IN HIGHER EDUCATION

A much more "hands off" approach to teaching and learning in higher education has been developing and will continue to grow in the next few years where there is a dire need for life-long learning. Taylor (1995) maintains that the emphasis on life-long learning within the context of a knowledge society is gaining widespread popularity and thus many higher education institutions are beginning to invest large sums of money on technology applications. Currently, the higher education sector of the world is grappling with dwindling funding for its existence as more and more governments are reducing their funding to tertiary institutions. Academics are being asked to teach more students with less money and even with the recent stabilisation of numbers the unit of resource is declining in real terms. This resource-driven rationale for CBA is the most widely accepted reason for its introduction in many individual modules, courses, and degree programmes. The increase in student numbers means that class sizes in first-year core modules can now easily reach 20,000-30,000 students in mega universities of the world. It is not possible to assess these numbers of students using the same traditional methods that were in place when the very same degree programmes had perhaps 50 to 80 students enrolled.

It is vital to be sure that if resource savings are the prime rationale, then the effort invested in development of CBA will pay off in the long run. Time is precious for academic staff when they are being asked to teach more students and do more and better research. Thus one of the prime motivations for developing CBA is to save time on tutor-based assessment.

Some Thoughts on How to Maximise the Time Savings

The potential time savings are attractive but the following parameters are need to be considered in order to maximise this benefit:

- **The timescale:** Considerable time savings are possible with CBA but are maximised over the longer term. Anybody thinking of developing a new CBA should first sit down and carefully and realistically evaluate how much time it will take to learn the software, write the questions and feedback, and sort out the practical difficulties of networking and obtaining computer time for students. On top of this the maintenance time once the system is up and running should be considered. The total time costs should be weighed up against the time saved on tutor-based assessment.
- **Are there any additional resources for development?:** Development costs can be very high for CBA. Many institutions and departments now have small funds, which can be used for assisting with this. It can be especially helpful to have someone assist with putting the questions on the system and providing graphics and so forth while some one else can concentrate on actually writing the text.
- **Changing course delivery and use of tutors:** Introduction of CBA may be part of a restructuring of the module delivery. It may be cost effective to employ tutors (typically postgraduate students) to help in the administration of CBA and other assessments.
- **Is there a wider departmental/institutional benefit?:** Development of materials for an individual module may not save time but if a broader departmental or even institutional strategy exists for CBA then the balance of the cost effectiveness equation may alter.
- **Are there existing assessments in the subject area?** The number of projects in different aspects of educational technology has grown enormously in the last 5 years. There are thus many existing assessments (question banks) which can be of direct use. This is obviously a major potential time saver.
- **Are there any question banks one could make use of?:** National or even international question banks are currently under consideration for some subject areas. Clearly this makes sense in disciplines where there is a large amount of subject material in common and bespoke assessments could then be constructed by question selection. The terms on which such question banks would operate are likely to vary but it seems probable that they will be available on payment by non-contributors and free to those who have contributed a minimum amount of material. The Geography, Geology, and Meteorology CTI centre is currently developing this idea. Given all the above considerations, a real-

istic cost-benefit analysis for developing a new CBA may give a negative answer and the real time savings for an individual will only be yielded over the longer term. In the light of this it is perhaps more realistic to stress the less well-advertised but far more important pedagogic rationale for using CBA in an assessment profile.

THE PEDAGOGIC RATIONALE FOR CBA

The focus on time savings which is often the main motivation for initial interest in CBA is sometimes an unfortunate distraction from the fundamental questions of why and what one wishes to assess. Many assessment practices in higher education (HE) are based on sound, but to some extent they are outdated principles. HE in South Africa, for example, until recently was geared towards the minority under apartheid and thus HE catered only for a small percentage of the population. Post apartheid has resulted in the masification for HE and thus university programmes were diversified and overhauled in many instances. It is within this framework that one ought to reexamine the types of assessment strategy that will employ within the university structure. In the past, assessments were designed to test a limited range of academic skills. Current modern university programmes are more diverse and need to cater for a greater diversity of student ability and produce graduates with a greater range of skills. Inherent is the change in the nature of degrees; there must be a corresponding change in the nature and type of assessment practices. Despite all this, academics take relatively little time to reevaluate what function they want the assessment to perform. Clearly CBA is likely to have considerable potential as a summative assessment process. Student responses can be regarded dispassionately and marked in an entirely consistent way. Assuming suitable questions can be designed, all the functions of summative assessment can be fulfilled with CBA. CBA is often perceived as only being useful for summative assessment of a wide range of low level skills in the curriculum of many degree programmes. While it is undoubtedly good at this, it is also becoming remarkably good at formative assessment, primarily due to technological advances which make assessments more interesting and interactive for students but also due to the fact that feedback is instantaneous, consistent, and infinitely repeatable.

THE SIGNIFICANCE OF CBA IN HIGHER EDUCATIION

There is a range of ethical values guiding design of good assessment outlined by Brown, Goma, and Gazzard (1996) which need to be considered. Computer based assessment has clear advantages over many other forms of assessment when some of these values are considered. Many of the applications of CBA are based on objective testing of one sort or another. They thus provide a much greater degree of reliability and equitability than almost any other form of assessment. The only variation in how different students are treated is as a result of the random choice of questions contained in many CBA systems. Otherwise there is exact repeatability at any time of day, with any number of repeat sittings. Computers don't get tired and they don't see or know individuals! Computer based assessment also contributes diversity to an assessment profile and thus contributes further to equitability, as different students perform better on different types of assessment. The nature of CBA also means that there is usually no effect of presentation or style on the judgment of quality so the validity is also assured. Of course this is one of the many limitations of CBA and a good reason why it is not suitable for every task but as long as it is an appropriate choice of skill to be tested and the questions are well designed, the computer will not be distracted by extraneous information. Finally, as discussed above, CBA may also be a highly efficient method of assessment particularly in the long term. Part of this efficiency arises from an ability to run repeated assessments at much more frequent intervals than many other forms of assessment normally allow. It thus makes timely assessment much easier.

FEASIBLITY OF CBA

So far, the general pedagogical rationale and advantages of CBA have been analysed. However, there are some significant limitations and drawbacks of using CBA which should be considered (Inoue, 1996), but that can be overcome or accommodated with a little forethought. These are mostly related to the provision of the infrastructure and equipment for the introduction of computers in assessment. Since these are often outside the immediate control of the individual, they will need to be resolved and acted upon at the departmental or institutional level if they are perceived to be serious enough to prevent development of computer assessments. There are key issues in assessment, which differentiates between the assessment system, or infrastructure, and the functions, types, and technology for delivery of assessment. In an individual situation, it is often the infrastructure which imposes the most significant obstacle to the introduction and development of new CBAs. Huges (1996) argues that special attention must be afforded to the following principles when considering CBA:

- Function of assessment
  - **Formative:** Course assessment
  - **Summative:** Diagnostic assessment
- Types of assessment
- The assessment process, for example, multiple-choice questions, use of pictures, problem solving, essays, practical work
- Technology
  - Using technology as a tool for assessment, for example, spreadsheet mode, hypertext "essay", and so forth
  - Using technology to assess, for example, delivering tests, marking tests
- Assessment system
- Collate/disseminate and analyse course and assessment information

STANDARDS

In the absence of a policy on standards, it is advised that Unisa adopt the IMS's question and test interoperability standard (IMS-QTI) for the CBA products that it wishes to acquire. It would be most valuable to adopt standards right from the onset as this would allow for conformance in attempting to develop CBA software either from open source or proprietary commercial products. It appears that the range of question formats is increasingly becoming more diverse and, in some cases, IMS standards have been extended in an ad hoc manner, suggesting the basic IMS standard is neither complete enough nor flexible enough for the needs of developers of the more complex products. This trend seems likely to continue, as IMS standards lag behind the products' capabilities which they aim to cover.

The Goals of IMS

- **Accessibility:** The ability to access instructional components from one remote location and deliver them to many other locations.
- **Interoperability:** The ability to use instructional components developed in one location with one set of tools on a platform in another location with a different set of tools or platform (there are multiple levels of interoperability).

- **Durability:** The ability to withstand technology changes without requiring redesign or recoding.
- **Reusability:** The flexibility to incorporate instructional design components in multiple applications.
- Restated in more concrete terms, these principles refer to the ability of a Web-based learning management system (LMS) to:
    - Launch content that is authorised by using tools from different vendors and to exchange data with that content.
    - Launch from different vendors the same content and exchange data with that content during execution.
    - Access a common repository of executable content and to launch such content.

IMS's Question and Test Interoperability Standard (IMS-QTI)

The IMS's Question and Test Interoperability standard (IMS-QTI) emphasises the following:

- To provide users with question items/banks regardless of virtual learning environment (VLE) deployed by the user
- To increase the ability to use question items/banks from various sources within a single VLE
- To supply support tools so as to develop new question items/banks in a consistent manner
- To increase the ability to report test results in a consistent manner

EQUIPMENT FOR WRITING CBA

Most individuals contemplating using CBA will have access to basic computing equipment but it is important to make sure that the:

- Hardware is capable of running software for designing CBA properly
- User can take full advantage of features such as the graphics facilities in order to make the assessments more attractive for students
- Uer have the right software for the creation and delivery of the assessments
- User is covered by the license to develop and deliver the assessments

EQUIPMENT FOR DELIVERY OF CBA

Most CBA will be delivered over a local network or, increasingly, the Internet. Some may be delivered locally on an individual machine but this is generally not very efficient except for small student numbers. Assuming the delivery is via a network, the following questions need to be asked:

- How much total access time (multiply students x number of assessments x average time for each assessment) will the assessments require?
- Are there sufficient computers available for this demand?
- Is all hardware comparable in performance (in timed assessments, it can be frustrating for students to experience different run times for tests!)?
- Do you have access to sufficient expertise to set up the network to deliver assessments, write the student answers to a suitable location and so forth?
- Do you have adequate control over the server where the assessments will be delivered? If the server is maintained by a section or service outside your immediate department, you may need to make sure they are aware of you and willing to provide the level of support you require.

INFRASTRUCTURE FOR LARGE-SCALE CBA

Besides the practical constraints of equipment, the infrastructure of the department, or more likely the institution, may present a serious impediment to wider development of CBA (Bull 1998). Heard (1997) constructed a protocol for the implementation of large scale usage of CBA for summative assessment, and he makes it clear that this is not a job to be done on a shoestring. Consideration of the facilities required for running examinations on a large scale is vital and the staff and support services must be in place and well organised for this to be successful. Good communication between academic and computing staff is essential and probably one of the weakest links in the chain in many institutions! However, the Universities of Aberdeen (Heard, 1997) and Luton (Zakrzewski, 1999; Bull, 1998) are good examples of institutions which run successful large scale computer examinations.

Gazzard (1998) maintains that it is probably not easy to implement measures such as the above on an individual or departmental scale without institutional backing. All of the issues considered here could be outlined as serious obstacles, but the key principle for success is to match the assessment design and ambitions to the resources available.

QUESTION CONSTRUCTION FOR CBA

It is already mentioned that CBA is more than just multiple-choice questions delivered on computer. Although these types of question do have a very important role to play in many examples of CBA, developments in software and in the ideas behind assessments on computers are producing a number of other approaches. This is probably one area where future advances in technology and software design will change the nature of CBA. This is not intended to be a manual on the detailed construction of CBA questions and there are a number of other sources of useful information on guidelines for writing objective questions (Lee, 1995; Partington, 1997; Jones, 1997) or for specific pieces of software (Heard, 1997). However, it is probably useful to highlight some of the key considerations here. In fact, although much can be learned from reading about the basic principles and from looking at other people's examples, perhaps the only successful way to develop question writing skills is to try it out. Only one can decide what to ask questions about and exactly how to ask them, although of course some trials with students and colleagues can be very worthwhile for guiding changes and additions once they are written.

QUESTION DESIGN

There is a common misconception that objective assessment is good at testing simple factual recall but not much else and many lecturers are indeed using it in this way. However, well-designed questions can do much more than this. Heard (1997) has based consideration of question design around Bloom's taxonomy which identifies six types of learning:

- Knowledge
- Comprehension
- Application
- Analysis
- Synthesis
- Evaluation

Objective tests are good at testing at least the first four of these and may occasionally be used for testing of synthesis and for evaluation too, although this can probably be done more effectively with other subjective assessments. It is vital to consider which of these skills the tutor wishes to test before deciding on an assessment strategy and certainly before writing the questions. However, there is a danger of encouraging surface learning strategies if a large proportion of the questions are simply testing factual recall (Scouller, 1994).

Multiple Choice

Multiple-choice questions are the simplest and best known types of objective assessment. The structure consists of a stem (question or incomplete statement) and a series of options, only one of which is correct. The other options are distracters, designed to provide a range of plausible but incorrect alternatives.

Up to five options are commonly offered and this is often a limitation on OMR-based tests, but with CBA you can provide as many options as you like. Equally, there is no reason why there should not be fewer options—true/false questions are the extreme of this. However, the marking scheme needs to take into account that with true/false questions guesses have a 50% probability of being correct: negative marking for incorrect responses is normally the best solution.

More complex question structures can be developed around multiple-choice questions, although it may be better to use multiple-response for such questions. If the tutor is limited to a multiple-choice structure, he/she can have a two-tier structure where the student selects a correct combination of options.

Assertion-reason questions can also be constructed and answered with a multiple-choice structure. These types of question tend to force greater thought and if well written, they can test much higher level skills than a basic multiple choice question. The normal phraseology of these questions gives five options:

- The assertion and reason are correct statements and the reason correctly explains the assertion.
- The assertion and reason are correct statements but the reason does not correctly explain the assertion.
- The assertion is correct but the reason is incorrect.
- The assertion is incorrect but the reason is correct.
- Both the assertion and the reason are incorrect.

In CBA, graphics can be incorporated into these types of question to test interpretation or identification skills.

Push Button

Push button questions are in effect a variant of multiple-choice questions, which computers make much easier to deliver than paper-based assessments. Instead of selecting an option by checking a box, the student actually clicks the answer itself. In principle, this could be a graphical representation as well as text but most software is limited to the use of text.

Multiple Response

Multiple-response questions have a similar structure to multiple-choice questions with a stem and a range of options but the student can select one or more responses. There are parallels of the two-tier multiple-choice question but multiple-response questions have the advantage that they can allocate marks for individual answers. For example, if there are three out of seven correct answers, then there is a possible mark of 3 maximum, but students may score 0, 1, or 2. With the two-tier multiple-choice structure, students can only be right or wrong even though they may have only part of the answer incorrect. Most software will permit either unlimited or limited numbers of choices. If unlimited choices are permitted then obviously there needs to be a penalty for incorrect answers. Likewise, students may or may not be told how many correct responses there are.

Hotspot Graphics

Graphical interaction is where CBA really has major advantages over other forms of assessment. The simplest approach is to have a graphic (which could actually be text as a graphic if necessary) where the student has to select an area or item in response to a question. The correct response area is defined by the question designer in advance but its definitions are hidden from the student. This often incorporates a moveable pointer which is picked up and placed by moving the mouse.

Text Matching

Questions involve completing a sentence or free entry in response to a question. The response is marked using pre-defined search strings. For example, if a question asked, 'Who is the president of the South Africa?,' the answers Thabo Mbeki, Mbeki, and Thabo might all be acceptable, so the string to search on would be 'Mbeki'. One might even consider it acceptable to have 'mbeki', in which case the answer could be made case insensitive. These question types need very careful design and testing, if they are to be used extensively in summative assessments.

Numeric

Numeric questions ask the student to enter a digital response to a question. They are often useful for assessment of end results or stages in a calculation series previously carried out. Again they need fairly careful design to ensure that rounding errors or precision does not affect the results. If in doubt, the precision required should be specified. Most software will accommodate ± ranges in recognition of these difficulties.

Other Question Types

The types of question discussed so far are standard approaches for CBA but there are a number of other more unusual techniques which have shown their worth in some circumstances. The best of these represent fruitful experimentation in new methods of CBA. Some further details on these are mentioned elsewhere but it is worth outlining some possibilities here to illustrate the broad potential of CBA.

- **Qualitative responses and self assessment:** This approach relies on qualitative input from the user and then uses that to decide on an overall grade.
- **Problem solving:** Problem solving is a difficult skill to assess automatically. It involves incremental steps and a consideration of the efficiency of a solution. The Mastertutor approach is a good example of how CBA can help with such complex assessments. Hughes (1996) also describes an assessment package which incorporates elements of problem solving and Leddo (1996) has developed a computer-based game for teaching scientific reasoning, although this involves no major elements of assessment.
- **Assessment of long text answers:** Text and numeric response questions can be longer than a few words but they will rarely go beyond a sentence. Marking computer programming is an example where much longer text responses can be assessed automatically.
- **Clinical decision making:** Computer simulation in teaching clinical skills often requires the use of multimedia. Testing of case-based clinical decision making has also been implemented by Scouller (1994).

QUESTION REVIEW

It is essential to review the questions after they have been written and during their use by students. The compilation of a decision tree is a useful initial way of reviewing questions and deciding whether they need changing. Equally, writing feedback to questions often points out problems with the question design itself. To counteract the above problems in the question review process, Brown et al. (1996) argue that the following principles should be adhered to:

- Ask colleagues to participate in the pilot of the questions to discover any initial errors or misleading questions.
- Check that distracters in multiple-choice questions are plausible.
- Try the test out on a group of students where the summative element is small or zero.
- Encourage students to respond to you on questions and feedback. If the CBA is networked, students can email responses directly to you while they are actually doing an assessment. Contrary to the above, an element of summative assessment encourages a better student response on the problems with the assessment, as they have an incentive for telling you if they think a question is wrong or misleading.
- After running an assessment, do at least some question analysis. Are there any questions that every student gets wrong? Are there particular distracters that almost all the students are choosing?

WRITING FEEDBACK

Provision of feedback is one the biggest advantages of CBA over paper-based objective tests. However, it has to be well thought out and useful without being long-winded. Similar considerations apply to any feedback to students but the following are useful things to bear in mind with CBA feedback:

- Make it clear whether the response is correct or not.
- Keep it as positive as possible—even if an answer is not correct, give credit for being nearly right and do not give the impression that the response given was completely idiotic. On the other hand, you do not need to be excessively effusive for correct answers unless they are the really difficult questions.
- Give feedback on correct responses too. Some answers may have been arrived at by the wrong reasoning or even by chance—even if the reasoning and answer were both correct, feedback can reinforce this.
- Make sure there is sufficient time for reading feedback. Formative assessments should not be strictly time limited if you are expecting students to make good use of feedback.
- Consider using graphics in feedback to make them more interesting.
- Explain why a wrong answer is not correct. Give the correct answer and explain how to derive it. If the response was only slightly wrong, try to suggest where the student could have gone wrong.
- Avoid over jovial responses but do keep them reasonably simple and friendly.
- Use feedback to adjust the possible responses. You may need to change the question once you come to write the feedback!
- Consider the timing of feedback—will it be given after each question or at the end of the assessment?
- This may affect the detail and form of feedback. Instant feedback on each question can be very effective but you need to bear in mind the time it takes for students to read and digest it.
- Give pointers to further learning opportunities and information such as page references in textbooks, course materials, and journal literature.

CONSTRUCTION OF THE ASSESSMENT: SOME CONSIDERATIONS

In some ways the overall design of CBA can be more complex than that for other assessments such as paper-based objective tests, simply because the range of options is very much larger. Computers are capable of selecting questions, managing student access times, and gathering and storing the information in many different ways. The widespread use of local networks and the World Wide Web means that this potential is now much greater than it ever has been, and there are therefore some fundamental aspects of assessment design which should be considered before writing and implementing a test.

The Random Element and Question Banks

Paper based assessments are fixed in advance, although it is possible to generate several different versions of an assessment either manually or in a semi-automated OMR system. Computers, however, can generate and deliver different sets of questions to different students or to the same student on different occasions. The variety of combinations depends on the size of the question bank from which the questions are drawn. Even a relatively small (say 2-3 times the number of questions delivered in any one test) question bank gives a good degree of variability. Randomisation of question order and of possible responses within multiple-choice and multiple-response questions gives a further degree of variability so that this is a very worthwhile feature even if the assessment delivers a fixed set of questions.

Libraries and Structuring Assessments

Randomisation of questions and answers is a tremendous advantage of CBA but it can lead to problems. If the question bank is large, there may be students who by chance only get assessed on a restricted part of the syllabus. One way to avoid this is to use libraries of questions within an assessment. Libraries are subsets of the full question bank, each of which contains a group of related questions. An assessment of 20 questions may be structured to take five questions randomly from within each of four libraries instead of taking all the questions randomly from the entire undivided question bank. More complex structures can be produced with some compulsory sets of questions and some randomly chosen questions. Here are some tips on structuring assessments and use of libraries:

- Think about what you would ideally like to do and then find out whether the software will allow this structure or a modification of it. Many poor CBAs are the result of restrictive thinking at the basic design stage.
- Build up question banks slowly. It is difficult to generate large numbers of good questions in one sitting. As the question banks grow, they can be subdivided into libraries.
- Think about the balance of the assessments. Are you assessing the full range of material? Are there areas that it is more important to assess than others? Do you need to ask all students one or two very important questions? Make sure your structure meets the needs of the curriculum and the intended learning outcomes.
- Have a look at other assessment structures. It helps to see how other people have approached this aspect. Beware though—not all assessments make use of these facilities as much as they might, so think about how they could be changed as well!
- Good control systems for access to question files are obviously important for ease of use as well as security. Perhaps, more important, however, is the careful management of answer files generated by students as these form

a resource not only for the summative scores but also for potential feedback to students themselves should they have difficulty in understanding where they have gone wrong from onscreen feedback or if they want to query results.

Again the network can facilitate all these activities by:

- Allowing students read access to question files when they need it but not to be able to corrupt or misuse these files
- Directing the "traffic" of incoming information to students from question banks and the outgoing answers from students to different locations if necessary
- Subsequent compilation of results and generation of data on student performance and use of questions provided as revision resources

ASSESSMENTS AS LEARNING RESOURCES

CBA has often been seen as an effective mechanism for large-scale summative assessment of students—for example the development of CBA examinations at Luton and Aberdeen (Heard, 1997; Zakrzewski, 1999; Bull, 1997). It can undoubtedly be used effectively in this way, but it has even greater potential for use as a learning resource, perhaps used for mixed formative/summative assessment or simply as formative assessment. However, if there is no summative element it may be difficult to achieve high usage by students. This has been partly covered by earlier sections, but this is an important consideration when designing assessment structures.

MANAGING ACCESS, TIME, AND PEOPLE

The widespread networking of computer systems in education has resulted in an additional level in the application of CBA. Stand-alone workstations can be a worthwhile way of delivering CBA to small numbers of students in closely controlled environments. Indeed, psychometric tests (where many CBA ideas initially sprang from) are still often delivered this way. However, networking allows close control of student access time and place automatically, as well as making it much easier to manage results and the assessments themselves. Using the network to manage access can be especially useful for:

- Controlling open access assessments which are essentially formative. Students can be limited to access times of a week or so, on a series of sequential assessments to help nudge them towards a balanced pattern of working.
- Overcoming limitations on workstation availability, by scheduling different sets of students at different times
- Allowing different kinds of access to students at different times. For example, assessments may be made available on a restrictive one-off basis for summative assessment, followed by open unrestricted access later in the module for revision purposes.
- Improving security, because it is a potential problem in allowing open and/or remote access to assessments. How can you be sure that students are not sitting the assessment for one another, even with password protection? If this is a major concern then supervised access is the only option (as in formal examinations). Clearly, it may be a difficulty, but it is certainly no more of a problem than dishonesty in conventional coursework assessments. Even if the main aim is to develop summative assessment,

consider using CBA to help prepare the students for the assessments. If you design a large enough question bank then it could even be made open access so that students can use it for revision.

MARKING SCHEMES

According to Gazzard (1996), any summative objective test must have a marking scheme which ultimately yields scores on a comparable scale to those being used by the institution to grade students. Objective tests frequently lead to a greater range of marks and since conventional marking schemes in the UK tend to bunch between <40% (fail) to >70% (first class) this can lead to problems. It is not such a contentious issue for pass/fail modules but it may still be difficult to design a marking scheme which yields appropriate final results.

DELIVERY OF CBA: CHOOSING THE SOFTWARE

It should be borne in mind that software is developing quite rapidly and further innovations are likely in the future, which will considerably expand the capabilities of CBA. The Internet is increasingly successfully used for delivery of assessments (Taylor, 1995; Dickinson, 1997) and offers great future potential as do other developments such as videodisc technologies (Mansen, 1996). However, although the Web is gaining its popularity as a mode of delivery for CBA, but it is not always fully taken advantage of and thus Littlejohn (1999) argues that appropriate steps must be taken to encourage HE academics to include all the possibilities of CBA when designing courses, modules, and programmes. The more the academics being to employ CBA, the more the costs will decrease.

Selection of Appropriate Software

Bull (1998) maintains that there is an increasingly wide range of software options which are available for developing CBA. Making a decision as to which software to use will probably depend on issues such as the following:

- **What is to be assessed?** Decide whether the assessment should be formative or summative or both and what the feedback requirements should be.
- **Group size:** How many students will there be and does this affect how/when the CBA will run?
- **Mode of delivery:** For example CD-ROM, floppy disk, Web, and so forth. Thought needs to be given as to how and when the assessment is to be undertaken. For example, it may be that you want the test to run at a prescribed time or for it to be available during a flexible time slot.
- **Security:** Decide on the level of security. For example, you may want password protection. Additionally, you will probably need to store question and answers files at differing places on the file server or drive directory.

Advantages

Internet based CBA methods have the following advantages:

- **Access time:** Can be accessed at any time so students are not dependent on University timetables or opening hours.
- **Access location:** Can be accessed from any location with an Internet connection. It is, therefore, especially useful for distance learning. It could also be seen as a means of reducing pressure on University resources.
- **Flexibility:** The Web is platform independent. The same CBA can be accessed

using PCs, Macintosh, or other types of computer.

Disadvantages

However, they may have some disadvantages:

- **Dependent on external connections and networks/hardware:** Local Networks may be more stable than external connections and networks.
- **Access control:** it is potentially harder to control access.
- **Security:** Due to potential access, security may be harder to ensure.
- **Available expertise:** What particular areas of expertise are already available (computer programming, networking skills, etc.) for developing and supporting the CBA?
- **Resources and facilities:** Rooms, networked PCs, Web access, program software, and so forth.

GENERIC SOFTWARE

With the services of an experienced programmer (this could be expensive if it requires to recruit or buy from someone the required program) a generic program language such as 'C,' 'C++,' 'FORTRAN,' 'COBOL,' or 'Basic' could be used to develop CBA. However, in most cases this would be rather like hitting a small pin with a sledge hammer! Microsoft's 'Visual Basic' is useful for less experienced programmers, of which the latest releases could arguably be considered authoring software. Recently available software like Java, J2EE, and php, which are open source and may be taken into consideration while developing CBA.

AUTHORING PACKAGES

An approach that many developers of CBA have chosen is to make use of authoring software (often used for creating multimedia applications). Most of these programs enable to build applications in an environment where text, graphics, audio, and video can be placed into an interface whose properties can be set by program coding. Manufacturers often claim that their authoring software products are easy to use and that "virtually" no training is required; applications are relatively quickly created by clicking icons and/or by cutting and pasting from "script libraries."

Creating a CBA application using an authoring software program usually follows a similar process:

- Design and create the interface using a toolbox provided (usually a set of icons which enable the user to draw, add boxes, push buttons, images, text, etc.)
- Set properties of the interface (give a value to a checkbox, provide a name to a box that may be used to enter text, etc.)
- Write the code to respond to the user's action (when the user clicks the mouse over a checkbox, etc.). As mentioned earlier, these program codes are often already written or can be assembled by clicking icons. However, regardless of what the manufacturers say about the need for virtually no training, some understanding of the principles of computer programming is an advantage especially for developing a sophisticated CBA, for example, opening up other external applications, or playing video or audio, and so forth.

Many authoring software programs are able to produce applications which can be used on the Web (mostly by the use of 'plug-ins'—a program usually available by downloading from the Web which allows Web browsers to run certain applications). Not only does the Web provide a wide delivery system but also provides advantage so that end-users do not have to use the same computer system platform (in other words a CBA developed on a PC version of the authoring software can often be used on a Mac or UNIX system over the Web). There is a growing number of authoring software tools on the market, many of which have an educational version. Probably the most familiar are Asymetrix's "IconAuthor" and "Toolbook" along with Macromedia's "Authorware". These programs are expensive, but the manufacturers claim that the pay-off comes in the time saved by easy development for non-experienced users. Asymetrix also market "CBT Express", which is a considerably cheaper product and is aimed specifically for training and educational institutions. This application has no programming or scripting language to worry about; applications are developed from template libraries or by the use of "Wizards", for example to create exercises or build tests. A few such applications are given below:

- **Asymetrix:** http://www.asymetrix.com
- **Macromedia:** http://www.macromedia.com
- **Digital Workshop 'Illuminatus':** http://www.digitalworkshop.co.uk
- **Linotype-Hell 'Dazzler':** http://www.linotype.co.uk
- **Matchware 'Mediator':** http://www.mwin.com
- **General:** http://www.mcli.dist.maricopa.edu/authoring/.

HARDWARE REQUIREMENTS

Depending on the intensity of the multimedia materials (graphics, video, audio, etc.) , for the development of CBA using authoring software it probably needs at least 16Mb of RAM, a Pentium processor, a 2Mb graphics card, and a fast hard disk of at least 1Gb capacity. For delivery of CBA the requirements may not need to be so high. However, conflicts can often occur especially when using a variety of differing specification end-user machines. The best advice here would be to test out the CBA exhaustively on all computers and networks (if appropriate and where practical) that are intended for use in delivery before final release of the application.

OPEN SOURCE SOFTWARE (OSS) / FREE SOFTWARE

This section deals with several open source software that are used to develop CBA.

CASTLE (Computer Assisted Teaching And Learning) http://www.caacentre.ac.uk/

This is an excellent free package to begin with CAA for formative assessment. CASTLE is a funded project which provides an authoring shell for the delivery of Web-based MCQs. Users need only minimal IT skills (being able to use a browser) and need not know any other CGI, HTML, or other scripting or markup language.

Clyde Virtual University Assessment Engine: http://www.caacentre.ac.uk/

CVU offers free tools, which enable the delivery of online assessments and evaluation forms. The assessment engine allows users to design Web-based tests and the assessment system displays tests to users, marks the responses, and offers immediate feedback. Question types available include multiple choice, multiple response, free text entry, and list ordering.

TRIADS: http://www.caacentre.ac.uk/

The University of Derby/Tripartite Interactive Assessment Delivery System is a toolkit for users of Authorware Professional to develop CAA using a wide range of sophisticated question types. The site contains an impressive demonstration of TRIADS materials.

WWWAssign: http://www.caacentre.ac.uk/

A free tool for the delivery of Web-based assessments, WebAssign, is developed by Larry Martin of North Park University, Illinois.

WebTest: http://www.caacentre.ac.uk/

WebTest (Heriot Watt University) is a system for creating, delivering, and marking Web-based assessments. Features include the randomisation of questions, ability to display mathematics and scientific formulae, and the creation of diagrams and graphs on the fly.

EXAMINE: http://www.caacentre.ac.uk/

Enables users to develop CBA for delivery with Windows. A range of question types is available, including those which incorporate graphics and multimedia. The software, which is free to HEIs, was developed under the United States Information Training Technology Initiative (USITTI), which has now ended.

COMMERCIAL PROPRIETARY CBA SOFTWARE

There are a number of commercially available software programs designed specifically for computerised assessment, of which Question Mark's "Designer" claims to be the world leader (King, 2001). The popularity of these packages suggests that they strike the right balance for many users since they offer considerable flexibility of assessment design.

QuestionMark: http://www.caacentre.ac.uk/

The most widely used commercial provider of CAA software in UK is HE. Visit the site for a demonstration of QuestionMark Designer (Windows) and QuestionMark Perception (Web-based).

WinAsks Professional: http://www.caacentre.ac.uk/

Software for the development of secure tests and questionnaires delivered using Windows 3.1 or higher.

WebMCQ: http://www.caacentre.ac.uk/

WebMCQ is an assessment service based in Australia, which hosts and maintains Web-based tests.

EQL: Interactive Assessor

EQL was developed as a system for creating question banks, generating customised tests, running tests, and providing automatic results analysis.

Hot Potatoes Web Authoring Tools

The Hot Potatoes suite is a set of six authoring tools, created by the Research and Development team at the University of Victoria (Australia) Humanities Computing and Media Centre.

CyberExam

CyberExam was originally constructed by Virtual Learning Technologies as an enabling software tool that allows schools, universities, and corporations to create online testing instruments that will automatically grade and analyse results.

Quiz Factory 2

Quiz Factory 2 has been developed by the LearningWare Inc. Quiz Factory 2's powerful, easy-to-use interface allows tutors to quickly design and save a quiz with sounds and movies, and to present formative assessment.

Test Pilot

Test Pilot has been designed to provide for the easy creation and deployment on online formative assessment using the latest Internet technologies. It has been vigorously time tested in the most demanding of corporate education, national university, primary, and secondary educational environments.

STRATEGIC PLANNING ISSUES

Assessment policy

It is recommended that every institution of higher learning must develop an assessment policy. The policy should spell out clearly the framework within which assessment design, assessment strategies, frequency of assessment, types of assessment, and so forth, that academics must adhere to. In essence the assessment policy must be aligned to the tuition policy of the institution.

Learner Management Systems

Institutions have to make judicious selection of which learner management system (LMS) to use as a platform for assessment. Choices have to be made with regard to proprietary and non proprietary learner management systems. The choice of a learner management system must be based on the ease of navigation, the ease with which information can be archived, and also the ease with which the functionality can be used by the user. In essence the LMS must allow for multiple assessment strategies such as portfolio assessment, multiple choice assessment, essay type assessment, and many more.

Working in Teams

There is no doubt that academics have to work with teams in order to develop sound computer automated assessment systems. Academics will provide the pedagogical inputs in so far as content to be assessed while ICT specialists will design the appropriate system and interface for the assessment process/event.

Budgeting

There has to be scope for prior budgeting for computer based assessment as in all other forms of assessment. In many instances budgets have to cater for licensing of software, hiring of computer programmers, instructional designers for e-Learning, and others.

CONCLUSION

Current assessment procedures in higher education are long overdue for a rethink. They are particularly ill suited to the digital age in which using information is more important than remembering it, and where reusing material should be viewed as a skill to be encouraged, not as academic plagiarism to be despised. Many online courses are leading the way in devising assignments and assessment procedures which reflect the call for HE to embark on lifelong teaching and learning within an outcome-based framework. This approach supports the call for developing competencies and skills among learners who will be able to participate in society as informed and critical citizens. The issue of critical thinking, working in groups, team work, peer supported learning, and so forth are all relevant to the new knowledge society that is currently besieging the educational landscape.

Computer based assessment offers enormous possibilities and it is hoped that HE institutions will start to appreciate the various advantages by adopting it into their assessment strategies. Academics can rest assured that the time-consuming task of pencil and paper marking can be radically diminished if they can incorporate CBA into their course/module design. Of course, the initial time outlay may appear to be great, however, the long-term time saving can be a huge positive factor that must be considered. The argument that initial costs in developing software for CBA is huge, is with the advent of open source software, proving to be false and thus one cannot hide behind the issue of cost.

While academics can play their role in assimilating CBA, it is really left with the decision makers within the HE institutions to drive the awareness campaign in terms of the significance and advantages of integrating CBA into the degree, diploma, and certificate programmes. University administrators and managers ought to realise the potential that CBA offers and thus begin to appreciate the cost saving in terms of the employment of tutors for marking assessments. This position chapter provided an in-depth rationale for the adoption and integration of CBA into the assessment policies of HE institutions and thus it is hoped that more thought would be afforded to this ever growing field of CBA. The ease with which students can now submit assignments electronically and take self-tests and even take examinations online, has led many institutions to exploit the technology to globalise their courses and in some cases to relieve tutors of the more tedious aspects of marking. Customised CBA software for education offers easy access to the creation and design of multiple-choice tests, short answer tests, essays, and so forth. Finally, the advantages of CBA could be summarised very briefly as:

- Affording the possibilities of undertaking more frequent formative and summative assessment
- Allowing staff to be alerted sooner so as to adapt their teaching
- Reducing the amount of time spent on marking
- Allowing for self-assessment whereby students can undertake assessment tasks in their own time, at their own pace and when and where they are ready
- Increasing student confidence, resulting in reduced costs—fewer tutors can be employed for marking
- Allowing for synchronous feedback which students always welcome

FUTURE RESEARCH DIRECTIONS

Having investigated the policy frameworks, implementation plans and advantages of computer based assessment in academic evaluation, the researcher believes that the following topics are worthy of consideration for future research.

Mainstreaming and Inclusion

The researcher concurs with Taylor (1995) when he argues that not much has been written about the impact of computer based assessment on students with learning difficulties and students with disabilities. There is currently a strong shift towards the inclusion of students into mainstream teaching and learning institutions and thus research should be undertaken to investigate how such students cope with "on demand" computer based assessments. Topics for future research should include:

- How do students with disabilities handle computer based portfolio assessment?
- What pedagogical and infrastructural support do students with learning difficulties and disabilities need to undertake computer based assessment?
- How do students with disabilities respond to "on-demand" performance based computer assessments?
- How do students with disabilities cope with the completion of projects within a specified time frame based on computerised assessment?
- How should performance and assessment standards be or not be adjusted for students with disabilities?

Impact of Computer Based Assessment on Teaching Strategies and Student Learning

Way back in 1999, Mansen (1999) argued that further research must be more specific in terms of evaluating the extent to which particular computer based assessment formats promote the acquisition of particular skills and knowledge. Unfortunately, Mansen's (1999) concern still persists today. The researcher believes that future research must investigate whether the acquisition of certain skills and knowledge precludes the acquisition of certain other skills and knowledge within a given domain applying computer based assessment formats. Indeed, research of this nature must be tailored to specific computer based assessment systems that are carefully chosen to represent different formats but within the same content areas. In the same vain research priorities within computer based assessment must focus on the development of models of assessment that cater for cross disciplinary assessment practices within higher education academic evaluation.

Professional Development and Support

Professional development and support needed to argument academic's expertise in applying and designing new and different types of computer based assessment applications should be given priority in future research. There is currently a high proliferation of both open source and commercial/proprietary computer based assessment software. It is therefore incumbent that research efforts be geared towards the investigation of skills, competencies, and paradigms that will enable academics to design and implement appropriate computer based assessment applications specific to their contexts.

The Relationship Between Long Term Benefits and Long Term Costs

The researcher believes that future research should focus on establishing workable funding models within which frameworks could be teased out on investigating relationships between long term costs and long term benefit analyses of engaging in computer based assessment practices. While it may prove to be difficult to pin point exactly the type of funding framework that could be used generically amongst educational landscapes of the world, it nevertheless may highly basic principles of funding that must be taken into account when undertaking computer based assessment.

REFERENCES

Bennett, R. (1998). *Reinventing assessment.* Princeton: Educational Testing Service.

Brusilovsky, P., & Miller, P. (1999). *Web-based testing for distance education.* New York: Kogan Page.

Bull, J. (1997). *A glimpse of the future in computer-assisted assessment in higher education.* London: Kogan Page.

Brown, J.R., Goma, G., & Gazzard, S. (1999). *Assisting student learning using Web-based assessment: an overview of the WebMCQ system*. Poster presentation at the 15th Annual Conference of the Australasian Society for Computers in Learning in Tertiary Education. Retrieved December 01, 2002, from http://www.webmcq.com/

Burstein, J.C., Wolff, & Lu, C. (1999). *Using lexical semantic techniques to classify free-responses. The depth and breadth of semantic lexicons*. London: Pergamon Press.

Burstein, J., Kukich, K., Wolff, S., Chi Lu, Chodorow, M., Braden-Harder, L., & Harris, M. (1998). *Automated scoring using a hybrid feature identification technique*. Retrieved March 21, 2003, from http://www.webmcq.com/

Dickenson, J. (1997). *Using WebMCQ for formative and summative assessment*. Proceedings of university science teaching and the Web workshop. Sydney: Uniserve Science. Retrieved December 12, 2002, from http://www.webmcq.com/

Gazzard, S. (1998). *ComDesign principles for next wave software: The development of the WebMCQ system*. Proceedings of the Fifteenth Annual Conference of the Australasian Society for Computers in Learning in Tertiary Education. Retrieved November 23, 2002, from http://www.webmcq.com/

Heard, B. (1997). A correlation of mutiple-choice and essay assessment measures. *Research in Education, 56,* 73-84.

Hogg G. (1997). About multiple choice questions, multiple choice testing: Green Guide, 16, *The Higher Education Research and Development Society of Australasia*. Cambeltown: Associated Press.

Huges, TK. (1996). An introduction to latent semantic analysis. *Discourse Processes, 25*(2), 259-284.

Inoue, A. (1996). *Developing and integrating a Web-based quiz into the curriculum*. Proceedings of WebNet97, World Conference of the WWW and Internet. Toronto: Association for the Advancement of Computing in Education.

King, G. (1994). *Developing adaptive tests for school children*. NJ: Lawrence Erlbaum Associates.

Leddo, H.J. (1996). *Computer-based features of the junior astronomy course at the University of Sydney*. Proceedings of University Science Teaching and the Web Workshop. Sydney: Uniserve Science.

Littlejohn, A. (1999). The virtual university as a conceptual model for faculty change and innovation. *Journal of Interactive Learning Environments, 7*(2/3), 209-226.

Mansen, J. (1999). *Innovations in computerised assessment*. NJ: Lawrence Erlbaum Associates.

Robinson, J. (1999). *Computer-assisted peer review*. New York: Kogan Page.

Scouller, A. (1994). Test writing on computers: an experiment comparing student performance on tests conducted via computer and via paper and pencil. *Education and Policy Analysis Archives, 5*(30), 21-25.

Sly, L., & Rennie, L. (1999). Computer managed learning as an aid to formative assessment in higher education. In S. Brown, J. Bull, & P. Race (Eds.), *Computer-assisted assessment in higher education*. London: Kogan Page.

Taylor, S. (1995). *Handbook of modern item response theory*. New York: Springer-Verlag.

Whittington, C.D., & Sclater, N. (1998). Building and testing a virtual university. *Computers and Education, 30*(1/2), 41-47. Retrieved February 21, 2003, from http://i.am/davewhittington/

Zakzewski, M. (1999). A linguistic perspective on multiple choice questioning. *The Higher*

Education Research and Development Society of Australasia Newsletter.

ADDITIONAL READINGS

Banta, T., & Associates (2002). *Building a scholarship of assessment.* San Francisco: Jossey-Bass.

Barker, K. (2002). *Canadian recommended e-learning assessment guidelines (CanRegs).* Retrieved May 20, 2004, from the LICEF Research Center Web site http://www.licef.teluq.uquebec.ca/fr/ pdf/CanREGs%20Eng.pdf

Benjamin Franklin Institute of Global Education (2001). *Distance education: Quality assurance initiative.* Retrieved May 5, 2004, from http://www.academyWeb.com/DEQA

Bennet, R. (2003). Using electronic assessment to measure student performance. *Issue Brief, NGA Center for Best Practices,*

Bennett, B.R. (2001). How the Internet will help large-scale assessment reinvent itself. *Education Policy Analysis Archives, 9*(5).

Brennan, J., & Shah, T. (2000). *Managing quality in higher education: An international perspective on institutional assessment and change.* Buckingham: Open University Press.

British Association for Open Learning. (2002). *The development, implementation and use of the BAOL Quality Mark: A report to the Department for Education and Skills.* Retrieved May 5, 2004, from http://www.baol.co.uk/PDF/qmrepdfes.pdf

Brown, R. (2000). The new UK quality framework. *Higher Education Quarterly, 54*(4), 323-342.

Carnevale, D. (2000). Assessing the quality of online courses remains a challenge and educators agree. *Chronicle of Higher Education,* p. A59.

Carnevale, D. (2002). Accreditors offer views on distance programs. *Chronicle of Higher Education,* p. A36.

Caudron, S. (2001). Evaluating e-degrees. *Workforce, 80*, 44-48.

Cleary, T. S. (2001). Indicators of quality. *Planning for Higher Education, 29*(3), 19-28.

Hooker, M. (1997). The transformation of higher education. In Diana Oblinger & Sean C. Rush (Eds.), *The learning revolution.* Boston: Anker Publishing Company, Inc. Retrieved May 5, 2004, from http://horizon.unc.edu/projects/seminars/Hooker.asp

Kingsbury, G.G., & Hauser, C. (2004, April). *Computerized adaptive testing and No Child Left Behind.* Northwest Evaluation Association

Kompf, M. (2001). ICT could be the death knell of professoriate as we know it. *CAUT-ACPPU Bulletin Online.* Retrieved May 5, 2004, from http://www.caut.ca/english/bulletin/2001_sep/commentary.asp

Lairson, T. D. (1999). Rethinking the "course" in and online world. *Campus-wide Information Systems, 16*(5). Retrieved May 5, 2004, from http://fox.rollins.edu/~tlairson/online.html

Milliron, M. D., & Miles, C. L. (2000). Education in a digital democracy: Leading the charge for learning about, with and beyond technology. *EDUCAUSE Review*, 50-62. Retrieved May 5, 2004, from http://www.educause.edu/pub/er/erm00/articles006/ perm0064.pdf

Noble, D. F. (2001). *Digital diploma mills: The automation of higher education.* New York: Monthly Review Press.

Pascarella, E. T. (2001). Identifying excellence in undergraduate education: Are we even close? *Change*, 19-23.

Phipps, R. A., & Merisotis, J. P. (2000). *Quality on the line: Benchmarks for success in Internet-based education.* Retrieved May 5, 2004, from the Institute for Higher Education Policy Web site http://www.ihep.com/Pubs/PDF/Quality.pdf

Pond, W. K. (2002). Distributed education in the 21st century: Implications for quality assurance. *Online Journal of Distance Learning Administration, 5(2).* Retrieved May 5, 2004, from http://www.westga.edu/%7Edistance/ojdla/summer52/pond52.html

Quinn, J. B. (2001). Services and technology: Revolutionizing higher education. *EDUCAUSE,* 28-37. Retrieved May 5, 2004, from http://www.educause.edu/ir/library/pdf/erm0141.pdf

Rabinowitz, S., & Brandt, T. (2001). *Computer-based assessment: Can it deliver on its promise?* Vancouver: Kogan Page.

Russell, M., & Haney, W. (2000). The gap between testing and technology in schools. *Statements, 1,* 2, National Board on Educational Testing and Public Policy, January 2000.

Sims, R., Dobbs, G., & Hand, T. (2002). Enhancing quality in online learning: Scaffolding planning and design through proactive evaluation. *Distance Education, 23*(2), 135-148.

Widrick, S. M., Mergen, E., & Grant, D. (2002). Measuring the dimensions of quality in higher education. *Total Quality Management, 13*(1), 123-131.

Section III
E-Concepts

Chapter XI
Creating Competitive Advantage:
The Emergence of a New Business through Collaborative Networks – An Empirical Case Study in the ICT Sector

Arla Juntunen
Helsinki School of Economics, Finland

ABSTRACT

This chapter addresses collaborative business networks at the level of industry/cluster networks, which is important and relevant from the strategic management perspective in several industries. This chapter examines two current "hot topics": the emergence of a new Internet driven business group, and its construction through a series of intentionally developed collaborative networks. Networks are seen to offer firms collective benefits beyond those of a single firm or market transaction. The author of this chapter aims to contribute to the development of the emerging theory of network management by integrating notions from the industrial network approach and the resource-based view of the firm, in particular its extensions into dynamic capabilities and learning. The starting premise is that the characteristics of the task that organizations try to accomplish through forming a specific collaborative network influence, the management of that network, and thus, the capabilities developed and required. This chapter is based on a longitudinal case study in the ICT-sector.

INTRODUCTION

This study examines two current "hot topics": the emergence of a new Internet driven business group, and its construction through a series of intentionally developed *collaborative networks*. This is a highly relevant topic from a number of perspectives. First, relatively little academic research into so-called *e-business* has been published, in spite of its huge impact on economy (Amit & Schott, 2001). Many Scholars argue that network organizations are expected to take the leading role in economic and social innovations in the world of increasing globalization, connectivity, and knowledge intensity (Castells, 1996; Grabher, 1993; Jarillo, 1993; Parolini, 1999; Thompson et al., 1994). Second, in spite of the few seminal studies (Alajoutsijärvi et al., 1999; Lundgren, 1991; Håkansson & Lundgren, 1995; Håkansson & Waluszewski, 2002) we have no more than a rather scant understanding of the dynamics of the emergence of business networks, and especially about the role of individual actors in this process.

Furthermore, the companies operating in a network context need to learn to master both the social and business networks in order to succeed in their business networks (e.g., Lechner & Dowling 2003; Uzzi, 1997) and use the knowledge and information shared in these networks to create competitive advantage. The partners and relationships in networks have also been studied in social networks studies (e.g., Griffith & Harvey, 2004). Relationships in collaborative networks can be divided as follows: Firstly, the relationships involved in research and development activities in software and hardware (platform) technology development and their production, including R&D organizations, universities, and research laboratories (Lee et al., 2001; Anderson & Jack, 2002), or as Lechner and Dowling (2003) stated, knowledge, innovation, and technology (KIT) relationships and networks. Secondly, relationships involved in marketing and distribution, including in after-sales services and maintenance, distribution and marketing networks, such as advertising agencies, marketing consultancies, and various distribution channel partners that enable access to new or current markets, and provide new market information and expertise in commercializing a product-service offering (Larson, 1991; Lechner & Dowling, 2003; Rocks et al., 2005). Thirdly, network actors facilitating or supporting business processes include business associations (Vanhaverbeke, 2001) and financial partners (Birley, 1985).

This chapter aims to show how an intentional networking among different actors can benefit a focal actor in developing a new business group based on emerging technologies, and thus, create competitive advantage. This study is based on a longitudinal case study in the ICT sector (Juntunen, 2005), and it takes the viewpoint of a focal actor. The author looks at how new breakthrough technologies (i.e., xDSL, multimedia, mobile technology, Internet) and the combination of formerly separate technologies and businesses are transformed into a viable business—the so-called *home commerce* business group—during the time period of 1990-2003. *Home commerce* refers to a set of Internet-based services targeted for consumers and accessible at home via different terminals. This study also examines how the focal actor tries intentionally to construct this new business group by developing of a series of interlinked strategic alliances and networks of organizations. Using the insights gained during this process the author tries to identify the capabilities that the focal actor develops, giving particular attention to network capabilities. The study of network *capabilities* is a rapidly growing area of research, which combines aspects of industrial network theory and the dynamic capability view of the firm. The basic idea is that companies must develop specific organizational capabilities, which the author calls network capabilities, in order to operate and survive in a network context (Möller et al., 2003; Möller &

Svahn, 2003, 2006) and to create competitive advantage (Ulaga & Eggert, 2005). This view is supported by the work of Gemunden and Ritter, who speak of network competence (Gemunden & Ritter, 1997; Ritter, 1999).

Academic interest in different forms of inter-organizational collaboration has increased considerably during the last decade. The author has adopted the following theoretical approaches in the empirical study: industrial network theory, and the resource-based view of the firm, in particular its extensions into dynamic capabilities and learning. As the emphasis in this study is on the empirical analysis, only a brief discussion of these approaches is provided.

The chapter is organized as follows. The author starts by discussing the conceptual background for the study. The methodological solutions are then described and arguments for them are presented. The fourth section of the paper is the case analysis, where periodic description of the emergence of the home commerce business group (HCB) is covered. The types of collaborative networks that are considered to be critical in this emergence are also discussed, with special attention being paid to identifying the network capabilities, through which the focal actor was created. The role played by these network capabilities is analyzed. A discussion of the theoretical and managerial conclusions and suggestions for future research conclude the chapter.

BACKGROUND

Academic interest in different forms of inter-organizational collaboration has increased considerably during the last decade. The author has adopted two theoretical approaches in the empirical study: industrial network theory and the resource-based view of the firm (RVB), in terms of dynamic capabilities and learning. The study emphasizes on the empirical analysis, and a brief discussion of these approaches has been provided.

Industrial Network Theory

The industrial network or "markets-as-networks"-approach (INT) pursued by the industrial marketing and purchasing group (IMP) view industrial markets as networks of inter-firm relationships. Industrial network theory (INT) provides the basic conceptual tools, the actors who carry out value-activities through the resources, and capabilities they command (Håkansson & Snehota, 1995). It stresses the importance of business relationships as a coordination mechanism on a balance with markets and hierarchies, and their role in supporting learning and innovation in industrial systems (e.g., Håkansson, 1982, 1987, 1989; Lundgren, 1991). Networks represent a complex system in which different interdependencies between actors are characterized by both competition and cooperation, continuously constituting and reconstituting business fields (Alajoutsijärvi et al., 1999; Halinen et al., 1999; Halinen & Törnroos, 1998; Håkansson & Waluszewski, 2002; Mattson, 1985). A development of new technical innovations and solutions (e.g., Håkansson, 1989; Waluszewski 1990; Lundgren, 1991) require continuity in relationships (i.e., long-term relationships). Also, they claimed that interdependence between actors and technologies is apparent (e.g., Håkansson, 1989; Lundgren, 1991; Waluszewski, 1990). Moreover, any single firm has only limited control over its relationship portfolio and network position (Håkansson & Ford, 2002). The position describes how the firm is connected to the other actor that is to what external resources it has access, with which actors it competes and includes the actor's internal resources (Mattsson, 2003). In an intentionally created business network, a focal actor is presumed to have some kind of power to control and mobilize other participants in the business network (Juntunen, 2005).

The firm negotiates through a network of relationships within networks of organizations. First, it is important to make a distinction between a network of organizations and a network organiza-

Vignette 1. The industrial network approach offers some useful concepts that are important in this study.

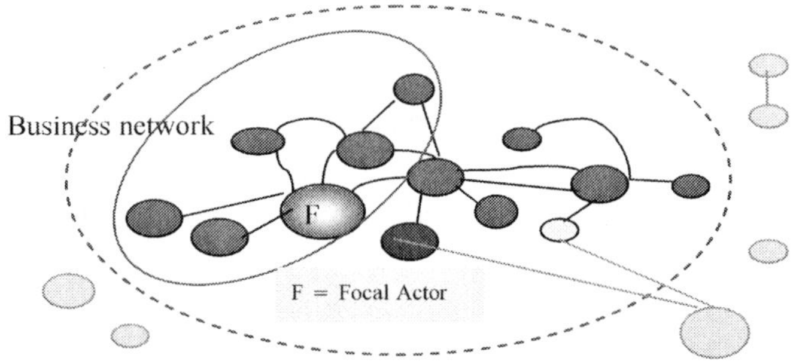

A business network refers to the overall configuration of interconnected actors around a focal actor. A network environment, a macro network, refers to the overall configuration of interconnected organizations around a focal actor. The surrounding macro network includes such entities as a market area, economy and society that the focal actor cannot openly influence. The network environment in this study incorporates many other companies which are not visible to the focal actor and to a particular net, and which cannot be directly influenced by the focal actor.

tion. Network of organizations refers to any group of organizations or actors that are interconnected by relationships (Möller & Svahn, 2003, 2004). According to Axelsson and Easton (1992) and Håkansson and Snehota (1995) any market can be described as a kind of a macro network. This markets as networks view also compares to that held by scholars of economic sociology (Granovetter, 1973, 1985; Knoke, 2001; Uzzi, 1996), and prevailing in the network externalities discussion in economics (Katz & Shapiro, 1985) as well as in the recent marketing studies of "network markets" (Frels et al., 2003; Srivastava et al., 1998).

Another relevant aspect is provided by the concept of embeddedness, which refers to an organization's relations with and dependence on the different types of networks (institutional and political networks, technological systems, and institutions), which form an organization's "environment." All these simultaneously improve the organization's actions, its potential relationships, and the outcomes it may achieve (Halinen &Törnroos, 1998; Lundgren, 1991). In other words, actors in a network should be viewed as important change agents and not only as passive adaptors to a faceless, changing environment.

When industrial network theory primarily focuses on the general characteristics of organically-evolved networks, for example, on their structure and development processes (Möller & Halinen, 1999), the emerging strategic network approach is primarily interested in the creation and management of intentionally-formed network organizations featuring a specific set of actors (Jarillo, 1993; Normann & Ramirez, 1993; Parolini, 1999).

Resource-Based View, Dynamic Capabilities and Organizational Learning

Many practitioners have opined that strategy depends on learning, and that learning depends on capabilities (e.g., Prahald & Hamel, 1990; Mintzberg et al., 1998, p. 213). Unique company resources and capabilities are, therefore, seen as a primary basis of profitability and the basis for formulating its longer-term competitive advantage. This organizational emphasis on restructuring, reengineering, outsourcing, and forming alliances to build unique capabilities became known as the resource-based view (RBV) (Barney, 1991; Penrose, 1959; Rumelt, 1974; Zollo & Winter, 2002). RBV perspective suggests that a firm is a collection of heterogeneous resources (Wernerfelt, 1984) and also, that a company's performance is related to differences in its resources. Resources are not easy to imitate and they form the basis for competitive advantages (Amit & Schoemaker, 1993; Barney, 1991).

The RBV-approach and its knowledge-based extension of inter-organizational collaborations conceptualize the firm as a collection of resources, of which knowledge and core capabilities are critical elements in achieving, maintaining, and renewing competitive advantage. Companies differ in how they can control resources that are necessary for implementing strategies. The core behavioral assumption is that companies are able to learn by acquiring, assimilating, sharing, and dissimilating knowledge within the organization and between organizations. RBV sees the knowledge, capabilities, organizational culture, and management as sources of competitiveness. The resource-based argument of alliance formation suggests that firms use alliances to establish the optimal resource configuration that maximizes the value of their resources. Companies use alliances and other types of cooperative forms to develop a value creating resource base that a single company could not have formed (e.g., Das & Teng, 1998; Möller & Rajala, 1999; Blomqvist, 2002).

The concept of dynamic capabilities became an addition to the RBV approach in the middle to late 1990's.The main thrust in the discussion of dynamic capabilities has been how firms integrate, reconfigure, renew, and transfer their own resources and the resources they control. Although in relevance of exploiting external resources (Teece et al., 1997), the importance of "alliance and acquisition routines that bring new resources into the firm from external sources" (Eisenhardt & Martin, 2000, p. 1108) and the "ability to integrate efforts of different actors" (Grant, 1996) have been mentioned, the challenges involved in operating in a complex network remain fairly unarticulated (Kenis & Knoke, 2002; Park, 1996)

The RBV and IMP approaches both view a company as an actor in a web of relationships which influence the firm's conduct, survival and success. While the RBV approach focuses on the constraints that a dependence on exchange partners poses and emphasizes the firms' pursuit for independency by increasing the firm's resource base, and thus, productivity (e.g., Rumelt, 1984; Wernerfelt, 1984), IMP theory focuses on accumulating benefits and effectiveness via a web of relationships (e.g., Ford et al., 2003; Håkansson 1987,1989). The dynamic capability -view adds the dimension of seizing business opportunity via the creation of new capabilities dynamically suitable for the changing situation within both a network and a business context.

In the construction of a new business group—which is often based on the ability to combine several technologies and coordinate the resources and capabilities of various actors coming from different fields—both knowledge and learning can be expected to play core roles. The author draws on studies of "knowledge management" (Larsson, Bengtsson, Henriksson, & Sparks, 1998; Nonaka, Toyama, & Konno, 2001) which suggest that inter-organizational relationships allow possibilities for collaboration, knowledge transfer, knowledge combination, and knowledge exploitation. For a more in depth discussion of inter-organizational learning, see Holmqvist (2003); and for a discussion of the role of knowledge and learning in different types of strategic networks, see Möller and Svahn (2003).

As a summary, a potential criticism of this study may concern the way these multiple research traditions were utilized for providing concept for analyzing the empirical material. The purpose of this was to have a selection of different, but interrelated, conceptual perspectives, to match the complexity and multi-layered characteristics of the phenomena under study. While it can be argued that using a single research tradition may have provided a more focused analysis, such a study would have been able to only partially cover the multidimensional and comprehensive findings achieved with the multi-theory approach. Another important aspect is the layered character of the new business emergence. In order to achieve a holistic view of the complex causal relationships between technological evolutions, strategic intentions of major actors, and strategic networks, study designs with multiple level of analysis is recommended.

RESEARCH SETTING AND METHODOLOGY

This study also argues that a better theoretical and empirical understanding of the mechanisms by which a focal actor constructs its business and networks is needed. Currently, it is not clear which specific factors in the various collaborative forms determine the positive outcomes and competitive advantage for a focal actor.

This study argues that an inter-organizational collaboration is needed to construct a new business. The goals of this study, for example, to identify, describe, and analyze how a new business group is constructed in an environment of dynamic change, exert a strong influence on the choices of methodology. In order to capture the construction dynamics, a longitudinal study is required in a field that is characterized by technological and commercial change and uncertainty (Huber & Van de Ven, 1995). The explicit focus of this study is the potential implication for the formulation of theoretical propositions regarding the emergence of a new business, and what capabilities and managerial processes are required in the development of business networks.

Furthermore, as the author wishes to examine the interaction between a focal actor and the networks in which it is embedded, an approach was needed to capture how events unfold in a specific context: this requirement is matched by the strong aspects of a longitudinal case study (Pettigrew, 1997; Van de Ven & Poole, 1990). As examining change in a longitudinal study is very intensive, It was decided that the focal actor in this study should be a single organization, *Elisa* (the 2nd largest tele-operator in Finland, see *www.elisa.com*), and more specifically one of its business groups *the home commerce business group* (HCB).

Development of the home commerce business (HCB) in Elisa in Finland based on several emerging technologies and its commercial development has been influenced both by the "e-hype" period of the late 1990's and the bursting of this "bubble" in the early 2000. The study was temporally limited to the time period from 1990 to 2003. The time period was long enough to capture the developmental process of a new business, home commerce business (HCB). The selection of a single corporation from which the multiple cases were chosen naturally brings limitations concerning the generalization of the results of the study. On the other hand, a multiple-case study within a single corporation with excellent data access made it possible to understand the phenomenon under study more profoundly. This decision was also guided by Elisa's cooperative attitude shown towards the study.

Elisa Corporation is a nationwide telecommunications group whose core business areas are ElisaCom, Elisa Mobile, Elisa Networks in Finland, and Elisa Kommunikation in Germany. HCB was part of ElisaCom. The business services HCB provided, as well as the technologies used, differed from those used earlier in the Elisa, and the processes, functions, activities, business requirements, and marketing and business strategies were seen to be different to those of traditional telecommunication business (i.e., fixed line telephone business). HCB integrated several technologies, capabilities, and resources from a number of actors in different industries to create services for consumers and communities, (e.g., Kotiportti™ which consists of various subscriber connections and community services; e.g., Kaasinen, 2001, p. 51); household monitoring and security systems; Efodi -learning space: IS-services for publishers, schools, groups, students and teachers. The development process during the years 1990-2003 was explored and evaluated, and an attempt was made to understand the factors and events affecting the development of business networks and the capabilities required.

The present study concentrates on the focal actor and presents the business and network development from the focal actor's point of view. In view of the richness of the material from the case corporation, it was decided to limit the perspective to the focal actor. The focal actor was one of Elisa's business groups; the home commerce business group (HCB) integrated several technologies and resources from a number of actors to create services for consumers and communities. Home commerce, in this instance, refers to a set of Internet-based services targeted toward consumers and accessible from home using various terminals. Internationally, the concepts of "*future home,*" "*smart home,*" "*intelligent home,*" "*digital home,*" "*networking home,*" "*smart environments,*" "*internet home,*" "*automated home,*" *and "smart housing*" are used for platforms and services in this business (Masala, December 16, 2003). The focal actor, HCB, uses the terms "Future Home" and "Smart Home" in the same context (Masala, December 16, 2003).

Research documentation in this case study consists of both information about the telecommunication industry and the case corporation, Elisa, and HCB's development during the period under study. About 20 persons representing various business units and subsidiaries of Elisa and cooperating partners were interviewed during the period of 1999-2003 (see reference list). Data gathered consists of articles, project documents, e-mails between members of the projects, memorandums concerning strategy and business plans, and annual reports from the years 1990-2001. As the development of the home commerce business (HCB) primarily took place through inter-organizational projects the documents used in the case were arranged in chronological order by project. Even if majority of the material came from HCB-based sources, this does not diminish the value of the conceptualized findings, however. It can be claimed that the conceptualized findings have more general relevance, particularly for firms operating in dynamic, rapidly changing fields characterized by several interlinked technologies. It is also claimed that this relevance reaches beyond the limited historical period.

The research design is a logical sequence that connects the empirical data to the initial research questions (Miles & Huberman, 1994; Yin, 1994). Since there are only a few seminal studies in e-business and on the emergence of business networks and the focal actor's role in the development process this study could be labeled as explorative but it also has some explanatory characteristics (Yin, 1994). The qualitative methods make it possible to develop a thorough understanding of a complex and a multi-dimensional phenomenon in a specific context. The case design here is a multiple-case study but conducted within one corporation.

The reason for choosing a case study approach lies in the in-depth knowledge needed regarding the evolution of different collaborative forms and their managerial processes. Information about these phenomena requires good access to an organization in order to be able to identify them (Heide & John, 1995; Yin, 1994). While a case study is claimed to increase understanding of the complexity of process and change (Cassel et al., 1994; Pettigrew, 1990; Van de Ven & Poole, 1990). Hartley (1994) emphasized the tailor-made nature of a case study because it allows observation within the context of a real-life situation. Furthermore, the longitudinal case study approach was selected so this case study can be described as cases of on-gong processes to increase the general understanding of the phenomena studied (see Carson et al., 2001, pp. 94-95).

Emerging Home Commerce Business Networks: Development Process During 1990-2003

First, this chapter provides an overall view of the development process of Elisa's *home commerce* business group (HCB) through the period 1990-2003. The types of networks through which HCB mobilized this development are then examined together with a discussion of the capabilities required and created during this emergence process. The networks start as technology project networks and develop towards strategic networks with strategic intentions and collective network strategies.

The Period of Innovating 1990–1998

European telecommunication networks of the 1980s and early 1990s were constructed under monopoly conditions (Beardsley, Bray, & van Rooijen, 1995, p. 157). During the first years of this study period, the main drivers of change were technology (e.g., broadband, multimedia), the deregulation in 1995 in Europe, and globalization of the telecommunication markets. Once the economic recession of the early 1990s in Finland was over, Elisa began investing in new R&D projects which led eventually to organization of the MuMe-team (this was the Multimedia team which later became HCB). This team participated in R&D, technology assessments and multimedia, Internet, mobile, and xDSL-based pilot projects. Figure 1 is a broad view of the whole process and the main projects. During this phase, development was strongly guided by the early vision of a single core individual in HCB (located in HTC, Helsinki Telephone Company, later Elisa). The capabilities acquired during this phase were technology-and project-related (For more thorough description of projects see Juntunen, 2005).

Figure 1. An overview of HCB's different collaborative forms from 1990 to 2003

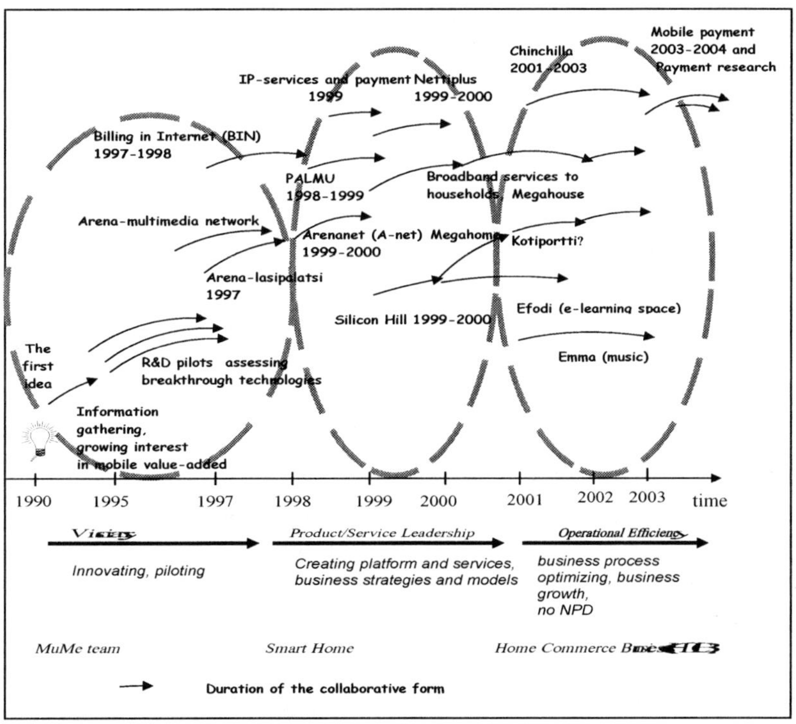

The central MuMe-team worked closely with Comptel, a company owned by the case corporation, which was its former IT-department. Cooperation with Comptel was essential when learning to integrate new services and products into existing architecture and platforms and also in product maintenance (i.e., on-going rationalizations of products, services, and IS/IT solutions). Benefits of this strategic dyad relationship were that both parties learnt and acquired new information and knowledge about the daily changes taking place and innovations being made inside the case company. Comptel was also able to use the knowledge gathered from this relationship in its other projects, including international ones, and this benefited the whole organization. When other resources or capabilities were required to development new services and products, external partners were engaged solely on temporary, project-based contracts. Towards the end of this period, in addition to its close cooperation with Comptel and related partners, HCB also became an active member of national and global cooperative forums in which the competing and collaborating companies from ICT-sector participated, developing new standards and assessing the possibilities offered by new technologies.

In the "e-hype" years of 1997-1998, the general assumption was that a large part of the fixed-line network would be replaced by broadband access technology in the following 5 years. The Internet had also shown its potential as a platform for telecommunication, telephone, and other Web-based services. One important development step was PALMU (see Figure 2, Picture 1), a technology project funded by TEKES (The Finnish Technology Agency) which focused on creating secure billing via the Internet (Hölttä, 1998a, 1998b, Palmu-projektisuunnitelma, Palmu-kokousmuistio). A user-friendly and secure method of e-payment was a prerequisite for new ways of selling goods in Web-based markets. It was expected that flexible payment methods and well-functioning customer interfaces would encourage rapid adoption of e-services by customers. The possibility of differentiating and integrating services in new ways would open opportunities to introduce self-service and the provision of services on demand (HPY Research, 1998). In this network, the case company was an equal partner with the other participants. The primary goal was to find a way of creating efficient web-based services and payment methods in the new network environment (Isomäki & Jäntere, 1998; Palmu, 1999). Both technological and business visions drove this kind of R&D technology-cooperation and technology piloting.

The results of this wider research into enabling technologies were transferred into the more business-oriented strategic dyad with Comptel and its surrounding network, for further development. HCB´s relationship with Comptel was very useful as each actor was a specialist and had its own support network, while there was sufficient common ground and mutual experience to co-create new solutions. As an IS/IT provider, Comptel maintained a portfolio of relationships from its business point of view while HCB initiated relationships from its perspective of starting to develop the first ADSL-based services while maintaining the vision of new home commerce business. Both actors could, at the same time, have the roles of a product developer and an IS/IT provider. During this period, Comptel, having gained significant knowledge in providing mediator-software services, embarked on rapid global growth.

The Period of Product/Service Development 1998-2000

Developments in 1990-1998 resulted in the creation of a technological platform of enabling Internet and broadband technologies, which could be used in a more business-oriented development of the home-commerce area. In Finland, this coincided with the deregulation of the telecommunication sector in 1998. During the period 1998-2000, the IT boom and rapid changes in competitive environment caused organizational changes, while new business and related projects received a higher proportion of resources and investments than they had before. R&D projects completed in 1990-2000 provided insight and direction suggesting the types of new knowledge, resources, and capabilities required when constructing architecture based on the new enabling solutions.

In Elisa, it was clearly understood that gaining competitive advantage in the new Internet and multimedia-driven business required additional technology, business, and managerial capabilities. This capability gap was perceived to be growing wider until Elisa's top management decided in 1998 to start building capabilities by establishing a product/service-oriented business unit called Smart Home (which later became HCB) and provided funding for further network projects.

The first 'real' business network aimed at new multimedia-technology-based business was Arenanet (A-net) (see Figure 2, Picture 2). The driving vision behind Arenanet, a coalition of culture and technology triggered by the City of Helsinki, European Cultural City of 2000 project, was to offer broadband-based services to every single home in Finland (Lehmus, 1999). The ambitious objectives of "Helsinki Arena 2000" were: to create a virtual Helsinki, a next generation multimedia network environment for Finland's capital city, to provide a forum for citizens, and a place where culture and business could meet (Salmi, 1999; Tenhovuori,1996). By joining this network, smart home group (HCB) was aiming to create competitive advantage in

Figure 2a. Picture 1: Equal partners in a network, (PALMU): Technology project-network with cooperation with competitors and institutions and international forums

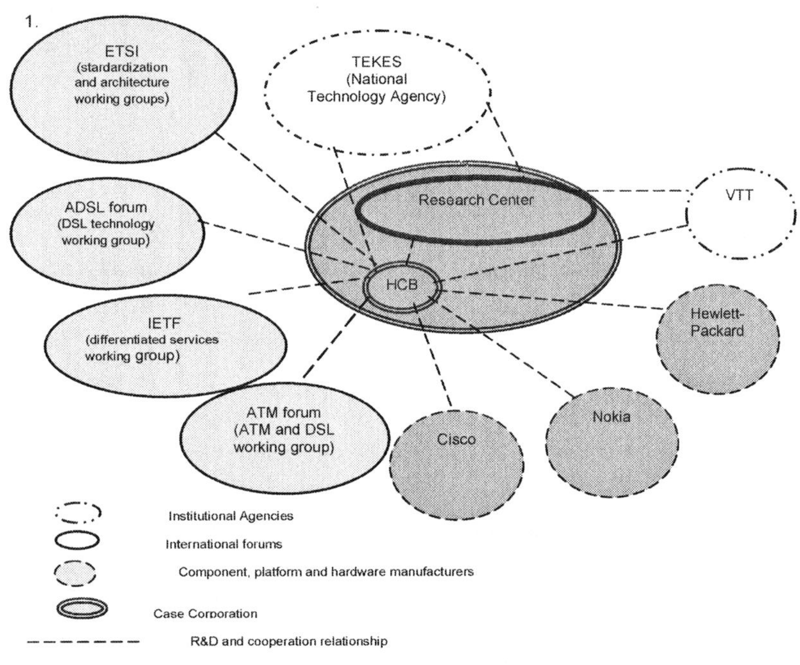

Picture 2b. Arenanet (A-net): A coalition of technology and culture

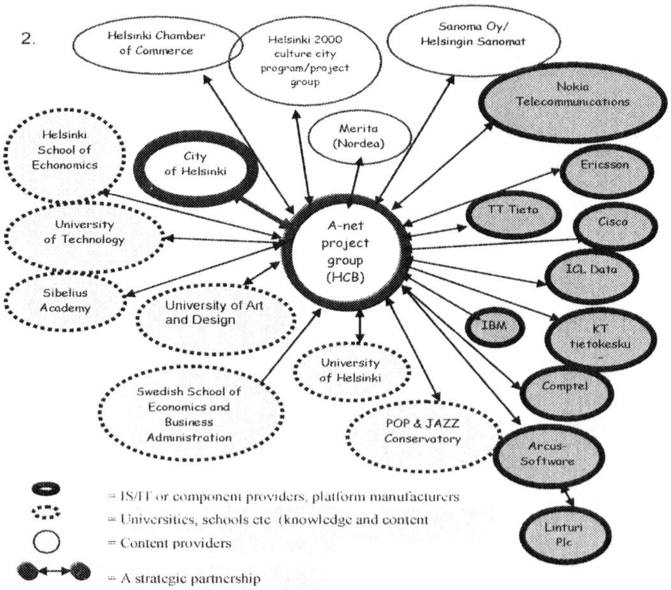

this area of e-business, to strengthen both its role and its image, and to expand its customer base. It was a deliberate strategic move to combine Smart Home's (HCB's) technological knowledge with new business development.

Actors in Arenanet (A-net) were chosen primarily for the purposes of creating a broad-ranging service and information base for an "information age city." The actors in this network were both global players like Nokia, Cisco, and Hewlett Packard as well as national players like Elisa, Linturi, and the city of Helsinki. Relationships in this network were controlled and coordinated by the A-net project group in the case company and this group later became part of HCB. A-net also initiated a change from the strong emphasis on the relationship with Comptel towards increased networking with other businesses. A-net was a complex, temporary business network attempting to combine the knowledge and capabilities of several actors, but there was no clear view on how to manage network relationships or how to maintain the actors' interest in the cooperation. Smart home (HCB) and A-net project group itself lacked a clear vision of where this business development might lead, and especially whether and how it could be made profitable.

The exploration shows that rather than being a real business network, A-net was still more akin to R&D, a project for gaining experience of how to be in the multimedia and Internet-based business. On the other hand, A-net proved invaluable as a means of learning and establishing relationships. The hub company gained important knowledge of this type of networking and cooperation, which involves issues such as the types of resource and capability required, how to continue in the multimedia and Internet-based business, and determining the types of service that would interest consumers. The A-net experience helped the key members of the smart home (HCB) realize that a network is an entity to be taken care of and that in business networks, there are different actors and different roles.

One important realization was that the primary reason for some actors' participation in the network was to gain knowledge and experience, which they could use in their own projects and in creating competitive services. Although a management board to control and coordinate the development of A-net existed, it did not effectively manage the roles of actors involved and the network's future development. A-net was also the first project in which actors from several different industries were involved and doing business together with HCB. The network was closed down in the end of 2000.

Even if the A-net was not a success from the business point of view, there was a consensus that networking is essential when dealing with multiple technologies and platforms within several industries. The experience confirmed that, in a context where knowledge is sophisticated, expanding and widely dispersed, the innovation and commercialization must be carried out through collaborative networks. Based on experience resulting from A-net, the case company initiated the *megahouse* and *megahome* project networks targeted at further development of the technology architecture and basic services, as well as new home commerce services. They became so called mega concept. What differentiated the mega-concept from previous collaborative networks were the advantages resulting from being the first actor in this area and its foundation on risky and bold resource commitments. However, the successful integration of technologies, marketing, and innovative business models did not automatically translate into a sustainable strategy and advantage, the construction of organizational-specific dynamic capabilities was required.

In the mega-concept's value net, HCB's roles were as the hub, for example, the intermediator and the value chain integrator. HCB integrated the value chains of several industries. It also developed and provided the collaborative platform. HCB designed the net's architectural structure. It created a virtual community solution for households (Kotiportti^{TM}) and for teachers and students (Efodi—the e-learning space). In accordance with this, HCB provided the security solutions and services because Datatie, a subsidiary of Elisa, had provided those services previously. Those services were now integrated to the same interface and platform by HCB. The roles of the other actors were as software, component, device, application,

Figure 3. Overview of the home commerce business (HCB) and industries involved in mega-concept

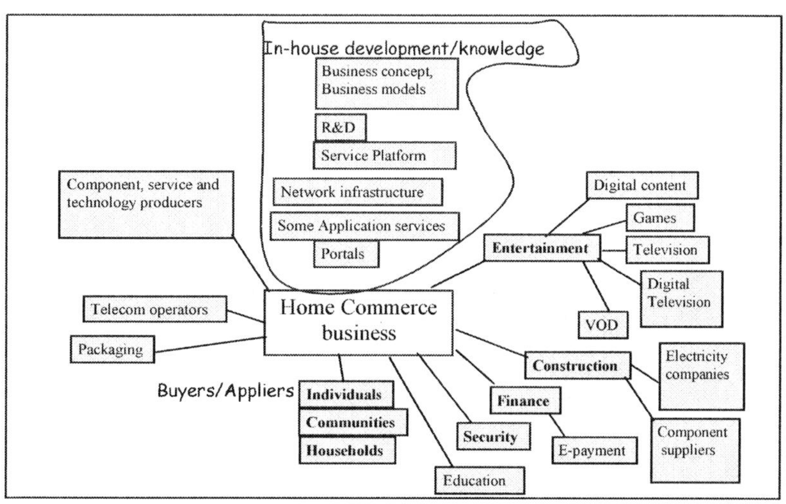

content and service, access and network providers (i.e., Internet, broadband, fixed, or wireless network), as well as infrastructure and platform manufacturers. When comparing mega-concept to A-net, the main difference was that in A-net, HCB was not capable of preventing opportunistic behavior. Figure 3 shows the different industries involved in mega-concept.

The number of competitors in the future home market was few during the early phase but there was a proliferation of competitors by the beginning of the 21st century. The basis of competition became more defined. Initially, it was a question of attracting customers and building customer base. The vision for the future home and home commerce business was to maintain its position as the pre-eminent provider of home commerce services and information. To achieve that goal, HCB had to continue to enhance both content and site services and further diversify revenues, achieving cost savings through process redesign in order to remain dynamically efficient in the future.

Learning new technologies, integrating and implementing them into a business context underlay the main learning race within the ICT-sector during the 1990s. The capabilities developed around these new technologies and the businesses based on them were the source for HCB's competitive advantage. Moreover, the results show that an innovative actor has to be able to develop capabilities in network management. Without these managerial capabilities the actor is not capable of influencing, orchestrating, and mobilizing the other net members.

In addition to the strong competence learning, as well as service technology and concept development, the 1998-2000 period can also be characterized as the top management team of the HCB 'giving promises' to the corporate management and other constituencies. This relates to the question of communicating the potential growth and profitability of multimedia and Internet-based business. This intra-organizational agenda "selling" resulted in HCB being given the resources it required to participate and mobilize all the development networks, and resulted in HCB achieving leadership in ADSL connections and the leading position in smart-home/future-home solutions (e.g., Kotiportti™, household monitoring and home security systems, digital television services) in Finland.

The Stabilization Phase 2001-2003

The years 1998-2000 were time of the 'IT-boom' and "e-hype" in Finland as well as in Europe. During that time, Elisa's R&D was allowed to test and fine-tune their products to be perfect in every way possible.

However, the years 2000-2003 showed unpredictable economic situation in the ICT-sector, and R&D investments were expected to have a short pay-back time and there was no extra time to spend in perfectionism in product designs. The years 2001-2003 showed the impact of exogenous factors, such as the burst of the "IT-bubble" and associated declining profits in the ICT-sector, in Elisa in many ways. The impacts seen included:

- **At a business level:** termination of unpredictable business, concentration on core and profitable products and services, the pruning of non-core business, the analysis of future business possibilities, and the investment in promising technological and business areas
- **At the organizational level:** the retrenchment of employees, the restructuring of the oganization
- Customer service was considered to be more important than in the years of rapid growth and expansion seen during 1998 to 2000.

The reduction of investment in R&D affected HCB to the extent that the new product development was terminated and HCB instructed to concentrate on their core business and key products like ADSL and Kotiportti™. Processes and interfaces of services underwent tuning to increase efficiency. Senior management's key motivation in investing in the new home commerce related business projects was because they saw these as a way to enhance current businesses and create new business opportunities. Organizational restructuring affected so that a couple of new emerging businesses were merged to HCB.

The formation of internal networks to focus on certain R&D areas of the business had changed within the case corporation. Whereas, in the 1990s, R&D was focusing on single technology pilots, in 2001-2003, the focus was on business issues and technologies related to that business.

In addition to networking within the case corporation, external partners were more precisely assessed, chosen, and classified not only for today's business needs but also with regard to future business needs. HCB's management needed to focus its attention to dynamic resource flows between actors in a business network. The managerial capabilities were needed to steer external resources toward supporting network's objectives. HCB retained only those activities to itself in which they excel over any other players in the markets. The management in Elisa was also required to have collaborative process design and management tools such as Elisa tried to develop in its knowledge management (KM) platform in late 1990s. The competition was not only between companies but also a competition of resources capable of adding the most value to the network.

Technological knowledge, the fast pace of technological developments and the business possibilities offered by the integration of technologies and businesses required more resources than HCB had originally envisaged. The concentration of business activity in an area, in which they performed optimally, including broadband, home automation and future home solutions, was sufficient for HCB.

HCB had realized that the formulation of a business plan and the collective strategy of a network involved understanding the structure of each industry engaged in the network, finding the competitive advantage and determining how a business house will compete in the markets. HCB had learnt that growing a new business involved the establishment of partnerships with prominent companies, seeding the market and encouraging partners to participate in the creation process of a new business.

In HCB's internal service development, its role initially changed from product and service developer to platform developer and provider, eventually moving towards a role as service environment–provider in which the service environment included homes, buildings in general, and communities.

Role of Networks in Creating a Business and Management Capabilities

The descriptive analysis of the construction of a radically new business group highlights the relevance of networks in this process. It also provides insights into the building of dynamic network capabilities through collaborative networks. The nature and goals of networks and networks changed in a systematic manner as HCB developed during 1999-2003.

In an entirely logical manner, the early years (1990-1997) were characterized by participating in and mobilizing networks for the development of technological knowledge. The strategic alliance with Comptel, the multimedia projects and R&D networks, and participation in the Finnish Technology Agency (TEKES) financed e-payment project network, and all were crucially import in helping HCB to create the technological platform it needed to start offering Internet and broadband-based services to businesses and consumers. Through this collaborative networking, HCB's core team learned that, in addition to the specific technological capabilities, it is essential to have networking capabilities, such as partner evaluation and selection, multiparty-project participation, and management. Because of the number and continuity of these primarily project-based networks (see Figure 1), HCB was able to build routines and a managerial infrastructure that turned primarily into people-embodied competencies with dynamic organizational capabilities. In conceptual terms, this process is termed as the experience accumulation and knowledge articulation learning mechanisms as suggested by Zollo and Winter (2002)

Furthermore, participation and the coordination of the A-net (the network coalition aimed at turning Helsinki into an e-city) was a turning point where the emphasis in HCB's network mobilization shifts from technology cooperation to business development. The A-net experience sensitized the HCB management team to the necessity of developing a network management capability; the careful selection of network members, and the provision of comprehensible roles with specific responsibilities, direction, and coordination. Development of these capabilities was carried out in the large, multi-firm and multi-industry networks (*megahouse* and *megahome*), which further emphasized aspects of network mobilization and management capability such as negotiating skills and the creation of new forms of contracts.

Another major learning experience that was already perceived during late R&D project networks, but really accentuated by the disappointment of not achieving any direct business objectives with the A-net, was the realization of the crucial role played by new business models in home-centered, Internet and broadband-enabled e-business. Without a clear and jointly agreed view about the earning modes and shares of each network member (e.g., operator, software provider, content providers, portal owner) the network management functions faced continuous conflicts and inefficiency. In brief, a good network mobilization and management capability was highly dependent on the focal actor's competence in designing effective and attractive business models.

What enables a management team to design a good business model in a fast developing business sector? Accumulated experience gained from several networks and from participation in industry-wide forums provided the HCB team with a visioning capability for the unfolding business sector. This allowed them to assess the business opportunities from the perspectives of potential network members and led to viable and mutually attractive business models. This "framing capability", which makes it possible to create an overarching view of the architecture of a business group in which the value-systems of different business sectors merge, is very hard to achieve. On the basis of the case evidence, it appears that the HCB team was able to accomplish this.

As a team, HCB was considered more of a profit and innovation center than a cost center and they were encouraged to develop through internal training programs and through different technology networks. Also, they did their own market research.

This case of HCB could also be seen as a case of intrapreneurship—fostering entrepreneurship within an established organization. HCB's management team put together team of enthusiastic volunteers, sponsored by Elisa's senior management. It was a way for an established organization to find new markets and new top products. HCB's team could be considered as inventors as they stepped out of their traditional roles within the established business, they were creative and took risks. Moreover, they worked towards minimizing the risks using the collaborative networks to more efficient and effective product production and sales. The success of HCB also showed a way to increase the speed and cost-effectiveness of technology transfer from research and development to the marketplace. It also seemed that success was based on the organizations ability to take on concurrently multiple and contradictory business elements through an organizational managerial structure that combined a mix of self and centralized government.

CONCLUSION

This study makes several contributions to the evolving theory of network management. It provides empirical evidence for the crucial roles played by different types of networks and networks in both the emergence of a radically new business group and in learning the managerial capabilities required to not only survive in this context but also to create it.

Different alliances, project networks, R&D forums, and large-scale project networks provided different types of learning experiences. In brief, they provided the case company with different resources and knowledge; in other words, they fulfilled the different functional needs that evolved during the evolutionary process. The author contends that a core management capability in this type of environment is the ability to utilize the available networks and mobilize one's own networks to meet the resource and learning needs that exist. This is not, however, an organizational capability which is inherent. It presumes two aspects—visioning and networking resources and capabilities. Firstly, a comprehensive or "architectural" view of the field, which enables a company to envision its development, at least a few steps ahead. Through visioning, an actor can anticipate the technologies and other capabilities it must develop. Visioning alone is not enough; the resources to carry out networking must also exist. In the phase where an actor wishes to mobilize its own strategic network, it must also be able to offer an attractive development agenda or business model if it is to engage partners with cutting-edge knowledge in their own areas of operation.

This "functionality" perspective on the role of networks suggests that firms with an "architectural" vision and adequate resources can, through network relationships, purposefully attempt to create an extended pool of resources and capabilities that matches their current and foreseeable needs. If this attempt is successful, a success circle may result.

In addition to this functionality aspect of network collaboration and network management, the case analysis highlights the relevance of understanding the process-like nature of capability development. Only after exposure to a variety of continuing networking experiences, the management team of the case firm was able to identify and systematically start developing network-management capabilities. This provides additional support to the experimental learning view of dynamic capabilities (Zollo & Winter, 2002). As seen in this case, the intelligence in a network structure is its ability to share, modify, create, and distribute information, knowledge, and resources among the different parties, and thus develops and creates network capabilities that can change and dynamically develop over time (see also Möller et al., 2003; Möller & Svahn, 2006). Creating competitive advantage with the capabilities developed and acquired, the case company succeeded in gaining the leading position in ADSL-services at the end of the 1990s which supports the earlier studies of creating competitive advantage (e.g., Penrose, 1959; Rumelt, 1974; Wernerfelt, 1984; Barney, 1991). On the other hand, since network participation also has the character of an investment, as emphasized by the industrial network theory (Håkansson & Snehota, 1995), creating a variety of resources and gaining the knowledge cannot be achieved without an associated cost. Taken together, these findings provide support for and expand the extent of discussion on network capabilities.

These findings have important theoretical and methodological implications. First, they support the industrial network approach (Håkansson et al., 2004) and the Resource Based View in emphasizing the key role of combining the heterogeneous resources controlled by various actors in order to be able to create new technological and business solutions.

Moreover, knowledge of processes, particularly in relation to tuning R&D processes, in late 1990s made it easier for HCB to create a competitive advantage. The corporate vision provided a continuity of direction, which allowed HCB to create unique skills and capabilities and to build a strong reputation of superior technical knowledge in certain new technology areas (cf. Porter, 2001, p. 71).

Cognitive aspects of social capital can be seen as human-related aspects and motivation for collaboration (e.g., Nahapiet & Ghoshal, 1998; Anderson & Jack, 2002). From the focal actor's perspective, this study shows that in the relationships involving actors and networks related to the supply-side; for example, research and development activities, as well as in/out-sourcing of components and platforms required in software and platform development and production, the cognitive aspects of social capital providing value to software vendors include capabilities in both current technology, as well as in emerging future technology. In addition, innovativeness, flexibility, and adaptability seem to be linked to cognitive capabilities of the network actors. These relationships involve actors and networks facilitating and supporting business; for example, those partners who do not directly contribute to the value-creation process, cognitive capabilities are related to business strategy, marketing, and management.

As a conclusion, the emergence of a new business is clearly a combination of the evolution of technological and market factors which are not controllable by any one actor, rather it is accumulation of the intentional strategy of developing collaborative forms—alliances, partnerships, and strategic networks—by the focal actor. First, this study shows that a focal actor can clearly promote its business by developing its business and social networks as seen in this case study. Without this strong intention or strategy, the focal actor would not have been able to develop its position in the emerging new Internet and mobility related services. Second, the results show that an innovative actor has to be able to develop capabilities in network management. Without these it cannot influence and orchestrate the other actors whose resources and competences are required in creating the technological platforms underlying new business services nor the services themselves. Network management was seen to be composed of such strategic level competencies as being able to design differentiated strategies for different types of networks, perceiving the networks and partnerships as an interrelated portfolio of collaborative forms, and being able to develop different contractual forms for different type of collaborative forms.

Finally, the global ICT-policy making addresses primarily the removal of all obstacles that might stand in the way of the unconstrained operation of the various international ICT actors on markets. Scope and direction of national ICT-strategies are influenced by the emerging global systems of the ICT-sector's governance. A successful national ICT-strategy as a part of a global ICT development can assist in the development of flexible and decentralized industrial production and R&D business networks, and thus, improve the competitive position of local ICT-suppliers in national markets against the global competitors. Through business networks, new global communities are being established. Increasingly, organizations in various countries are integrated into these webs of horizontal and vertical information and knowledge exchange. However, the biggest challenge for both the national and global ICT-strategies are to ensure the sustainable growth and the development of the ICT-sector and ICT-related capabilities and skills. This sustainable development will not be possible only through technology development but by development of national and international policies. Business networks can offer new sources of innovations and knowledge, and thus, assist in developing the skills and capabilities needed in national ICT-development and assuring the competitive advantage locally or internationally.

Future Research Directions

ICT has been recognized as a major catalyst for socioeconomic development, therefore, the experiences and information of the usage of ICT are of key importance. ICT can be used in service of broader strategic objectives in a number of development sectors, for example: health, military, education, e-commerce, and e-government. Many innovative partnerships between governments, businesses, and private companies have been formed to build collaborative networks and deploy advanced ICT applications both nationally and internationally. However, we still lack the analytical and political tools to better conceptualize the business-university-government networking, and the role of each partner in the strategic ICT-development. Due to this networking, we also need tools to evaluate the processes and working methods cross different industry and organizational boundaries. Interaction in ICT-developmental projects between organizations is undertaken with intention to achieve an economic goal, with strategic objectives being implemented to obtain final economic rewards. Further, economic goals involve some overlapping of self and collective network-level interest. And probably, there is some disparity between interacting organizations on their combinations of expected self and collective network-level interest in the collaborative networking.

In addition, we need new frameworks to monitor and evaluate ICT-enabled projects and their role in national, economic, and/or business development. Measuring the benefits and failures of ICT development is particularly interesting because available data is limited. Collaborative networking in ICT-projects is difficult to track and there is a lack of exact data to serve as the basis for wide-scale policy decisions.

REFERENCES

Alajoutsijärvi, K., Möller, K., & Rosenbröijer, C.J. (1999). Relevance of focal nets in understanding the dynamics of business relationships. *Journal of Business-to-Business Marketing, 6*(3), 3-35.

Amit, R., & Zott, Ch. (2001). Value creation in e-business. *Strategic Management Journal, 22*, 493-520.

Anderson, A.R., & Jack, S.L. (2002). The articulation of social capital in entrepreneurial networks: A glue or lubricant? *Entrepreneurship and Regional Development, 14*, 193-210.

Barney, J. B. (1991). Firm resources and sustained competitive advantage. *Journal of Management, 17*, 99-120.

Beardsley, S.C., Bray, W.D., & van Rooijen, M.C.W. (1995). The great European multimedia gamble. *McKinsley Quarterly ,3.*

Birley, S. (1985). The role of networks in the entrepreneurial process. *Journal of Business Venturing, 1*, 107-117.

Castells, M. (1996). *The rise of network society.* MA: Blackwell, Cambridge.

Gemunden, H. G., & Ritter, T. (1997). Managing technological networks: The concept of network competence. In H. G. Gemunden, T. Ritter, & A. Walter (Eds.), *Relationships and networks in international markets* (pp. 294-304). Oxford: Elsevier Science.

Grabher, G. (Ed.). (1993). *The embedded firm. On the socio-economics of industrial networks.* London: Routledge.

Griffith, D.A., & Harvey, M. G. (2004). The influence of individual and firm level social capital of marketing managers in a firm's global network. *Journal of World Business, 39*, 244-254.

Halinen, A., Salmi, A., & Havila,V. (1999). From dyadic change to changing business networks: An analytical framework. *Journal of Management Studies, 36*(6), 779-94.

Halinen, A., & Törnroos, J.Å. (1998). The role of embeddedness in the evolution of business networks. *Scandinavian Journal of Management, 14*(3), 187-205.

Heide, J.B., & John, G. (1995). Measurement issues in research on interfirm relations. In K. Möller & D. Wilson (Eds.), *Business marketing: An interaction and network perspective* (pp. 531-554). Boston: Kluwer Academic Publishing.

Holmqvist, M. (2003). A dynamic model of intra- and interorganizational learning. *Organization Studies, 24*(1), 93-121.

Huber, G.P., & Van de Ven, A. H. (Eds.). (1995). *Longitudinal field research methods. Studying processes of organizational change.* USA: SAGE Publications Inc.

Håkansson, H., & I. Snehota (Eds.). (1995). *Developing relationships in business networks.* London: Routledge.

Håkansson, H., & Waluszewski, A. (2002). *Managing technological development. IKEA, the environment and technology.* Routledge Advances in Management and Business Studies. London: Routledge.

Jarillo, J. C. (1993). *Strategic networks: Creating the borderless organization.* Bodmin, Cornwall: MPG Books Ltd.

Juntunen, A. (2005), *The emergence of a new business through collaborative networks: a longitudinal study in the ICT sector.* Acta Universitatis oeconomicae Helsingiensis. Printed in HSEPrint. A-series, no 256, ISSN 1237-556X. ISBN: 951-791-957-3/1.

Kaasinen, K. (2001). *Describing business concept based on partnership network.* Thesis. Lappeenranta University of Technology.

Kenis, P., & Knoke, D. (2002). How organizational field networks shape interorganizational tie-formation rates. *Academy of Management Review, 27*(2), 275-293.

Larson, A. (1991). Partner networks: Leveraging external ties to improve entrepreneurial performance. *Journal of Business Venturing, 6,* 173-188.

Larsson, R., Bengtsson, L., Henriksson, K., & Sparks, J. (1998). The interorganizational learning dilemma: Collective knowledge development in strategic alliances. *Organization Science, 9*(3), 285-305.

Lechner, C., & Dowling, M. (2003). Firm networks: External relationships as sources for the growth and competitiveness of entrepreneurial firms. *Entrepreneurship & Regional Development, 15,* 1-26.

Lee, C., Lee, K., &. Pennigs, J.M. (2001). Internal capabilities, external networks and performance: A study on technology-based ventures. *Strategic Management Journal, 22,* 615-640.

Lundgren, A. (1991). *Technological innovation and network evolution. The emergence of industrial networks.* EFI, Stockholm School of Economics, Stockholm.

Möller, K., & Halinen, A. (1999). Business relationships and networks: Managerial challenge of network era. *Industrial Marketing Management, 28,* 413-427.

Möller, K., Rajala, A., & Svahn, S. (2005). Strategic business nets–Their types and management. *Journal of Business Research,* 1274-1284.

Möller, K., & Svahn, S. (2003). Managing strategic nets: A capability perspective. *Marketing Theory, 3*(2), 201-226.

Möller, K., & Svahn, S. (2006). Role of knowledge in value creation in business nets. *Journal of Management Studies, 43*(5), 985-1007.

Nahapiet, J., & Ghoshal, S. (1998). Social capital, intellectual capital, and the organizational advantage. *Academy of Management Review, 23*, 242-266.

Nonaka, I., Toyama, R., & Konno, N. (2001). SECI, Ba and leadership: A unified model of dynamic knowledge creation. In Nonaka & Teece (Eds.), *Managing industrial knowledge. Creation, transfer and utilization* (pp. 13-43). London: Sage.

Normann, R., & Ramírez, R. (1993). From value chain to value constellation. Designing interactive strategy. *Harvard Business Review, 71*(4), 65-77.

Park, S. H. (1996). Managing an interorganizational network: A framework of the institutional mechanism for network control. *Organization Studies, 17*(5), 795-824.

Parolini, C. (1999). *The value net: A tool for competitive strategy.* UK: John Wiley & Sons

Porter, M. E. (2001, March). *Strategy and the Internet.* Harvard Business Review.

Penrose, E. (1959). *The theory of growth of the firm.* London: Basil Blackwell.

Pettigrew, A.M. (1997). What is a processual Analysis? *Scandinavian Journal of Management, 13,* 337-348.

Ritter, T. (1999). The networking company. *Industrial Marketing Management, 28*(5), 497-506.

Rocks, S., Gilmore, Audrey & Carson, David (2005). Developing strategic marketing through the use of marketing networks. *Journal of Strategic Marketing, 13*, 81-92.

Rumelt, R. F. (1974).*Strategy, structure, and economic performance in large industrial corporations.* Boston: Harvard University Press.

Teece, D. J., Pisano, G., & Shuen, A. (1997). Dynamic capabilities and strategic management. *Strategic Management Journal, 18*(7), 509-533.

Thompson, G., Frances, J., & Levačič, R., & Mitchell, J. (Eds.). (1994). *Markets, hierarchies & networks: The coordination of social life.* Thousand Oaks, CA: SAGE Publications.

Ulaga, W., & Eggert A. (2005). Relationship value in business markets: The construct and its dimensions. *Journal of Business-to-Business Marketing, 12*(1), 73-99.

Uzzi, B. (1997). Social structure and competition in interfirm networks: The paradox of embeddedness. *Administrative Science Quarterly, 42*, 35-67.

Van de Ven, A.H., & Poole, M. S. (1990). Methods for studying innovation development in the Minnesota Innovation Research Program. *Organization Science, 1*(3) 313-335.

Vanhaverbeke, V. (2001). Realizing new regional core competencies: Establishing a customer-oriented SME network. *Entrepreneurship & Regional Development, 13*, 97-116.

Wernerfelt, B.(1984). A resource-based view of the firm. *Strategic Management Journal, 5*, 171-180.

Zollo, M., & Winter, S.G. (2002). Deliberate learning and the evolution of dynamic capabilities. *Organization Science, 13*(3), 339-351.

UNPUBLISHED REFERENCES

HPY Research (1998). *Palmu-project. 5.2.1998. Palmu-projektin kick-off tilaisuus.* Helsingin Puhelin Oyj. Helsinki, Finland.

Hölttä, P. (1998a). *Lisäselvitys Palmu-projektin tuotekehityshakemukseen.* 30.4.1998. Helsingin Puhelin Oyj. Helsinki, Finland.

Hölttä, P. (1998b). *Palmu-projektin väliraportti ajalta 1.1.1998–30.9.1998.* Helsingin Puhelin Oyj. Helsinki, Finland.

Isomäki, M. & Jäntere, K. *PALMU working draft. Laajakaistainen DSL Liittymä–Business-mallien ja niihin liittyvien teknisten toteutustapojen vertailu.* Helsinki, Finland.

Lehmus, P. (1999). *Strategiasuunnitelma 2000-2003. Arena.* Helsingin Puhelin Oyj, Liiketoimintaprojektit. 14.4. (1999). Helsinki, Finland.

Palmu-kokousmuistio. 13.1. (1999). Helsingin Puhelin Oyj. Helsinki, Finland.

Palmu-projektisuunnitelma HPY.(1999). Helsingin Puhelin Oyj. Helsinki, Finland.

Palmu (1999). *Palvelujen Hallinta Mobiilissa Internetissä.* Helsingin Puhelin Oyj.13.1.1998. Helsinki, Finland.

Salmi, P. (1999). *Helsinki Arena 2000-esittelymateriaali.* Helsingin Puhelin Oyj. Helsinki, Finland.

Tenhovuori, T. (1996). *Helsinki Arena 2000–palveluverkko. Tavoitesuunnitelma Helsinki Arena 2000.* Helsingin Puhelin Oyj. 2.12.1996. Helsinki, Finland.

ADDITIONAL READING

The following list contains some useful and interesting books, reports, statistics, and articles of the ICT development, business networks, and value creation in networks:

Global Information and Communication Technologies (2006). Information and Communications for Development 2006. GlobalTrends and Policies (IC4D 2006), available at http://www.worldbank.org/ict/

Global Information and Communication Technologies (2006). The Evolution of Telecoms in Emerging Markets, available at http://www.worldbank.org/ict/

Global Information and Communication Technologies (2005). E-Development: From Excitement to Effectiveness. Available at http://www.worldbank.org/ict/

Helander, Nina (2004). *Value-creating networks: An analysis of the software component business.* Acta Universitatis Ouluensis, G 16. Doctoral dissertation.

Forsström, B. (2005). *Value co-creation in Industrial buyer-seller partnerships–Creating and exploiting interdependencies*, Åbo Akademi University Press: Åbo. Doctoral dissertation.

Gabrielsson, P., & Gabrielsson, M. (2004). Globalising internationals: Business portfolio and marketing strategies in the ICT field. *International Business Review, 13*(6), 661-684.

Gabrielsson, M., & Gabrielsson, P. (2003) Global marketing strategies of born globals and globalising internationals in the ICT field. *Journal of Euromarketing, 12*(3/4), 123-145.

Gabrielsson, M., & Kirpalani, V.H.M. (2004) Born globals: How to reach new business space rapidly. *International Business Review, 13*(5), 555-571.

Gemunden, H.G., Ritter,T., & Heydebreck, P. (1996). Network configuration and innovation success: An empirical analysis in German high-tech industries. *International Journal of Research in Marketing, 13*(5), 449-462.

International Telecommunication Union (2007). *Measuring the Information Society 2007/ICT Opportunity Index and World Telecommunication/ICT Indicators.* Available at www.itu.int

International Telecommunication Union (2007). *Core ICT indicators.* Available at http://www.itu.int/ITU-D/ict/partnership/material/CoreICTIndicators.pdf

Kirpalani, V.H. M., & Gabrielsson, M. (2004). Worldwide Evolution of Channels Policy: Impact of Globalization, Societal Problems and Research

Needs. *Journal of Marketing Channels, 11*(2/3), 123-134.

Lindgreen, A., & Wynstra, F. (2005). Value in business markets: What do we know? Where are we going? *Industrial Marketing Management, 34*(7), 732-748.

Möller, K., & Svahn, S. (2004). Crossing East-West boundaries: Knowledge sharing in intercultural business networks. *Industrial Marketing Management, 33*, 219-28.

Möller, K., & Svahn, S. (2005). Managing in emergence: Capabilities for influencing the birth of new business fields. In Sanchez & Freiling (Eds.), *Research in competence-based management, Issue 1: The Marketing Process in Organizational Competence*. Elsevier Science.

Ritter, T., Wilkinson, I. F., & Johnston, W. J. (2002). Measuring network competence: Some international evidence. *Journal of Business and Industrial Marketing*, 119-138.

Rusten, G., & Skerratt, S. (Eds.). (2007). *Information and communication technologies in rural society.* Routledge Studies in Technology, Work and Organizations. Routledge, UK.

Steinbock, D. (2001). The Nokia Revolution. *The story of an extraordinary company that transformed an industry.* AMACOM.

Steinbock, D. (2006). *The Mobile Revolution: The making of worldwide mobile markets.* UK: Kogan-Page.

Taisch, M., Thoben, K-D., & Montorio, Marco (Eds.). (2007). *Advanced Manufacturing. An ICT and Systems Perspective.* UK: Routledge.

Tuominen, M., Rajala, A., & Möller, K. (2004). Market-driving versus market-driven: Divergent roles of market orientation in business relationships. *Industrial Marketing Management, 33*(3), 207-217.

Törnroos, J.Å., & Halinen, A. (1998). The role of embeddedness in the evolution of business networks. *Scandinavian Journal of Management, 14*(3), 187-205.

INTERVIEWS

Name (Interviewee)	Position	Organization	Interview	Time
Ahlstrand, Klaus	Business Development Manager	Datatie	in person e-mails	March 29th 2000,February 11th 2001,October 31st 2001,November 1st 2001, November 12th 2001, December 5th 2001
Arhi, Mika	BI, Analyst	Elisa Corporation	in person e-mail	May 2001,July 2002
Buuri, Marko	Product Development	Elisa, HCB	in person, cooperation meeting	September 2001
Hakanen, Seppo	Marketing Manager	Elisa, PCS	in person, cooperation meeting	March 30th 2000,April 17th 2000
Hedberg, Nina	Voice Services, Manager	Elisa, Traffic and Subscriber connections, Elisa, PCS	in person, cooperation meeting	June 1999, March 30th 2000,April 17th 2000
Hietanen, Petri	Development Manager	Elisa, HCB	in person, cooperation meeting	September 2001
Hölttä, Pertti	Research Center	Elisa, Research Center	in person, cooperation meetings	January 31st 2000,January 25st 2001
Jäntere, Kirsi	Development Manager	Elisa, PCS	in person, cooperation meetings	March 2000,April 2000
Kaasinen, Katariina	Student, R&D developer	Elisa, HCB Nokia, R&D	in person ,e-mails	November 30th 2001,August 4th 2003
Lehmus, Pasi	Director	Elisa, PC	in person, cooperation meeting	January 15th 2002
Malmberg, Juha	Director	Elisa, Traffic and Subscriber connections Elisa, PCS	in person	May 1999,March 30th 2000,April 17th 2000, November 20th 2001
Masala, Sami	Business Development Manager	Elisa, HCB	in person ,e-mails	April 17th 2000, ,March 30th 2000,April 17th 2000, November 26th 2002, December 12th 2002, January 23rd 2003,February 27th 2003, November 13th 2003,December 16th 2003
Peltola, Hannu	Product Development	Elisa, Traffic and Subscriber connections	in person, cooperation meeting	May 20th 1999
Rasia, Olli	Department Manager	Elisa, PCS	in person, cooperation meetings	May 1999,March 30th 2000,April 2001,August 2001

continued on next page

continued

Simula, Timo	Head of Development, HCB's Manager	Elisa, HCB	in person, cooperation meetings, e-mails	March 30th 2000,April 17th 2000, November 7th 2001, November 26th 2002, December 12th 2002, January 30th 2003,March 14th 2003, September 9th 2003, October 13th 2003, November 3rd 2003
Tirkkonen, Piia	Student, R&D	Elisa, HCB, Nokia	in person, e-mail	November 22nd 2002, July 22nd 2003
Vainionpää, Sami	Multimedia Access, Development Manager	Elisa, PCS	in person, cooperation meeting	January 15th 2001
Viitala, Erkki	Customer Service Manager	Comptel	in person	December 2001, July 7th 2003
Vuolteenaho, Petri	Product Development, Manager	Elisa, PCS	in person, cooperation meeting	March 30th 2000 April 17th 2000

Chapter XII
Implementation of E-Governance:
Some Technical Issues and Challenges

Lalitsen Sharma
University of Jammu, India

ABSTRACT

This chapter introduces the term e-governance and puts on view that e-governance is not merely the computerization of government processes but a government process reengineering (GPR). E-governance is not only about implementing technologies and supplying services but that it provides solutions. The initiatives taken in this regard should not only remain limited to informational and transactional level but also reach complete transformational level. Some success stories of e governance initiatives have been reported with positive impact and value addition in citizen's services. However, low percentage of successful e-governance projects compels to identify and address the various issues accountable for it. It is argued that technical issues are equally imperative to be addressed along with socio-economic and political issues for implementing interoperable e governance solutions. An attempt has been done to identify such technical issues and provide solution so as to go for successful ICT strategy.

INTRODUCTION

The World Bank, in 2000, organized a forum called 'Voices of the Poor', which got feedback from 60,000 people in 60 countries and concluded that people wanted access to knowledge and opportunities instead of charity to fight conditions leading to poverty[1] (Nath, 2001).

The need to have access to information and knowledge by the citizen about the governmental programmes has been comprehended worldwide. In India, the government realized that the citizens must have right to information regarding government affairs towards development programmes and put forward the Freedom of Information Bill—December, 2002 which stood up for:

To provide freedom to every citizen to secure access to information under the control of public authorities, consistent with public interest, in order to promote openness, transparency, and accountability in administration and in relation to matters connected therewith or incidental thereto.

The above bill was enacted as *Right to Information* act from October 12, 2005 (The Gazette of India, 2005). The Information Technology Act was already enacted from October 18, 2000 that facilitated legal validity of transactions on the Internet and paved the way for the citizens to conduct business with the Government online rather than to be in line.

The integration of information and communication technology (ICT) in governance has been envisaged as a tool to accelerate the flow of information between different nodes and to cut down the time lag and reduce the probability of information loss and distortion; to widen the domain of information recipients and providers, leading to greater diffusion of information; to create information exchange interfaces that cuts across the traditional and organizational barriers to information flow such as hierarchy and bureaucracy (Baharul Islam, 2003). The transformation of government processes and services that make the use of most innovative information and communication technologies, particularly Web-based Internet applications, is termed as electronic governance or e-governance (Fang, 2002).

The European Commission (Alabau, 2005), defined electronic governance as "the use of information and communication technologies in public administrations combined with organizational change and new skills in order to improve public services and democratic processes and strengthen support to public policies."

Bill Gates wrote in his book "Business @ The speed of Thought"—E-governance is a method by which governments are empowering citizens to act for themselves without having to go through bureaucracy (Gates & Hemingway, 1999). Government, perhaps more than any other organization, can benefit from the efficiencies and improved services that stem from digital processes.

Electronic governance (e-governance) is not merely the computerization of government processes (Kumar, 2005) but also refer to the government process re-engineering (GPR), which has an innovative, technical, and managerial role. It refers to the use of technology to transform (not to translate) and enhance the access to and delivery of government services to benefit citizens, business partners, and employees (Torre, 2003).

This chapter deals with some technical issues and challenges in implementing e-governance at the national level. It also discusses different stages of implementations of e-governance system and recommends on developing successful ICT strategies incorporating comprehensive e-governance action plans.

STAGES OF IMPLEMENTATION OF E-GOVERNANCE

World Wide Web, shortly known as Web is the most ubiquitous and least expensive service of Internet and a medium for disseminating information. Due to its global interactivity and universal connectivity beyond organizations that helps creating new business models, processes, and strategies, it is regarded as technology innovation (Kim & Galliers, 2004). It has also emerged as a software technology and most of the applications are being developed Web based. Gartner Group predicted that by 2007, 70% applications would be Web Services[2] enabled (Lim, Sun, & Vila, 2004). The applications based on a Web Services framework promise to be a feasible solution and a standard way of setting up communication between applications. Irrespective of technology, the progress in implementation of an e-governance project is divided in the following three stages (Palanisamy, 2004).

Informational

This stage is again divided into three sub-stages namely; *emerging Web presence*, *enhanced Web presence,* and *interactive Web presence*; which covers three out of five phases identified by United Nations Public Administration Network (UNPAN) for assessing progress of a country towards e-governance development (Raghavan & Nair, 2004). In *emerging Web presence*, a country may have a single or few national Web sites that offer static legislative information to the citizens. The rules and regulations, laws, guidelines, handbooks, directories, and so forth, are published on the Internet. In *enhanced Web presence*, the number of government Web pages increases as information becomes more dynamic with citizens having more options. In *interactive Web presence*, the citizens can interact online with government bodies. The examples include seeking expert opinion about a subject matter, interacting with minister or an administrative officer, filing FIR or income tax return, and so forth. This phase also covers the first phase of Layne's four phased e-governance model that he named as *cataloguing* (Layne & Lee, 2001) and stage 1 and stage 2 named as *Web* and *data Web* respectively of four-stage model given by Australian National Audit Office (ANAO) in its Audit Report No.18 (1999-2000).

Transactional

The e-governance efforts mature at this stage. This covers the fourth phase of five phases identified by United Nations Public Administration Network (UNPAN). The security and privacy features become integral part of the Web sites. Business transactions can be fully completed on the Web by electronic filing and subsequent transfer of funds. The efforts at this stage remain to the level of translation; however, restructuring of government becomes essential at this stage. The e-governance efforts of Government of Andhra Pradesh, an Indian state, have reached this stage (Gupta & Sahu, 2005). This phase also covers the second phase of Layne's four phased e-governance model that he named as *transaction* (Layne & Lee, 2001) and stage 3 named as *service Web* of four-stage model given by Australian National Audit Office (ANAO).

Transformational

This stage is also termed as *fully integrated Web presence,* and represents the complete integration of all government services through one-stop-shop portals (Bagla, 2005). It integrates people, processes, information, and technology to meet government objectives. This stage completes the CIRCLE of e-governance by connecting the citizens; Informing the citizens; representing the citizens; consulting the citizens; listening to the citizens; and engaging the citizens. This covers fifth phase out of five phases identified by United Nations Public Administration Network (UNPAN). This phase also covers the third and fourth phases of Layne's four-phased e-governance model that he named as *vertical integration* and *horizontal integration* (Layne & Lee, 2001) and stage 4 named as *cooperation Web* of four-stage model given by Australian National Audit Office (ANAO). The vertical integration links the local systems to higher level systems within the same functionality where as horizontal system integrate systems across different functions providing real one-stop-shop portal for citizens.

The initiatives in e-governance, namely, government to citizens (G2C), government to business (G2B), government to government (G2G), government to employees (G2E), government to foreign trade (G2X), and citizen to citizen (C2C) are fully met at this stage.

CONCEPT AND CHALLENGES

The conceptual design of an e-governance application moves around two factors, namely, information integration and process integration with interoperability as a frame of reference and security and privacy of transaction as essential parts.

Within information integration, the guiding vision is to attain an inter-organizational information flow between information producers and consumers (which may be organizations, humans, or machines). For this purpose, the critical aspects of analysis and modeling include:

- Data definitions and information models (which may include process related data)
- Classification of information (e.g., through ontologies[3])
- Quality of information (reliability, access rights, etc.) (Klischewski, 2004)

Within process integration, the main objective is to achieve an inter-organizational workflow between different process participants (which may be organizations, humans, or machines). In order to achieve this objective, the critical aspects of analysis and modeling include:

- Identification and classification of performing entities according to role, function, or service within their organizational structure (e.g., front office versus back office)
- Description and representation of items to be processed and shared
- Indicators for process control (e.g., conditions, events, service levels)
- Process models (business processes, workflows with reference to informational resources) and process reference models (e.g., classification/types of administrative processes) (Klischewski, 2004)

There are number of success stories of e-governance initiatives in India that significantly reduced intermediation, corruption, exploitation (Chandrashekhar, 2005). Bhoomi in Karnataka, Computerized Interstate Check Post in Gujrat, and registration of property in Andhra Pradesh, Maharashtra and Karnataka are a few cases. For instance, people go to nearby public information kiosk, lodge their complaint or submit service request at nominal fee of Rs., 10 to 50 depending upon the type of complaint or service and get the complaint rectified or service done within a week's time. The OPEN project in Korea is widely known for reducing corruption[4]. E-citizen in Singapore (http://www.ecitizen.gov.sg) and Centerlink in Australia (www.centrelink.gov.au) speak the success stories of e-governance.

Quite obvious, to have access to e-governance services the working knowledge in computer is also essential. The findings of *hole-in-the-wall* experiment conducted in 1999 by Dr. Sujata Mitra, a pioneer Indian scientist suggested that:

Using the minimally invasive education method, six-to-thirteen year-olds can teach themselves to use computers regardless of their social, economic, ethnic, and even linguistic status (PTI, April 5, 2005).

For a successful implementation of e-governance programme, the "bottom up" approach has been found a worthy (Grundén, 2005) ICT strategy. A "bottom up" approach implicates grassroots feedback of citizens and is a biggest challenge. Most of the Canadian Web sites provide "top down" flow of informationand has turned out to be at saturation stage. There has to be a good number of forums, discussion groups, list servers, or interactive sites for public reaction to policy (Jenkins, 2002).

Countries may have number of social, economic, and political issues, which shape the implementation of e-governance programmes. Rajiv Rattan Sharma and Lalitsen Sharma (Sharma & Sharma, 2003) have discussed some socio-economic issues regarding the use and implementation of Information and Communication Technology in India. These are economy and infrastructure, language barrier, cultural diversity, and government attitude. However, technical capacities also appear to be significant factors for the advancement of e-governance (Moon & Norris, 2005). The various significant technical issues have been discussed below.

TECHNICAL ISSUES

According to Gartner Research Group, the transition from governance to e-governance takes place in four phases, namely, presence, interaction, transaction, and transformation. India is in early phases of its transition (Gupta & Sahu, 2005). In India, Andhra Pradesh is an action leader in e-governance initiatives, is still at transactional stage. There are a few projects that reached complete transformational stage. USA, Canada, and Singapore have reached this stage in few sectors like e-procurement (Contenti, et. al., 2003). According to an estimate, on an average, only 15% of e-governance projects are successful all over the world (www.dqindia.com, 2005). The development of e-government in most countries is still primarily aimed at developing electronic services that customers can access via Internet. However, continued development requires the development of an integrated government portal and reengineering of back-office processes (Kunstelj & Vintar, 2004).

A survey, based on observation and analysis of Web sites and portals of each of the 467 local authorities in the UK, has been undertaken by Society of IT Managers (SOCITM) in corroboration with the UK portal survey in the year 2001, 2002, and 2003. The SOCITM 2003 survey's results showed that only 2% sites have reached *transactional* stage. This low percentage is mainly due to the lack of interaction in front-office and back-office, difficulties in systems integration and lack of middleware development capability (Musgrave, 2005).

The present technological solutions are sufficient enough to develop and implement e-governance applications to the transformational stage and it has positive impact in delivering government services (Kochha & Dhanjal, 2005). Despite that, the low percentage of success of e-governance applications is a matter of apprehension. A completely technology driven approach will equally fail as a soft system methodology. Instead, designing e-governance systems requires a balanced approach which integrates the many facets of public administration for example, a combination of strategic, organizational, technical, functional, cognitive, social, judicial, and cultural aspects (Wimmer & Krenner, 2001). The various technical issues accounted for the low percentage of successful e-governance initiatives are described below.

Management of Change

While practicing e-governance the most fundamental issue to be addressed is management of change. The Brown University, USA, analyzed 1197 national government Web sites in 198 countries using various possible criteria and reported that services offered by 88% of the Web sites were not executable online, 86% of the Web sites had no privacy policy, 91% had no security policy, and 75% did not respond to an e-mail responsiveness test (Raghavan & Nair, 2004). Technology is not to be blamed here. In fact, acclimatization to new culture is very much required by the service providers and service consumers. E-government implementation is affected by the perception of ICT by civil servants, who must see it as a tool rather than a threat in order to adapt it as a bridge between citizen and government (Joia, 2005).

Dutch parliamentary committee in 2001, affirmed that '*information and communication technology (ICT) is not a supporting technology, but coincides with the primary process and touches government at its core*' (Bourn, 2002)[5]. The government processes has to be reengineered in context with ICT developments (Grabow, Drüke, & Siegfried, 2004). For a successful project, change management and process re-engineering are found to be nearly 80% accountable. To a certain extent reengineering of processes relies heavily on capable IT enabler. In the absence of this enabling, reengineering of processes is unavoidable to fail. Outsourcing effectiveness is minimal, confusing, and highly subjective. It saves money but no improvement in quality at all (Joia, 2005).

Reengineering involves innovative delivery of public services and facilities, rural and urban development schemes through electronic data interchange (EDI), procedural and legal changes in the decision and delivery making processes for Internet, and other IT based technologies, fundamental changes in Government decision management, changes in the decision making procedures in terms of decision making levels, and delegation of authority, mandatory changes in legal provisions and administrative reforms to give effect to the technology objectives (Kothari & Bansal, 2005). The efforts are required to be made to institute *process-led initiative* rather than *champion-led initiative*.

Taking a case of India, a cultural difference has to be removed between Central and State governments. Organizational structure and hierarchy has to be the same. The processes should not vary a large and be suitably customized as per location and work specific requirements (Budhiraja, 2003). A simple and comprehensive Web portal should be designed with uniform look and must comply with the guidelines issues by the Ministry of Information Technology, Govt. of India. A memorable nomenclature of official web sites is required. As every state has been assigned two-letter code for registration of vehicles, it is opined to use this code as a generic name to derive the address of official Web sites like http://www.ap.gov.in for Andhra Pradesh; http://www.hp.gov.in for Himachal Pradesh; http://www.pb.gov.in for Punjab, and so forth.

Interoperability

Interoperability is the ability of a system or process to use information and/or functionality of another system or process by adhering to common standards. Interoperable designs play important role in addressing implementation issues and providing flexibility in e-governance models. According to the New Zealand e-government strategy document (E-government Interoperability Framework Version 2, published in December 2001), interoperability has to be achieved in the five elements namely; business process interface, service delivery, access, information sharing, and interconnection[6].

In European Union, member states started a project to provide electronic identification and authentication of citizens through the use of electronic identity (e-ID). The electronic identity in Europe was introduced by Finland in 1999 followed by Italy (March 2001) and Belgium (April 2003). In less than 3 years Estonia has equipped 45% of its population with e-ID cards (Europa News, 2004). But, in year 2005, Austria and other member states of EU started rolling out the e-ID card, as the interoperability of these cards was not given due importance by the member states while developing their individual systems. As a result, the citizens could no longer use their e-ID card for identification and authentication with government and private sector of another member state of EU (Hayat, Posch, & Leitold, 2005).

Many e-government projects like eGOV (http://www.egov-project.org), EU-PUBLI.com (http://www.eu-publi.com), OntoGov (http://www.ontogov.org), and FASME (http://www.fasme.org/index-org.html) are being developed in European Union and various approaches have been proposed for the design and the development of an interoperable architecture to deliver e-government services to citizens.

Web services is the latest software technology in the industry to address the system interoperability problems. It is based on the technologies governed by the World Wide Web Consortium (W3C). W3C is an international standard body that looks all Web-related technologies with a support from major IT companies such as Microsoft, IBM, Sun, HP, and Oracle. As such, it is projected as a strong technology and many IT organizations will investigate and adopt if proven viable. Gartner Research Group compared Web services with earlier technologies and stated that this time things may be different because "*With Web services, all the major vendors are on board with their support*" (Lim, Sun, & Vila, 2004).

Legacy Applications

The integration of various old applications (legacy applications) has been identified as a major attribute of second-generation portal products (Musgrave, 2005). Such integration can be achieved by the adoption of common standards and the use of Web Services for sharing data between different systems and implementing service oriented architecture (Musgrave, 2005).

Linking legacy systems and their attached databases to new Web-based applications and distributed databases seems to be a natural way of evolution. The main areas to deal with the legacy systems are given hereunder. A careful selection of these can lead to better cost benefit trade off.

Migration/Replacement

It refers to phasing out the old systems and structures, building new systems, porting the old databases into the new environment, rewriting the application code on basis of renovated or reengineered processes. Microsoft .NET framework provides extensive facility to support the legacy source code migration (Gowthman, Mustafa, & Khan, 2005).

Taking a case of Department of Social and Family Affairs (DSFA), Republic of Ireland[7], this formulates and delivers various citizen-friendly schemes such as Old Age Pensions and Child Benefits. The delivery is managed by various applications. These applications were built using pre-.NET technologies. Due to the design constraints enforced by these technologies, the applications have lots of code duplication and many joins at the database level thus making them slow. Hewlett Packard, Ireland with Trisoft Systems as the partner migrated application to the .NET platform. Migration of application resulted in a re-architecture, reengineering cum migration exercise that ultimately resulted in a 70% increase in overall performance of the application.

Encapsulation

It refers to shielding legacy system and freezing them in terms of further developments. This approach involves building front-ends that serve as gateways between old and new systems. Common Gateways Interface (CGI) plays an important role in connecting old systems with Web-based front end.

Researchers at Virginia Tech and Purdue University have developed project for Indiana's Family and Social Services Administration (FSSA), which encapsulated the legacy applications into Web services. Web services enabled the agency to provide value-added services by defining a new service that uniformly handled privacy issues; and standardized the description, discovery, and invocation of social programs (Medjahed, et al., 2003).

Expansion

It refers to adding Web/Internet capable components to legacy systems in a fashion that the legacy systems can directly be accessed from Web-based application and databases and vice versa.

Regional Language Support

E-governance strategy had been a top-down, techno-centric exercise that neglected intended client groups and resulted in failure. The language is a main issue (Thomas, 2004). The access to information must be permitted in the language most comfortable to the public user, generally the local language.

Taking a case of India, a limited number of Indian language based applications are developed and deployed in e-governance domain. Most of these applications still have English interface. Mostly, the static reports are generated using Indian language template. Few applications are enabled at the level of data input-output in local languages. E-governance application penetration in various States is at various levels with the governments of Andhra Pradesh, Karnataka, and Madhya Pradesh being the first few to take major initiatives in this area (Dash, Ray, & Surinder Kumar, 2005). Reason of non-proliferation of applications is the lack of appropriate language interface (Chopra & Swaran Lata, 2004). There is a need to build quick translation tools, which aid the human translators to meet the large translation requirement of the e-governance domain.

Accessibility

According to UN World Public Sector Report 2003, E-government at the Crossroads, October 2003, "The potential of e-government as a development tool hinges upon three prerequisites—a minimum threshold level of technological infrastructure, human capital, and e-connectivity for all"[8]. E-government readiness strategies and programmes will be effective and will include all people only if all are connected and have access to the Internet. The primary challenge of e-government for development therefore, is how to accomplish e-connectivity through commonly used communication devices.

Accessibility refers to the extent to which the Web site and its contents are available to a wide range of users with varied levels of physical capabilities/skills and technologies. Accessibility should be regardless of disability and independent of technology. The contents should be in all forms, viz. text, video, audio, Braille, and so forth, and accessible through all available devices such as computer, PDA, mobile, and so forth.

The World Wide Web Consortium's (W3C) Web accessibility initiative (WAI) is an internationally agreed recommendation for Web site accessibility for people with special needs and government Web sites are expected to adhere to these standards. The government Web sites under evaluation should be studied and tested to check how easy/difficult it is for the common citizens (even those with physical disabilities) to search and access them and whether the portals appear and function properly on the commonly available browsers. The presence of metadata information, which is vital to ensure a good visibility of the site in major search engines, should also be adequately tested. The response speed of the sites is another important criterion for testing the accessibility and checking how much time it takes for a common citizen with an average speed of

Internet Connectivity to access the information given on a government Web site (Verma, Kalra, & Mohandas, 2005).

Digital divide is not a major concern because those who do not have Internet access at home can always travel to community technology center or information kiosk to go online. Indeed, telecenters are one of the driving forces in combating the digital divide, particularly because they are often seen as trusted institutions within the community where residents unfamiliar with ICTs may interact with fellow community members to learn how to use them.

Standardization in Data Encoding

Existing developments in data encoding in regional languages are being carried out on a number of vendor dependent non-standard platforms, which has resulted into non-interoperability of stored data. Hence, implementation of e-governance solutions using standards in data encoding is of utmost priority.

Taking a case in India, IS13491:1991, Indian standard code for information interchange (ISCII) is all set for use, but a number of vendors, while using ISCII still store data in their own font codes. Due to the limited market size, many vendors implement their proprietary codes to protect their market segment. In the wake of globalization, multi-nationals around the globe formed a Unicode consortium to define the Unicode standard representing the languages of the world. The latest version of Unicode is 4.0, brought out in 2004. Since most of the e-governance applications are Intranet/ Internet based, the need to migrate to Unicode is being felt, so that data can be ported and accessed across the platforms (Chopra & Swaran Lata, 2004). The data encoded using nonstandard and vendor dependent fonts need to be converted to Unicode for easy interoperability.

Database Reengineering

Migrating data to new environment is another hard challenge. Studies show that data migration costs can often escalate to 10 times the original budget or cause scope contraction and therefore poor cost-benefit return (Fryman, 2005). Database reengineering consists of transforming a legacy database according to new technical requirements while keeping the information contents unchanged. Substituting a modern data management system (relational DBMS for instance) for an outdated manager (typically standard file manager) or improving the logical schema to gain better performance are popular scenarios. Transformational engineering that defines processes as chains of transformations has proved to be both an elegant and efficient approach to perform these processes as semantics-preserving transformations (Cleve, Henrard, & Hainaut, 2005).

Data warehousing becomes imperative for transformational e-governance. It provides architectures and tools to systematically organize heterogeneous data and bridging information gap so as to understand and use data to make strategic decisions.

RECOMMENDATIONS

This has become obvious from above discussion that e-governance applications demand interoperability at all the technical levels and domains. Interoperability in process, data and information flow is considered to be a decisive factor for the success of e-governance initiatives and is envisaged as a successful ICT strategy. The e-governance application must comply with the standard interoperability framework that governs accessibility and adaptability irrespective of user diversity. The European interoperability framework (EIF) is worth mentioning, which defines organizational, semantic, and technical interoperability (Muller, 2005). Organizational interoperability concerns with thc rcquirement of the user and varies from organization to organization. The technical interoperability is being achieved through open standards like TCPIP, SMTP, XML, ebXML, and Web Services. The semantic interoperability is addressed by the application designer with service ontologies. Semantic Web services provide good support to designer to put into practice the semantic interoperability. The Service Oriented Architecture is advocated here which not only integrate the legacy applications but also useful for software re-use. The centrally agreed XML schemas should be re-used for interchanging data. For the interoperability in data interchange Unicode is the best choice.

CONCLUSION

The success stories of e-governance conclude that it not only adds to the value of services, but also maintain disintermediation and minimizes corruption. It is essential that a bottom-up approach with the participation of stakeholders should be adopted. Interoperability of e-governance application enhances the scope and the viability of the services and hence this issue should be addressed at the very beginning of the design of the application. The integration and the compatibility of legacy application should be kept in mind while developing a new e-governance model. Web services are the best suitable software technology to address the most of the issues of e-governance applications. Accessibility and other standards should also be complied. E-governance refers to government process reengineering (GPR); it is a transformation not translation. It mixes up innovative, managerial, and technical role and appropriate accommodation of these processes will lead into successful e-governance system at all levels.

FUTURE RESEARCH DIRECTIONS

The technical issues have been discussed here contemplating interoperability that is highly demanded in any transformational e-governance model. Application interoperability has been well addressed with Web services and XML and is widely implemented (Miguel, 2003). Semantic interoperability and its link with enterprise architecture represent the core technological research area. In this regard much research work is needed in the development of data definitions, structures, and models to allow the great variety of data to be integrated and rich pattern matching (Cabral et al., 2004). Semantic Web services is an emerging area of research and will have attention of industry and academia for next few years (Davis, Fensel, & Richardson, 2004).

According to a report of the E-government interoperability workshop held in Brussels on 1st March 2004, an issue was raised by several participants regarding the development of e-government services that should be event-driven and self-adaptive[9]. Thus, a qualitative research is required to be carried out in connecting and orchestrating independent processes in a macro process that supports self-management and self-healing. AgrentEclipse, a real-time, event-based billing system for communications companies from Argent Networks (www.argentnetworks.com) is an example of event-based system (Langdon, 2003).

The primary research focus is required to be made on proactive notification of events that has to be generated in heterogeneous environments. The infrastructure may consist of a publish/subscribe notification mechanism and a reactive functionality (Cilia, Bornhövd, & Buchmann, 2005). The events should originate from heterogeneous sources together with their ontology-based context metadata to make them self-describing beyond the boundaries of the event source (Yagüe, Maña, & Lopez, 2005).

REFERENCES

Alabau, A. (2005). The European Union and its e-government development policy. *Vodafone Spain Foundation*, (p.17).

Bagla, P. (2005). Booming computer sector seen as a mixed blessing. *Science, 310*, 1754. Retrieved from www.sciencemag.org

Baharul Islam, K. M. (2003, May 6). Information age government: Success stories of online land records & revenue governance from India. *Executive Summary, United Nations Economic and Social Council.*

Bourn, J. (2002, April 4). Better public services through e-government. *Report by the Comptroller and Auditor General HC 704-III Session, National Audit Office.*

Budhiraja, R. (2003). *Electronic governance-A key issue in the 21st century.* Retrieved from http://mit.gov.in

Cabral, L., Domingue J., Motta, E., Payne, T., & Hakimpour, F. (2004). Approaches to Semantic Web services: An overview and comparisons. *Computer Science*, *3053*, 225-239.

Chandrashekhar, R. (2005, January 14). e-Governance: Unlocking India's latent potential. *ASSOCHAM–World Bank 2nd Summit*

Chopra, P., & Lata, S. (2004). E-governance-Localization issues. *Vishva Bharat@tdil.* Retrieved from www.mit.gov.in/tdil/Oct_2004/egov-li-16.pdf

Cilia, M., Bornhövd, C., & Buchmann, A. P. (2005). Event handling for the universal enterprise. *Information Technology and Management*, *6*, 123-148.

Cleve, A., Henrard J., & Hainaut, J-L. (2005). Co-transformations in information system reengineering. *Science Direct (Electronic Notes in Theoretical Computer Science)*, *137*, 5-15.

Contenti, M., Termini, A., Mecella, M., & Baldoni R. (2003). An e-service-based framework for inter-administration cooperation. *LNAI, 2645*, 13-24.

Dash, S. S., Ray, D., & Surinder, K. (2005). Developing interoperable information system. *e-gov*, *1*(2), 10-13. Retrieved from www.csdms.in/egov/march05.pdf

Davis, N. J., Fensel, D. & Richardson, M. (2004). The future of Web services. *BT Technology Journal*, *22*(1), 118-130.

Europa News (2004, September). *Belgium close to solving electronic ID cards distribution deadlock.*

Fang, Z. (2002). E-government in digital era: Concept, practice, and development. *International Journal of The Computer (The Internet and Management)*, *10*(2), 1-22.

Fryman, H. (2005). Data migration: Lowering cost and risk to move to new systems and applications. *Informatica, What Works, 19*(5). Retrieved from http://www.tdwi.org/Publications/WhatWorks/display.aspx?id=7524

Gates, B., & Hemingway, C. (1999). *Business @ the speed of thought.* Warner Books.

Gowthman, K., Mustafa, K., & Khan, R. A. (2005). Legacy source code migration to .NET architecture. *Journal of Computer Science*, *1*(1), 45-52.

Grabow, B., Drüke, H., & Siegfried, C. (2004). Factors for success for local community e-government. *German Journal of Urban Studies, 45*(2). Retrieved from http://www.difu.de/ index.shtml?/publikationen/dfk/en/04_2/04_2_kubicek_wind.shtml

Grundén, K. (2005, September 13). Implementation of "Vision of E-Government 2007" at the county administration of Sweden. *eGovernment Workshop '05 (eGOV05),* Brunel University, West London.

Gupta, M. P., & Sahu, G. P. (2005). Need fruits of e-governance? *e-gov., 1*(2), 6-9.

Hayat, A., Posch, R., & Leitold, H. (2005). *Identifying obstacles in moving towards an interoperable electronic identity management system.* IDABC Conference.

Jenkins, G. (2002). Lessons from the trenches of e-Gov: Part 2. *Ubiquity: An ACM IT magazine and forum.* Retrieved from http://www.acm.org/ubiquity/views/g_jenkins_2.html

Joia, L. A. (2005). A framework for developing regional e-government capacity building networks. *MIT Press Journal, 2*(4), 61-73.

Kim, C., & Galliers, R. D. (2004). Toward a diffusion model for Internet systems. *Internet Research, 14*(2), 155-166.

Klischewski, R. (2004*). Information integration or process integration? How to achieve interoperability in administration.* Retrieved from http://is.guc.edu.eg/uploads/egov2004/klischewski.pdf

Kochha, S., & Dhanjal, G. (2005). E-governance report card. *Yojana, 49*, 60-65.

Kothari, S., & Bansal, R. (2005). Innovative thinking in government. *Yojana, 49*, 34-39.

Kumar, K. (2005, February). Strategic framework for effective e-governance. *PC Quest*, 68-69.

Kunstelj, M., & Vintar, M. (2004). Evaluating the progress of e-government development: A critical analysis. *Information Polity, 9*(3-4), 131-148.

Langdon, C. S. (2003). Information systems architecture styles and business interaction patterns: Toward theoretic correspondence. *Information Systems and e-Business Management, 1*, 283-304.

Layne, K., & Lee, J. (2001). Developing fully functional e-government: A four stage model. *Government Information Quarterly, 18*, 122-136.

Lim, B.B.L., Sun, Y., & Vila, J. (2004). Incorporating WS-security into a Web services-based portal. *Information Management & Computer Security, 12*(3), 206-217.

Medjahed, B., Rezgui, A., Bouguettaya, A., & Ouzzani, M. (2003, January/February). Infrastructure for e-government Web services. *IEEE Internet Computing*, 58-65.

Amutio-Gómez, M.A. (2003). The construction of a Pan-European e-government: The IDA Programme. *UPGRADE, the European Journal for the Informatics Professional, 4*(2), 52-62.

Moon, M. J., & Norris, D. F. (2005). Does managerial orientation matter? The adoption of reinventing government and e-government at the municipal level. *Information Systems Journal, 15*, 43-60.

Muller, B. (2005). The European interoperability framework: An industry perspective. *European Review of Political Technologies, 3.* Retrieved from www.politech-institute.org/review/articles/MULLER_Benoit_volume_3.pdf

Musgrave, S.J. (2005). Community portals –The UK experience a false dawn over the field of dreams? *The Journal of Community Informatics, 1*(2), 32-44.

Nath, V. (2001). Empowerment and governance through information and communication technologies: Women's perspective. *Intl. Inform. & Libr. Rev., 33*, 317-339.

Palanisamy, R. (2004). Issues and challenges in e-governance planning. *Electronic Government, 1*(3), 253-272.

PTI, April 5, 2005. www.agencyfaqs.com, April 5, 2005. Indo-Asian News Service. Retrieved April 3, 2005, from http://www.infochangeindia.org/Itanddltop.jsp? section_idv=9#1917 as accessed on 22/08/2005

Raghavan, G. V., & Nair, V. S. M. (2004). e-Governance: Leveraging IT for reinventing the government. *Manorama Yearbook 2004* (pp. 130-138).

Sharma, R. R., & Sharma, L. (2003). Use of information technology in education: Some issues and challenges. In B. Shah, Pub., K. Mahal, & A. (Eds.), *Information technology and education: Vision and prospects* (pp. 44-57).

The Gazette of India (2005, June 21). Extraordinary Part II, Section I. Mnistry of Law and Justice (Legislative Department). Govt. of India. New Delhi.

Thomas, J. (2004). Strategically transforming governance for the new global economy. *Strategies for Competitiveness in the New Global Economy.* Presented at the 4th International Conference on Quality Management, Teheran, Iran. Retrieved at www.qmconf.com/Docs/0097.pdf

Torre, N. P., & Ponce de la (2003). e-Government and the "network society". *UPGRADE, the European Journal for the Informatics Professional, 4*(2), 14-20. Retrieved from http://www.upgrade-cepis.org

Verma, N., Kalra, S., & Mohandas, V. S. (2005). Assessing citizen centricity of government websites. *e-gov (www.csdms.in/egov/july05.pdf), 1*(4), 29-31.

Wimmer, M., & Krenner, J. (2001). An integrated online one-stop government platform: The eGOV Project. In *Proceedings Hofer, Chroust IDIMT-2001 9th Interdisciplinary Information Management Talks* (pp. 329-337) (ISBN 3-85487-272-0).

Yagüe, M. I., Maña, A., & Lopez, J. (2005). A metadata-based access control model for web services. *Internet Research, 15*(1), 99-116.

ADDITIONAL READINGS

Harris, R., & Rajora, R. (2006). *Information and communications technology for governance and poverty reduction—A study of rural development projects in India.* New Delhi, India: Reed Elsevier India Private Limited.

Jones, S. (Ed.). (1999). *Doing Internet research: Critical issues and methods for examining the net.* India: Sage Publications Pvt. Ltd.

Kumar, R.J. (2006). *IT in governance: An introduction.* Hyderabad, India: ICFAI Press.

Manorma Year Book, 2004.

Rajagopalan, S. (2006). *The digital divide: An introduction.* Hyderabad, India: ICFAI Press.

Traunmüller, R. (Ed.). (2002). *The e-Business challenge.* Boston: Kluwer Publishing.

Yojana (2005, August). *Planning commission,* Govt. of India.

ENDNOTES

1. Available at www1.worldbank.org/prem/poverty/voices/
2. The W3C Web services Description Working Group defines Web service "a software application identified by a URI [1-IETF RFC 2396], whose interfaces and binding are capable of being defined, described, and discovered by XML artifacts and supports direct interactions with other software applications using XML based messages via internet-based protocols". Further information at <http://www.w3.org/ 2002/ws/desc/>.
3. Explicit and formalized specifications of conceptualization
4. Case study available at http://www1.worldbank.org/publicsector/egov

[5] Available at http://www.governmenton-theweb.org/downloads/papers/Cultural_Barriers.pdf

[6] Available at http://www.e-government.govt.nz/docs/e-gov-strategy-dec-01/

[7] http://www.third-i.trisoft.net/casestudies/DSFA_Case_Study.pdf; as accessed on Oct. 11, 2005

[8] Available at unpan1.un.org/intradoc/groups/public/documents/UN/UNPAN012733.pdf

[9] Available at www.egovinterop.net/ SHBlob.asp?WCI= ShowD&I=176&name= 040301 %20 Research %20 Workshop %20 report.pdf

Chapter XIII
Integration Strategies and Tactics for Information Technology Governance

Ryan R. Peterson
Information Management Research Center, Spain

ABSTRACT

Amidst the challenges and changes of the 21st century, involving hyper competitive market spaces, electronically-enabled global network businesses, and corporate governance reform, IT governance has become a fundamental business imperative. IT governance is a top management priority, and rightfully so, because it is the single most important determinant of IT value realization. IT governance is the system by which an organization's IT portfolio is directed and controlled. IT governance describes (a) the distribution of IT decision-making rights and responsibilities among different stakeholders in the organization, and (b) the rules and procedures for making and monitoring decisions on strategic IT concerns. The objective of this chapter is threefold. First of all, to describe past developments and current challenges complex organizations are facing in governing the IT portfolio of IT applications, IT development, IT operations and IT platforms. Based upon the lessons we've learned from the past, one of the key objectives is to move beyond 'descriptives', and discuss how organizations can diagnose and design IT governance architectures for future performance improvement and sustained business growth. The final objective of this chapter is to provide a thorough understanding and holistic picture of effective IT governance practices, and present a new organizing logic for IT governance.

INTRODUCTION

One morning in 1997, Ralph Larsen, CEO of Johnson & Johnson, called his controller, JoAnn Heisen, to his office for a meeting. Ralph had just launched a corporate-wide cost-cutting campaign to help finance a drive into highly competitive and costly new drug markets. That morning, Ralph wanted Johnson & Johnson's IT organization to be a bigger part of all of that - and to get smarter about how the company was using IT.

Johnson & Johnson was spending millions annually on IT, yet business executives and customers weren't getting the business information they needed, and the business value they wanted. Hospitals, for example, were asking Johnson & Johnson to help them cut their stashes of supplies, but Johnson & Johnson didn't have the Web-based tracking systems needed to deliver on that request. The electronic networks that did exist suffered frequent break-downs. JoAnn recalls "Nobody was talking to each other. And why should they? Nobody asked the business units to talk with each other before, and no one had asked IT how much we were spending on the business." Ralph told JoAnn he wanted to cut IT costs dramatically, but he also wanted to oversight reform.

That morning JoAnn left Ralph's office with a new job—as CIO—and a mission to standardize systems, cut IT costs, and align the IT organization with business strategies, while simultaneously acknowledging the decentralized culture of Johnson & Johnson's numerous business divisions across different countries.

Does the Johnson & Johnson chronicle (Alter, 2001; Scheier, 2001) seem familiar to you? This real-life case illustrates many of the problems and challenges large complex organizations have been facing for over a decade. It's almost cliché that chief executives across the board have experienced many failures and disappointments with IT-enabled business transformations. Expecting strategic value innovation, executives have faced project cancellations, business disruptions, rising customer churn, decreasing shareholder value, and many other disappointments, including loosing their jobs.

In fact, executives today are less concerned about getting 'Amazon-ed', than about getting 'Enron-ed'. Corporate responsibility, business sustainability and governance reform are currently high on the strategic agenda in many companies. The growing scrutiny over shareholder interests, lingering economic growth and corporate performance, have now also prompted renewed soul-searching and interest into the governance of information and Internet-based technologies. Amidst all these changes and challenging responsibilities, governing IT for sustaining business value has become a fundamental business imperative for thriving in the 'old new' economy.

Re: Information Technology Governance

Boards and business executives have come to recognize that whereas traditionally they could delegate, avoid, or ignore IT decisions, today they cannot conduct marketing, R&D or HR without depending on IT at some point in time. Metaphorically, a 'Speak-See-Hear No Evil' attitude towards IT governance is no longer viable in today's business landscape (Figure 1).

With the dawning of the 21st century, organizations are experiencing a global digital revolution with profound impacts on their business models and electronic business processes, wherein the interdependency between business and IT is intensely reciprocal. This e-business genesis coincides with a business landscape, in which the intensity, unpredictability and diversity of change has accelerated to create a condition of hypercompetition (D'Aveni, 1994), in which there is no stable competitive position, bureaucratic hierarchies become a competitive liability, organizational boundaries are being redefined, core competencies develop into core rigidities, and strategic fit is fleeting, all against a backdrop of global economic turmoil and market volatility.

Figure 1. IT governance: No more 'monkey business'

Figure 2. Top management priorities in 2002 (GartnerEXP, 2002).

Rank	Priority
1	Strategizing for business-IT alignment
2	Providing leadership and guidance for the Board of Directors and senior executive
3	Demonstrating business value of IT
4	Developing IT senior management team
5	Reducing total IT costs
6	Strengthening program management, and project prioritization and selection
7	Tightening security and privacy safeguards
8	Developing electronic business architectures
9	Attracting, nurturing and sustaining IT resources
10	Reducing IT complexity

The growing infusion of e-business technologies in and between organizations, and the 'e-wakening' from the dot.com frenzy, has made both business and IT executives recognize that getting IT right this time will not be about technology, but about governing IT. GartnerEXP's 2002 survey of its 1,500-member CIO community, indeed confirms that the top management priorities are all within the domain of IT governance (Figure 2). As business models and IT become virtually inseparable, managing their integration and co-evolution involves putting the right people in the right place to understand and take direct responsibility for making sure the organization meets its strategic goals, and that all efforts—including IT—are directed toward that end.

Nevertheless, how to govern IT for sustained value innovation, remains, an enduring and challenging question. In the case of Johnson & Johnson, questions the new CIO needed to answer were:

- How can IT best support a complex organization composed of diverse, globally operating business units?

- What and how much should we standardize? And who should be involved in making these strategic decisions?
- How does a company choose the best arrangement for IT governance? And what is the 'best' arrangement for our company?

At the heart of Johnson & Johnson's journey, as in many other organizations, has been a need to find answers to tough, almost timeless, questions of governance: how to organize for diversity and differentiation, while preserving integration and unity of direction? How to promote local innovation, yet reap the benefits of scale and scope in a global economy? In terms of IT governance, the general question is how should we be governing our IT activities in order to manage the imperatives of the global, hyper competitive, digital economy? More specifically, how to direct, organize and control IT in order to realize value for the enterprise, and its key constituencies? In this chapter, these and other questions, and their answers, are explored and discussed.

Objectives and Organization of this Chapter

The purpose of this chapter is, first of all, to describe past developments and current challenges complex organizations are facing in IT governance. Based upon the lessons we've learned from the past, one of the key objectives is to move beyond 'descriptives', and discuss how organizations can diagnose and design IT governance solutions for future improvement and growth. The final aim of this chapter is to provide a thorough understanding and holistic picture of effective IT governance practices, and present a new organizing logic for IT governance.

This chapter is organized as follows. In Section 2, a review of previous studies and developments is presented, and contrasted with current changes and challenges in contemporary businesses. The fundamental problem underlying IT governance is described in Section 3, along with a discussion of design strategies and integrative solutions. In Section 4, the results of a multi-client study are summarized, and a diagnostic model for IT governance is outlined. Several lessons for future architecting of IT governance are presented in Section 5, in which a 'new' logic for IT governance is described. This chapter concludes with a discussion of the major implications and directions for practice and research on IT governance.

INFORMATION TECHNOLOGY GRASPING AT GOVERNANCE

In this section, the concept of IT governance is discussed, and a review of previous studies is presented, and contrasted with current business environments. Specifically, this section addresses the body of knowledge on IT governance (Section 2.1). A review of previous studies is presented (Section 2.2), and compared to state-of-the-art developments in contemporary organizations (Section 2.3). This section concludes with a discussion of existing IT governance principles, and questions the suitability of the current paradigm (Section 2.4).

Information Technology Governance: Definitions and Myths

IT governance has been the subject of much debate and speculation over the past decades, yet it remains an ephemeral and 'messy' phenomenon, emerging in ever-new forms with growing complexity. Despite more than 30 years of empirical research, and management theory and practices, there are still many breaches to be bridged.

Although questions and concerns regarding IT governance have been around since the introduction of IT in companies, currently there is no consistent, well-established body of knowledge and skills regarding IT governance. This is partly attributed to the simultaneous enduring and evolving nature of IT governance, but partly also due to the specialization and disconnectedness between globally-dispersed IT governance interest communities. For instance, the differentiation and dissonance among these professional communities has led to the use of various definitions and models on IT governance (Figure 3), which consequently has not been beneficial to the development of a cumulative body of knowledge and skills on IT governance[1].

Similar to corporate governance, IT governance is a topic that has recently been re-discovered, and as yet, is ill-defined and consequently blurred at the edges. To cite Burt (1980), the rich vocabulary emerging from the literature is like a terminological jungle in which any newcomer may plant a tree. Considering the foregoing definitions, and consistent with Cadbury and in line with the OECD's definition of corporate governance, the following definition of IT governance is used in this chapter (Peterson, 2001):

IT governance is the system by which an organization's IT portfolio is directed and controlled. IT governance describes (a) the distribution of IT decision-making rights and responsibilities among different stakeholders in the organization, and (b) the rules and procedures for making and monitoring decisions on strategic IT concerns.

IT governance thus specifies the structure and processes through which the organization's IT objectives are set, and the means of attaining those objectives and monitoring performance. IT governance is a second order phenomenon, i.e., governance is the set of decisions about *who* and *how* decisions on strategic IT concerns are made. The abovementioned definition also (implicitly) addresses several 'IT governance myths' that have existed for a long time and still persist, yet, which need to be debunked (explicitly) if business and academic communities are to move forward in their IT governance thinking and IT governance practices.

Figure 3. IT governance definitions

What is Information Technology Governance?

IT governance is the organizational capacity exercised by the board, executive management and IT management to control the formulation and implementation of IT strategy and in this way ensure the fusion of business and IT.

- Van Grembergen, 2002.

IT governance describes a firm's overall process for sharing decision rights about IT and monitoring the performance of IT investments.

- Weill & Vitale, 2002

IT governance is the responsibility of the board of directors and executive management. It is an integral part of enterprise governance and consists of the leadership and organizational structures and processes that ensure that the organization's IT sustains and extends the organization's strategies and objectives.

- IT Governance Institute, 2001

IT governance defines the locus of enterprise decision-making authority for core IT activities

- Sambamurthy & Zmud, 2000

IT governance refers to the patterns of authority for key IT activities.

- Sambamurthy & Zmud, 1999

IT governance is the degree to which the authority for making IT decisions is defined and shared among management, and the processes managers in both IT and business organizations apply in setting IT priorities and the allocation of IT resources.

- Luftman, 1996

IT governance describes the locus of responsibility for IT functions.

- Brown & Magill, 1994

IT Governance Myth: IT Governance is the Responsibility of the CIO

While IT governance is certainly an essential element of a CIO's portfolio, the CIO is *not* the primary stakeholder. Still too often, corporate executives and business managers assume that the CIO is taking care of IT governance. Moreover, abdication of responsibility and accountability by the business, and 'pointing the finger' at IT will not resolve the 'IT productivity paradox', nor the many misalignments between business and IT.

More then a decade ago, Rockart (1988) argued that business management needs to take charge of IT. Remenyi (1997) also indicates that line leadership is an absolute necessity, and placing business management as the principle stakeholder, repositions the locus of responsibility for realizing IT value squarely where it should be, i.e., with the business. IT governance effectiveness is only partially dependent on the CIO and other IT executives, and should be viewed as a shared responsibility and enterprise-wide commitment towards sustaining and maximizing IT business value (Figure 4).

Developing business-IT leadership and IT competencies in business management is pivotal for IT governance effectiveness and realizing business value with IT (Peterson, 2001; Weill & Broadbent, 1998). In essence, IT governance relies on the capability of business executives at all levels to set the strategic business—including IT—agenda, understand the business capabilities—not the technicalities—of IT, and monitor business value appropriation from IT.

IT Governance Myth: IT Governance is Concerned with Organizing the IT Function

Traditionally, the IT function has been regarded as a single homogenous function. However, given the widespread proliferation and infusion of IT in organizations, involving e.g., technical platforms, shared IT services centers, and local business-embedded applications, *the notion of a single homogenous IT function is obsolete.*

Weill & Broadbent (1998) indicate that contemporary organizations consists of a portfolio of different interdependent business functions and technical capabilities, some of which are allocated to different levels in the enterprise, and/or to third-party vendors. In discussing 'the new infrastructure', Weill & Broadbent (1998), graphically summarize the different IT functions in a *portfolio* of IT capabilities (Figure 5). IT capabilities are the organizational skills and knowledge that are essential to IT-enabled business transformation,

Figure 4. Primary and secondary stakeholders in IT governance (Peterson, 1998)

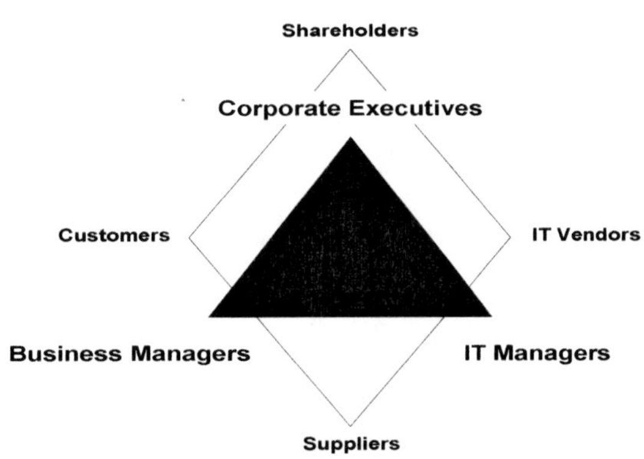

Figure 5. IT Portfolio (Adapted from Weill & Broadbent, 1998)

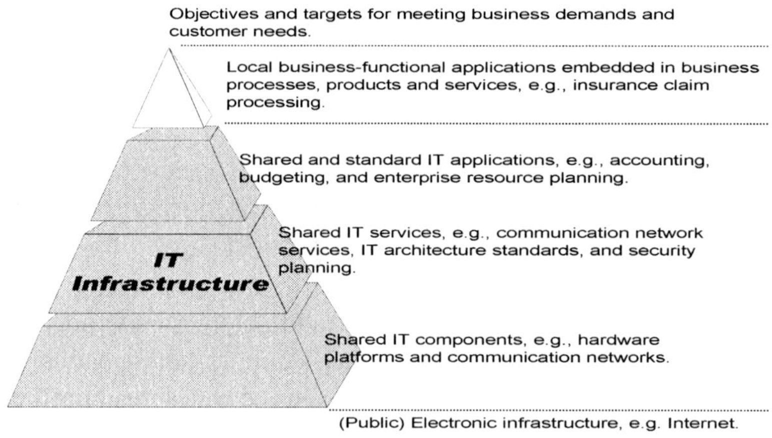

Figure 6. IT governance and IT management (Peterson, 2001)

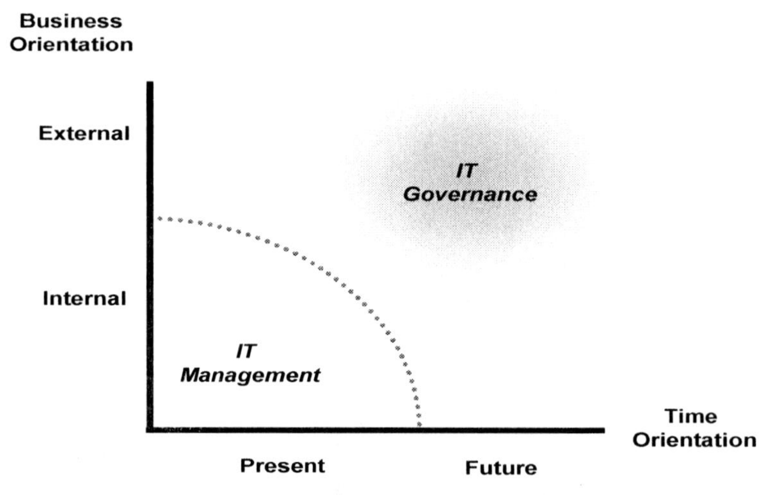

and include IT infrastructure, and the competency to partner with business clients, executive management, dispersed and/or external IT specialists and vendors (Peterson, 2001).

The IT infrastructure is the base foundation of IT capabilities, delivered as reliable shared services throughout the organization, and centrally directed, usually by corporate IT management. The purpose of the IT infrastructure is to enable organization-wide data sharing and cross-business integration. The infrastructure capability describes the degree to which its resources are shareable and reusable. In contrast, local business-embedded applications are concerned with product/service specific needs in order to meet the changing demands of the business and its customers, usually directed by local business management. These applications utilize the infrastructure services and are built on the shared technical platforms.

IT Governance Myth: IT Governance is a New Form of 'Old-School' IT Management

Due to the enduring nature of IT governance, and the perennial, often intractable problems associated with IT value delivery, some may draw the conclusion that IT governance is simply a new form of 'old-school' IT management. However, although there may be a mere thin dotted line separating IT governance from IT management, there is a fundamental difference between IT governance and IT management that goes well beyond theory, which has profound implications for the design and effectiveness of IT governance in practice.

Whereas the domain of IT management focuses on the efficient and effective supply of IT services and products, and the management of IT operations, IT governance faces the dual demand of (a) contributing to present business operations and performance, and (b) transforming and positioning IT for meeting future business challenges (Figure 6). This does not undermine the importance or complexity of IT management, but goes to indicate that IT governance is both internally and externally oriented, spanning both present and future time frames. One of the key challenges in IT governance is therefore how to simultaneously perform and transform IT, in order to meet the present and future demands of the business and the business' customers in a satisfying manner.

The difference between IT governance and IT management is akin to Parker & Benson's (1988) discussion of enterprise-wide information management, i.e., a shift in emphasis away from information *technology*, to the *business/information* relevance of IT. Whereas elements of IT management and the supply of (commodity) IT services and products can be commissioned to an external IT provider, IT governance is organization-specific, and direction and control over IT can not be relegated to the market.

IT Governance Myth: IT Governance Focuses on the (De)centralization of IT

Acknowledging the rebuttal of the previous myth, often leads to a fourth myth, i.e., IT governance focuses on the locus of IT control, or where IT decision-making authority is allocated in the organization. The discussion on the formal allocation of IT decision-making as vested in organizational positions has led to much rhetoric and speculation on the 'best way' to organize IT governance, and in the process has rekindled the classical 'centralization versus decentralization' debate. Yet, scholars have recently questioned whether the concept of IT governance is simply about centralization or decentralization (Peterson et al., 2000; Sambamurthy & Zmud, 2000; Vitale, 2001; Whetherbe, 2001).

The keen reader will recognize that by building forth on the discussion of the previous three myths, the terms centralization and decentralization provide a dichotomy that is meaningless

Figure 7. Example of federal patterns in IT governance (Peterson et al., 2000)

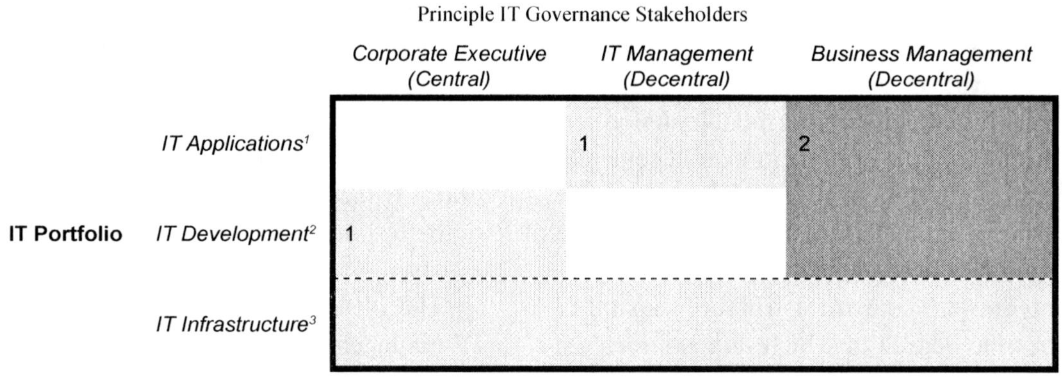

Figure 8. Tradeoffs and the best of both

Drivers	IT Governance	IT Governance	IT Governance
Synergy	+	-	+
Standardization	+	-	+
Specialization	+	-	+
Customer responsiveness	-	+	+
Business ownership	-	+	+
Flexibility	-	+	+

when employed as a generality to IT governance. The centralization or decentralization can be applied to each of the main IT capabilities in the IT portfolio, yielding eight distinct patterns in IT governance[2]. In the first pattern, the corporate executive is responsible for IT infrastructure and IT development decisions, whereas (decentral) IT management is responsible for IT application decisions (Figure 7). In the second pattern, however, (decentral) business management is responsible for IT development and IT application decisions.

The discussion whether to centralize or decentralize IT governance is based on a rational perspective of the organization, in which choices are reduced to one of internal efficiency and effectiveness (March & Simon, 1958). This view assumes a system of goal consonance, and agreement on the means for achieving goals, i.e., rational and logical trade-off between (a) efficiency and standardization under centralization, versus (b) effectiveness and flexibility under decentralization (Figure 8). In general, it is assumed that centralization leads to greater specialization, consistency, standardized controls, while decentralization provides local control, ownership and greater responsiveness and flexibility to business needs (Brown & Magill, 1998; King, 1983; Rockart et al., 1996). However, flexibility under decentralization may lead to variable standards, which ultimately result in lower flexibility, and specialization under centralization incurs risks, due to bounded rationality and information overload (Mintzberg, 1979; Simon, 1961).

A *socio-political view* of IT governance, suggests that the debate concerning centralization versus decentralization is used to further the goals of specific organizational actors in a *satisficing* manner (Simon, 1961), in ways that might not help to meet organizational goals (Cyert & March, 1963). There are important differences among factions within organizations, leading to the presence of conflict and disagreement over organization goals and the means for achieving them. Consider, for instance the notorious strategic conflicts between business and IT stakeholders. These stakeholders represent groups or individuals that influence, and are influenced by strategic decisions regarding IT, and may thereby have different, often competing stakes in IT. Power struggles and cultural clashes are endemic to IT governance, and the question is often '*who's way is it going to be?*', rather than, '*which way is the best?*'.

The potential risk in contemporary business environments is that either centralization or decentralization fit the organization into a fixed structure. The challenge is therefore to balance the benefits of decentralized decision-making and business innovation, and the benefits of central control and IT standardization. As Mintzberg (1979) points out, centralization and decentralization should not be treated as absolutes, but as two ends of a continuum.

Over the past decade, organizations have set out to achieve the 'best of both worlds' by adopting a federal IT governance structure. In a federal IT governance model, IT infrastructure decisions are centralized, and IT application decisions are decentralized (Brown & Magill, 1998; Rockart et al., 1996; Sambamurthy & Zmud, 1999; Weill & Broadbent, 1998). The federal IT governance

Figure 9. Examples of leading companies adopting federal IT governance

Companies Adopting Federal IT Governance	
Verizon Communications Inc.	Cisco Systems Inc.
United Technologies Corp.	Royal Dutch Shell Company
Chevron Texaco Corp.	British Petroleum PLC
General Motors	ABN AMRO
Avnet Inc.	Royal Dutch Airlines KLM
Siemens AG	HCA Inc.

model thus represents a hybrid model of both centralization and decentralization. Across the Atlantic, a number of prominent companies in different industries have been actively experimenting with this relatively new model for IT governance (Figure 9).

Recall, for instance, the CIO's mission at Johnson & Johnson: to standardize systems, cut IT costs, and align the IT organization with business strategies, while simultaneously acknowledging the decentralized culture of the business divisions across different countries. The competing drives towards improving cost-efficiencies, but also business responsiveness, and IT standardization, yet also IT innovation, led the CIO at Johnson & Johnson to adoption and use of a federal IT governance model (Alter, 2001; Scheier, 2001).

While the implementation of the federal IT governance model has paid off for Johnson & Johnson (e.g., significant cost reduction, cheaper maintenance costs, eliminated duplicate IT developments, enhanced pharmaceutical R&D, improved time-to-market for new products, profit growth), and they have been able to develop unprecedented levels of cooperation between, traditionally independent business units, it was a perilous and painstaking transformation.

Earlier attempts to (re-) centralize IT failed due to cultural barriers and business' resistance to change, and relinquish of IT control. The CIO indicates that the federal IT governance model is tricky to manage, and after commencing new IT governance regime the CIO states (Alter, 2001; Scheier, 2001):

It's hard getting all of Johnson & Johnson's businesses to go along with some of even the most benign changes in policy. Originally, we hoped to create a single, centralized strategy, but soon realized that only a federalized approach would work. We are too complex and independent from one business unit to the next to devise one strategy. I get '190 land mines' in any given day. Some business units, for example, try to convince me they can't adopt some corporate technology standards, or kick in their share of the cost for upgrades in infrastructure.

In sum, the federal IT governance model is often easier contemplated than implemented. A federal approach towards IT governance chal-

Figure 10. Determinants of IT governance (Peterson, 1998).

DETERMINANTS	Centralized Model	Federal Model		Decentralized
Business strategy	Cost-focus	←	→	Innovation-focus
Business governance	Centralized	←	→	Decentralized
Firm size	Small	←	→	Large
Information-intensity	Low	←	→	High
Environment stability	High	←	→	Low
Business competency	Low	←	→	High

lenges managers in local business units to surrender control over certain business-specific IT domains for the well-being of the enterprise, and to develop business-to-corporate and business-to-IT partnerships. As the Johnson & Johnson case illustrates, this a feat that many companies struggle with, especially considering the 'IT governance legacy' large and complex organizations face.

Determinants of IT Governance: Reviewing the Empirical Evidence

While the previous section discussed some of the myths regarding IT governance, and presented anecdotal cases concerning contemporary IT governance practices, this section takes a more scientific view of IT governance by reviewing the empirical data to date. Previous studies have sought an answer to the 'best way' of designing IT governance, recognizing that this 'best way' is contingent upon internal and external factors (Brown & Magill, 1994; Sambamurthy & Zmud, 1999).

These studies examined the influence of various determinants, including *organization size* (Ahituv et al., 1989; Brown & Magill, 1994; Clark, 1992; Ein-Dor & Segev, 1982; Sambamurthy & Zmud, 1999; Tavakolian, 1989); *business strategy (*Brown & Magill, 1994; Peterson, 2001; Sambamurthy & Zmud, 1999; Tavakolian, 1989); and *business governance structure (*Ahituv et al., 1982; Brown & Magill, 1994; Peterson, 2001; Sambamurthy & Zmud, 1999; Tavakolian, 1989).

The cumulative of these empirical findings indicate that (Figure 10):

a. Central IT governance is associated with small-sized organizations, following a cost-focused business (competitive) strategy, and characterized by a centralized business governance structure, environmental stability, low information-intensive business products/services, and low business experience and competency in managing IT.

b. Decentral IT governance is associated with large, complex organizations, following an innovation-focused business (competitive) strategy, and characterized by a decentralized business governance structure, environmental volatility, and high information-intensive business products/services and processes, and high business experience and competency in managing IT.

But what type of IT governance arrangement should a company adopt in the following (though hypothetical, certainly not unrealistic) scenario's?

Company A is a large bank, focused on operational excellence, and new product and service development, and whose business management, however, lacks the professional skills and knowledge to manage IT.

Company B is a small pharmaceutical company, focused on product excellence, rapid innovation and quick commercialization, and whose business management is leading in IT innovation.

Company C is a large insurance company focused on achieving cost-efficiencies and providing customer value through customized products, and whose business management has developed the required competencies to manage IT projects.

Each of these scenario's describes a situation in which multiple contingency factors are active in determining IT governance, yet provide conflicting solutions. Should Company A adopt decentral IT governance because of its size and strategy? Yet, how does it reconcile this with the lack of business competency? Should Company B adopt central IT governance because of its size? Yet, how does it combine this with its focus on innovation and business leadership? Should Company C decentralize due to size and customization, or should it centralize due to a focus on cost-efficiencies?

Although previous studies have increased our understanding of IT governance determinants, they have failed to assess the multiplicity of these determinants. In practice, IT governance models are influenced by many factors simultaneously, and determining the right structure is a complex endeavor. While traditionally IT governance focused on either efficiency or flexibility, often in a sequential manner - recall the continuous pendulum swing of centralization and decentralization - currently, IT governance faces the dual demands for (a) flexibility and speed, and (b) efficiency and reliability. The latter concern is of long standing, in which IT governance was concerned with efficiency and cost reduction, often directed at the operational level. Subsequently, IT governance focused on managing IT as a strategic resource, in which the primary aim was to align IT with the business strategy, in order to gain competitive advantage. Having emerged from both practices with ambiguous results and experiences, executives are recognizing the need to meet the demands for both (a) delivering customized, high quality IT products and services, and (b) compressing costs, risks and time, in order to meet business needs in an efficient, reliable and effective manner.

Consider, for instance, the case of Johnson & Johnson. The competing drives towards improving cost-efficiencies, but also business responsiveness, and IT standardization, yet also IT innovation, led Johnson & Johnson to adopt a federal IT governance model. The complexity (size and span) of this organization would dictate a decentral approach, which was the traditional IT governance approach. Yet, the need to cut costs, standardize IT, and improve IT performance led Johnson & Johnson to centralize IT infrastructure decisions. Thus, the case of Johnson & Johnson illustrates how IT governance is subject to the pulls and pressures of multiple, rather than singular, contingency forces.

Recent empirical evidence indicates that organizations adopt federal IT governance when pursuing multiple competing objectives (Peterson, 2001; Sambamurthy & Zmud, 1999). Due to the relentless pace and unpredictable direction of change in contemporary business environments, it should be no surprise that many organizations need to focus on both standardization and innovation, and in the process have adopted a federal IT governance model.

Strategic Flexibility: The New Enterprise Logic

Business has embarked on a new era of competition that is faster, more turbulent, and increasingly global and digital, simultaneously requiring relentless cost-efficiencies, as well as the flexibility to find new ways to innovate and create value. It is no secret that the contemporary business landscape is characterized by (D'Aveni, 1999; El Sawy et al., 1999):

- Time and cost compression in product-life and design cycles
- Accelerating technological advancements;
- Fickle customer loyalty
- Tailored, knowledge-intensive products and services
- Unexpected entry by new competitors, and repositioning of incumbents
- Redefinition of industry and organizational boundaries
- Global market volatility

As the mosaic of these developments transpires simultaneously in unpredictable patterns, organizations face significant uncertainty and ambiguity in determining their strategic direction. Organizations operating in turbulent fields, experience competing goals and performance demands, including, e.g., pressures to innovate and customize products and services, improve levels of responsiveness and speed, and increase productivity and efficiency (Daft, 1998; Mintzberg, 1979). These competing demands cause conflicting contingencies, which are endemic to complex open social systems, and require both exploitation and exploration strategies (March, 1991).

Exploitation strategies involve taking advantage of what is already known, i.e., cashing in on the investments made in existing capabilities. *Exploration* strategies, on the other hand, involve the creation of new knowledge and capabilities. Whereas exploitation builds forth on the efficient supply of extant products and services, exploration is geared at developing new products and services, in order to meet changing and ambiguous environmental demands (March, 1991). Organizations must determine the proper balance between these competing strategies for developing capabilities, and preserving organizational conditions. Thus, turbulent environments require the ability to explore new opportunities effectively, and exploit existing opportunities efficiently.

Contemporary organizations do not have single goals, and face multiple, often conflicting, contingencies. The 'low-cost versus differentiation' dichotomy (Porter, 1980) is currently fallacious, as organizations are effectively pursuing both strategies simultaneously, in order to meet the competing demands of volatile global electronic markets. D'Aveni (1999) concludes that under these conditions, there is no sustainable competitive advantage based on either a low-cost strategy or a differentiation strategy. Companies can only build temporary advantages in order to sustain strategic momentum through a series of initiatives, rather than achieve 'fit' with the external environment.

Instead, companies need to adopt *simultaneous strategic thrusts*, in rapid and surprising manners, in order to offset competitors, and satisfy customer needs (D'Aveni, 1999). Adopting simultaneous strategic thrusts requires an organization to focus on multiple value-creating drivers, involving excellence in business operations, product development, *and* customer service delivery (Figure 11). Market leaders focus on *at least* one value driver, and meet a minimum threshold of competence in the other two (Treacy & Wiersema, 1995).

Avnet Inc. is a perfect example of how market leaders are transforming towards meeting multiple value drivers (El Sawy et al., 1999). In 1991, Avnet Inc., then Marshall Industries, realized that customers wanted products and services at the lowest possible cost, highest possible quality, greatest possible customization, and fastest possible delivery time. Avnet Inc. recognized that

Figure 11. Complementary value-creating drivers (Adapted from Treacy & Wiersema, 1995)

Value Driver	**Operational Excellence**	**Product Excellence**	**Service Excellence**
Description	Providing reliable products or services at competitive prices, and delivered with minimal inconvenience	Offering customers leading edge products and services that consistently enhance the customer's use of the product	Segmenting and targeting markets precisely and tailoring offerings to match exactly the demands of those customers
Focus	Competitive prices, minimize overhead costs, reduce transaction costs, optimize business processes across functional boundaries, improve reliability, streamline enterprise workflows	Innovation, speed, state-of-the-art technologies, commercialization, short cycle times, time-to-market, selective controlled experimentation, directed 'grass root' innovation	Customization of products and services, build customer loyalty and life-time value to the company, responsiveness, cultivate relationships and understanding, satisfy unique customer needs
Threshold	Operational Efficiency	Product Enhancement	Customer Responsiveness

Figure 12. Strategic flexibility: The new enterprise logic (Peterson, 2001)

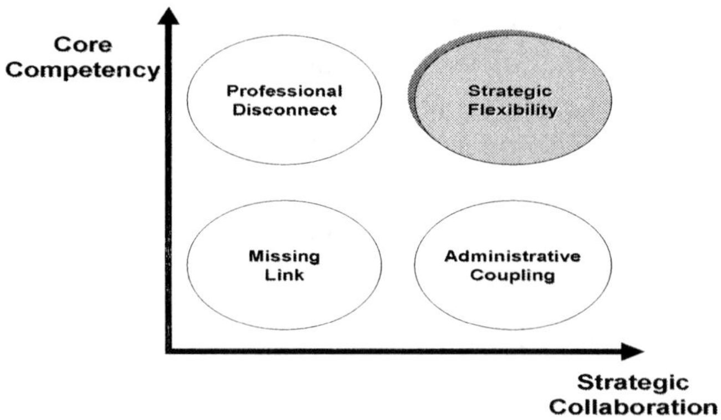

they needed to focus on all three value drivers. They realized that it was no longer adequate to think of operational excellence in terms of cost of individual transactions, and expanded that to include the total cost of value-added services (El Sawy et al., 1999). Avnet Inc. also focused on product excellence by enhancing product features, customization, and the anticipation of future needs. Service excellence was also improved by 7x24 services, reduced delivery time, and reduced time-to-market for customers' products (El Sawy et al., 1999).

Empirical studies indicate that value-creating organizations have learned to master more than one value driver (Boynton, 1993; Buenger et al., 1996; Khandwalla, 1976; Miller & Friesen, 1978; Peterson, 2001; Quinn & Rohrbaugh, 1983; Quinn et al., 2000; Treacy & Wiersema, 1995). Organizations learn to manage the conflicting pulls by becoming 'bi-focal', and recognize the requisite *complementary* - not competing - nature of value creation.

In order to satisfy complementary value-creating drivers, an organization should have a variety of capabilities at least as great as the demands and disturbances in the environment (Ashby, 1956; Hitt et al., 1998). Consequently, organizations develop a repertoire of competencies to respond to, and influence their external environment. Whereas, in the 70's and 80's organizations would integrate vertically, today organizations have recognized the need to focus on their core competencies. Yet, continuous differentiation leads to fragmentation, unless a corresponding process of integration complements it. The uncertainty and ambiguity associated with the complex of external demands and differentiated capabilities creates the need for integration to achieve clarity of direction and unity of purpose in responding decisively and swiftly (Lawrence & Lorsch, 1969; Hitt et al., 1998; Venkatraman, 2000).

The concentration on core competencies in the early 1990s, and the recent focus on (inter- and intra-) organizational collaboration are exemplary of the need to both differentiate and integrate in turbulent environments. Focusing on core competencies alone leads to a situation of professional disconnectedness, whereas collaboration without core competencies only provides for administrative coupling (Peterson, 2001). Value-creating organizations thus focus on both their core competencies and strategic collaboration (Figure 12). The degree to which organizations can achieve this is a measure of an organization's

Figure 13. Primary IT value drivers (Peterson et al., 2000)

Value Driver	Service Delivery	Solution Integration	Strategic Innovation
Description	Providing reliable IT operations and services, delivered with maximum reliability and availability	Offering business leading-edge products and services that consistently enhance the business' use of products and services	Targeting business value drivers, and tailoring offerings that supersede the demands of the business
Focus	Provisioning of utilities, provide reliable, cost-effective, and secure IT services, manage cross-unit synergies across the corporation, manage IT infrastructure standards, establish IT infrastructure flexibility and scalability.	Strategic analysis of business needs for IT, deciding on the necessary applications, and delivering them either through internal development, external contracting or packaged software. Focus on ensuring timely and cost-effective delivery of IT applications.	Focus on ways in which IT can be used to strengthen business competencies, customer relationships and business partner networks. Ensure IT applications have a value focus (operational excellence, product excellence, service excellence).
Threshold	Infrastructure	Integration	Impact

Figure 14. IT value drivers: An integrated perspective (Peterson et al., 2000)

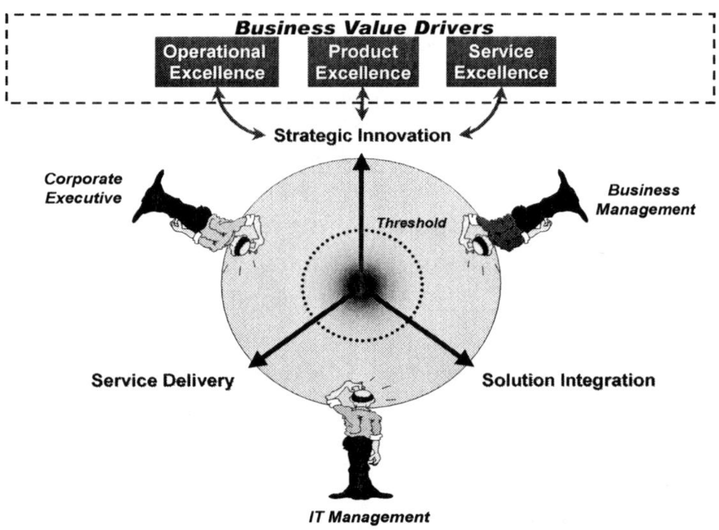

strategic flexibility, i.e., developing differentiated capabilities to pro-act in an integrated manner to unanticipated changes (Hitt et al., 1998). Organizations improve their chances of success and survival in a turbulent environment by creating strategic flexibility that gives them the ability to pursue alternative course of actions in response to unexpected environmental conditions.

Strategic flexibility in a turbulent and competitive business environment requires organizations to be dynamically stable (Marchand, 2000). Organizations need to simultaneously develop a variety of differentiated capabilities in order to serve a range of changing customer and market demands, and integrate these for developing joint expertise and providing direction (Hitt et al., 1998; Nadler & Tushman, 1998). The underlying enterprise architecture is an interconnected network of differentiated strategic capabilities, in which IT is an integral part, wherein the interdependency between IT and business is intensely reciprocal. Consequently, the efficiency *and* flexibility with which IT capabilities are infused in the enterprise architecture become business critical. The business imperatives of strategic flexibility and dynamic stability thus require the business of IT to:

a. Develop and deliver applications that facilitate business responsiveness to customer demands in a rapid and efficient manner;
b. Provide cost-effective, scalable infrastructures and operations that enable cycle time improvement and streamlined, enterprise-wide business processes, and;.
c. Add value to the enterprise by focusing on operational excellence, product excellence, and service excellence.

Similar to the business enterprise, IT needs to develop distinct strategic capabilities in order to satisfy multiple value drivers (Figure 13). The primary value drivers for IT focus on service delivery, solution integration and strategic innovation (Feeny & Willcocks, 1998; Sambamurthy & Zmud, 2000; Peterson et al., 2000).

Value-adding IT organizations focus and channel energy at excelling in a specific dimension, yet maintain threshold standards on the other performance dimensions. The strategic importance of measuring different performance criteria for IT has also been recognized in the development of a 'balanced scorecard' for IT (Van Grembergen & Bruggen, 1997). IT organizations today cannot afford to focus on service delivery at the expense of solution integration, or vice versa. Furthermore, strategic innovation is difficult, if not impossible, to achieve without some base-line performance in service delivery and solution integration.

In terms of cause-effect relationships, strategic innovation is the value link between IT performance and business performance. Whereas, service delivery and solution integration provide the foundation for strategic innovation, the latter provides the strategic IT capability for operational, product and/or service excellence. Moreover, it is pivotal that key stakeholders involved in IT governance recognize the multiplexity of value drivers, and develop a shared understanding of the value of IT (Figure 14).

Similar to the business, the new enterprise logic for the IT business is based on developing the agility and flexibility to sense and respond to existing and emerging demands in an integrated and pro-active manner (Peterson, 2001). The challenge for IT governance is then, as discussed in Section 2.1, how to simultaneously perform and transform IT, in order to meet the present and future demands of the business in a satisfying manner. Thus, the IT business also needs to manage the balance between exploitation and exploration.

Consequently, the IT business also needs to transform. Analogous to the business, IT requires a dynamically stable organizational architecture to deliver value to the business, i.e., stable enough to exploit IT capabilities for service delivery and solution integration, and sufficiently dynamic

Figure 15. Transformation of the IT organization (Peterson, 1998)

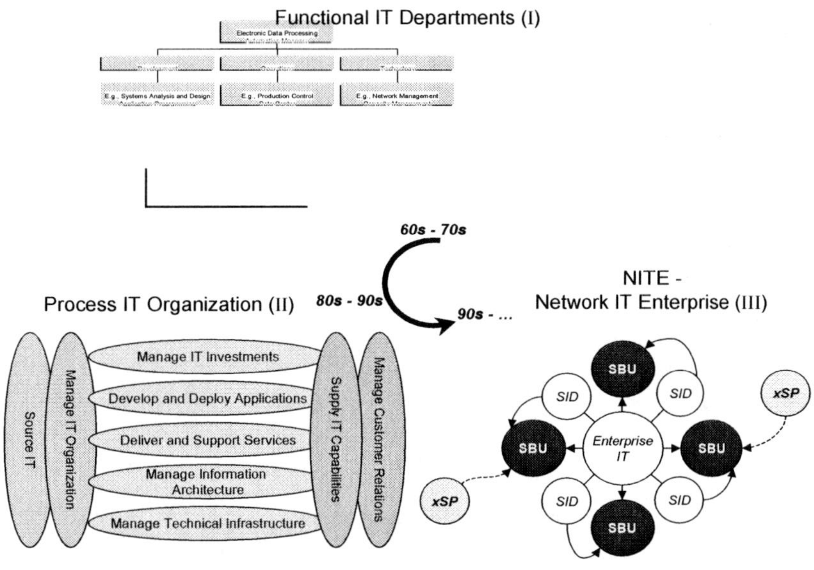

to explore new value innovation opportunities. Consistent with 'law of requisite variety' (Ashby, 1956), the IT business should have a variety of capabilities at least as great as the demands and disturbances in the business enterprise environment.

The business of IT has indeed experienced a fundamental transformation over the past few decades. In the 1950s, accounting and controller departments mainly used electronic computers. Some of these departments were centralized, while others were decentralized, and decision-making defaulted to those departments that made use of the technology. The benefits gained from automating the business's processes and functions led most organizations to integrate and centralize their technologies in the 1960s and 1970s. Specialists skilled in a number of hardware and software fields were required, and often reported to a central IT group. This central or corporate IT group primarily served a manufacturing role in a function-oriented organization with different functional IT departments, e.g., development, operations and technology (Figure 15).

By the mid 1980s, both the business and IT environments had changed significantly. Business markets became more complex and competitive. Businesses adopted divisional structures, each with their own products and market services, and concurring responsibilities and accountabilities. The proliferation of IT also became more complex with the dispersion of IT to the business units. Local IT managers were given authority over IT in order to respond to the local needs of the business in the competitive environment.

In the early 1990s, as companies experienced the demise of traditional geographic and business boundaries, and the emergence of 'the new infrastructure' (Weill & Broadbent, 1998), IT was again resorted to a central IT group. The title and function of CIO emerged on organizational charts, and many organizations were characterized by both a corporate IT department, often led by a CIO or IT director, and several local IT departments, often still functionally organized. During this period, IT organizations also began to see the benefits of employing a process-based IT organization (Figure 15), and reengineering IT around value added process, both upstream (e.g., source IT, manage IT investment) and downstream (deploy IT applications, deliver IT services, manage customer relations).

Consider, for instance, the transformation of British Petroleum IT's function. In 1989, the IT function set out to transform itself, in line with its mission to "...become the best in class in its support of BP's goal to be the best upstream company in the world" (Cross et al., 1997). They supported this mission by changing their direction, with the explicit purpose of moving from "system provider" to "infrastructure planner". By 1994, the IT function was truly an "infrastructure planner", overseeing technical integrity and value creation through information sharing. Although some IT resources remain in the local businesses, the top IT management team provides global vision for infrastructure planning (Cross et al., 1997)

Today, a new transformation is transpiring. IT organizations are transforming toward a 'core-peripheral' organization design, in which the IT infrastructure is directed by Corporate IT management, and local business applications are managed by business or IT management. At the 'core' of the organization, IT infrastructure decisions are centralized and allocated to the corporate unit, whereas IT application decisions are decentralized and allocated to the different operational business units at the organization 'periphery'. This network IT enterprise (NITE) is built around solution integration delivery (SID) teams that focus on the needs of the business, and contracts with external service providers (xSPs). The (internal) SID and (external) SP components 'hover' around a center that manages and provides core IT competencies (e.g., IT services & IT skills). NITE resembles the enterprise architecture in contemporary organizations, and embodies the principles of strategic flexibility and dynamic stability.

A good example of NITE is found at Verizon Communications Inc., one of the world's largest investors in global communications markets. Verizon Communications, formed by the merger of Bell Atlantic and GTE, is a leading provider of communications services. Verizon companies are the largest providers of wireline and wireless communications in the US, with nearly 134 million access line equivalents and over 29 million wireless customers. Verizon is also the world's largest provider of print and online directory information. With more than 247,000 employees and $67 billion in 2001 revenues, Verizon's global presence extends to 45 countries in the Americas, Europe, Asia and the Pacific.

In 1996, Verizon (then Bell Atlantic) adopted a core-peripheral design in molding a network IT enterprise. The NITE at Verizon reflects a 'centers of excellence' (CoE) approach (Figure 16), in which roles and processes are differentiated for conceptualizing IT applications, delivering IT services and building the requisite IT competencies (Clark et al., 1997). The implementation of NITE has enabled Verizon to develop a flexible IT organization focused on IT-based value innovation.

Discussion

In retrospect, comparing the empirical evidence on IT governance (Section 2.2), with the new business imperatives of strategic flexibility and dynamic stability (Section 2.3), it is clear how and why the federal IT governance model has emerged as the dominant design in contemporary organizations. Already forecasted by Zmud et al. (1986), different studies indicate that this federal model is the dominant IT governance practice in contemporary organizations (Feeny et al., 1989; Hodgkinson, 1996; Sambamurthy & Zmud, 1999). The federal model is also propagated as the best model, 'capturing the best of both - centralized and decentralized - worlds' (Von Simson, 1990). Rockart et al. (1996) describe the federal model as one of the fundamental imperatives of IT in the late 1990s, and urge organizations to adopt the federal model, regardless of organizational

Figure 16. Verizon's NITE (Adapted from Clark et al., 1997)

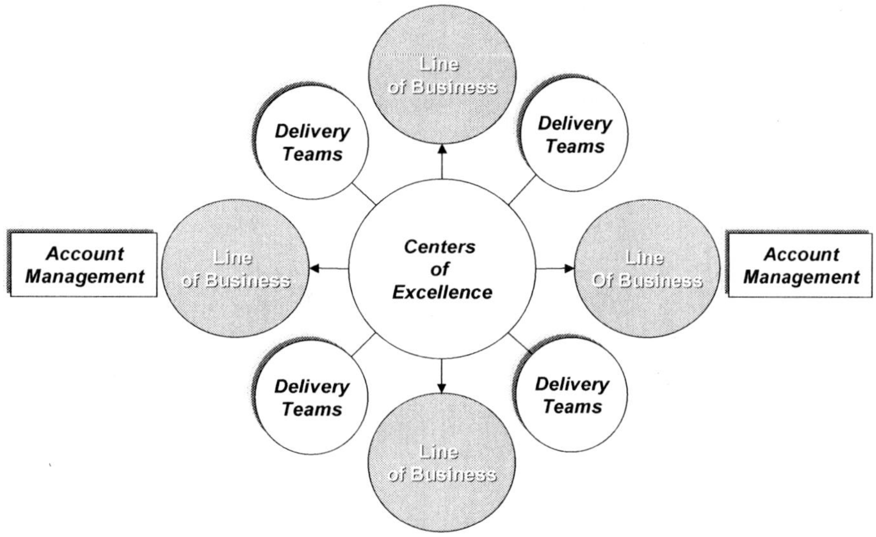

contingencies. Recently, Earl (2000) argues that every company needs to build a degree of IT federalism.

From a strategic perspective, both business and IT are facing multiple, competing objectives to reduce costs, standardize, and innovate and provide customer value. From an organizational perspective, the enterprise architecture is characterized by a dynamic network of integrated business-IT capabilities (Figure 17). The federal IT governance model, by dividing direction and control over IT between central and local offices, across different business and IT constituencies, creates a structure that is consistent with the enterprise architecture, i.e., both stable and dynamic, and enables IT-based strategic differentiation of the business.

However, some researchers have recently questioned whether 'the truth' regarding IT governance is still out there. Is the federal IT governance model really where the buck stops? Is dividing the locus of IT decision-making authority the answer in our quest for IT value?

Peppard & Ward (1999) state that while appealing at present, the federal IT governance model is more of a theoretical construction, than a direct practical solution. The problem, according to Sauer & Yetton (1997), is that on the surface, the federal model seems to provide a solution to the competing demands of corporate efficiency and business flexibility. However, the structure it creates depends on IT managers' ensuring that activities of the different business units, which have competing interests, are integrated across the organization. The reality is that local IT units tend to be 'captured' by the goals of the business unit, while the central IT unit tends to become divorced from the business (Sauer & Yetton, 1997).

Sambamurthy & Zmud (2000) argue that there are some signs that the accumulated wisdom on IT governance might still be inadequate. Further, Peterson (1998; 2001) states that our understanding of IT governance is not only inadequate, it is incomplete. To cite Simon (1960), the current federal IT governance model is like designing an unreliable ship with risky steersmanship:

...there is no use in one group of experts producing the hull, another the design for the power plant, and a third the plans for the passenger quarters, unless great pains are taken at each step to see that all these parts fit a seaworthy ship".

Figure 17. Evolution of value drivers, enterprise architectures and the governance of IT (Peterson, 2001)

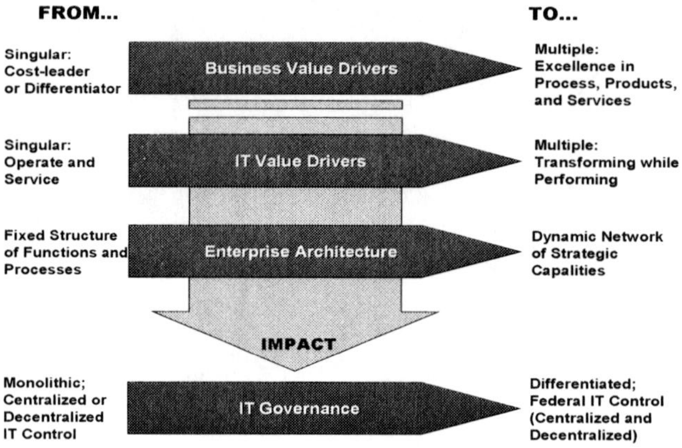

DESIGNING EFFECTIVE IT GOVERNANCE ARCHITECTURES

In this section, the fundamental IT governance problématique is addressed, and architectural solutions are discussed. Specifically, this section describes how and why our current framing of IT governance is incomplete, and needs a fundamentally different approach (Section 3.1), and how strategic risks can mitigated through effective IT governance architectures (Section 3.2), and integration strategies and tactics (Section 3.3). This section concludes by discussing the need to align IT governance architectures with value drivers and enterprise architectures (Section 3.4).

Federal IT Governance: What is Wrong with This Picture?

In *The Blind Men and the Elephant*, American poet John Godfrey Saxe (1816-1887) retells an ancient Indian fable of six blind men who visit the palace of the Rajah and encounter an elephant for the first time (Figure 18). The first blind man put out his hand and touched the side of the elephant: "How smooth! An elephant is like a wall". The second blind man put out his hand and touched the trunk of the elephant: "How round! An elephant is like a snake". The third blind man put out his hand and touched the tusk of the elephant: "How sharp! An elephant is like a spear". The fourth blind man put out his hand and touched the leg of the elephant: "How tall! An elephant is like a tree". The fifth blind man reached out his hand and touched the ear of the elephant: "How wide! An elephant is like a fan". The sixth blind man put out his hand and touched the tail of the elephant: "How thin! An elephant is like a rope".

Although a well-known, simple story, the morale of this ancient Indian fable applies equally well to IT governance, and in particular, the federal IT governance model. Much like an elephant, or any other large living organism, the federal IT governance model is a complex system, involving different stakeholder constituencies with specific perceptions, views and motivations. Similar to the blind men, these stakeholders have specific stakes and responsibilities in governing IT. This is especially the case with federal IT governance, in which corporate executives are focused on creating enterprise-wide synergies, standardizing IT, and controlling the IT infrastructure, whereas local business and IT managers are concerned with improving flexibility and responsiveness to busi-

Figure 18. Framing federal IT governance

ness-specific problems and opportunities. Though each constituency may be correct in pursuing their own strategic objectives, their 'blinded' focus impedes effective governance of IT.

While research insists on the dominance and importance of the federal IT governance model, previous studies have failed to specify how and which of the eight distinct patterns of federal IT governance are adopted, and under what circumstances these distinct patterns are effective. The federal model is not a single design, but consists of different patterns in dividing IT decision-making authority (Figure 18). Furthermore, the federal IT governance model introduces a 'new division', in which the decision-making actions of individual units are divided, thus becoming interdependent, thereby requiring coordination, especially considering the dynamic task environments. Previous studies, however, assume that once IT decision-making is allocated to corporate executives, business management, and IT management, coordination will follow automatically through hierarchical lines of reporting. Thus, while previous studies focus on the differentiation of decision-making for IT, they do not address the integration of decision-making for IT.

In terms of Mintzberg (1979), previous studies address the division of responsibilities, but fail to take into account coordination to accomplish activities. Lorsch & Lawrence (1970) state that the real difference between centralization and decentralization is much more complex than patterns of decision-making:

Another shortcoming in the traditional views about centralization and decentralization is a failure to recognize that the issue is really one of a vertical division of labor and coordination. Therefore, it is not just a question of dividing responsibility up and down the hierarchy, but it is also a question of organizing the flow of information and coordinating devices. If these labels are to capture the realities of how complex organizations operate, they must refer to systems of organizational variables which include division of work and differentiation; the integration among divisions and the headquarters; the types of integrative structural devices used, as well as the information flows and decision-making processes operating within the organization.

Although this discussion may seem utterly theoretical and outdated, consider the symptoms and strategic risks associated with the lack of integration in companies today (Figure 19), involving not only internal IT troubles, but moreover, major business tribulations.

Figure 19. Symptoms and strategic risks associated with the lack of integration (Peterson, 2001)

Symptoms and Strategic Risks	
• Lack of IT prioritization	• Business management takes no responsibility for IT-enabled business change
• Wasted IT investments	• IT organization fails to realize business value
• IT organization fails to meet commitments	• Increasing customer dissatisfaction
• IT management does not understand the business	• Less revenues
• Unresolved conflicts between business and IT management	• Loss of market share
	• Weakened competitive position
• Executives do not support IT	

Figure 20. Design logic at different levels of governance

Level of Governance	Design Logic	
Inter-Organizational	Core Competency	Strategic Collaboration
Intra-Organizational	Differentiation	Integration

To create value for the business, and mitigate strategic risks, IT governance should have a variety of capabilities at least as great as the demands posed by the multiplicity of value drivers. Consequently, IT governance requires a fusion of different business and IT competencies, involving both corporate executives, and business and IT management. However, the differentiated fusion of business and IT competencies leads to fragmentation, unless there is a corresponding balanced process of integration. The complexity associated with the multiplicity of value drivers and stakeholder constituencies, creates the need for integration to achieve clarity of direction, and unity of purpose in governing IT effectively (Peterson, 2001).

The organizing logic is similar to that of the enterprise (see Section 2.3), albeit at a different level of governance. Whereas the enterprise logic focuses on core competencies and strategic collaboration at the inter-organizational level, the organizing logic for IT governance focuses on the differentiation and integration of IT decision-making and monitoring at the intra-organizational level (Figure 20).

In the past, however, there has been a strong bias toward defining and designing IT governance in a disjoint manner, i.e., reductionism. This predisposition is often instilled upon us—both executives and researchers—through years of professional education and indoctrination. We have been thought, especially in Western cultures, that when faced with a complex problem, to divide the problem into smaller pieces, and solve each piece separately. This 'divide-and-conquer' mentality prevails in much problem-solving in both business and information systems communities (e.g., IT business investment processes, IT program management, algorithm design). However, solving each piece of the problem does not address the complete problem, i.e., addressing sub-systems sequentially or in parallel is no guarantee that the supra-system will function effectively (Von Bertalanffy, 1968). The classic Indian parable thus illustrates the need for a holistic systems view of IT governance.

Designing Effective IT Governance Architectures

A holistic systems[3] view emphasizes the need to view IT governance as a complex open social system, interacting with its environment, and consisting of a set of interdependent subsystems that produce a purposeful whole (Peterson, 2001). Complex systems, are characterized by reciprocal interdependence (Thompson, 1967), in which decisions made by sub-units are mutually dependent and influential, thereby increasing the need to exchange information. In complex governance systems, each decision-making unit presents direct decision contingencies for every other unit (Lorsch & Lawrence, 1970). Interacting subsystems in a social system imply that stakeholders are interdependent, and need to work together in a coordinated fashion to achieve objectives. A systems thinking approach towards IT governance acknowledges its complex and dynamic nature, and underscores the importance of personal mastery and mental models, and team learning and shared vision (Senge, 1990).

The systems logic underlying governance is the division and coordination of decision-making, in order to direct the operational system towards the realization of the goals of the organizational system (Simon, 1961). This systems logic is the foundation for Organization Design Economics[4]. The design logic of organizing revolves around the processes of (a) *division* of labor into various tasks to be performed, and (b) *coordination* of these tasks to accomplish the activity, and achieve the organization's goals. Thompson (1967) describes the basic logic of organizing in the following manner:

By delimiting tasks, responsibilities and other matters, organizations provide their participating members with boundaries within which efficiency may be achieved. But if structure affords numerous spheres of bounded rationality, it must also facilitate the coordinated action of those interdependent elements.

Figure 21. Example of vertical coordination and coordination by plan in IT governance

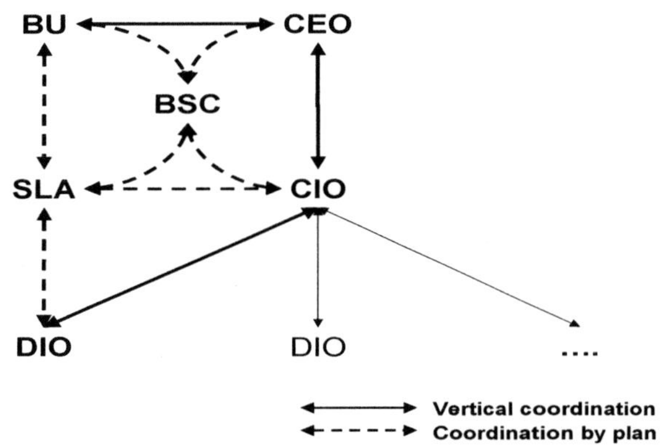

In terms of IT governance, and specifically the 'new division' of federal IT governance, this involves centralizing and decentralizing specific aspects of the IT portfolio. In order to achieve the organization's objectives, IT governance also needs to integrate these different aspects of the IT portfolio across the principle stakeholders, i.e., corporate executives, IT management and business management.

The manner in which responsibilities and accountabilities for the IT portfolio are organized and integrated is defined as an *IT governance architecture.* An IT governance architecture describes the *differentiation and integration of strategic decision-making for IT* (Peterson, 2001). The IT governance architecture specifies the strategic policies and business rules that provide direction to strategic IT decision-making, and plots a path for achieving business objectives (Weill & Broadbent, 1998). Designing an effective IT governance architecture is thus dependent on both the differentiation and integration of strategic decision-making for IT (Peterson et al., 2000). However, whereas executives and scientists have been keen to adopt and propagate a federal model of IT governance, i.e., differentiate IT decision-making, following through on integration has been somewhat undermined and still remains a challenge.

Traditionally, organizations and IT governance relied on the hierarchy and standardization for coordination. *Hierarchy* or *vertical coordination* describes the hierarchical referral of infrequent situations for which standardized programs have no solution (Galbraith, 1973). The hierarchy achieves coordination by having one person (e.g., CIO-Chief Information Officer) take responsibility for the work of others, issuing instructions and monitoring actions (Figure 21). If the hierarchy gets overloaded, additional levels or positions can be added to the hierarchy (e.g., DIOs-Division Information Officers). *Standardization* or *coordination by plan*, on the other hand, describes the use of standard programs, formal rules and procedures, and the specification of outputs, goals and targets (Galbraith, 1973; March & Simon, 1958). The adoption and use of balanced scorecard methodologies (BSC) and service level agreements (SLAs) are typical examples of how contemporary organizations coordinate by plan (Peterson & Ribbers, 2002).

Vertical coordination and standardization, however, only provide limited coordination capability in complex and uncertain environments (Daft, 1998; Galbraith, 1973, 1994; Mintzberg, 1979). With the profusion of electronically-enabled, globally-operating organizations, characterized by a multiplicity of value drivers in a dynamic network of strategic capabilities, the best CIOs, DIOs, BSCs and/or SLAs, will not suffice in designing effective IT governance architectures. Instead, IT governance needs to focus on *lateral coordination capabilities.*

Traditionally described as 'informal mutual adjustment' (March & Simon, 1958; Thompson, 1967), lateral coordination and horizontal relationships represent the most significant contemporary development in practice and theory. Lateral coordination has recently been rediscovered as a strategic organizational capability for competing in contemporary hyper competitive environments (Galbraith, 1994; Hitt et al., 1998). In turbulent environments, performance is driven by an organization's resources that are valuable and unique (Collis & Montgomery, 1995). From a resource-based perspective, lateral coordination is a resource that is hard to imitate, cannot be purchased, and is time dependent and socially complex (Hitt et al., 1998; Powell, 1992).

How then can organizations develop this requisite lateral coordination capability, and what (horizontal) integration mechanisms should be used for governing IT effectively?

Integration Strategies and Tactics for IT Governance

Integration strategies for IT governance can be classified according to two dimensions. Vertically, integration mechanisms can focus either on integration structures or integration processes, whereas horizontally, a division is made between formal positions and processes, and relational networks and capabilities (Peterson, 2001). Collectively, this provides four types of integration strategies for IT governance (Figure 22):

a. Formal integration structures
b. Formal integration processes
c. Relational integration structures
d. Relational integration processes

Formal integration structures involve appointing IT executives and accounts, and institutionalizing special and standing IT committees and councils. The use of account and/or relationship managers aid IT managers to develop an improved understanding of business needs, and aid in proactive - versus reactive - behavior by IT managers (Peterson, 2001). Committees and/or executive teams can take the form of temporary task forces—e.g., project steering committees—or can alternatively be institutionalized as an overlay structure in the organization in the form executive or IT management councils (Figure 22). Committees vary in the degree to which they have an advisory function or have formal decision-making authority. Advisory steering committees are also referred to as advisory, review or guidance committees. Contrary to specialized task forces, steering committees and advisory boards bring together different stakeholders on a relatively permanent basis for resolving - Business/IT - decision-making questions and problems.

Formal integration processes describe the formalization and institutionalization of strategic IT decision-making-/monitoring procedures and performance. Formal integration processes vary with levels of (Peterson, 2001):

a. **Comprehensiveness:** The degree to which IT decision-making/-monitoring activities are systematically and exhaustively addressed, i.e., (i) the identification and

Figure 22. Integration strategies and tactics for IT governance (Peterson, 2001)

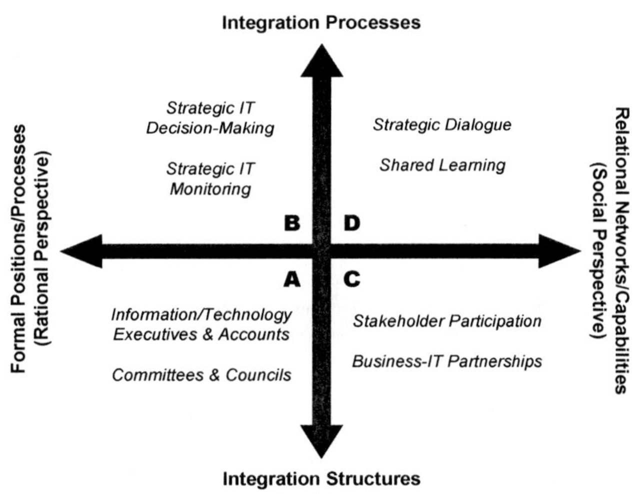

formulation of the business case and/or 'business rationale' for IT decisions, (ii) the prioritization and selection of business-IT course-of-action, and (iii) monitoring and evaluating of IT decision-implementation and IT performance (Parker & Benson, 1988; Luftman & Brier, 1999; Willcocks, 1996).

b. **Formalization:** The degree to which IT decision-making/-monitoring follows specified rules and standard procedures, often embedded in formalized decision-making methodologies and management frameworks (e.g., BITS-Balanced IT Scorecard, SAM-Strategic Alignment Model, CSF-Critical Success Factors, and/or IE-Information Economics, etc.).

c. **Integration:** The degree to which business and IT decisions are integrated, i.e., (i) administrative integration, in which budgets and schedules are pooled between business and IT; (ii) sequential integration, in which business decisions provide directions for IT decision-making; (iii) reciprocal integration, in which business and IT decisions are mutually influential; and (iv) full integration, in which business and IT decisions are made concurrently in the same process (Teo & King, 1999)

Whereas the foregoing formal integration mechanisms tend to be mandatory, tangible, and often implemented in a 'top-down' manner, relational integration mechanisms are 'voluntary actions', which cannot be programmed and/or formalized, and which are often intangible and tacitly present in the organization (Peterson, 2001). Kahn and McDonough (1997) refer to integration as a composite of interaction and collaboration. Interaction describes the formal structures and processes used for information-

Figure 23. Operationalization of integration strategies and tactics for IT governance (Peterson et al., 2000)

Integration Strategy	(a) Formal Integration Structures	(b) Formal Integration Structures	(c) Relational Integrationt Structures	(d) Relational Integration Processes
Tactics	IT Executives & Accounts Committees & Councils	Strategic IT Decision-Making Strategic IT Monitoring	Stakeholder Participation Business-IT Partnerships	Strategic Dialogue Shared Learning
Mechanisms	- CIO on Board - IT program managers - IT relationship managers - IT executive councils - eBusiness advisory board - eBusiness task force - IT standing teams	- Balanced (IT) Scorecard - Critical Success Factors - Scenario analysis - SWOT analysis - Strategic Alignment - Information Economics - Service Level Agreements - IT benefits management	- Active participation by principle stakeholders - Collaboration between principle stakeholders - Partnership rewards and incentives - Business/IT co-location - Business/IT 'virtual connection'	- Shared understanding of business/IT objectives - Active conflict resolution ('non-avoidance) - Cross-functional business/IT training - Cross-functional business/IT job rotation
	Low	← Lateral Coordination Capability →		High

exchange and communication, whereas collaboration is described as the affective, participative and shared element of integration, corresponding to a willingness to work together. Galbraith (1994) and Malone & Crowston (1994) describe these levels as a layered system of successively deeper levels of coordination, i.e., higher levels of lateral coordination capability (Figure 23). Recent evidence indicates that while formal integration mechanisms are necessary, they are insufficient for designing effective IT governance architectures in competitive environments (Peterson et al., 2000).

Relational integration structures involve the active participation of, and collaborative relationships between corporate executives, IT management, and business management. Central to relational integration is the participative behavior of different stakeholders to clarify differences and solve problems, in order to find integrative solutions. The ability to integrate relationally allows an organization to find broader solutions, and unleashes the creativity involved in joint exploration of solutions that transcend functional boundaries and define future possibilities (Peterson, 2001).

Relational integration structures are characterized by their participative and shared nature. Participation is a process in which influence is exercised and shared among stakeholders, regardless of their formal position or hierarchical level in the organization. Active stakeholder participation balances the involvement of business and IT communities in information processing, decision-making and problem-structuring/-solving (Peterson, 2001). Collaboration integration refers to a close, functionally interdependent relationship in which organizational units strive to create mutually beneficial outcomes. Henderson (1990) describes this as a strategic partnership that reflects a working relationship of long-term commitment, a sense of mutual collaboration, and shared risks and benefits. Mechanisms that facilitate relational integration include joint performance incentives and rewards, co-location of business and IT managers, and the creation of 'virtual meeting-points' for business and IT managers (Peterson et al., 2000; Ross et al., 1996; Weill & Broadbent, 1998).

Relational integration processes describe strategic dialogue and shared learning between principle business and IT stakeholders. Strategic dialogue involves exploring and debating ideas and issues in depth prior to decision-making or outside the pressure of immediate IT decision-making (Van der Heijden, 1996). A strategic IT dialogue incorporates a wide range of initially unstructured business perspectives and IT views, and involves rich conversation and communication to resolve diverging perspectives and stakeholder conflicts. Conflicts are resolved through the use of active and passive resolution strategies. Active conflict resolution involves confrontation and competition strategies, whereas passive conflict resolution involves avoidance and smoothing-over strategies, i.e., conflicts remain and are not explicitly resolved (Lawrence & Lorsch, 1969; Robbins, 1994).

Shared learning is defined as the co-creation of mutual understanding by members of organizational sub-units of each other's goals and objectives (Peterson, 2001). The essence of organizational coordination capability is the integration of domain-specific expertise and tacit knowledge. Shared learning is developed when people in close collaboration enact a single memory, *with differentiated competencies* and *responsibilities* (Weick & Roberts, 1993). Shared learning resides in specialized relationships among stakeholders, and in particular, the information flows and decision-making processes that shape their dealings with each other (Lorsch & Lawrence, 1970). Parker et al. (1997) argue that identifying acceptable solutions to ambiguous problems in complex and dynamic environments, requires the collaboration of different stakeholders, working with different paradigms, and offering different insights. Shared learning is inherently dynamic, and results in coordinated decision-making and collaborative relationships, which are particularly relevant and beneficial when the need for reliability is high, and decision-making is non-routine, involving interactive complexity, i.e., the combination of complex interpersonal interactions with a high degree of interdependence (Weick & Roberts, 1993).

Research indicates that when business and IT managers understand each other's perspectives in IT decision-making, they can accurately interpret and anticipate actions, and coordinate adaptively (Peterson, 2001). Within the context of IT governance, shared learning describes the mutual understanding of business and IT objectives and plans by business and IT executives (Reich & Benbasat, 1996; Weill & Broadbent, 1998). Mechanisms that support shared learning include strategic dialogues between business and IT executives, cross-functional business-IT training, and cross-functional business-IT job-rotation/-transfers (Luftman & Brier, 1999; Peterson et al., 2000; Ross et al., 1996; Weill & Broadbent, 1998).

Discussion

Designing an effective IT governance architecture is dependent on both the differentiation and integration of strategic decision-making for IT. Whereas differentiation focuses on the distribution of IT decision-making rights and responsibilities among different stakeholders in the organization, integration focuses the coordination of IT decision-making/-monitoring processes and structures across stakeholder constituencies. The notion of an IT governance *architecture* emphasizes the need to define and control the interfaces between the separate components of the IT governance system. Designers of IT governance architectures thus need to consider and implement integration strategies and tactics for governing IT effectively.

Contrary to the IT governance evolution described in Section 2 (e.g., Figure 17), the governance of IT has evolved not only from a monolithic towards a differentiated structure, but moreover, towards an integrated system of differentiated business and IT capabilities. Similar to the enterprise architecture, but then at a 'micro-level', an IT governance architecture resembles a dynamic network of strategic capabilities for directing and controlling IT in line with the business, and the business' objectives. Embedding the IT governance architecture in the enterprise architecture, and aligning IT governance with IT value drivers, enables an organization to realize its business value drivers (Figure 24). Effective IT governance architectures should be consistent with the enterprise architecture and IT value drivers, whereas enterprise architectures and IT value drivers should be in sync with business value drivers (Peterson, 2001). Given the multiplexity of value drivers and the dynamic stability of the enterprise architecture, aligning value drivers, enterprise architectures and IT governance architectures is inherently dynamic.

The key question which then arises is, of course, how do we—executives and researchers—know whether an IT governance architecture is effective, and what measures can be taken to redesign and improve the effectiveness of the IT governance architecture. In the next section, this challenging question is addressed.

DIAGNOSING IT GOVERNANCE EFFECTIVENESS

In this section, the design of a measurement system for diagnosing IT governance effectiveness is discussed. In the following sections, the rationale and foundation for an IT governance diagnostic is described (Section 4.1). This section presents the IT governance diagnostic model, and discusses how it can be used to diagnose and track IT governance effectiveness (Section 4.2).

Figure 24. Aligning value drivers, enterprise architectures and IT governance architectures (Peterson, 2001)

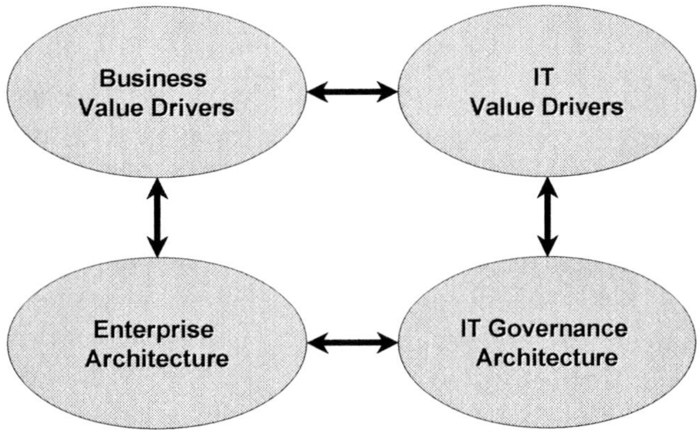

Motivation & Methodology

A critical part in designing effective IT governance architectures is devising a measurement system to assess and affect IT governance effectiveness. However, while our descriptive knowledge of IT governance has increased dramatically over the past three decades, prescriptive actions with regard to designing effective IT governance architectures are scant. Recently, Weill & Vitale (2002) indicate that executives moving into e-business should examine their IT governance and determine whether IT governance encourages the necessary communication and control of e-business initiatives. Diagnosing IT governance effectiveness, and establishing an IT governance measurement system that link cause and effect is consequently strategically important and highly relevant (Weill & Woodham, 2002).

Drawing upon the results of a longitudinal study on the design and effectiveness of IT governance (Peterson, 1997, 1998, 2000, 2001), an IT governance diagnostic model is developed and implemented in several large, multi-business-unit organizations across different industries in Europe. This comprehensive multi-client study was based on several in-depth case studies conducted in different sectors of the economy, including, Banking, Insurance, Manufacturing, Health Care, Telecom & IT, and Travel & Leisure. Data was collected and analyzed on (a) the strategic challenges and IT value drivers in business transformations, (b) the locus and distribution of IT decision-making/-monitoring responsibilities, (d) the integration strategies and tactics, and coordination capability and maturity of IT governance, and (d) the impact of IT governance on IT value realization.

More than 100 interviews were conducted with senior executives, directors and managers in both business and IT organizations and departments. These interviews were complemented by the collection and analysis of strategic company reports (e.g., IT investment analysis, Service Delivery Audits), and industry trends publications. In several of the client-organizations, executive workshops were organized to (a) present the main findings of the study, and (b) discuss the implications for the design and effectiveness of IT governance in their organization. In the next section, the results of this comprehensive study are described.

The IT Governance Diagnostic Diamond

The 'IT Governance Diagnostic Diamond' (Peterson, 2001) describes a holistic, high-level assessment model of IT governance design and effectiveness. The model is theoretically inspired by Organization Design Economics, specifically, the congruency model of organizations (Galbraith & Lawler, 1993; Nadler & Tushman, 1998), a balanced strategic management system (Kaplan & Norton, 1992, 1996), and conversion effectiveness (Weill & Broadbent, 1998). Each of these concepts and underlying models are used extensively in practice, and have been empirically validated.

The congruency model directs our attention towards the contextualistic nature of effective designs, emphasizing the interplay between context, design and effectiveness. The balanced strategic management system reminds us that different (types of) measures should be used to assess performance, and that these measures are embedded in a process linking strategic objectives with adaptive actions (i.e., strategic flexibility). In contrast to the micro-economic 'black-box' model of IT value, conversion effectiveness, recognizes that realizing IT value is based on the level of Value Conversion Process Maturity - VCPM -, which is characterized by (Peterson, 2001):

a. Senior (business) executive input and involvement (not 'abdication')
b. Collaborative relationships and shared learning (not 'political clout')
c. Integrated business-IT decision-making/-monitoring (not 'management-by-budget')
d. Business-IT competence (not 'ignorance')

Note: high VCPM is primarily dependent upon the business organization, and its executives, and only secondly on the IT organization. Furthermore, high VCPM is directly related to the level of lateral coordination capability, i.e., the adoption and use of relational integration structures and processes (see Figure 23).

The IT governance diagnostic diamond is organized into two axes (Figure 25). Vertically, the model depicts the IT governance architecture consisting of the:

a. *Differentiation of IT decision-making*, i.e., who has what authority and responsibility to make decisions regarding the IT portfolio?;
b. *Integration of IT decision-making*, i.e., what integration strategies, tactics and mechanisms are used to coordinate IT governance?

Answering the first question regarding differentiation provides (a) a specific federal profile of IT governance (see Figure 8), and (b) improves transparency into the distribution and allocation of IT decision-making authority and responsibility. Answering the second question regarding integration provides (a) a description of the integration strategies and tactics used, and (b) an assessment of the current level of lateral coordination capability (see Figure 23).

Horizontally, the IT governance diagnostic diamond distinguishes between *strategic direction* and *strategic control*, i.e., what are the value drivers and strategic intents, and to what extent have these value drivers been realized, or is IT contributing to business performance? The latter is the ultimate measure of true IT governance effectiveness.

In assessing the value drivers, organizations focus on the value propositions for IT (i.e., Strategic Innovation, Solution Integration, and Service Delivery), and the operationalization of these value drivers into operational goals, and specific measures and targets. Moreover, it is vital that executives also indicate *what* initiatives are currently underway, and planned for the future, that address this value driver, and *who* are involved and affected by these programs.

In determining IT value realization, the IT governance diagnostic diamond recognizes the dilution of IT impacts, and consequently, distinguishes four generic measures for assessing IT value realization, i.e.:

a. **IT services:** Consistent with 'Service Delivery' value driver (see Figure 14, Section 2.3);
b. **IT solutions:** Consistent with 'Solution Integration' value driver (see Figure 14, Section 2.3);
c. **Business impact:** Addressing the business value drivers, i.e., 'Operational Excellence', 'Product Excellence', and 'Service Excellence', and the respective foci and operationalization (see Figure 12, Section 2.3);
d. **Business value:** Addressing (future) financial performance, e.g., revenue growth, sales growth, return on assets, market share, shareholder value.

Similar to strategic direction, assessing strategic control involves the specification of the measures and targets involved, and to what extent the specific initiatives and programs have achieved their targets and objects, including level of stakeholder satisfaction.

Diagnosing IT governance effectives, and applying the IT governance diagnostic diamond follows a 'left-to-right' logic in a step-wise process. The following four steps are followed for both the '*Ist*' (current 'as-is' position), as well as for the '*Soll*' (desired 'to-be' position) situation.

1. Describe and assess the *current/future* business and IT value drivers for organization:
   a. *What are the main business value drivers?*
   b. *What are the main IT value drivers?*
   c. *What specific goals and targets, and initiatives and programs are underway that address the IT value drivers and business value drivers?*

2. Describe and assess the current/future differentiation of IT decision-making authority for the IT portfolio:
   a. Who is responsible for IT-business applications decisions, IT innovation decisions, and/or IT infrastructure decisions?
   b. How clearly are these responsibilities formulated, and transparent to the organization, especially other key stakeholders?
   c. Is the differentiation of IT decision-making in line with the value drivers?

3. Describe and assess the current/future integration of IT decision-making:
   a. *What integration strategies, tactics and mechanisms are employed by the organization?*
   b. *What is the level of lateral coordination capability, and value conversion process maturity?*
   c. *Are IT governance integration strategies, tactics and mechanisms in line with the type value drivers and IT governance differentiation employed by the organization (now and in the future)?*

4. Describe and assess the current/future IT value realization:
   a. What is the contribution of IT to improved business performance?
   b. What are the main business impacts, and how do these relate to the business value drivers, goals and measures?
   c. How is the IT organization performing on service delivery, solution integration, and strategic innovation?

Applying the IT governance diagnostic diamond, and following this step-wise approach for both the current and the desirable situation, provides a diagnosis of the suitability of the existing IT governance architecture, and identifies

Figure 25. IT governance diagnostic diamond (Peterson, 2001)

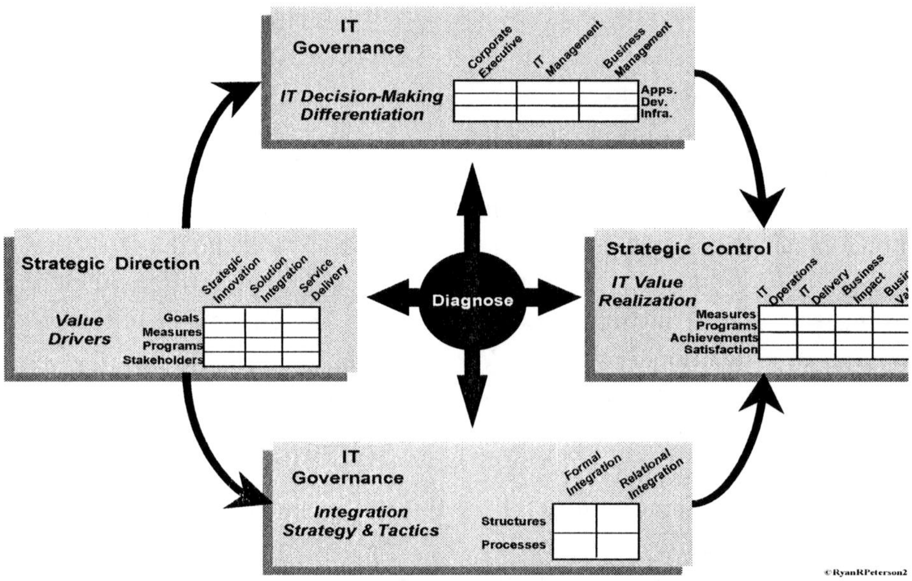

strategic discrepancies with the future, desirable position, and measures to redesign and improve the IT governance architecture.

CONCLUSION: TOWARDS A NEW IT GOVERNANCE PARADIGM

This section concludes the chapter on 'Integration Strategies and Tactics for IT Governance'. In this final section, the main lessons we have learned are summarized (Section 5.1), and based upon the previous sections, a new IT governance paradigm is presented (Section 5.2). In the final section, directions and opportunities for practitioners and researchers of IT governance are outlined (Section 5.3).

What Have We Learned? Implications for Management

Despite the many bridges that still remain to be breached in the (theories and practices) of IT governance, there are a number of important lessons we have learned over the past decades, which were addressed in this chapter. When contemplating and/or studying IT governance, both executives and researchers would do well to consider these important lessons learned.

Amidst the challenges and changes of the 21st century, involving hyper competitive market spaces, electronically-enabled global network businesses, and corporate governance reform, IT governance has become a fundamental *business* imperative. IT governance is a top management priority, and rightfully so, because it is the single most important determinant of IT value realization.

IT governance is the system by which an organization's IT portfolio is directed and controlled. IT governance describes (a) the distribution of IT decision-making rights and responsibilities among different stakeholders in the organization, and (b) the rules and procedures for making and monitoring decisions on strategic IT concerns. IT governance is thus corporate governance focused on IT, and it is not (only) the responsibility of the CIO or the organization of the IT function, in which an 'old new' IT management form is introduced to choose between centralization or decentralization.

The terms centralization and decentralization provide a dichotomy that is meaningless when employed as a generality to IT governance. The centralization or decentralization can be applied to each of the main IT capabilities in the IT portfolio, yielding distinct patterns of federal IT governance. The federal IT governance model is currently the dominant model in contemporary organizations, as they seek to combine the benefits of synergy, standardization and specialization, with the advantages of autonomy, innovation and flexibility.

The dominance of the federal IT governance model is a strategic response to the needs of strategic flexibility and dynamic stability in contemporary organizations and markets, in which both business and IT organizations are adopting multiple, complementary value drivers. However, it should be emphasized that redesigning for federalism challenges managers in local business units to surrender control over certain business-specific IT domains for the well-being of the enterprise, and to develop business-to-corporate and business-to-IT partnerships.

An IT governance architecture recognizes this coordination challenge, and emphasizes the need to take a holistic systems view of IT governance. An IT governance architecture is defined as the manner in which responsibilities and accountabilities for the IT portfolio are organized and integrated, and describes the differentiation and integration of strategic decision-making for IT (Peterson, 2001). Four types of integration strategies, each consisting of several integration tactics and integration mechanisms, can be employed to assure the level of coordination capability in the IT governance architecture. Embedding the IT governance architecture in the enterprise architecture, and aligning IT governance with IT value drivers, enables an organization to realize its business value drivers.

Consequently, diagnosing the IT governance architecture, and developing a balanced strategic measurement system to assess and track IT governance effectiveness is essential. Applying the IT governance diagnostic diamond, and utilizing the proposed step-wise approach, provides a diagnosis of the suitability of the existing IT governance architecture, and identifies strategic discrepancies with the future, desirable position, and measures to redesign and improve the IT governance architecture in terms of strategic flexibility and dynamic stability.

Towards a New IT Governance Paradigm

Until recently, efficiency was the keyword in designing IT governance. This made sense in a world characterized by a stable placid environment, in which neither the core technology nor the markets in which companies were operating changed drastically over time. Organizations could afford to use a 'command-and-control' structure to govern IT. However, with the business imperatives and new enterprise logic of strategic flexibility and dynamic stability, this 'old' IT governance paradigm no longer seems viable, nor prudent.

Rather than being a system of command-and-control, focusing on the locus of IT decision-making authority, this chapter concludes that effective IT governance in contemporary organizations is more likely to resemble a network of multiple business-IT collaborative relationships based on competencies and flexibility (Figure 26). IT is less about who is hierarchically positioned to be in control, and more about the complementary—business and IT—competencies an organization possess, and how it can integrate these, in order to develop the required strategic flexibility for realizing and sustaining business value from IT in a complex and dynamic environment. The emerging IT governance paradigm is based on principles of collaboration, competency and flexibility – not control, authority and efficiency.

These principles of the emerging IT governance paradigm underscore and reaffirm the importance of flexible management systems in complex and uncertain environments. The organizing logic in the emerging IT governance paradigm is characterized by a collaborative network structure, where communication is more likely to be lateral, task definitions are more fluid and flexible—related to competencies and skills, rather than being a function of position in the organization—and where influencing of business-IT decisions is based on expertise rather than an individual (or group's) position in the hierarchy.

In collaborative relationships between business and IT stakeholder constituencies, managers work together to understand business and IT competencies, opportunities, risks and benefits. This collaborative relationship demands that both business and IT managers take responsibility for business operations and IT innovation, which is achievable only when stakeholder constituencies share their unique expertise and competencies. This emerging paradigm for IT governance is based on a 'philosophy' of collaboration, where the need for distinct competencies are recognized and developed, and shared adaptively across functional, organizational, cultural and geographic boundaries.

Directions and Opportunities for Future Research

In Spanish there is saying "*el camino se hace al caminar*", which roughly translates into English as the road is paved as you go. While we have come a long way in understanding and implementing strategies for IT governance, and designing effective IT governance architectures, the real journey still lays ahead of us. Specifically, there are (at least) five areas of theoretical and empirical study that seem particularly fruitful for future research and practice of IT governance. These are:

Figure 26. The emerging IT governance paradigm (Peterson, 2001)

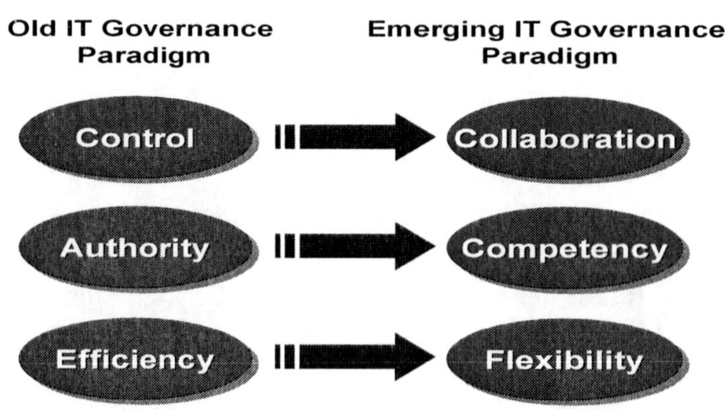

1. The development and integration of multiple complementary business and IT value drivers, e.g., how do organizations and networks of organizations develop and integrate the multiplicity of business and IT value drivers? How do organizations develop the requisite capabilities to become strategically flexible and dynamically stable?
2. The development of lateral coordination capability and value conversion process maturity, e.g., what are the interrelations between the different integration strategies and tactics? Is relational integration dependent upon formal integration, or does it enable formal integration? Are certain integration mechanisms 'multi-modal', i.e., do they facilitate more than one integration strategy?
3. The alignment of the IT governance architecture with IT value drivers and the enterprise architecture, e.g., how do organization evolve and maintain such a complex and dynamic alignment? Are there different paths to aligning the IT governance architecture?
4. The application of the IT governance diagnostic diamond, e.g., what are the experiences of (other) organizations applying the IT governance diagnostic diamond? What actions does an organization take after diagnosing IT governance effectiveness?
5. The emerging IT governance paradigm, e.g., how can the new principles be operationalized for research and practice? Is this emerging paradigm typical for contemporary firms in Europe, or are these principles also emerging in other continents? And what about the 'dot.coms'?

These five directions provide ample opportunities to expand our knowledge on strategies for IT governance, and will enable us - executives and researchers - to improve IT governance effectiveness.

REFERENCES

Ahituv, N., Neumann, S., & Zviran, M. (1989). Factors affecting the policy for distributing computing resources, *MIS Quarterly, 13*(4), 389-401.

Alter, A. (2001). Thinking out loud: Interview with JoAnn Heissen, *CIO INSIGHT*, December.

Ashby, W. R. (1956). *Introduction to cybernetics*. New York, John Wiley & Sons.

Boynton, A. C. (1993). Achieving dynamic stability through information technology, *California Management Review, 35*(2), 58-77.

Brown, C. V., & Magill, S. L. (1998). Reconceptualizing the context-design issue for the information systems function, *Organization Science, 9*(2), 176-194.

Brown, C. V., & Magill, S. L. (1994). Alignment of the IS function with the enterprise: Toward a model of antecedents, *MIS Quarterly, 18*(4), 371-403.

Buenger, V., Daft, R. L., Conlon, E. J., & Austin, J. (1996). Competing values in organizations: Contextual influences and structural consequences, *Organization Science, 7*(5), 557-576.

Burt, R. S. (1980). Models of network structure. *Annual Review of Sociology, 6*, 79-141

Clark, T. D. (1992). Corporate systems management: An overview and research perspective, *Communications of the ACM, 35*(2), 61-75.

Clark, C., Cavanaugh, N., Brown, C. V., & Sambamurthy, V. (1997). Building change-readiness IT capabilities: Insights from the Bell Atlantic experience. *MIS Quarterly, 21*(4), 425-455.

Collis, D. J., & Montgomery, C. A. (1995). Competing on resources: strategy in the 1990s. *Harvard Business Review, 73*(4), 118-128.

Cross, J., Earl, M., & Sampler, J. L. (1997). Transformation of the IT function at British Petroleum. *MIS Quarterly, 21*(4), 401-423.

Cyert, R. M., & March, J. G. (1963). *A behavioral theory of the firm*. Englewood Cliffs, NJ: Prentice Hall, Inc.

Daft, R. L. (1998). *Organization theory and design* (6th ed.). Cincinnati, OH: South Western College.

D'Aveni, R. A. (1999). Strategic supremacy through disruption and dominance, *Sloan Management Review, 40*(3), 127-135.

D'Aveni, R. A. (1994). *Hypercompetition: Managing the dynamics of strategic maneuvering*. New York: Free Press.

Earl, M. (2000). Local lessons for global businesses. In D. A. Marchand, T. H. Davenport, & T. Dickson (Ed.), *Mastering information management*, London: Financial Times/Prentice Hall.

Ein-Dor, P., & Segev, E. (1982). Organizational context and MIS structure: Some empirical evidence. *MIS Quarterly, 6*(3), 55-68.

El Sawy, O., Malhotra, A., Gosain, S., & Young, K. (1999). IT-intensive value innovation in the electronic economy: Insights from Marshall industries. *MIS Quarterly, 23*(3), 305-335.

Feeny, D. F., & Willcocks, L. (1998). Core IS capabilities for exploiting information technology, *Sloan Management Review, 39*(3), 9-21.

Feeny, D. F., Edwards, B. R., & Earl, M. (1989). *Complex organizations and the information systems function*, Oxford Institute for Information Management, Oxford, RDP 87/7.

Galbraith, J. R. (1994). *Competing with flexible lateral organizations*. MA: Addison-Wesley Pub.Co.

Galbraith, J. R., & Lawler, E. E. (1993). *Organizing the future: The new logic of managing complex organizations*. San Francisco: Jossey-Bass.

Galbraith, J. R. (1987). Organizational design. In J. W. Lorsch (Ed.), *Handbook of organizational behavior*, Englewood Cliffs, Prentice Hall, Inc.

Galbraith, J. R. (1974). Organization design: An information processing view. *Interfaces, 4*(3), 28-39.

Galbraith, J. R. (1973). *Designing complex organizations*. MA: Addison-Wesley.

Henderson, J. C. (1990). Plugging into strategic partnerships: The critical IS connection. *Sloan Management Review, 31*(3), 7-18.

Hitt, M. A., Keats, B. W., & DeMarie, S. M. (1998). Navigating the new competitive landscape: Building strategic flexibility and competitive advantage in the 21st century, *Academy of Management Executive, 12*(4), 22-42.

Hodgkinson, S. L. (1996). The role of the corporate IT function in the federal IT organization. In M. Earl (Ed.), *Information management: The organizational dimension* (pp. 247-268). Oxford, Oxford University Press.

IT Governance Institute (2001). Board Briefing on IT Governance. Retrieved from http://www.itgovernance.org

Kahn, K. B., & McDonough, F. (1997). An empirical study of the relationships among co-location, integration, performance, and satisfaction, *Journal of Production Innovation Management, 14*, 161-178.

Kaplan, R. S., & Norton, D. P. (1996). Using the balanced scorecard as a strategic management system. *Harvard Business Review, 74*(1), 75-85.

Kaplan, R. S., & Norton, D. P. (1992). The balanced scorecard: Measures that drive performance. *Harvard Business Review, 70*(1), 71-79.

King, J. L. (1983). Centralized versus decentralized computing: Organizational considerations and management options, *Computing Surveys, 15*(4), 319-349.

Khandwalla, P. N. (1976). The techno-economic ecology of corporate strategy. *Journal of Management Studies, 13*(1), 62-75.

Lawrence, P.R., & Lorsch, J.W. (1969). *Developing organizations: Diagnosis and action*. Addison-Wesley Publishing Company Inc.

Lawrence, P.R., & Lorsch, J.W. (1967). Differentiation and integration in complex organizations, *Administrative Science Quarterly, 12*(1), 1-47.

Lorsch, J. W., & Lawrence, P. R. (1970). *Studies in organization design*. Georgetown, Ontario: Irwin-Dorsey Limited.

Luftman, J., & Brier, T. (1999). Achieving and sustaining business-IT alignment, *California Management Review, 42*(1), 109-122.

Malone, T. W., & Crowston, K. (1994). The interdisciplinary study of coordination, *ACM Computing Surveys, 26*(1), 87-118.

March, J. G. (1991). Exploration and exploitation of organizational learning, *Organization Science, 2*, 71-87.

March, J.G., & Simon, H.A. (1958). *Organizations* (2nd ed.). [s.l.], Wiley.

Marchand, D. A. (2000). Hard IM choices for senior managers. In D. A. Marchand, T. H. Davenport, & T. Dickson (Ed.), *Mastering information management* (pp. 295-300). London: Financial Times/Prentice Hall.

Marchand, D. A. (2000). How to keep up with the hypercompetition. In D. A. Marchand, T. H. Davenport, & T. Dickson (Ed.), *Mastering information management* (pp. 120-126). London: Financial Times/Prentice Hall.

Miller, D. (1988). Relating Porter's business strategies to environment and structure: Analysis and performance implications, *Academy of Management Journal, 31*(2), 280-308.

Miller, D., & Friesen, P. H. (1978). Archetypes of strategy formulation, *Management Science, 24*(9), 921-933.

Mintzberg, H. (1979). *The structuring of organizations*. Prentice-Hall: Englewood Cliffs.

Nadler, D., & Tushman, M. L. (1998). *Competing by design: The power of organizational architecture*. Oxford: Oxford University Press.

Peppard, J., & Ward, S. (1999). Mind the gap: Diagnosing the relationship between the IT organization and the rest of the business. *Journal of Strategic Information Systems, 8*, 29-60.

Parker, M., Benson, R., & Trainor, H. (1988). *Information economics: Linking business performance to information technology*. NJ: Prentice Hall, Englewood Cliffs.

Parker, M., Trainor, H., & Benson, R. (1989). *Information strategy and economics: Linking business performance to information technology*. NJ: Prentice Hall.

Parker, M.M., & Benson, R.J. (1989). Enterprise-wide information management: State-of-the-art strategic planning. *Journal of Information Systems Management*, 14-23.

Parker, M.M., Ribbers, P.M., & Parker, S.R. (1997). *Critical IT Issues for Business Transformation*. Handboek BIK, D 1005-1, Tilburg University.

Peterson, R., & Ribbers, P.M.A (2002) *Service Level Agreements: Noodzakelijk en toereikend?* .ego. Retrieved from http://www.sbit.nl/ego/archief.php

Peterson, R.R. (2001). *Information governance: An empirical investigation into the differentiation and integration of strategic decision-making for IT*. Tilburg University, The Netherlands

Peterson, R.R. (2001). *Configurations and coordination for global information governance: Complex designs in a transnational European Context*. Proceedings of the 34th HICSS Conference, Hawaii.

Peterson, R.R. (2000). *Examining ambidextrous designs for information governance: Evidence from a longitudinal comparative case study in Financial Services*. Conference proceedings of the European Conference on IT Evaluation 2000, Dublin, Ireland.

Peterson, R.R. (2000). *Emerging capabilities for IT governance: Exploring stakeholder perspectives in Financial Services.* Conference Proceedings European Conference on Information Systems 2000, Vienna, Austria.

Peterson, R.R., O'Callaghan, R., & Ribbers, P.M.A. (2000). *Information technology governance by design.* Conference proceedings of the International Conference on Information Systems 2000, Brisbane, Australia.

Peterson, R.R., Smits, M., & Spanjers, R. (2000). *Exploring IT-enabled networked organizations in health care: Emerging practices and phases of development.* Conference Proceedings the European Conference on Information Systems 2000, Vienna, Austria.

Peterson, R.R. (1998). Successful exploitation of information technology in financial services. The role of IT governance. In *Proceedings of the International Conference on Information Systems (Doctoral Consortium)*, Helsinki, Finland.

Porter, M. E. (1980). *Competitive advantage.* Harvard Business Press.

Powell, T. C. (1992). Organizational alignment as competitive advantage, *Strategic Management Journal, 13*, 119-134.

Quinn, R. E., DeGraff, J., & Thakor, A. (2000). Creating shareholder value - and dispelling some myths. In T. Dickson (Ed.), *Mastering strategy.* London: Pearson Education Limited.

Quinn, R. E., & Rohrbaugh, J. (1983). A spatial model of effectiveness criteria: Towards a competing values approach to organizational analysis, *Management Science, 29*(3), 363-377.

Reich, B. H., & Benbasat, I. (1996). Measuring the linkage between business and information technology objectives. *MIS Quarterly, 20*(1), 55-81.

Remenyi, D. (2000). The elusive nature of delivering benefits from IT investment. *Electronic Journal of Information Systems Evaluation,* 1. Retireved from http://www.iteva.rug.nl/ejise/index.html

Robbins, S. P. (1994). *Essentials of organizational behavior.* London: Prentice Hall International.

Rockart, J. F., Earl, M., & Ross, J. W. (1996). Eight imperatives for the new IT organization, *Sloan Management Review, 38*(1), 43-55.

Rockart, J. F. (1988). The line takes the leadership: IS management in a wired society. *Sloan Management Review, 29*(4), 57-64.

Ross, J. W., Beath, C. M., & Goodhue, D. L. (1996). Develop long-term competitiveness through IT assets. *Sloan Management Review, 38*(1), 31-42.

Sambamurthy, V., & Zmud, R.W. (1999). Arrangements for information technology governance: A theory of multiple contingencies. *MIS Quarterly, 23*(2), 261-290.

Sambamurthy, V., & Zmud, R.W. (2000). Research commentary: the organizing logic for an enterprise's IT activities in the digital era: A prognosis of practice and a call for research, *Information Systems Research, 11*(2), 105-114.

Sauer, C., & Yetton, P. W. (1997). The right stuff: An introduction to new thinking about IT management. In C. Sauer, P. W. Yetton, & Associates (Eds.), *Steps to the future: Fresh thinking on the management of IT-based organizational transformation* (pp. 1-25). San Francisco: Jossey-Bass Publishers.

Scheier, R. (2001). Central intelligence: Johnson & Johnson case study. *CIO INSIGHT,* December.

Senge, P. M. (1990). *The fifth discipline.* New York: DoubleDay.

Simon, H.A., & Barnard, C.I. (1961). *Administrative behavior: A study of decision-making processes in administrative organization.* New York: The Macmillan Company

Simson, E. M. v. (1995). The recentralization of IT. *Computerworld, 1*(11), 2-7.

Simson, E. M. v. (1990). The centrally decentralized IS organization. *Harvard Business Review, 68*(4), 158-160.

Tavakolian, H. (1989). Linking the information technology structure with organizational competitive strategy: A survey. *MIS Quarterly, 13*(3), 309-317.

Teo, T. S. H., & King, W. R. (1999). An empirical study of the impacts of integrating business planning and information systems planning. *European Journal of Information Systems, 8*, 200-210.

Thompson, J. (1967). *Organizations in action: Social sciences bases of administrative theory.* New York: McGraw-Hill Book Co.

Treacy, M., & Wiersema, F. (1995). *The discipline of market leaders.* MA: Addison-Wesley.

Treacy, M., & Wiersema, F. (1993). Customer intimacy and other value disciplines. *Harvard Business Review, 71*(1), 84-93.

Van Grembergen, W., & Van Bruggen, R. (1997). Measuring and improving corporate information technology through the balanced scorecard technique. In *Proceedings of the 4th European Conference on the Evaluation of Information Technology* (pp. 163-171). Delft.

Van Grembergen, W., & Saull, R. (2001). Information technology governance through the balanced scorecard. In *Proceedings of the 34th Hawaii International Conference on System Sciences (HICSS).* IEEE.

Van Grembergen, W. (2002). Introduction to the minitrack: IT governance and its mechanisms. In *Proceedings of the 35th Hawaii International Conference on System Sciences (HICSS).* IEEE.

Van der Heijden, K. (1996). *Scenarios. The art of strategic conversation.* New York: John Wiley & Sons.

Venkatraman, N. (2000). Five steps to a dot-com strategy: How to find your footing on the web. *Sloan Management Review, 41*(3), 15-28.

Von Bertalanffy, L. (1968). *General system theory: Foundations, development, applications.* New York: George Braziller.

Vitale, M. (2001). *The dot.com legacy: Governing IT on Internet time.* Working Paper, Information Systems Research Center, University of Houston, Houston, Texas.

Weick, K. E., & Roberts, K. H. (1993). Collective mind in organizations: Heedful interrelating on flight decks. *Academy of Science Quarterly, 38*, 357-381.

Weill, P., & Broadbent, M. (1998). *Leveraging the new infrastructure: How market leaders capitalize on information technology.* Harvard Business School Press.

Weill, P., & Vitale, M. (2001). *Place to space. Migrating to eBusiness models.* Harvard Business School Press.

Weill, P., & Vitale, M. (2002). What IT infrastructural capabilities are needed to implement e-business models. *MIS Quarterly Executive, 1*(1), 17- 34.

Weill, P., & Woodam, R. (2002). *Don't just lead, govern: Implementing effective IT governance.* CISR Working Paper No. 326. Sloan School of Management, Cambridge, Massachusetts.

Whetherbe, J. (2001). *Achieving the high-performance, information-based networked organization.* Working Paper, Information Systems Research Center, University of Houston, Texas.

Willcocks, L.P., Feeny, D.F., & Islei, G. (1997). *Managing IT as a strategic resource.* London: McGraw-Hill Companies.

Willcocks, L. (1996). *Investing in information systems: Evaluation and management.* London: Chapman & Hall.

The Blind Men and the Elephant. (1963). McGraw-Hill Book Company.

Zachman, J.A. (1987). A framework for information systems architecture. *IBM Systems Journal, 26*(3), 276-292.

Zmud, R. W., Boynton, A. C., & Jacobs, G. C. (1986). The information economy: A new perspective for effective information systems management, *Database,* Fall, 17-23.

ENDNOTES

1 Two exceptions and noteworthy initiatives undertaken to address the development of a cumulative body of knowledge and skills are the organization of a special 'IT governance research track' at the 35th Hawaii International Conference on System Sciences (2002), and the IT Governance Institute™, established by the Information Systems Audit and Control Association® (ISACA™) in 1998.

2 Theoretically, there are eight distinct federal patterns in IT governance (excluding the completely centralized and the fully decentralized IT governance model).

3 IT application decisions address applications prioritization and planning, budgeting, and the delivery of application services.

4 IT development involves blending knowledge of business processes and functions with IT infrastructure capabilities along the complete IT systems development life cycle.

5 In a federal IT governance model, enterprise IT infrastructure decisions are always allocated to a central corporate IT office.

6 The systems view (Ashby, 1956; Von Bertalanffy, 1968) is based on cybernetic principles (Wiener, 1956), which draws upon the Greek word *Kubernesis* – steersmanship, the task of keeping a ship on its course in the midst of unexpected changing circumstances.

7 Organization Design Economics is an interdisciplinary field of study based on the work of, e.g., March & Simon, 1958; Cyert & March, 1963; Lawrence & Lorsch, 1967, 1969; Thompson, 1967; Galbraith, 1973, 1994; Daft & Lengel, 1984; 1986; Williamson, 1996; Malone & Crowston, 1994; Hitt et al., 1998; Nadler & Tushman, 1998.

This work was previously published in Strategies for Information Technology Governance, edited by W. Van Grembergen, pp. 37-80, copyright 2004 by IGI Publishing, formerly known as Idea Group Publishing (an imprint of IGI Global).

Chapter XIV
The Paradigm of E-Commerce in E-Government and E-Democracy

Bernd Carsten Stahl
De Monfort University, UK

ABSTRACT

During its relatively short history e-commerce, the use of information and communication technology (ICT) in business has been more successful and glamorous than e-government or e-democracy, the comparable use of ICT in governments and administration. This may be the reason why many government initiatives try to emulate the success of e-commerce by using concepts, processes, technologies, and approaches pioneered by businesses. This paper analyses the relevance and limits of this use of e-commerce as a paradigm in government. For this purpose it starts out by distinguishing between e-government and e-democracy. In the following step the paper discusses which factors have led to the success of e-commerce and might therefore be applicable as parts of the paradigm. It then discusses the strengths and the weaknesses of e-commerce as applied to government. The conclusion will be that there are good reasons to use the commercial paradigm in e-government and e-democracy. However, this may lead to an unintended shift towards e-government. E-democracy may even be weakened by the paradigm which might turn out to be detrimental to the democratic legitimacy of e-government and e-democracy.

INTRODUCTION

Many governmental initiatives aimed at promoting the use of information and communication technology (ICT) for the purposes of government and administration try to transfer ideas from the area of e-commerce to the area of e-government. Most notably, one can find any number of initiatives world-wide that emphasis the idea of citizen-centeredness which is based on the example of customer-centeredness in e-commerce. Furthermore, governments try to take advantage of the strengths of e-commerce in order to improve their e-government initiatives. Such attempts to import successful examples from e-commerce into e-government refer to all sorts and aspects of information systems. On the one hand governments buy hardware and software that was originally developed for the private sector and apply it to their tasks. On the other hand, governments take over arguments and whole discourses from the commercial sector. Customer or citizen-centeredness is only one example of this. Others would be the ideas of efficiency, optimisation, or cost-benefit analysis. While these ideas are not confined to the commercial world, they have a strong association with it and during the last decades have mostly been developed in the context of private enterprises.

The attempts by governments to improve and optimise their services are usually met with approval. One of the central and frequently-voiced criticisms of governments is that they are slow, don't react to the demands of their citizens, and that they are generally bureaucratic and wasteful. The business world, on the other hand, does not seem to be bothered by these problems. Businesses are deemed to be efficient, quick, and responsive. Commercial entities that do not take their customers seriously are quickly replaced in the marketplace by those that do. A large portion of the criticism levelled at governments and the way they do their business can therefore apparently be taken care of by doing things the way they are done in business.

However, things may not be quite as simple as they seem. While citizens are the customers of governments to a certain extent, there are also limits to this analogy. The question that this paper will analyse is therefore where the limits of the application of commercial ideas to government are. In order to be able to discuss this question on a meaningful basis the paper will start out by discussing the concepts of e-government and e-democracy. In the following section the strengths of e-commerce are analysed insofar as they are relevant or translatable to e-government. After that the strengths and weaknesses of using the paradigm of e-commerce in e-government are discussed. The result of the paper will be that the commercial paradigm is useful for most service delivery tasks that we find in e-government but that it is quite less useful for most applications in e-democracy. At the same time the strength of electronic service delivery along the lines of e-commerce is such that it threatens to blend out e-democratic applications. If this is so, then the commercial paradigm may turn out to become a threat to the legitimacy of democratic e-government.

E-GOVERNMENT AND E-DEMOCRACY

In order to understand the impact of the commercial paradigm on e-government and e-commerce we will first of all have to clarify the concepts. This section will therefore start out with a brief introduction of the concept of a paradigm and will then continue to discuss e-government as well as e-democracy. The importance of the distinction between the two will be the reintroduction of the separation of powers which in discussions about e-government is often neglected.

Paradigms and Information and Communication Technology

The term "paradigm" will in this paper be used in accordance with Kuhnian epistemology and the current use in information systems literature. While paradigm originally means something like "example", the term has taken on a more specific meaning following the publication of Kuhn's (1996) classic of the philosophy of science: "The Structure of Scientific Revolutions". For Kuhn, a paradigm is a framework for understanding the world, that is on the one hand "sufficiently unprecedented to attract an enduring group of adherents away from competing modes of scientific activity" (Kuhn 1996, 10). On the other hand it must be "sufficiently open-ended to leave all sorts of problems for the redefined group of practitioners to resolve" (ibid.). Examples are Aristotle's Physica or Newton's Principia and Opticks. These are works that shape the perception of researchers and provide them with a theoretical framework to carry out their work. Kuhn's writings seem to suggest that only great scientific developments constitute frameworks but the term has since evolved to be used for specific ways of approaching research.

In the area of information systems, the term has been used to denote "the most fundamental set of assumptions adopted by a professional community that allow them to share similar perceptions and engage in commonly shared practices" (Hirschheim & Klein 1994, 108). Paradigms in information systems are thus the lenses which are used to perceive and understand reality. They are contained in the narratives that are used to make sense of technology and its impact on social relationships (cf. Pentland 2003). There is a multitude of paradigms in most academic disciplines and information systems are no exception. Where the term "paradigm" is used explicitly it usually refers to questions of methodology and epistemology. However, it can also be used to analyse the use of ICT within organisations and here it may be used to explain why certain technologies are successful or fail. It can also be used to give explanations for problems of change management and changing perceptions. Finally, it has also been applied to the area of e-government and the understanding of developments in there (Wastell, 2003).

In this paper we will regard e-commerce as a paradigm in e-government and e-democracy. This is not to say that e-commerce is nothing but just that, a paradigm. E-commerce can be seen as an important part of the economy, as a particular channel of doing business, as an academic discipline, and many other things. For us the main point of interest here is whether and in what way e-commerce can influence or determine the perception of another, albeit related, field, namely e-government. This means that we will necessarily have to abstract from the totality of phenomena in e-commerce and try to distil those aspects that render it useable as a paradigm, as a lens for perception or a framework for understanding. Before we return to the question of why e-commerce is a paradigm and what the specifics of the e-commerce paradigm are, we should first clarify what the area is that is supposedly affected by the paradigm, namely e-government and e-democracy.

E-Government

In order to understand where and how the paradigm of e-commerce can be useful and also where its limits are when applied to public administration and political processes it is useful to distinguish between e-government and e-democracy. It will later become clear that e-commerce may be rather useful as a paradigm in e-government but it may be less so or even dangerous in e-democracy. To make this argument transparent we need to define the terms more clearly.

E-government will in this paper be understood as those aspects of public administration that have to do with the tasks of the executive. When these tasks are discharged with the help of ICT we speak of e-government. Typically these are administrative tasks, service delivery, but they may also include other executive duties such as the interpretation or enforcement of laws. E-government in this sense can aim at internal processes, meaning the use of ICT for the optimisation of internal processes. It can also aim at international processes, where computers and networks can be used for all sorts of international political matters from the exchange of statistics over cultural collaboration to combining efforts in crime prevention and detection. Finally, and this is the aspect that tends to be most clearly emphasised, e-government can have to do with the interaction of the administration with its citizens. Here, most of the examples revolve around service delivery where certain governmental or administrative functions are discharged with the use of computers.

It should be noted that this paper concentrates exclusively on the use of ICT in democratic governments. This is a non-trivial limitation of the scope of the topic because it implies several aspects. Democratic governments are supposed to represent their citizens and to act in their best interests. Democratic governments have to adhere to legal processes, they are subject to checks and balances and they have to adhere to the underlying ethical expectation that they do "the right thing", whatever that may mean in any specific situation. This limitation of the concept of e-government is also important because e-government is just as well imaginable in non-democratic environments. In fact, Orwell's "1984" is a good example for potential use of ICT in an administration without a democratic background.

Another interesting aspect that will be largely ignored for the rest of this paper is the interaction between technology and administration. The central question of this paper is whether e-commerce is a useful paradigm for e-government and where the limits of this approach are. The role of technology in this area is not discussed in any depth. However, it should be conceded that there can be a close connection and mutual influence of technology and e-government. On the one hand, governments can be large or even the largest users of ICT and thus shape the market for this technology. On the other hand, governments and administrations usually set the technical and legal framework within which technology is developed. Decisions about technology development can play a central role in governments (Weiser & Molnar 1996). Finally, existing technology from other realms can influence which types of policies or service deliveries are deemed possible. "[...] policy is often tightly coupled with, or biased by, the technology it applies to, and vice versa" (Reagle 1996, 18).

To come back to e-government, one can conclude that it mainly occupied with administrative and bureaucratic tasks. When we hear of e-government initiatives then these tend to concentrate on innovative and better ways to provide citizens with services. At the same time e-government also has a deep influence on the internal processes in administrations. External service delivery and internal process optimisation go hand in hand. The bureaucratic nature of e-governmental service delivery may be one of the reasons why computerisation appeals to administrative decision makers. Following Wiener, Postman (1992) argues that computers are the technology of command and control and that they need something to control. Given that bureaucracies are there to control and supervise, they may have a natural affinity to ICT that other aspects of democratic governments lack.

E-Democracy

Many authors define the notion of e-government much wider than we do here and include other political elements as long as they are related to ICT. On a local level, Wastell (2003), for example, sees three functions of government, namely to provide the mechanisms of local democracy, to be the focus for public policy making and to provide a range of public services, mostly in the social domain. If one defines the scope of government this wide, then e-government, consequentially, is the use of ICT in these three domains. Given the classical division of powers in executive, legislative, and judiciary, however, it seems useful to distinguish between the different aspects by introducing different terms. In this paper we will therefore distinguish between e-government which stands for the use of ICT for the purposes of the executive branch of government and e-democracy, which represents the use of ICT in all other aspects of political processes in democracy. This choice of terms is not perfect because it may be misunderstood to imply that e-government is not democratic. Also, it may be too limiting because it subsumes all non-executive functions of government under one term and the different applications of ICT in the legislative branch as well as all other sorts of democratic processes may require a further distinction. For the purposes of this paper, however, the dichotomy between e-government and e-democracy will suffice because it highlights the two most important branches of the use of ICT in government and public administration.

There is another reason why the distinction between e-government and e-democracy is of importance. Looking back at the idea of the division of power since Montesquieu, there has always been the belief that different parts of government should be responsible for making the rules, for enforcing them, and for judging breaches. The division of power was supposed to put checks and balances to the powers and thereby avoid misuses of power. In this sense, e-democracy can be seen as a possible check on the powers of e-government.

Most factual uses of ICT in government are examples of what we have named e-government instead of e-democracy. However, one should note that e-democracy has at least a strong theoretical influence. Ideas of computer use for the purposes of democratic discourse and decision making have long accompanied the rhetoric surrounding computers and particularly the Internet. Johnson (2001, 211) summarises the argument as follows: "(1) Democracy means power in the hands of individuals (the many); (2) information is power; (3) the Internet makes vast quantities of information available to individuals; (4) therefore, the Internet is democratic." The internet can thus be called a "democratic technology, suggesting that it is inherently democratic" (Johnson 2000, 181). And, indeed, the idea of democratic participation was one of the inspiring factors that led Al Gore to his political efforts that promoted the development of the Internet and the World Wide Web as we know it today (cf. Gore 1995).

There is a number of possible ways in which ICT and particularly the Internet might be beneficial to democracy. On the one hand, technology might allow the formation of democratic interest groups and the development of open democratic discourses. Electronic deliberation could allow the incorporation of a multitude of voices and stakeholders into the process of decision making (cf. Lévy 1997). The new technology might "free us to build a better world, promote democratic equality, improve the quality of education, and create new economic opportunities for underdeveloped nations" (Stichler & Hauptman 1998, 1). Optimists envisage a world where the increase in bandwidth will eliminate the difference between haves and have-nots in terms of information access and telecommunication will form the backbone of society (Meeks 2000). On the other hand ICT can have a positive effect on the individual and thereby improve the functioning of democratic societies. ICT has frequently been described as a means of emancipation and empowerment (cf. Hirschheim & Klein 1994). If it helps people live up to their psychological and intellectual potential and helps them optimise their organisations then this would again strengthen the basis of democracy.

Problems of E-Democracy

While e-democracy is a central aspect of the use of ICT in democratic societies, it is not without drawbacks. There are a number of areas where e-democracy either fails to live up to its expectations or where it may even have negative effects. One of the fundamental critiques of e-democracy is that instead of expanding dialogues and discourses it may do the opposite and result in a restriction of information flows. It may allow ICT-literate elites to separate themselves from other parts of society (Breen 1999). The question of power distribution does not have an unambiguous answer with regard to ICT. While the ideal developed in the last section is one of emancipation, decentralisation, and empowerment, ICT can be used for the exact opposite as well. Stallman (1995) argues that the computer system and the societal system can be designed in such aw way as to keep the elite in power (cf. Weizenbaum 1976). Similar power issues can appear in other social settings such as commercial organisations. Ischy & Simoni (2002) emphasise that many of the problems of ICT development are linked to power struggles that link them to politics, be it organisational politics or politics on a wider scale. Another argument against e-democracy is that despite the promising rhetoric that accompanies it, the examples we see of it so far do not seem to deliver on the promises. For example, computerisation does not seem to empirically promote decentralisation of power and public communication which can be seen as conditions of functioning democracy (cf. Yoon 1996).

A fundamental problem of e-democracy that might materialise in case e-democracy is introduced successfully is populism. The fundamental idea of e-democracy seems to be some kind of direct democracy, of electronic plebiscites based on the idea of direct democracy as we know it from the Greek Polis. It is debatable, however, whether and in what form this would be applicable and useful in complex modern societies. One could argue that e-democracy might lead to a shallow exchange of phrases. Instead of developing a political will it could lead to a simplification of issues and to non-optimal solutions. Instead of a democratic utopia we might end up with anti-democratic plebiscites (cf. Ess 1996). Instead of including larger groups or even all of society, e-democracy can lead to the creation or petrification of special interest groups (cf. Paletz 2000), to lobbying, and to a skewed idea of public interest.

A final problem of e-democracy that needs to be mentioned here because it can have an influence on which effects the commercial paradigm may have is that of the change of society. This is a highly complex problem with many different aspects that cannot be analysed comprehensively in this paper. It is based on the fact that the ubiquitous use of ICT changes not only the way we can use democratic institutions but affects most parts of society. Furthermore it seems to do so without being subject to the intentional steering of any one actor and it seems to take away power from the nation state. In the literature this development is often called globalisation. Globalisation is at least partly based on the use of ICT and many of its aspects are not possible without this. The networking of international financial and information markets, for example, are dependent on a functioning ICT infrastructure. Globalisation can be seen as a chance for e-democracy because it may enable international networks and international consultations on political issues. An interesting aspect here is that the opponents of globalisation are using the very technologies that it is based on to organise their resistance to it as could be seen in the anti-globalisation protests in Seattle, Milan, and other places in the last few years.

However, globalisation also seems to threaten the classical political basis of democracy, namely the nation-state. ICT allows the creation of international networks from global trading networks over internationally working NGOs to international crime syndicates. Most individual nation-states are powerless to control or oversee these institutions. Worse, the power of globalised developments forces individual states to adhere to certain rules without their having an equitable voice in the development of these rules (cf. Johnson 2001).

Finally, e-democracy and globalisation may just be aspects of a fundamental change in the fabric of our states and societies. The nation-state may evolve into networks and parts of wider networks as Castells (2001) predicts. In this network of networks e-democracy may take on new forms and meanings. We may be witnessing the evolution of a truly new form of society, the network society, or even a new age, the information age (Castells, 2000). All of this forms the background to our understanding of e-government and e-democracy and it is important for seeing how the paradigm of e-commerce can affect the development of these notions. In the following section we will now analyse what exactly the paradigm of e-commerce stands for and which of its aspects can be of relevance to e-government and e-democracy.

THE PARADIGM OF E-COMMERCE

This section will attempt to give an overview over e-commerce with regards to its use as a paradigm. It does not claim to be conclusive and exhaustive of the topic. Rather, it will attempt to isolate those factors that might affect our perception of the use of ICT in government. For this purpose it will try to capture the positive aspects frequently used to explain the success of e-commerce but also the characteristics which might become problematic when applied to democratic processes.

Characteristics of E-Commerce

In this paper we will take the term "e-commerce" to denote the buying and selling of products, services, or other commodities through the use of information and communication technology. Furthermore, we will include internal processes of companies that are geared toward the support and facilitation of such commercial exchanges. That means that we will ignore the difference between e-commerce and e-business as external and internal aspects of commerce using ICT. This is justified by the fact that we are looking for the possibility of transfer from the economic sphere to the administrative sphere and the boundaries between internal and external may not be comparable between the two. On the basis of this wide and inclusive definition of e-commerce one can note that it is not a radically new phenomenon. Companies have exchanged business data over a variety of communication networks for a number of years (Currie, 2000). However, the rapid expansion of the Internet and its use for commercial purposes have increased the volume of e-commerce immensely. In some areas this has led to completely new business models, while in many cases old business models have been adapted to be able to make use of the technology. It is debatable how deep the changes introduced by e-commerce really are, whether we are looking at a revolution of the business world or just a gradual change of some aspects. Independent of the answer to this question, there are some aspects of e-commerce that could be observed during the last few years which are important enough to speak of a new paradigm when applied to e-government.

The use of the Internet to buy and sell goods has introduced a new form of competition to many markets. While competition is at the heart of capitalist economic models and generally recognised as a positive part of market economies, competition in traditional markets is often limited. Individual consumers may have a choice of shopping in supermarket A or B but they rarely have the necessary information and resources to get an overview over markets and make economically rational decisions. Markets in economic theory tend to have a number of characteristics such as an infinite number of participants, complete transparency, and infinite reaction speed, that are not even approximated by most real markets. E-commerce has in many cases led to an evolution of markets in the direction of perfect markets of economics. The use of the Internet as a market platform facilitates access to information and reduces barriers to access (Spinello 2000). The individual market participant can get an overview over markets more easily and technology allows access to vendors and customers far beyond the regional area of traditional markets. A customer who wants to buy a book, for example, can now choose between hundreds of booksellers located all over the world. She can easily compare prices and conditions and make a better-informed decision. E-commerce thus realises the promises of market economies by providing better service to customers.

E-commerce does not only optimise existing markets but it also creates new ones. There are a number of goods and services for sale online that are only possible through the use of ICT in commerce (Schiller, 1999). Examples might be online information data bases, outsourcing of company tasks to web-based companies, and a whole host of services related to information as a commodity. Again, this may not be radically new (Stichler 1998) but it has taken on new meaning through the Internet and related technologies.

The ease of obtaining information on the Internet has brought with it the disadvantage of information overload. The amount of information available to each and every Internet use is greater than the capacity for processing it. This has produced the side effect that attention has become more important than clear information. Even the best information is useless if it does not reach potential users. Information suppliers on the internet are therefore concentrating more on grabbing potential customers' attention than on providing them with useful information. The Internet and e-commerce have thereby created a sort of attention economy where attention is the scarce good instead of goods, services, or information (Zerdick et al., 2001; Liebl, 1999).

The most important positive aspect of e-commerce, the aspect that may have facilitated the success of the Internet economy and the one that renders it desirable to politicians and administrators is that it is supposed to save costs. Shin (2003, 127) quotes three reasons why e-commerce can be less costly than traditional exchanges: the use of ICT decreases information processing costs, it decreases the costs of product selection, and it allows decreasing inventories, thereby saving capital costs. Another view of the cost-saving capacities of e-commerce concentrates on transaction costs. These costs that appear when transactions are prepared and realised consist of different costs, including search costs, information costs, bargaining costs, decision costs, policing costs, and enforcement costs (Welty & Becerra-Fernandez, 2001, 68). The use of computers and networks does not necessarily affect all of these components but it does bring down the overall transaction costs which can form a considerable part of transactions. At the same time, the use of ICT within businesses can also produce cost savings by optimising structures and processes, by allowing to react to customer preferences etc. These aspects carry different weight in different industries but they are probably the most important reason why e-commerce was and is successful.

Advantages of E-Commerce

The above characteristics are a collection of reasons why e-commerce may make sense as a business institution. In order to understand why it might be tempting to use the ideas in e-government it is helpful to translate them into manifest advantages for the customer. This section will therefore describe why e-commerce is good for customers (mainly working with the model of business to consumer (B2C) e-commerce) because this explains best why politicians and bureaucrats believe it to be good for citizens. The point of this section is therefore to translate the economic advantages named above into more general advantages for the customer.

The central theme of the last section was that e-commerce makes good business sense by improving efficiency and decreasing costs. For companies this translates into higher earnings which, given the supposed purpose of companies, namely to create profits, does not need further justification. But why should the consumer care? The answer comes from fundamental assumptions about the nature and functioning of markets. In functioning markets (and we have seen that e-commerce is supposed to create these) producers and service providers will not be able to retain the gains they make through efficiency and productivity gains but they will eventually have to pass them on to customers. Customers can therefore save money which then translates into a greater freedom of choice. Cost savings thus take on an ethical meaning which can easily be translated into something a democratic government can identify with. Following this line of reasoning one can see other ethical benefits in e-commerce which stem from its realisation of functioning capitalist markets. Cost savings not only increase freedom of choice but they also allow to produce more with the same amount of resources and they are therefore a necessary condition of redistribution and thus of justice (Kreikebaum 1996). The ethical advantages of e-commerce are thus of a fundamental nature and they touch deep-seated ethical concerns such as freedom and distribution. While most of these are of course double-edged swords, it is nevertheless easy to make a case for e-commerce in these terms. Another example might be tele-working which can also be understood to be one aspect of e-commerce. This, too, can be seen as an increase of freedom of the employees. It can also have negative consequences (McCalman 2003) but for the moment we are more interested in positive sides as these may be used as arguments for the use of the commercial paradigm in e-government.

One big advantage for customers that e-commerce promises is that it facilitates a more personalised service. The use of ICT allows vendors and service providers to accumulate great amounts of information on the customers and this information allows them to cater to individual preferences and needs. An integral part of many e-commerce applications is the so-called customer relationship management (CRM). CRM systems can be bought as complete packages or they can form a part of other e-commerce applications. Their use varies between industries and sellers but the overall defining feature is that the customer's views are taken seriously. Again, this is a very important point for governments, and it is the reason why the use of the term "customer-centred" which is closely linked to e-commerce and CRM is transformed to "citizen-centred" in e-government.

THE PARADIGM OF E-COMMERCE IN E-GOVERNMENT AND E-DEMOCRACY

Having discussed the concepts of e-government and e-democracy as well as the positive sides of e-commerce we are now in a position to discuss the advantages and disadvantages of the transfer of the concept of e-commerce to democracy and administration. In the first part of this section we will look at the reason why governments may want to use the commercial paradigm. The second part will be dedicated to the limits of the commercial paradigm.

Reasons for the Adoption of the Commercial Paradigm in E-Government and E-Democracy

The reasons why governments and administrations might want to use ideas and concepts from e-commerce for their own activities should have become quite clear from the enumeration of the advantages of e-commerce. Fundamentally, one can summarise the advantages of e-commerce as those of a functioning capitalist market economy and those advantages should be transferable to the activities of government. Among these positive points we have found efficiency, which should allow governments to deliver the same or better services at lower costs. This idea of efficiency also mirrors a hope that government bureaucracies, which are traditionally seen as inefficient and reluctant to change, could be accelerated and streamlined. Here, the commercial paradigm tends to aim at motivational structures, at the fact that e-commerce companies have found it possible to become much more flexible than traditional companies. The hope is that bureaucracies may find it possible to become as flexible by using processes and motivational measures copied from the commercial sector.

But efficiency in service delivery is no end in itself. In e-commerce efficiency gains are supposed to maximise profits, whereas in government they have different justifications. These can best be understood when the state is seen as a representative of its citizens which requires the state to act in the sense of the citizens and to do what they believe to be right. Saving costs may be something citizens desire but that only counts in the context of the other wishes citizens have. The ethical advantages of e-commerce therefore play a central role for the adoption of the paradigm in government. Efficiency and cost savings can then be seen as measures that increase the citizens' freedom. E-government can thus be seen as an aspect of liberty, which is something that democratic states are supposed to provide their citizens with. This liberty includes the freedom of choice, and again, the use of ICT can improve this by providing information. In the extreme, e-government could be seen as an introduction of competition between governments because the increased information flow might allow citizens to make an informed decision in which jurisdiction they want to live.

The adoption of the concept of customer-centeredness in the form of citizen-centeredness should also be understood in this context. Since, fundamentally, the state is there for the citizens it should be focused on the citizen anyway. If governments try to become more citizen-centred, then this is an expression of the reflection of the government on their original purpose. As such, it is clearly positive if e-government following e-commerce focuses the awareness of administrations on their *raison d'être*.

Limits of the Commercial Paradigm in E-Commerce and E-Democracy

As we have just seen, there are numerous good reasons for trying to extend the ideas of e-commerce to e-government and e-democracy. However, there are also limits to how far this transfer of ideas can go. These will be discussed in this section, starting with the differences between customers and citizens, proceeding with the limitations of economic analogies in government and administration and ending with genuine political problems caused by the adoption of e-commerce as a paradigm.

The starting point of this discussion of the limits of the commercial paradigm is the difference between customers and citizens. As we have seen above, the idea of customer-centeredness, which is closely linked to e-commerce is quite attractive to administrators in public bodies. Just like companies have to satisfy their customers to receive orders and survive, governments should satisfy their citizens. Accepting the commercial paradigm should lead to desirable developments such as speedier service and more efficient processes. However, there are limits to this analogy. The role of citizens in state and government is fundamentally different from the role of customers in a company. Companies exist for purposes that are defined by their owners. These may include profit generation, power exertion, reputation enhancement, and many others. Customers are important for companies as means to achieve these ends but they carry no intrinsic value. This is different for governments which exist for the citizens. The citizens are at the same time the owners and decision makers of governments. Governments have to realise the collective will of citizens. Without citizens governments would cease to exist. To return to the commercial paradigm, citizens are not only customers, they are also the shareholders. This means that while some aspects of the analogy of the customer apply to citizens, these have a richer meaning in governments. Citizens are the ultimate sovereign and they should have a voice in decision making. Some companies try to give their customers a voice in decision making or product design, but again, the motivation for doing so is different. Governments are there for their citizens in a fundamental ontological sense, whereas companies are only there for their customers as long as it suits their needs.

Another group of limits of the commercial paradigm results from the dissimilarity of states and markets. Markets are self-organising entities whereas states are directed and led. Furthermore, markets are not natural occurrences but they require a framework of rules, regulations, and enforcement, that can only be supplied by political entities (cf. De George, 1999; Hayek, 1994). Markets are thus fundamentally different entities from states and governments and, consequentially, some of the characteristics and advantages of e-commerce may not be transferable from one to the other.

One example of this is the idea of competition which is so central to success of e-commerce. There are several reasons why competition may not be equally desirable in government and administration. First, competition, by definition, produces winners and losers. Not everybody can survive in competition and the threat of losing is one of the main motivators in markets. This very idea is not suitable to the way governments treat their citizens. Citizens retain their status as citizens independent of their personal abilities and the state cannot accept a system which necessarily leads to some people losing out. Second, competition is only relevant for those players who have market power, meaning who have financial resources. E-commerce may be highly customer-centred but this applies only to those customers who can afford the products. Again, governments cannot copy this aspect, as their citizens have rights to services, independent of their financial means. In fact, those citizens who have little financial means are the same ones that need the services of the state most. Finally, competition does not apply to states and governments. Citizens do not generally have a choice between who they want to be governed by and ICT does little to change this. In order for the e-commerce paradigm to develop all of its potential, there would have to be competition between different e-government agencies or initiatives and this is impossible because of the nature of governments as natural monopolies.

Another problem is the concept of efficiency. Our description of the advantages of the commercial paradigm has emphasised efficiency by using a common sense understanding of efficiency. However, looking closer at the term shows that it is quite difficult to define efficiency in such a way that it reflects the common sense understanding. Because of this difficulty economics defines efficiency as Pareto-optimality (Hausman & McPherson, 1996; Sen, 1987). This definition means that an economic state is efficient if there are no more possible exchanges between two agents that are mutually advantageous. This is fundamentally problematic because it means that an economic state is efficient when one agent owns everything and nobody else owns anything at all. By definition this would be efficient but it would not be something that people and governments strive for.

A further problem of the commercial paradigm is that it may not be applicable to the goods that states are responsible for. Arguably one of the more important tasks we need states and governments for is the allocation of public goods. Public goods are those that belong to the community and can be used by all or some members thereof. They are highly problematic because the benefits of using them can be reaped by individual users whereas the costs tend to be socialised. They therefore create incentives for anti-social behaviour where individuals optimise their benefits and minimise their costs. Since these incentives are similar for all users, public goods can be overused and destroyed, to the detriment of all users. Pure market tools are incapable of dealing with this sort of situation. The situation requires state intervention and thus goes against the commercial paradigm. Interestingly, the information age seems to create new cases of public goods. The entire network structure of the Internet, for example, is a public good. It is questionable whether this can be managed by principles of capitalist business (Chapman & Rotenberg, 1995) even though it is the basis for most of today's e-commerce activities. Furthermore, the idea of information itself also seems to be a public good. Democracies need some sort of information to be workable. Applying the principles of private ownership to this may be harmful to the very idea of democracy (Blanke 1998).

Finally, there is the character or e-commerce as an attention economy. In terms of e-government and e-democracy this might be taken to mean that only those topics are taken serious that are at the top of current attention. While this is the case to some degree in any democracy, and maybe even stronger in modern media democracy, it is not necessarily a desirable development. If the increased use of ICT leads to an overflow of information and to citizens who are disoriented because of too much rather than too little information then e-commerce may be better used as a bad example than as a paradigm that one should follow.

Apart from these points where there is a lack of fit between the paradigm of e-commerce and the reality and requirements of government and democracy, there are some issues of genuine political importance which stand against the use of e-commerce ideas. Generally, these issues can be summarised by saying that the introduction of ICT into government, administration, and democracy can have political results that are not desirable.

The first point here is that ICT can lead to a redistribution of power. By their very nature, government and democracy have to do with power. However, the idea of democracy is to render the distribution and use of power transparent and understandable. ICT may lead to more or less subtle power shifts which are not transparent and which are not coupled to institutions of accountability. Much of the literature on how ICT changes power structures starts with the works of Foucault and analyses the power relationships within organisations and companies (cf. Healy & Iles, 2002; Introna, 2001). The same processes of power shifts might take place within governments and between different stakeholders of governments. One aspect of this is that of access, where e-government and e-democracy will favour those who have access to technology. A similar argument can be made in international relationships as well. The increasing use of ICT in commerce can be seen as a facilitator of international trade but it can also easily become an expression of cultural imperialism (Weckert, 2000). The rich western countries force the developing world to subscribe to their values and rules because it allows them to increase profits.

The most serious political problem produced by the use of e-commerce as a paradigm is that it may promote a particular ideology under the guise of addressing technical problems. When we look back at the advantages of the commercial paradigm then these could be summarised as saying that ICT can improve processes, mostly without changing the substance of activities. One could hold against this that e-commerce is not value-neutral but rather that it is deeply entrenched in the ideology of liberalism. The use of the paradigm in e-government would consequentially lead to a shift toward liberalism. This is in itself not a negative thing but the problem is that this might happen masked as a technical change. In fact, it has been noticed that digital technology, including personal computers, networks, and the Internet, and the mindset of those who developed this technology are quite closely related to liberalism or even libertarianism (Fagin, 2000). Many of the aspects of e-commerce that allow it to produce its positive effects are based on the ideas of free markets and can be related to neo-liberalism and the Chicago School of economics (Winner, 2000). These ideas which originally aim mostly at the economic sphere also have consequences in the political system. Many of the aspects and advantages of what is sometimes called "information democracy" are close to or originate from liberal thoughts. The free flow of data assumes independent and equal individuals with a sufficient amount of knowledge and freedom to act according to the information. This is, of course, the very starting point which liberalism takes (cf. Kester, 1998). It should thus be noted that e-government and e-democracy, as long as they work on the basis of this view of humanity, are using presuppositions which move them closely to liberalism.

The use of the term "ideology" here should not be misunderstood to be something entirely negative. An ideology can be seen "as a set of assumptions of which we are barely conscious but which nonetheless directs our efforts to give shape and coherence to the world" (Postman, 1992, 123). As Postman himself points out in the next sentence, according to this view, language itself is pure ideology. It is impossible to exist without ideologies as these are the building blocks of our world. In this they are very close to the concept of a paradigm which is so central to this paper. This part of the argument should thus not be misconstrued to be overly critical of liberalism. Rather, it is meant to show that the adoption of one paradigm, e-commerce, may lead to or be influenced by another set of fundamental assumptions, namely liberalism. This is nothing bad in and of itself. But it can become politically problematic it is not admitted openly and made subject of discussion and political decisions.

CONCLUSION

The question of this paper was whether the success of e-commerce could somehow be emulated or imported into state, government, and democracy. The paper discussed why e-commerce might be used as a paradigm as well as the strengths and weaknesses of this approach. The result was somewhat ambivalent. Some of the aspects of e-commerce can be used and applied in administration and democratic decision making whereas others seem to run counter to the idea of democracy. So where does this leave us, what should decision maker try to achieve?

The answer to this question becomes a bit clearer when one looks at the areas where e-commerce as a paradigm displays strengths and weaknesses. As a general rule, one can say that the success of the paradigm is the greater the closer the government application is to e-commerce. That means that in those areas where governments provide goods and services for the citizens, where citizens can thus justly be seen as customers, e-commerce may provide a useful role model. The further government applications move away from this service provision model into the genuine tasks of democratic politics, the less useful e-commerce will be. Democratic decision making including elections, representation, parliamentarianism, etc. have few or no equivalents in the business world. E-commerce can therefore not provide governments with suggestions how ICT can be used in these areas.

Returning to our distinction between e-government and e-democracy one could now say that e-commerce is a useful paradigm for the former but less so for the latter. At the same time one can observe a tendency of many of the organisations and institutions charged with using ICT in government to adopt the language of e-commerce and thus presumably the paradigm (cf. Remenyi & Bannister, 2003). This can be seen as a good sign because it means that the advantages of e-commerce may be realised in government. At the same time it can also mean that the emphasis of governments will move toward service provision and e-government and away from the politically more important e-democracy. This may lead to a growth of the power of the executive to the detriment of the legislature. If this is so then it might be deeply damaging to democracy as we know it.

This paper was meant to draw attention to this hidden danger. It was not intended to say that we should not make good use of positive experiences in the business world and use them to improve the workings of our states and administration. However, we should realise that there are fundamental differences between democratic government and business. Neglecting to take these differences seriously may in effect do more harm than good by weakening the participative basis of democracy.

REFERENCES

Blanke, H.T. (1998). Librarianship and public culture in the age of information capitalism. In R.N. Stichler & R. Hauptman (Eds.), *Ethics, information and technology: Readings* (pp. 184-199). Jefferson, NC: MacFarland & Company.

Breen, M. (1999). Counterrevolution in the infrastructure: A cultural study of techno-scientific impoverishment. In L.J. Pourciau (Ed.), *Ethics and electronic information in the 21st century* (pp. 29-45). West Lafayette, IN: Purdue University Press.

Castells, M. (2001). Epilogue: Informationalism and the network society. In P. Himanen (Ed.), *The hacker ethic and the spirit of the information age* (pp. 155-178). London: Secker & Warburg.

Castells, M. (2000). *The information age: Economy, society, and culture* (Volume I). The Rise of the Network Society. 2nd edition. Oxford: Blackwell.

Chapman, G., & Rotenberg, M. (1995). The national information infrastructure: A public interest opportunity. In D.G. Johnson & H. Nissenbaum (Eds.), *Computers, ethics & social values* (pp. 628-644). Upper Saddle River: Prentice Hall.

Currie, W. (2000). *The global information society.* Chichester: John Wiley & Sons.

De George, R.T. (1999). *Business ethics* (5th ed.). Upper Saddle River, NJ: Prentice Hall.

Ess, C. (1996). The political computer: Democracy, CMC, and Habermas. In C. Ess (Ed.), *Philosophical perspectives on computer-mediated communication* (pp. 197-230). Albany: State University of New York Press.

Fagin, B. (2000). Liberty and community online. In R.M. Baird, R. Ramsower, & S.E. Rosenbaum (Eds.), *Cyberethics - Social and moral issues in the computer age*. New York: Prometheus Books.

Gore, A. (1995). Global information infrastructure. In Johnson & Nissenbaum (Eds.), *Computers, ethics & social values*. Upper Saddle River: Prentice Hall.

Hausman, D.M., & McPherson, M.S. (1996). *Economic analysis and moral philosophy.* Cambridge: Cambridge University Press.

Hayek, F. A. von (1994). *The road to serfdom.* Fifthieth Anniversary Edition. Chicago: The University of Chicago Press.

Healy, M., & Iles, J. (2002). The impact of information and communications technology on managerial practices: The use of codes of conduct. In Alvarez, et al. (Eds.), *The transformation of organisations in the information age: Social and ethical implications. Proceedings of the sixth ETHICOMP Conference*, Lisbon, Portugal. Lisbon: Universidade Lusiada.

Hirschheim, R., & Klein, H.K. (1994). Realizing emancipatory principles in information systems development: The case for ETHICS. In *MIS Quarterly, 18*(1), 83-109.

Introna, L. (2001). Workplace surveillance, privacy, and distributive justice. In R.A. Spinello & H.T. Tavani (Eds.), *Readings in cyberethics.* Sudbury, MA: Jones and Bartlett.

Ischy, F., & Simoni, O. (2002). Representations as factor of organizational change. In Alvarez, et al. (Eds.), *The transformation of organisations in the information age: Social and ethical implications. Proceedings of the Sixth ETHICOMP Conference*, Lisbon, Portugal. Lisbon: Universidade Lusiada.

Johnson, D.G. (2001). *Computer ethics* (3rd ed.). Upper Saddle River, NJ: Prentice Hall.

Johnson, D. G. (2000). Democratic values and the Internet. In D. Langford (Ed.), *Internet ethics* (pp. 181-196). London: McMillan.

Kester, G.H. (1998). Access denied: Information policy and the limits of liberalism. In R. N. Stichler & R. Hauptman (Eds.), *Ethics, information and technology: Readings* (pp. 207-230). Jefferson, NC: MacFarland & Company.

Kreikebaum, H. (1996). *Grundlagen der Unternehmensethik.* Stuttgart: Schaeffer-Poeschel

Kuhn, T.S. (1996). *The structure of scientific revolutions.* 3rd edition. Chicago; London: The University of Chicago Press.

Lévy, P. (1997). *Cyberculture.* Paris: Editions Odile Jacob.

Liebl, F. (1999). What system have you announced? (Impressionen aus einer Ökonomie der Ankündigung). In Liebl (Ed.), *e-Conomy - Management und Ökonomie in digitalen Kontexten.* Marburg: Metropolis Verlag.

McCalman, J. (2003). What can we do for corporate nomads? IT and facilities management. In Joia (Ed.), *IT-based management: Challenges and solutions* (pp. 130-142). Hershey: Idea Group Publishing.

Meeks, B. N. (2000). Better democracy through technology. In Baird, Ramsower, & Rosenbaum (Eds.), *Cyberethics - Social and moral issues in the computer age.* New York: Prometheus Books.

Pentland (2003). Panel contribution to: Ackerman, Mark, Pentland, Brian T., Qureshi, Sajda & Yakura, Elaine K. (2003): Visual Elements in the Discourse on Information Technology. In: Wynn, Eleanor; Whitley, Edgar; Myers, Michael D. & DeGross, Janice (Eds.) (2003): *Global and Organizational Discourse About Information Technology*: Ifip Tc8/Wg8.2 Working Conference on Global and Organizational Discourse About Information Technology, Barcelona, Spain. Dordrecht: Kluwer Academic Publishers.

Paletz, D. L. (2000). Advanced information technology and political communication. In Baird, Ramsower, & Rosenbaum (Eds.), *Cyberethics - Social and moral issues in the computer age* (pp. 285-287). New York: Prometheus Books.

Postman, N. (1992). *Technopoly - The surrender of culture to technology.* New York: Vintage Books.

Reagle, J.M. Jr. (1996). Trust in electronic markets. In *First Monday* (1:2). Available www.firstmonday.dk

Remenyi, D., & Bannister, F. (Eds.). (2003). In *Proceedings of the European Conference on e-Government,* Trinity College Dublin.

Schiller, D. (1999). *Digital capitalism: Networking the global market system.* Cambridge, MA / London: MIT Press.

Sen, A. (1987). *On ethics and economics.* New York: Basil Blackwell.

Shin, N. (2003). Productivity gains from IT's reduction of coordination costs. In Shin (Ed.), *Creating business value with information technology: Challenges and solutions* (pp. 125-145). Hershey, PA: Idea Group Publishing.

Spinello, R. (2000). *Cyberethics: Morality and law in cyberspace.* London: Jones and Bartlett.

Stallman, R. (1995). Are computer property rights absolute? In Johnson & Nissenbaum (Eds.), *Computers, ethics & social values*. Upper Saddle River: Prentice Hall.

Stichler, R. N. (1998). Ethics in the information market. In Stichler & Hauptman (Eds.), *Ethics, information and technology: Readings*. Jefferson, NC: MacFarland & Company.

Stichler, R. N., & Hauptman, R. (Eds.). (1998). *Ethics, information and technology: Readings*. Jefferson, NC: MacFarland & Company.

Wastell, D.G. (2003). Organizational discourse as a social defense: Taming the tiger of electronic government. In Wynn, Whitley, Myers, & DeGross (Eds.), *Global and Organizational Discourse About Information Technology: Ifip Tc8/Wg8.2 Working Conference on Global and Organizational Discourse About Information Technology*, Barcelona, Spain. Dordrecht: Kluwer Academic Publishers.

Weckert, J. (2000). What is new or unique about Internet activities? In Langford (Ed.), *Internet ethics*. London: McMillan.

Weiser, M., & Molnar, K. K. (1996). Advanced telecommunications infrastructure policies - A comparative analysis. In *Proceedings of the Americas Conference on Information Systems*.

Weizenbaum, J. (1976). *Computer power and human reason*. San Francisco: W. H. Freeman and Company.

Welty, B., & Becerra-Fernandez, I. (2001). Managing trust and commitment in collaborative supply chain relationships. *Communications of the ACM* (6), 67-73.

Winner, L. (2000). Cyberlibertarian myths and the prospects for community. In Baird, Ramsower, & Rosenbaum (Eds.), *Cyberethics - Social and moral issues in the computer age*. New York: Prometheus Books.

Yoon, S-H. (1996). Power online: A post-structuralist perspective on computer-mediated communication. In C. Ess (Ed.), *Philosophical perspectives on computer-mediated communication* (pp. 171-196). Albany: State University of New York Press.

Zerdick, A., et al. (2001). *European Communication Councel Report: Die Internet-Ökonomie: Strategien für die digitale Wirtschaft*. 3rd edition Berlin, Heidelberg: Springer.

This work was previously published in Electronic Government Strategies and Implementations, edited by W. Huang, K. Siau, & K.K. Wei, pp. 1-19, copyright 2005 by IGI Publishing, formerly Idea Group Publishing (an imprint of IGI Global).

Section IV
Strategic ICT Applications

Chapter XV
Delivering More Effective Community Consultation and Support for Regional ICT Programs

Lynne De Weaver
Southern Cross University, Australia

Allan H. Ellis
Southern Cross University, Australia

ABSTRACT

This chapter looks at the role of language and the community consultation process in overcoming the digital divide by facilitating the uptake of information and communications technologies (ICT) in small regional communities in Australia. It focuses on one of the 'telecentre' programs funded by State and Federal governments in Australia—the Community Technology Centre at New South Wales (CTC@NSW) program. The authors look at some of the key issues that emerged in communities that applied for CTC@NSW grants when a more culturally relevant consultation process, based on Maslow's Hierarchy of Needs, was developed and implemented. This consultation process was successfully used in targeted regional communities to build social capital, facilitate regional economic development, and empower communities through the use of ICT. The chapter also includes projects that demonstrate the diversity of ICT usage in the communities that received funding to establish a CTC.

INTRODUCTION

Federally funded programs in Australia, such as Networking the Nation, have provided millions of dollars to Australian regional communities to help them gain access to a range of ICT products and services. The Community Technology Centre at New South Wales (CTC@NSW) was one such program, with grant funding being made available to communities in partnership with the NSW State Government. A new consultation model was developed by De Weaver in order to increase the understanding and uptake of information and communications technology (ICT) products and services in small regional communities in New South Wales. Using the CTC@NSW program as a case study, the chapter will focus on this consultation process and how it was used to achieve successful outcomes for some of the communities that participated in the program. It will also highlight underlying language and technology issues that arose when small rural and regional NSW communities applied for CTC@NSW grants.

One of the major challenges of facilitating effective community consultation at a statewide level was in presenting information about the benefits of ICT products and services in a way that was relevant to all communities. In many instances, 'Communities don't know what they don't know'. This proved to be the case in many of the communities that sought grant funding through the CTC@NSW program. A recent report from a regional research forum (2004, p. 8) stated, "...regional development policy in Australia is 80 years old, but despite this, and the research issues in the last 40 years in particular, what we know is substantially less than what we don't know." Hearn et al. (2004, p. 7) identified lack of ICT knowledge in regional Australia as a further complicating factor and states, "Advocates of ICT projects can mis-specify and under resource the social and human infrastructure required. It is necessary that citizens are able to not only access that technology but that they also possess the skills to use and the resources to access them."

Finally, the chapter looks at the consultation process and methodology used by De Weaver, in her role as regional coordinator for the CTC@NSW program, in achieving successful ICT strategies and outcomes for the regional communities in which she worked. It also highlights the importance of using 'plain English' in the consultation process and in dealing with the underlying resentment that may exist in various stakeholder groups when their needs are misunderstood or inadequately addressed.

BACKGROUND

The CTC@NSW began its operational phase in March 2001. Its original charter, as shown on its Web site (July 20, 2004), was "...to establish up to 55 Community Technology Centres (CTCs) in regional NSW communities, with populations of less than 3000, to help them gain access to a range of technology services for local businesses, students, community organizations, individuals, and identified special groups within these communities." The specific goals of the CTC@NSW program were to:

1. Assist local communities to establish a network of sustainable CTCs across rural and remote NSW
2. Promote community development and networks of interests in rural and remote NSW
3. Encourage community demand for information technology (IT) services in rural and remote communities
4. Encourage existing telecentres to join the CTC support network

By June 2003, the CTC@NSW program had achieved its goals with a total of 83 fully funded CTCs serving over 100 NSW communities. The

CTC network included 50 newly funded CTCs with a further five CTCs in the planning stage, plus 33 telecentres that had been funded by an earlier federal government program. The momentum for establishing CTCs came from the communities themselves, and involved a lengthy community consultation process. The first step in this process was to help communities determine their eligibility through the preparation of an expression of interest (EOI). Following approval of an EOI, a regional coordinator was assigned to the 'approved' community so work could commence on their application. CTC applications were for seed funding rather than an out right grant. Communities had to determine how much they could apply for *after* they had completed their 3 year business plan. This plan was also the key component against which applications would be evaluated by the Canberra based CTC Funding Panel that met three times a year. Consultation with communities usually lasted from 4 to 6 months, although for some, it was an even longer process if their first application proved unsuccessful.

A wide range of motivational factors needed to be incorporated into the community engagement and consultation process in order to ensure the best possible outcomes for applicant communities. However, as Pigg and Crank (2003, p. 25) stated in their paper, which looked at similar programs in rural American communities, "Community capacity to successfully capture the advantages engendered via the deployment of ICTs is not particularly unique to the technology. Key agents of change must work together with local residents to actively support and engender an atmosphere for collective action towards improvement for their rural communities." Maslow's five-stage Hierarchy of Needs, as re-framed by Chapman (2002, p.1), provided an excellent model to promote collective action as his concept of self-actualization related well to the challenges and opportunities that ICT presented to the small, regional communities involved in the CTC@NSW program. It was this reframing that provided the conceptual framework for the new Hierarchy of Community Needs and offered the structure, understanding, and the methodology by which collective action could be taken to meet the identified ICT needs of individual communities as they sought funding for a CTC.

While there are many definitions for the term 'community', the authors have used the definition given by the American Society of Association Executives (2001, p. 2), that defines communities in two related ways:

- Community as social glue—providing 'stickiness' or connection
- Community is also what people do for each other—to create a Web of social relationships.

De Weaver used Action Research (AR) as her methodology throughout the 2 years of fieldwork she undertook as the Northern Rivers Regional Coordinator for the CTC@NSW program. It was the cyclical nature of AR, for example, planning, action, and review that enabled her to achieve better outcomes in each of the communities in which she worked, as she was able to build on the insights gained from the critical reflection component of AR when she commenced work in new communities. Action Research was especially effective in this regard because as Dick states (2001, p. 21) "Action Research... pursues both action (change) and research (understanding) outcomes. It achieves change through its participative approach often in conjunction with other change processes."

THE HIERARCHY OF COMMUNITY NEEDS CONSULTATION MODEL

Although much has been written about community consultation per se, when the consultation incorporates explaining the benefits of ICT, identification of needs, and dealing with technical usage issues, a whole new dimension of complexity is added to the process. Aslin and Brown (2004, p. 3) state:

Good community engagement can be achieved by using a wide range of tools and techniques, not by applying a simple recipe or prescription applicable to all situations ... The engagement process is part of furthering another process: decision-making for a particular purpose.

This proved to be the case in the process of applying for a CTC grant and it was Maslow's Hierarchy of Needs in its re-framed five-stage model that formed the basis of the new consultation model that evolved. Maslow's 'Needs', as explained by Chapman (2002, p. 4), consist of:

Biological and Physiological needs including basic life needs. Safety needs including protection, order, etc. Belongingness and love needs including work groups. Esteem needs including status, responsibility, reputation and self-actualization including personal growth and fulfillment.

While Maslow's theory is often closely linked to pedagogy or psychology it also has important applications in understanding what motivates communities to apply for capacity building grants such the CTC@NSW program. This view was reflected by Chapman (2002, p. 6), when he stated that:

Abraham Maslow developed the Hierarchy of Needs model in 1940-50's USA, and the Hierarchy of Needs theory remains valid today for understanding human motivation, management training, and personal development ... we must satisfy each in turn, starting with the first, which deals with the most obvious needs for survival itself. Only when the lower order needs are satisfied are we concerned with higher order needs.

However, for community consultation in the CTC@NSW program to be effective, it was necessary to reframe Maslow's model while adhering to his core values so that it became a new hierarchy of community ICT needs consultation model as shown in Figure 1.

Just as in Maslow's Hierarchy of Needs, the most basic need in many small regional communities is for survival itself as banks, businesses, and governments withdraw services from them in the name of economic rationalism. These communities are then faced with the need to explore alternatives and find out if ICT has a role to play in helping them build capacity and overcome the loss of face-to-face and shop front services that have had such a negative impact on their quality of community life. While Plowman et al. (2003, p. 135) believe that:

Figure 1. Hierarchy of community needs consultation model (Derived from De Weaver's unpublished monthly progress reports prepared in her capacity as regional coordinator for the CTC@NSW program from 2000–2002.)

Community Actualization
Access to identified ICT products & services
Growth of community business planning & implementation skills
Identification of key community stakeholder groups & their ICT needs
Information about the benefits of ICT products & services presented in plain English

Governments have started to respond to the strong expressions of dissatisfaction of rural people concerning the decline of social and economic fabric in rural communities attempting to develop programs for the reversal, or at least the amelioration, of the observed trends....governments are increasingly calling on the presumed self-help capacities of rural communities for them to become 'partners' with government in developing and delivering programs for rural communities...however the depth and breadth of resources, particularly the human and social capital, may not be sufficient to allow communities to respond to, and participate effectively in, such programs.

ASSESSING NEEDS REQUIRED INFORMATION IN PLAIN ENGLISH

It is important to consider how a community's ability to access government funded programs to satisfy its most basic 'survival' needs is often challenged by the bureaucratic and technical nature of the language used in grant applications. De Weaver and Lloyd (2005, p. 4) noted that "...applicant communities had to come to grips with the technology driven and corporatized language of ICT applications, a 'literacy *for* technology' issue emerged that needed to be addressed." They also needed to understand how access to ICT funding could help their communities survive in a global marketplace. In one community, when a survey was circulated to determine the type of ICT services that residents required, some of the answers gave a clear indication that they first had to deal with a 'literacy *in* technology' issue. For some, ICT meant being able to have public access to a facsimile machine in order to fax in their benefits claim forms while for others it meant having a public pay phone outside the general store or being able to get mobile telephone coverage. But, for the more sophisticated, access to high-speed bandwidth was a key driver because it meant more cost effect access to national and international markets for their products and services.

In order to understand the implications of what ICT might offer a community, members of that community had to be able to articulate their needs. Wolstenholme (1995, p. 27), provided insight into this stage when she stated, "Action and interaction by members of a community in the pursuit of shared interests is probably the strongest force within a community. It is a way of achieving things for the community...Community action is the important source of power for any change which is likely to promote the well-being of a community." Information about the CTC@NSW program was therefore imparted and gained in a variety of ways including: media releases, community surveys, free information sessions to explain how the Internet worked, how small businesses could take advantage of online services to do their banking, pay taxes, seek regulatory assistance, and more. The use of the Internet to help high school and tertiary students with their studies by providing access to a vast array of information databases was often seen as a highly desirable outcome for applicant communities. The aim of all these activities was not only to raise awareness of greater access to ICTs but also to explain ICT features in terms of consumer benefits using plain English.

The importance of using plain English to empower people was expounded by Fairclough (Cope & Kalantzis, 2000, p. 163) when he stated, "We are living through a period of intense social and cultural change which is pervasive and universal in its global, national, and local effects...these fundamental changes are changes in languages, different dialects, different genres, and different discourses." Fairclough believes that the language issues of marketization and globalization also arise out of these changing discourses. He also says (Cope & Kalantzis, 2000, p. 164), "The processes of marketization are in part linguistic in nature: they involve a marketization of language, in the sense that the language of areas like public services are being colonized by the language of the market."

It was also interesting to look at the language issue from the perspective of social researchers. In this aspect, Carson and Gelber (2001, p. 5) commented on the community consultation process by stating that, "The achievement of the specific goal of achieving better communication and participation faces a number of obstacles, on a range of levels. At a macro level, those who decide which services should be provided are not the same people who actually provide the service." This was indeed the case in the CTC@NSW application, as it was written by bureaucrats who operated in a different paradigm than that of the communities. This implies that the language used in these applications often posed a significant challenge for the communities. In fact, of the 26 communities in the Northern Rivers region of NSW that had initial meetings with regard to the preparation of an EOI, only 7 went on to apply for grant funding, all the rest indicated that they either did not feel that they had the skill or time to deal with its complexity and that they found the wording in the documents too confusing.

For communities that went on to apply, not only did they have to deal with 'literacy' issues in order to apply for a CTC@NSW grant but they also had to deal with technical literacy issues as well. Luke and Gilbert (1993, p. 5) provided insight into this when they stated, "To refer to 'technology' in association with 'literacy' can signal a number of discourses concerning literacy, technology, and various combinations thereof...technology *for* literacy, literacy *for* technology, literacy *as* technology, and technology *as* literacy." For example, in the technology section of the CTC application, those communities that wanted to use Apple Macs had to deal with information such as:

It is recognized that there is a strong rationale for using Apple Macs particularly in the multimedia field...In recognition of this, and the fact that most CTCs will be running PC based infrastructure, the question then becomes how to integrate the Macs into a PC based network. Although there are a number of ways of doing this, and this has become somewhat simpler since the release of OS 10.1, we still recommend the Macs only be linked to the network via an IP address...

The above was perhaps one of the more glaring examples of the complexity of the language used in a CTC application. It was also one of many such instances. The authors concur with the findings of Shepherd (1998, pp.19-20) about other Australian regional development programs, "Project documents are peppered with buzzwords...required by financing organizations that must account for their spending or lending in the most up-to-date or fashionable terms. Buzzwords are often used by people with little understanding of what they might mean so they take on a hollow, artificial tone...language can also easily prevent understanding. Frequently outsiders and insiders think of a phenomenon in quite different ways. If the outsider has power to impose his or her thinking, unfortunate consequences many follow." All of which tend to impede a community's path to actualization and empowerment.

The culture of communities was also a major variable in achieving successful outcomes. In working with some of the applicant communities in Northern New South Wales, it was found that coastal communities, that had recently experienced an influx of well educated 'sea changers' had the greatest propensity to successfully complete their applications. On the other hand, many of the more rural, in-land communities that had been undergoing a steady outward migration of their younger, better educated, more productive citizens, for a variety of reasons, did not even complete their EOI. This outcome reflects Shepherd's earlier work (1998, p. 204), when he stated:

The likelihood is that the rich and non-marginalized will always be more articulate and dominant in any community or grouping. So outsiders can play an especially critical role in creating space for poor people, economically and environmentally, in terms of rights, access to services, and control over assets and services which the poor can provide to the rest of the community; and politically, by ensuring their inclusion in associations, helping them build networks and coalitions which will protect positions gained.

Effective community consultation needs to give people who live in rural and remote regions not only a voice but also the capacity to deal with issues arising out of inappropriate communication which can fuel misunderstanding and misinformation in these communities and prevent them from gaining access to the resources they need to ensure that their basic community needs for ICT are met.

IDENTIFICATION OF STAKEHOLDER GROUPS AND MEETING THEIR NEEDS

The establishment of a truly representative 'application committee', one that could speak for all the various stakeholder groups was another challenge for applicant communities. Unfortunately, it was found that the groups that had the most to gain were often the most inarticulate. Luke and Gilbert (1993, p. 43-44) spoke of this when they wrote about literacy issues in an Australian context and said "A group without access to literacy in our society is relatively powerless, unless trusted "brokers" exist to act on their behalf. By examining literacy pedagogy in cross-cultural contexts, it is possible to see a new approach emerging that challenges both traditional and progressivist theory and practice."

The consultation practice employed in applicant communities also needed to build trust so that communities were able to identify their needs in a non-confrontational, safe environment to enable them to move up to the next stage in becoming empowered. However, needs varied considerably from one community to the next, and the key to unlocking those differences was in identifying and engaging all the relevant stakeholders as well as building on their strengths. This view also reflects the earlier work of Wolstenholme (1995, p. 15) when she stated:

While traditional rural identity remains largely intact, there appears to be a move away from idealizing the more Spartan existence of traditional country life. People were questioning their rural ideology and comparing their lifestyles with those in the city. They wanted access to more resources ...

In this context, identifying the strengths of an applicant community as well as its key stakeholder groups posed a great challenge. But the greater challenge was in finding members from these groups who could work together as a committee to complete the application. This idea was also put forward by Aslin and Brown (2004, p. 4) in another regional Australian project, when they stated, "Recognizing that a wide range of stakeholders exists means facing the likelihood that local and more distant interests may be in conflict, as local communities often have to bear the personal and immediate consequences of decisions being made...Done well community consultation can feel real, committed, integrated, and influential." At the same time, further consideration must also be given to meeting the needs of all the various stakeholder groups so that linkages can be established to identify *their* needs in order for them to be integrated with those of the community at large. Furthermore, it is not enough to say that 'youth needs' should be met if they are not represented on the application committee. They need to be given a voice along with other stakeholder groups. The formation of sub-committees enabled diverse groups to have their own meetings and report back to the main Committee on a regular basis thus keeping applications on track.

It is of interest to note that Collits (2000, p. 30) also reminded that, "Communities have finite resources at their disposal for economic development projects and some difficult choices." This was evident in the CTC's final report, as those 'finite resources' extended to the skills and commitment of the people in the communities that participated in the CTC@NSW program. The CTC program's final report (2004, p. 3), highlighted, "Lack of available time from community members and other local issues, such as drought, often hindered progress with the application for seed funding and the development and establishment of Centres." It was the people from the various stakeholder groups that provided the membership of the CTC Application Committee and it was their skills and vision that ICT could serve as a catalyst for growth in their communities that proved to be the most significant determinant in the success of an application.

BUSINESS PLANNING AND IMPLEMENTATION SKILLS

The selection and composition of the CTC Application Committee and its sub-committees, proved to be the critical factor in achieving successful outcomes. While these committees needed to include people from the various stakeholder groups, they also had to have members with the requisite business acumen, marketing and computer skills as the business plan portion of the application was a complex spread sheet full of macros to reflect various equipment requirements and usages scenarios over three years. However, the committees that achieved the best outcomes all had a 'Dorothy' as a member. Dorothy said at the onset of the application process in her community, "I don't know much about ICT but I know how to make a good cup of tea, get people to attend meetings and do what they said they'd do." She was certainly right. What was Dorothy's motivation in wanting to get a CTC into her community? She wanted to do something to help keep the young people in her community from drifting away to Sydney or Brisbane and she saw ICT as a means of slowing their departure.

However, committees needed to prepare sound business plans and have access to the skills that would enable them to run a community owned and operated business. They needed to understand the implications of cash-flow on their business and what ICT could and could not deliver in their community. They also needed to take advantage of the many products and services that the CTC@NSW program could offer through its technical and business support staff. The right balance of skills on these committees enabled them to stay focused so that they could gain access to the ICT products and services needed in their communities

ACCESS TO IDENTIFIED ICT PRODUCTS AND SERVICES

The ICT products and services sought by the various communities varied significantly from one community to the next and were dependent on what had been identified in the consultation process. The other major variable was the actual delivery of broadband services to these communities as this was often limited by geographic location. This was also supported by Hearn et al. (2004, pp. 4-5), "...market failure occurs, especially in rural, regional, and remote areas where the costs of setting up and maintaining ICT services can be very high. These factors make government support of ICT initiatives necessary...in the current policy context, government funds are limited. Consequently, greater financial responsibilities are being foisted onto local communities." However, when a community decided to build on its identified strengths, it was far more likely to achieve its goals even with limited funding.

In a discussion paper produced by the Australian Government, *The Role of ICT in Building Communities and Social Capital*, (2005, p.1) it states:

As the use and impact of ICT increases, so does the prospect that ICT can play a role in shaping the nature of community development and contributing to the building of social capital... There is a greater variation in the needs and demands of communities and the role of ICT in meeting these.

However, this also suggests that, if government policy-makers determine that building social capital and community capacity should be seen as major outcomes of the uptake of ICT in small rural communities, why is ongoing federal funding and support so hard to come by? Communities will only remain empowered if their basic survival needs are met, because empowerment equates to greater social capacity with the flow-on benefit of diversity and innovation through the uptake of ICT. The following projects reflected the diversity of community needs that were met through the establishment of CTCs in several Northern Rivers communities.

Project One

Alive and Driving project was a program that one CTC initiated for learner drivers in its community. During the application process this community had identified lack of public transport as a major issue compounded by the fact that many young people were not very safe drivers. Moreover, there had been a horrific road accident that claimed the lives of three young people from this community that galvanized them into action. The CTC Committee instigated a driver-mentoring program that could give learner drivers access to a car and supervisor so that they could gain the 50 hours of driving experience they needed to sit for their NSW driving test. Learners paid a $50 membership fee for the first 10 hours access to a car as well as petrol costs during their driving session. After this they could pay a further $50 that enabled them to access as many hours as required until they got their license. This project was supported by funding received from the Australian Government under its regional partnership program, the Myer Foundation, and through partnerships with organizations and volunteers from the local community. But all the research to gain access to this funding was done using ICT, something that would not have been possible prior to the opening of its CTC.

Project Two

Another Northern Rivers community wanted to use the equipment in its CTC to create a virtual museum. They wanted to gain access to the technology that could assist them in preserving their community's rich and varied history. They used their funding to acquire the equipment they needed to scan and digitize old photographs and documents, record and preserve interviews with World War Two veterans. They also wanted to be able to catalogue and map the graves at the local cemetery and provide facilities for local residents to research their family histories. Currently they are seeking accreditation with museums Australia. Their virtual museum is now fully functional and they continue to work hard to enrich the content in their virtual museum.

Project Three

In early 2005, another CTC in the Northern Rivers was awarded a grant from the NSW Premier's Department Office for Women to conduct a training course specifically designed to help women learn filmmaking, computer, and marketing skills. The project, "Reeling in Women's Stories," saw 16 women of all ages create a 'Short Bytes Film' based on stories relating to womanhood. The participants undertook 14 free workshops to learn new computer and film making skills that included story boarding, filming, producing, and editing and marketing techniques. At the conclusion of the course, they held a movie night at the local cinema to showcase their combined efforts. The event was well attended by people of all ages from the local community.

Project Four

The Australian Taxation Office (ATO) has also worked with a number of CTCs to help deliver its regional and remote service delivery program. The online ATO products and services that are available through this service are designed to minimize time spent managing tax affairs as well as enable businesses to gain access to information regarding taxation by learning how to access information from the ATO Web site.

A Few Other Projects

CTCs also provide free access to the NSW government portal, which has been designed as a one-stop entry point for accessing information on a variety of government services as well as a place to pay bills to government departments.

During 2005 Senior's Week, 50 CTCs across NSW worked in partnership with the NSW Department of Ageing, Disability, and Home Care (DADHC), and the Country Women's Association of NSW to promote and organize activities for Seniors Week. The 'TechnoSeniors' project was designed to promote the use of ICTs by seniors throughout the week at participating CTCs, these activities included, amongst other things, beginners and intermediate computer classes. This partnership with DADHC saw nearly 3000 participants attend Senior Week activities at the CTCs and 'TechnoSeniors' is now set to become an annual event.

Although not all CTCs have access to videoconferencing facilities, many of those that did participated in youth week celebrations in 2005 through a *young people and work videoconference* sponsored by the Office of Industrial Relations and Work Cover. This Videoconference provided an opportunity for young people in rural communities to have their say on issues affecting them, such as work safety, pay, and conditions. Additionally, many CTCs also host a range of school holiday programs where kids can learn new computer skills, create videos and sound files, and access the Internet in a safe environment.

The CTC Association Web site, http://www.ctca.net.au can provide additional examples of other ICT based programs that are helping communities in the CTC network to remain empowered.

COMMUNITY ACTUALIZATION: SUCCESSFUL ICT OUTCOMES

While the communities mentioned have achieved only some of the goals articulated in their CTC applications, they all have built greater social capital in their communities through their CTC. Their CTCs have also played an important role in their ongoing regional economic development initiatives. The achievements of these communities are representative of what other small NSW communities have achieved through their continued participation in the CTC network. They also highlight the social implications of what access to ICT can facilitate in small regional communities. Hearn et al. (2004, p. 14) state, "...local needs and interests must be addressed...It can be extrapolated that success in attaining these goals necessitates the community having effective opportunities to be involved in deciding on the project(s) to be pursued...it is essential the community 'owns' the initiative.... the strengths and resources of the community should be identified and be used effectively to advance the community." This has also proved to be the case in many of the CTCs located throughout regional New South Wales.

ISSUES ARISING

In the chapter, *The Role of ICT in Building Communities and Social Capital* (2005, p. 4) published by the Australian government, it states; "ICT has a use in the development of community engagement and building social capital. Communities are at different stages of this process and do not always recognize the vast potential of ICT. When it is recognized, barriers may exist in the form of financial support, leadership, professional facilitation, technological development, and support and vision." It continues (2005, p .7) "Governments and communities have committed major resources to ICT projects and programs to foster community development and to assist with networking across groups. However, there has been comparatively little research in Australia of the actual and potential use of ICT as an enabler of community and social capability." An inhibitor of such research may be linked to the 3 or 4 year government funding cycles of ICT programs, like the CTC@NSW. This has been the case in the communities that were successful in their CTC applications, and are now expected to be able to manage their community owned and operated businesses, even though many of those staffing the CTC are volunteers. However, it is worth noting that if their original cash flow projections were not accurate, the first thing they usually cut is the salary of the CTC manager, and from that point onwards business productivity goes into decline and the long term viability of the CTC becomes a major issue. Although Plowman et al. (p. 8) states, "...No town has a 'right' to survive. To survive a town needs to be flexible and adaptable, to provide amenities and services to its members and to those outsiders which interact with it. Atrophy or decline awaits those towns that are not innovative or adaptable."

However, becoming or remaining an innovative town is a very big challenge indeed, as many observers and writers have pointed out, homogeneity; conservatism, and conformity exert a constant pull. Towns that strive to be innovative in their use of ICT need to be supported, and as shown in the previous examples from the CTC@NSW program, non-traditional uses of ICT can go a very long way towards building greater social capital in these communities. Gray and Lawrence (2001, p. 206) stated, "...new initiatives in telecommunications, transport, health, education, culture, and the environment alongside eliminations of the forces depleting social capital as they bolster inequality...will also be achieved by identifying mechanisms which aid and reward regions in their quest for sustainability."

SOLUTIONS AND RECOMMENDATIONS

It is important that successful communities remain empowered, as well as funded, if they are to obtain ongoing access to the many products and services that ICT can deliver. It also means remaining focused on the many challenges of running a small business, including cash-flow, meeting consumer demand and expectation, and building a loyal base of customers. In many cases, it may also mean seeking further assistance with funding not only for a CTC manager who can continue to work with all community stakeholders, but also to assist in upgrading equipment as technologies and applications change.

More creative funding strategies need to be implemented at every level of government because building social capital is not a one-off project, but something that needs to be continually nurtured. The whole process of applying for grants and/or additional 'top-up' funding should be re-thought with wider, more community oriented criteria applied because as Gray and Lawrence (2001, p. 112) stated, "It has been acknowledged that regional telecommunications users are disadvantaged. Service providers are reluctant to invest outside the more profitable metropolitan markets and existing government policies to redress this problem are not working."

Unfortunately, a creative opportunity was overlooked when the federal government's launched its *Clever Networks* program on August 31, 2006. The following was downloaded from the DCITA Web site (November 30, 2006), "The *Clever Networks* program, part of the $1.1 billion 'Connect Australia' package, will deliver innovative broadband services to communities in regional, rural and remote Australia. *Clever Networks* will fund broadband applications and leverage broadband infrastructure to foster innovative service delivery for communities. Broadband development activities will complement the delivery of these services by assisting communities to enhance their understanding and use of broadband. Delivering innovative broadband services to Australian communities will encourage economic growth, increase productivity, foster innovation and increase quality of life."

However, nowhere in the *Clever Networks* application is there opportunity or encouragement for 'top-up funding' from communities that had received earlier grants from the federal government to establish the CTC@NSW, and other similar state-wide programs such as the Western Australian Telecentre Network (WATN) and the Tasmanian Communities Online (TCO) program. The Western Australian Telecentre Network is the oldest of these networks and commenced operation in 1991 and is still growing. While the TCO Program, which was funded in 1998, continues to provide significant economic and social benefit to the 64 regional communities it serves.

Programs such as the CTC@NSW, TCO, and WATN could provide ideal venues for demonstrating broadband delivery of ICT. Federal funding, if cleverly used, could also promote the cross-fertilization of ideas and programs that use broadband delivery to facilitate the use of ICT in new and challenging ways. More importantly this type of ongoing funding would encourage small communities to continue broadening their approach to the uses of ICT.

Additionally, government funding should be delivered based on innovation and thus reward communities that continually strive to remain empowered through being innovative. Once communities become empowered by using ICT as a means of meeting their needs, they should be given the opportunity to share their ideas and programs with other 'networked' communities through a 'building on success' type program

FUTURE TRENDS

There will be continued erosion of face-to-face services in remote and regional areas due to the economic rationalist policies of government brought about by the range of forces associated with globalization. This means that there will be an ongoing demand by people living in regional areas to increase their skill base, so that they can have more equitable access to a greater range of products and services delivered via ICT. It also means that communities that have access to the latest ICT have the potential to become catalysts for regional growth and those communities that have the drive, as well as the capacity to survive, will be empowered and enabled to do so.

Empowered communities will continue to build on the strengths that lie in connecting communities to technology and information dissemination. They will continue to work with governments as well as explore and develop new e-business solutions for their communities. They will also explore opportunities to work with and support the micro and home based businesses that are a growing economic force in many regional and rural areas. CTCs in these empowered communities will continue to secure venue hire from both government and commercial enterprises in order to meet local needs. CTCs will also continue to promote and support computer literacy programs and serve as a one-stop shop for the delivery of more and more local, state, and federal government programs. New and innovative uses for ICT will come from these empowered communities as they use ICT to meet their needs, and at the same time remain connected to wider national and international communities.

CONCLUSION

Preliminary research indicates that a whole of government approach is essential if ICT is to serve as a catalyst for regional economic growth as well as assist in the building of social capital. De Weaver and Ellis (2006, p. 9) state, "Governments need to leverage the investment they have already made in programs such as the CTC@NSW if they are to pay more that just 'lip service' to the goal of regional economic development..."

It is also important to consider that while there are significant stages involved in delivering effective community consultation, it remains an ongoing challenge to ensure that communities that have become empowered as they gained access ICT products and services will remain committed to what they can achieve as an empowered community. In order to do this, the CTCs in each and every community must be run as financially sound businesses with management committees that can continue to provide the skills and support required to keep these community-owned and operated businesses viable. The long-term success *and* empowerment of communities that are part of the CTC and other similar Networks will also be predicated on continuing to build on the social capital that they have achieved to date.

Additional research undertaken by the authors will again use Action Research to assess whether or not the communities funded by the CTC program have remained viable.

FUTURE DIRECTIONS

The authors are also exploring the experiences of other communities in establishing similar ICT based programs.

Preliminary research has shown that as various government agencies and departments in Australia have rolled out other ICT programs designed to promote regional economic development in rural and remote communities through access to information and communication technology (ICT) programs, the language used in the documentation that supports these programs has often put up unnecessary barriers in the targeted communities and inhibited them from applying for funds.

How rural and remote communities actually engage in, and undertake, the process of applying for these programs is, to a large extent, dependent on their language and business skills. Not only are there social and political implications to be considered by governments in the design and roll-out of these programs, there are also technology and communications issues that need to be addressed in order to achieve more equitable access to these programs.

We believe that:

- The documentation that accompanies ICT based programs designed to deliver services in rural and remote areas should be written in plain English and be devoid of technical jargon. Where acronyms are used they should be clearly defined and examples given.
- Applied sociolinguistics principles need to be introduced and employed in all government grant applications programs for targeted populations.
- Glossaries of terms, which include appropriate examples, should be included in all documentation for funding programs that deal with services and/or products that are outside the usual frame of reference of people who live in rural and remote areas.
- Applications and documentation written for programs designed to build capacity in rural and remote communities should be trialed in selected communities, via focus groups

or a similar consultation process, before programs are rolled out.

- All such programs need to be supported by field facilitators who understand the needs of targeted communities and who can provide appropriate support and assistance.
- Programs that promote regional economic development in rural and remote communities also need to include an element of education and/or training in them in order to empower the community and allow members to deal affectively with grant applications.
- Outcomes should incorporate community expectations as well as that of government funding bodies.

REFERENCES

American Society of Association Executives Foundation (2001). *People make it work: The social fabric of community on the Web.* Retrieved April 20, 2006, from http://ww.asaenet/org/foundation

Aslin, H.J., & Brown, V.A. (2004). *Towards whole of community engagement.* Canberra: ACT Murray-Darling Basin Commission.

Beer, A., Maude, A., & Pritchard, B. (2003). *Developing Australia's regions: Theory and practice.* Sydney NSW: UNSW Press.

Carson, L., & Gelber, K. (2001). *Ideas for community consultation: A discussion on principles and procedures for making consultation work.* Sydney NSW: NSW Department of Urban Affairs and Planning.

Chapman, A. (2002). *Abraham Maslow's hierarchy of needs motivational model.* Retrieved May 11, 2006, from http://www.businessballs.com/

CTC Association Web site. Retrieved October 16, 2006, from http://www.ctca.net.au

CTC@NSW Web site. Retrieved July 20, 2004, from http://ctc.nsw.gov.au

Department of Communications, Information Technology & the Arts Web site. *Communications for Business Report.* Retrieved November 30, 2006, from http://www.dcita.gov.au/communications_for_business/funding_programs__and__support/clever_networks

Department of Local Government and Regional Development (2004). *Planning for the future report.* Perth, WA: Government of Western Australia.

De Weaver, L., & Ellis, A. (2006). The CTC @ NSW Program: Achievements and ongoing challenges. *International Journal of Education and Development using ICT, 2*(2),15-26.

De Weaver, L., & Lloyd, D. (2005, August). *The language of community engagement in a regional and indigenous context.* Paper presented at International Conference on Engaging Communities. Brisbane: QLD. Government of the State of Queensland.

Fairclough, N. (2000). Multiliteracies and language: Orders of discourse and intertextuality. In Cope & Kalantis (Eds.), *Multiliteracies: Literacy learning and the design of social futures* (pp. 162-181). South Yarra,Vic: Macmillan Publishers Australia Pty Ltd.

Hearn, G., Kimber, M., Lennie, J., & Simpson, L. (2004). *ICTS and regional sustainability: A critique and a way forward.* Paper presented at CIRN Conference and Colloquium, Prato, Italy.

Local Access Global Reach: Five Years of Tasmanian Communities Online 1998- 2002. Retrieved May 31, 2005, from http://www.education.tas.gov.au/tco

Luke, A., & Gilbert, P. (1993). *Literacy in contexts: Australian perspectives and issues.* St. Leonards, NSW: Allan & Unwin.

NSW Department of Commerce & Office of Information Technology (June 2004), *CTC Final Report,* Sydney: NSW.

Pigg, K., & Crank, L.D. (2003). *Rural economic development: The impact of information and telecommunications technology.* Paper presented at 5th International Information Technology in Regional Areas Conference, Caloundra, Qld.

Plowman, I., Ashkanasy, N.M., Gardner, J., & Letts, M. (2003). Main Report, *Innovation in rural Queensland: Why some towns thrive while others languish.* Brisbane, QLD: University of Queensland.

Department of Communications, Information Technology and the Arts (2004, February). *Regional Policy Research Forum Report.* Canberra: ACT.

The Role of ICT in Building Communities and Social Capital (2005, January). *A discussion paper.* Access Branch Department of Communications, Information Technology and the Arts.

Sankaran, S., Dick, B., Passfield, R., & Swepson, P. (2001). *Effective change management using action learning and action research.* Lismore, NSW: Southern Cross University Press.

Shepherd, A. (1998). *Sustainable rural development.* London: Macmillan Press Ltd.

Wollstenholme, R. (1995). *Understanding small rural communities and the consultation process.* Armidale, NSW: The Rural Development Centre, University of New England.

ADDITIONAL READINGS

While there appears to be a time lag regarding documentation and publications about other similar ICT programs, the following references should be of interest.

Fairclough, N. (1989). *Language and power.* London: Longman Group Ltd. Retrieved July 20, 2004, from http://www.ctc.nsw.gov.au/about/

MacLachlan, G., & Ian, R. (1994). *Framing and interpretation.* Melbourne: Melbourne University Press.

Ricento, T. (2000). *Journal of sociolinguistics, 4*(2), 196-213.

Romaine, S. (1994). *Language in society: An introduction to sociolinguistics.* New York: Oxford University Press.

Sorensen, T., & Epps, R. (Eds.). (1993). *Prospects and policies for rural Australia.* Melbourne: Longman Cheshire.

Stephens, T. (2003, November). Words fail us. *The Sydney Morning Herald,* 32.

Stilwell, F. (1993). *Reshaping Australia urban problems and policies.* Leichhardt: Pluto Press.

Trudgill, P. (1995). *Sociolinguistics: An introduction to language and society.* London: Penguin Group.

Watson, D. (2003). *Death sentence: The decay of public language.* Sydney: Random House Australia.

Chapter XVI
Policy Initiatives in Jamaica and their Impact on the Development of ICT

Sheryl Thompson
Lancaster University Management School, UK

David Brown
Lancaster University Management School, UK

ABSTRACT

The chapter explores the diffusion of information and communications technology (ICT) in the country of Jamaica, and within the context of the national strategic plan for information technology. Further, it utilizes the concept of institutional intervention to discuss some policies and initiatives undertaken in support of the plan. Within this framework, the impacts of managed interventions via collaborative projects between government and international agencies on the adoption of Internet technologies by small and medium-sized enterprises are also examined. The aim of the authors is to indicate that as far as developing countries are concerned, governments can affect the diffusion and adoption of ICT through policymaking but more positive outcomes can be realized through managed interventions.

INTRODUCTION

New technology offers an unprecedented chance for developing countries to 'leapfrog' earlier stages of development. Everything must be done to maximize their peoples' access to new information networks (Annan, 2000, heading III, para. 5).

Bilateral and multilateral assistance will continue to play a significant role in the building framework conditions for IT development. International financial institutions are well placed to contribute in this regard by formulating and implementing programs that foster growth, benefit the poor, as well as expand connectivity, access, and training (World Bank, 2001, p. 2).

The two quotes above set the agenda for this chapter. In the first, Kofi Annan the United Nations secretary-general in his millennium report singled out the crucial contribution of information and communications technology (ICT) to economic development. In the second, the World Bank in its contribution to the G-8 Okinawa Charter drew attention to the key role to be played by the global community in assisting developing countries to narrow the digital divide. These twin ideas of economic development using ICT with assistance from international institutions are explored in this chapter in the specific context of Jamaica. As a developing country in the Caribbean region, Jamaica is an interesting case as it is one of the first developing countries to prioritize information technology in this way. In 2000, the Jamaican Government introduced clear IT policies to assist economic growth and implemented these through initiatives funded in part through international assistance. Included in these initiatives were measures for smaller businesses—an important part of the Jamaican economy. In 2006 it is possible to reflect on these policies and initiatives in terms of progress made and future possibilities.

The background to the Jamaican Government's initiatives lies in the wider debate about technology, its contribution to economic growth, and the role of government. The advent of the Internet and its subsequent proliferation into most aspects of society and commerce has propelled information and communication technologies to the forefront of innovation. Its ubiquitous nature and the varied business opportunities it presents are potentially beneficial to all citizens and to all organizations irrespective of size or location. Therefore, it is not surprising that ICT is proposed as a key element to economic development and a means to the alleviation of poverty among the world's less developed countries. The debate is not one sided and opponents continue to argue that investment in information and communication technologies will have little impact on productivity and growth, if unaccompanied by institutional and economic reform, especially in the case of developing countries. Microsoft's Bill Gate's comment reinforces this reality:

Let's be serious. Do people have a clear view of what it means to live on a $1 a day?...There are things those people need at that level other than technology...About 99% of the benefits of having a PC come when you've provided reasonable health and literacy to the person who's going to sit down and use it (Steinberg, 2003, p. 46).

As in most situations, the truth lies somewhere between these two positions and is where this debate is situated. The pervasive nature of ICT demonstrates its value as a vehicle through which innovations are effectively diffused and enables organizations to be creative in the development of new products and services. Although the scale of its contribution to economic growth and productivity continues to be discussed, it is clear that ICT is accepted as an enabler and a driver of globalization. Governments and businesses are increasingly reliant on information and communication technologies, and the elimination of time-space distanciation, to conduct business internationally (Walsham, 2001). In today's world, geographical locations and differences in time zones are no longer a hindrance to business activities since the Internet can enable organizations to appear to be operating 24 hours per day and the use of e-mail, for example, provides a means for constant communication.

Though information and communication technologies will not singly spur economic growth, it is evident that ICT play an instrumental role in development. Strategic use of ICT can boost foreign direct investments (FDI) thereby resulting in economic growth as in the case of Costa Rica, who attracted 32 foreign electronic firms, including Intel, since 1995 and India, who is a leading outsourcing destination for firms in developed countries. Countries can also use ICT to service niche markets as in the cases of Singapore, Malaysia, and Taiwan in the electronics industry. The potential of information and communication technologies to disseminate information is sufficient to demonstrate that it is valuable to the development process. It facilitates the sharing of ideas and experience among international communities, as well as urban and rural communities; it increases access to education, facilitates the improvement of health services, and puts local organizations in touch with the global environment.

Irrespective of how a country decides to use ICT to foster development, having a national policy for the diffusion and adoption of the technology will support the process. A comprehensive policy can have significant effects on the transfer of technology and enable governments to encourage adoption by carrying out managed interventions targeting certain sectors, for example small and medium sized enterprises (SMEs) that may not otherwise be able to adopt the technology; this is the situation in Jamaica. As a part of its strategy for achieving economic growth and creating employment, the Government has made the integration of information technology (IT) into the Jamaican economy a high priority and a strategic imperative. The objectives of the Government of Jamaica (GOJ) are to promote the country as a Caribbean hub for IT activities and investment and to create a 'knowledge-based society'. With the advantage of an English-speaking labour force and low labour costs, the aim is to promote Jamaica as a viable outsourcing destination to companies in the United States, Canada, and Europe (GOJ, 2002).

The designation of ICT as the focus for attracting major international investments and generating jobs led to the establishment of a national strategic plan in the year 2000. *This chapter will explore:*

- The objectives outlined in the national strategic plan for information and communication technologies and what has been achieved since its inception in 2000
- Policies implemented and how these policies translate into initiatives that support the growth and use of ICT
- The role of government and aid-supported projects in the diffusion of ICT
- How managed interventions affect the adoption and implementation of Internet technologies by small and medium sized enterprises

To achieve this, the chapter is presented in six parts. In the first part, an overview of ICT in the context of economic development is provided. This is followed by a short profile of Jamaica and a description of its policy approach to ICT. In part three, the model of institutional intervention proposed by King et al. (1994) is used to discuss some initiatives set up to implement the ICT policy. In part four the chapter draws upon some recent primary research by the authors and looks at the initiatives in the important small business sector. Part five looks at some future issues and finally, in part six, a reflection on the Jamaican experience is offered with some wider implications for other developing countries.

ICT, DEVELOPMENT, AND INSTITUTIONAL INTERVENTION: AN OVERVIEW

The discussion on the diffusion and adoption of ICT can often be characterized by comparisons between the developed and developing economies in terms of their rate, depth, and reach of ICT adoption—the so called 'digital divide' and the role of institutional intervention to facilitate this phenomena. Twenty-five years ago developing countries had 20% of the world's telephone lines but in 2005 this figure had risen to 60% (World Bank, 2006). ICT has become pervasive in many developing countries during the last decade, but a significant divide still exists between the developed and developing economies. Basic ICT infrastructure has improved; however there is still a gap pertaining to more complex technologies. The lack of existence of secure Internet servers, an indication of advanced ICT such as e-commerce, exposes this fact. Whereas developed nations have more than 300 secure servers per 1 million people, developing nations have fewer than 2 (World Bank, 2006).

Statistics from the International Telecommunications Union (ITU) indicate that the 'digital divide' between developed and developing countries has narrowed since 1992. In 2002 the number of fixed and mobile telephone subscribers increased by 24% and 34% respectively in developing countries, accounting for 45 and 46% of the world figure, as against 21 and 12% in 1992. Moreover, in 1992, developed countries represented 83% of total telephone subscribers compared to the 17% for developing countries; in 2002 the numbers were 55% and 45% respectively, showing a very significant growth of 28%. In the case of the Internet, the positive growth trend was mirrored; 34% of Internet users were in developing countries in 2002, up from only 3% in 1992 (ITU, 2003).

While the figures presented above are remarkable, the growth in the use of information technology and the access to telecommunications in developing economies are not evenly distributed across countries. Asian countries such as Taiwan, Hong Kong, South Korea, Singapore, and Malaysia, classified as newly industrialized nations by Palvia, Palvia & Roche (2002), have utilized ICT strategically to mobilize their economies for growth and development. China has experienced the most significant increase in Internet users becoming the largest Internet-using country in the world (Pace, 2006), in addition to being the largest mobile telephone market (Twist, 2004). Elsewhere there is still a lot to be done in regions such as Sub-Saharan Africa, South and East Asia. However, the picture for Latin America and the Caribbean looks more promising; the region accounted for more than half of the world's foreign direct investments (FDI) in telecommunications during the period 1990–2003 (World Bank, 2006).

The commitment to bridge the 'digital divide' has resulted in a number of initiatives by international organizations responsible for development, regional and local governments, as well as private sector organizations. Facilitating improvement and growth in telecommunications is an essential antecedent for ICT development. Countries such as the United States and United Kingdom have proven that deregulation and liberalization of telecommunications resulting in increased competition, causes market forces to react, which can in turn drive down the cost of access to individuals and businesses. Peru, Chile, and Mexico are examples of the developing countries that have experienced positive impacts because of similar policies. Although a liberal telecommunication sector and national policies for ICT will support diffusion of Internet technologies to organizations, they are relatively blunt instruments. Access per se to the relevant infrastructure does not mean that businesses will automatically adopt Internet technologies. This is evident in developed countries where proper infrastructure exists but some companies, particularly small and medium-sized enterprises, remain non-adopters (Brown & Lockett, 2004).

In the case of developing countries, governments and donor agencies may have to push adoption by being change agents and instituting managed interventions from which these organizations are able to benefit.

Government intervention and initiatives, implementing national policies that allow local organizations to be competitive, investing in the proper infrastructure, and defining efficient standards can create an enabling environment for the successful technical transfer of technology to local organizations. According to Gilbert (1992), "creating a policy climate which nurtures international transfers of information technology is particularly relevant to policymakers in developing nations" (p. 422). The roles of institutions in the technology diffusion process are multifarious. Kraemer, Gurbaxani, and King (1992) in their study of the diffusion of computing in Asia-Pacific countries posited that while some fiscal and trade policies initiated by governments did affect the investment made in computing expenditures, a country's economic status was more likely to dictate the diffusion of information technology.

A later study by the authors, however, acknowledged *institutional intervention* as one of the keys to the successful adoption of information technology in developing countries (King et al., 1994). This thesis has been adapted and used to explain the diffusion of ICT in countries such as Denmark, Finland, Hong Kong, Chile, Peru, Costa Rica, and Ecuador (Damsgaard & Lyytinen, 2001; Montealegre, 1999; Silva & Figueroa, 2002). An institution is defined as any standing social entity that exerts influence and regulations over other social entities, and includes governments, international agencies, professional and trade associations, educational, financial and religious institutions, multinational corporations, and labour organizations. Institutions are thought to be able to stimulate supply by encouraging investment in production and promoting the transfer of foreign technology into a country, and stimulate demand by creating a need for innovative products and processes, or by enhancing the efficiency and effectiveness with which they are used (Montealegre, 1999).

In their influential article King et al. (1994) postulated a model purporting that "institutional intervention in IT innovation can be constructed at the intersection of the influence and regulatory powers of institutions and the ideologies of supply-push and demand-pull models of innovations" (p. 139). According to the model, there are six types of actions: knowledge building, knowledge deployment, subsidy, mobilization, standard setting, and innovation directives, that institutions can perform in an influential or regulatory role to produce the 'supply (technology)-push' or 'demand (need)-pull' context for the actions to be effective. Influential intervention is where the institutions use their powers, without being forceful, to affect attitudinal and behavioural changes from those governed. On the other hand, regulatory interventions are direct and forceful in nature where standards and rules are enforced (Henriksen & Andersen, 2004). International lending agencies typically intervene, in an influential way, by stipulating certain guidelines with which developing countries must comply in order to be eligible for loans; in many instances, this leads to changes in economic policies that may not necessarily be to the benefit of the country. Telecommunication liberalization is one example of the strategies propagated by international donor agencies without sufficient consideration of culture and traditions (Madon, 2000). Institutional intervention therefore can occur via the means of influence or regulation, and can either take the form of direct or indirect actions.

PROFILE OF JAMAICA AND ITS NATIONAL STRATEGIC PLAN FOR ICT

Jamaica is the largest English-speaking island in the Caribbean and the third largest in the region. In 2004 the population was 2.6 million and Gross National Income (GNI) US $3,300 per capita. It is classified as a developing country falling into the lower-middle income category and has a mixed, free-market economy consisting of a combination of state-owned entities and private-sector organizations. Tourism and mining are the two mainstays of the economy with agriculture also contributing significantly. The country has had a stable democracy since gaining independence from Britain in 1962. Two main political parties have shared rule of the country at various times during this period of relative political stability. The country has made remarkable social progress in the last 15 years. As evidence of this, the poverty headcount ratio has fallen from 30.5% in 1989 to 16.8% in 2001 and life expectancy has risen to 75.7 years in 2002 from 73.2 years in 1990, and total GNI has increased from US $7.6 billion in 2000 to US $8.7 billion in 2004. However, social problems particularly crime and high unemployment rates have had negative effects on the country. The unemployment level in October 2005 was 10.9%; with the age group 20-34 accounting for 79.9% of females and 57.6% of males (Statistical Institute of Jamaica, 2006). Some economic indicators for Jamaica are presented in Table 1. It is within this economic and social context that government policy, including ICT policy, has been formulated as part of the national strategic planning process.

National Strategic Plan for Information Technology

The impetus for the national strategic plan for IT came from the 1996 National Industry Policy that was created to outline a strategic plan for economic growth and development for Jamaica. The vision of the government is to use information and communication technologies to create an enabling environment for the evolution of other sectors in the economy. The notion is that using the capabilities provided by ICT will allow the country to become more competitive internationally and capture niche markets. The expectation is best summed in the following excerpt:

Table 1. Economic indicators for Jamaica (Source: World Bank data profile on Jamaica, 2006)

	2000	2003	2004
Population, total (million)	2.6	2.6	2.6
GNI (current US$ in billion)	7.6	8.1	8.7
GNI per capita (current US$)	2,940	3,090	3,300
GDP (current US$ in billion)	8.0	8.2	8.9
GDP growth (annual %)	0.7	2.3	0.9
Inflation, GDP deflator (annual %)	11.4	12.4	12.6
Industry, value added (% of GDP)	31.3	31.7	32.6
Services, etc., value added (% of GDP)	62.0	62.8	62.0
Exports (% of GDP)	42.6	40.2	40.6
Imports (% of GDP)	53.9	57.8	57.8
Foreign direct investment, net flows (BoP, current US$ in million)	468.0	721.0	601.6
Long-term debt (DOD, current US$ in billion)	3.9	4.6	5.3
Present value of debt (% of GNI)	....	....	89.0

The plan comes to terms with the realization that the technology is a ready tool to enhance the rapid development of all sectors in the economy and therefore it is truly national in its scope. High on its agenda is the massive provision of jobs—not only in the technology industry but also in all sectors that utilize the technology (GOJ, 2002, p. 12).

The Jamaican Government presented its five-year strategic plan for information technology in 2000. It projected that the IT sector would become a major source of job creation over the life of the plan. The sector would increase the variety of services on offer beyond the established data entry services. The overall objectives of the national strategic plan for information and communication technologies included:

- **Infrastructure and access:** Creating a competitively priced nationwide public information and communication technologies network
- **E-government:** Using information and communication technologies to provide efficient government services to the public
- **Economic development:** Using the Internet to facilitate growth in International trade
- **E-business:** Promoting the development of electronically transacted business

These objectives would be achieved by setting short and long-term goals geared towards improving telecommunications infrastructure and regulations, by policy implementation through initiatives and by adopting best practices. Human resource development—ICT skills—was another target area in order to harness the potential from the investments in the industry. The Central Information Technology Office (CITO) was established in 2000 to take ownership of the plan and coordinate its implementation by providing recommendations about appropriate policies, and liaising between local and overseas public and private sector organizations with a view to form partnerships.

The creation of CITO as a limited liability company was seen as strategic. The justification was that this separation from government would allow CITO the latitude needed to perform the roles it had been assigned. The then Ministry of Industry, Commerce, Science, and Technology also identified key agencies in government that would be mobilized to support the achievement of the goals outlined in the strategic plan. Finally, key policies for the liberalization of telecommunications, legislations and regulations, and e-government were implemented at the macro level to mobilize the plan and provide an enabling environment for its objectives to be achieved. As a result of these actions there has been US $700 million invested in the ICT sector (GOJ, 2006). These three key macro level policies and their impact are outlined below.

Liberalization of Telecommunications

The result of the first macro policy enacted was a major reform of the telecommunications industry with the passing of a new Telecommunications Act in 2000. This effectively saw the cessation of what was a monopoly for both fixed and mobile telephone services. This was a significant step because the then monopoly, Cable and Wireless Jamaica had been granted licences in 1988 for 25 years with an option to renew in 2013. A new agreement in 1999 between the government and the company led to a phased liberalization of the telecommunications industry over a three-year period. Subsequently, the industry has become more competitive and there are currently two additional mobile service providers. It is in this arena where there has been phenomenal growth; in 1999 there were approximately 90,000 mobile subscribers, by 2003 this number had increased to 1.2 million. At the end of 2004, there were 2.11 million telephone lines in Jamaica with mobile lines accounting for 1.7 million (Paulwell, 2005). Table 2 provides some ICT indicators for Jamaica depicting changes to the sector since the liberalization of telecommunications.

Similarly, licences have been granted to other companies to provide wired telephone services but growth has been minimal. The absence of growth in fixed lines may be due to persons opting to have mobile phones instead of the former, which is in keeping with the global trend. However, it may also be due to the financial outlay needed to construct the infrastructure for the provision of these services. Therefore, despite the fact that the number of players in the industry has increased, the former monopoly still has control because they own the infrastructure utilized by the other companies, and this has implications for provision of other types of services in the sector. The situation is likely to change soon since licences have been granted to two new companies to install undersea fibre-optic cables to provide additional access channels into the country.

These are positive steps taken towards the expansive deployment of broadband technology, which in turn is seen as crucial to the successful implementation of e-business by companies. Following through on plans to impose a tariff on telecommunication companies in order to create a fund to finance the implementation of broadband infrastructure, the Jamaican universal service fund was started with the collection of tariffs on calls coming into the island. The fund will be used to finance the E-learning project intended to ensure that Internet access is available in schools, libraries, and post offices. The push for broadband deployment puts the Jamaican Government in line with the European Union, which has promoted the importance of broadband connectivity for the economic development of member states through a dynamic e-business environment in its eEurope 2005 Action Plan (European Union, 2005).

Legislation and Regulatory Environment

In keeping with its mandate of creating an attractive investment environment for foreign companies, the Jamaican Government has also taken steps to improve its legislative and regulatory framework for new economy practices. An

Table 2. ICT indicators for Jamaica (Source: World Bank ICT at a glance, 2006)

	2000	2004
ICT Sector Performance		
Access		
Telephones main lines (per 1,000 people)	198	174
International voice traffic (min. per person; outgoing & incoming)	156	230
Mobile subscribers (per 1,000 people)	142	615
Population covered by mobile telephony (%)	80	95
Internet users (per 1,000 people)	30.9	403.5
Personal computers (per 1,000 people)	47	54
Quality		
Broadband subscribers (per 1,000 people)	....	9.5
International Internet bandwidth (bits per person)	28	43
Institutional efficiency, sustainability and ICT applications		
Total telecommunications revenue (% of GDP)	6.4	6.0
Total telecommunications investment (% of revenue)	32.5	24.9
ICT expenditure (% of GDP)	11.2	12.4
E-government readiness index (scale 0-1)	.....	0.41
Secure Internet servers (per 1 million people)	1.9	9.0
Schools connected to the Internet (%)	....	10

electronic transactions policy (ETP) was drafted 2003, which details the Government's intention to establish the appropriate policies to facilitate the growth of e-commerce:

Government must also ensure, as a priority, that legislation covering areas such as digital signatures which may hinder the growth of electronic commerce, be reviewed and revised to reflect the new technology…and the establishment of a transparent regulatory framework consistent with, and adaptable to, the emerging electronic business environment, covering areas such as privacy, protection of intellectual property, and digital signatures (GOJ, 2003, p. 5).

There have been other activities in the regulatory arena. The spectrum management authority (SMA) was established in 2001 with the responsibility of managing the radio spectrum of Jamaica; the Office of Utilities Regulations (OUR) and the Fair Trading Commission (FTC) have increased their focus as regulators in the telecommunications industry. Further, the Jamaica Intellectual Property Office (JIPO) was formed in 2002 with the purpose of ensuring Jamaica's compliance with international standards for Intellectual Property Rights and to administer intellectual property laws in support of other national policies. The Government, however, has not been timely in enacting all the necessary regulations to support an e-business environment.

E-Government

The Jamaican Government views its own use of technology as critical to the diffusion and adoption of Internet technologies by both businesses and individuals. As such, it aims to become a model user of the Internet. There have been a few e-government initiatives undertaken in its bid to provide online government services. The Public Sector Modernization Programme (PSMP) is a large-scale effort to improve public sector services and make the entities more efficient. The initiative is funded by GOJ with International Development Partners (IDP): the World Bank, European Union (EU), Canadian International Development Agency (CIDA), and the Department for International Development in the United Kingdom (PIOJ, 2005). Since the inception of PSMP, ICT-related tasks completed under the initiative include the computerization of operations for some ministries and agencies. The acquisitions of computer hardware and software, as well as the provision of technical support, have resulted in a few of the agencies offering online services.

Another pertinent initiative is the ICT project, jointly funded by the Government of Jamaica (GOJ) and the Inter-American Development Bank (IDB) that was launched in 2002. This GOJ/IDB ICT project was initially planned to last for 3 years with an IDB loan of US $17 million but the project has been rescaled and will continue for 5 additional years with a reduction in the loan component to US $8.5 million (Paulwell, 2005). The Project's main objectives are to accelerate Jamaica's e-readiness and help attain international standards of performance and efficiency in both the public and private sectors, while reducing the cost of services to customers, promoting convenient access, and enhancing their quality of life. A complementary policy environment that encourages growth in the ICT sector supports these initiatives. In 2002, the Government in a review of tax policy removed all duties and sales tax from computers. Development incentives such as an Encouragement Act for the ICT sector allow investors exemption from income and dividend taxes for up to 15 years, and exemption from import duties on ICT machinery and equipment during incentive period.

INTERVENTIONS VIA POLICIES AND INITIATIVES

The implementation of the national ICT strategic plan is proceeding through a series of Government initiatives. One of the first and most important initiatives was the establishment of the Information Technology Employment Creation and Development project (INTEC). INTEC was the main initiative devised to implement the strategic plan. The project, which began in the financial year 2000/2001 with an initial allocation of US $40 million, aimed to create 40,000 jobs within the sector over three years. In addition, as these jobs were linked to the establishment of call centres, an additional 250,000 square metres of factory space was to be constructed with a view to attract investment from local and overseas companies. Funds under the project were earmarked to provide the necessary training to expand Jamaica's pool of skilled IT workers. INTEC and other initiatives undertaken by Jamaica are discussed below using the King et al. (1994) typology of six categories introduced earlier—knowledge building, knowledge deployment, subsidies, mobilization, innovation directives, and standards setting. For the last category, a detailed example is provided.

Knowledge Building

Knowledge building can take place when institutions take actions to provide the scientific and technical knowledge that are essential for innovation and the diffusion of innovations. The obvious type of support for knowledge building would be in the form of sponsoring research and development for information systems and technology. While this is most important when countries are seeking to be producers of the technology, it can also be used to develop new ways of using existing technology. In Jamaica, there are no major government initiatives to build knowledge in the area of information technology hardware development. However, Fiscal Services Limited is a GOJ company that develops software solutions for government clients, and in this sense the learning is accumulative and shared within Jamaica. There is a considerable scope to share this knowledge and expertise in the use of IT solutions within the wider Caribbean region since it is easily adaptable to the cultural, traditional, and political environments. Further, the establishment of the Caribbean single market economy (CSME) brings even more opportunities for networking within the region and providing appropriate IT solutions.

Knowledge Deployment

A knowledgeable and well-trained population is essential for the successful transfer of technology. This is especially needed in the case of developing countries in order to encourage the diffusion and use of ICT. Four mechanisms for knowledge deployment typically exist. Firstly, governments can intervene in a regulatory capacity by ensuring that it is mandatory for information technology subjects to be taught in schools, particularly at the secondary level. The use of public institutions to provide basic information technology training to marginalized groups in society is the second mechanism that can be used by governments. Thirdly, partnerships with private and other social institutions are another way for governments to use their influence and assist in the provision of education to the general population, or to promote knowledge deployment by training a group of potential users of the innovation. Finally, knowledge is also deployed through multinational corporations or in the form of knowledgeable individuals migrating to the country.

In Jamaica, human resource development was a specific area targeted by the five-year national strategic plan for ICT in the effort to become a knowledge-based society. In the case of information technology, there was a lack of personnel possessing the requisite skills needed for promotion and growth of the sector, particularly at the higher end. Jamaica has suffered from the migration of skilled and tertiary level educated young people to countries like the USA, Canada, and the United Kingdom. The literacy rate for 2004 was 79.9%, and only about 19% of the population received a tertiary level education. These figures are significantly lower than those of developed countries and even some developing countries. Under the INTEC project, partnerships were formed with private sector organizations. For example, the Caribbean Institute of Technology was established to provide training and development in basic computing skills, telemarketing, call centre operations, Web development courses, and computer programming. The main vocational institute, Human Employment and Resource Training/National Training Agency (HEART/NTA) was the key public institution involved and received major funding to provide the training. The other local partners in this venture were the Ministry of Commerce, Science and Technology, University of the West Indies, Mona School of Business, and Montego Bay Free Zone. The international partners were Idusa Global and Furman University from the United States, and the International Development Consortium located in the United Kingdom. A CISCO training academy was also established through an alliance between HEART/NTA and CISCO, which provides certification up to the level of Certified Cisco Network administrator. In the overall context, however, the number of partnerships established under INTEC has been lower than that was anticipated.

At the primary and secondary levels, the national strategic ICT plan complements a well-set education policy. Information Technology is now a compulsory subject for students in grades 10 and 11 at secondary schools and the government will pay the fees for the IT qualifying examination to be taken at the end of their studies. Although the interventions by the Jamaican Government have resulted in more schools having IT labs, only a small percentage were connected to the Internet by the end of 2004 due to high broadband access cost. This is a reminder of the point made by King et al. (1994) of the importance of ensuring that policy acts, such as privatization and the liberalization of telecommunications, achieve the requisite effect of a decrease in the cost of access to the technology.

In developing countries it is highly unlikely for families that fall within the lower-middle income and lower income group to be able to purchase computers and expend a portion of their monthly income for Internet access. For these families, the only exposure that children will have to the Internet is in schools. With this in mind, governments can prioritize initiatives that promote the diffusion and use of IT to schools, especially those at the secondary levels. This can have long-term benefits; with the creation of a secondary level educated workforce that has been trained with low level IT skills, thus eliminating the need to spend on future projects providing training for secondary school leavers. A greater number of persons can then be trained in more highly skilled areas of IT, allowing more people to become specialist and increasing the pool of skilled IT personnel. This can be an attractive feature for foreign direct investments.

Subsidies

Subsidies are targeted actions by institutions to defray costs and risks to innovators and users. Subsidies can come in the form of: funding of prototype development and demonstration projects; encouraging the availability of funds for innovative activity, providing low-interest loan, provision of tax breaks, and purchasing innovations and locally produced goods and services containing these innovations. Developing countries have used the strategy of providing tax breaks and incentives to attract foreign direct investments. Under the national plan Jamaica has made significant incentives to companies investing in the ICT sector. To date these include tax exemptions, duty free imports, and free repatriation of profits and dividends.

Companies that have been granted a single entity free zone status have the option of applying for licenses to provide their own telecommunications. In addition, certified data processing/system development businesses will be eligible for a special capital expenditure allowance. Call centre employers receive a grant to offset the costs for each employee trained. These incentives are more in favour of foreign owned companies and new entrants in the sector, but government can extend incentive programmes to include existing companies and also to the locally owned companies. One option is to offer tax exemptions to existing companies in the sector that employ persons trained in ICT under government partnership schemes or by HEART/NTA.

Mobilization

In this setting, mobilization refers to actions taken to encourage stakeholders to think about an innovation in a particular way and hence to aid its diffusion and adoption. While this can be done using promotional and awareness campaigns, another method is by becoming a user of the innovation and acting as a change agent. In most countries, government ministries and agencies are the largest clientele for private sector organizations. This gives them some influence as customers, but also presents opportunities for them to be change agents for technology by becoming model users. Jamaica's intervention to aid the mobilization of the use of the Internet and information technologies comes in the form of an initiative jointly funded by the Inter American Development Bank (IDB) namely, the Government of Jamaica/Inter-American Development Bank Information and Communications Technology (GOJ/IDB ICT) project.

The project is focused solely on information and communications technologies and its components, includes: supporting e-government by facilitating the use of Internet technologies by some government agencies, institutional strengthening for the Ministry of Commerce, Science and Technology; establishing community access points for the Internet in marginalized communities, and training. The project is being funded through GOJ budgetary support in the amount of US $3.4 million, a loan of US $8.5 million from the IDB and US $2.3 million beneficiaries. This and other initiatives were undertaken to foster an enabling environment by integrating ICT within government departments and agencies so that services will be improved. The expectation is that stakeholders will identify the benefits of using the technology, especially the Internet, and this will propel them to adopt more quickly.

Innovation Directives

Innovation directives refer to the use of commanding actions to produce or use innovations, or to engage in activity facilitating the production or use of innovations. Regulations can be introduced in requiring organizations to invest specified amounts into research and development (R&D) activities leading to innovations. There are no known interventions by the Jamaican Government that illustrates the use of innovative directives in the area of ICT. The preferred approaches have been influence, provide incentives, and create partnership, such as evidenced in the Mona Research and Innovation Park planned in 2005 (Paulwell, 2005).

Standard Setting: A Detailed Example

For information and communication technologies the setting of standards can be used for diffusion of an innovation by making its use mandatory. King et al. (1994) suggested that standard setting could be used in the regulation of decentralized actors and organizations to bring them in line with larger social or institutional objectives. Interventions of this nature can involve the establishment of standards under which innovative activity might be encouraged, requiring that particular products or processes are to be used in any work for the institution, or requiring conformance with other standards that essentially mandates the use of particular products or processes. Standard setting is a powerful intervention that can be used by governments to act as a catalyst for stakeholders' attitude towards the use of ICT.

The national initiative that most exemplifies the use of standard setting to aid diffusion, is the use of electronic data interchange (EDI) technology between the Jamaica Customs Department and Customs Brokers. The system known as Customs Automated Services (CASE) was first launched in 1997 with manual and electronic features. It involved brokers completing electronic customs entries or forms and then transporting the information via floppy disks to the Customs Department to be uploaded onto their system. CASE has evolved through a phase where brokers were able to email the file containing their completed entries to the current stage where the Internet is used as the backbone to a completely electronic system that allows brokers direct access to the Customs Department. The result is that there is no practical alternative for the brokers. Compliance is now almost 100% and has resulted in more brokerage companies getting access to the Internet. An offshoot was that once the companies got Internet access, the next step was the exploration of how it could be further incorporated within their businesses. For some companies it has led to innovation and the creation of new services to be offered to their customers.

When CASE was initially implemented in 1997 the aim of the system was to decrease processing time and increase efficiency, the vision for utilizing the Internet and e-payment was absent. The compliance rate was low, with brokers shying away from using the system. In 2005, however, Fiscal Service Limited, developers of the application and the official provider of IT services to the Government of Jamaica, estimated the compliance rate to be 99.9%. This, albeit the absence of the appropriate legislation that makes digital signatures legally binding, making it still necessary for brokers to give a printed and signed copy of their entries to Customs. Needless to say the vision for CASE has changed since the development of the national IT plan and it is now integral to the e-government system linking the various government departments to form a portal for import and export trade.

From the perspective of the businesses, the introduction of CASE has served as somewhat of a catalyst for their venture into technology adoption. One brokerage firm that employs one full-time and three part-time staff acknowledged that they only recently adopted the Internet in response to the need to comply with CASE. The attitude towards technology has made a positive shift since, and there are plans to utilize IT more by automating daily processes and creating a Web site. In the case of another firm, RCM Customs Brokers, which had its first computer since 1989, the use of IT has always been a way of life. This is due in great part to the fact that the Managing Director, also owner of the company, has a positive attitude toward and a high awareness of IT. In fact this individual was instrumental in the development of CASE and provided a vital stakeholder perspective to the government. The newer version of CASE with direct access to the Customs Department has enabled RCM to be even more innovative by using data from Customs to provide new services to their clients. CASE is an example of one of the ways in which the Government of Jamaica has successfully used the Internet to provide more efficient services. It also illustrates how policymakers can use intervention to aid the diffusion and adoption of ICT.

IMPACT OF INTERVENTION ON SMALL- AND MEDIUM-SIZED ENTERPRISES

Small and medium-sized enterprises (SMEs) are the backbone of the economy for many countries and make significant contributions to the economy through employment and productivity. The United Kingdom is typical where SMEs account for 99.1%

of all businesses and 56% of employment, contribution to GDP is similar to that of large firms and in some sectors they out-perform larger companies in terms of productivity (DTI, 2002). In Jamaica it has been estimated that SMEs employ 60% of the work force and contribute 40% annually to Jamaica's national income. These figures confirm the importance of micro, small, and medium sized companies to the Jamaican economy.

Developed and developing countries have attested to the value of SMEs to their economy and are striving to make sure that they can survive in the digital economy by providing the proper infrastructure. Despite the lauded advantages to be gained by adopting information and communications technologies and specifically Internet technologies, SMEs have been cautious in their explorations beyond the simple applications of e-mail and Web sites (Brown & Lockett, 2004). This inertia among SMEs has propelled regional and national governments to develop policies and programmes that are geared towards encouraging small firms to adopt Internet technologies and engage in e-business. The EU plan mentioned earlier is evidence of this.

It has been suggested that the decision to adopt Internet technology by SMEs is opportunistic, rather than strategic (Mehrtens, Cragg, & Mills, 2001; Poon & Swatman 1999; Sadowski, Maitland, & van Dogen, 2002). In some cases firms do not adopt because of the belief that there is little or no benefit to be gained for their type of product or service. As with the case of the Custom Brokers, this perception can be simply wrong. For other potential adopters, a lack of both human and financial resources is a barrier, or in some cases the absence of suitable and affordable services from ICT providers to SMEs.

In the above context, interventions by governments, associations, and international agencies in the forms of mobilization, creating awareness about the technology, knowledge deployment, and training in the use of the technology, can lead to changes in the attitudes of SMEs. It is important, however, to ensure that participating organizations are ready to absorb the adoption of the technology. This will help to create sustainability of the adoption and safeguard against the process being abandoned after the initial implementation. Organizational readiness refers to the level of IT sophistication and financial resources of the firm. Small firms can often be faced with financial deficiency and in most cases this will dictate the type of financial risks that will be undertaken by the firm. Therefore, firms exhibiting higher levels of financial and technological readiness are better prepared and are more likely to sustain the adoption (Chwelos, Benbasat, & Dexter, 2001; Iacovou, Benbasat, & Dexter, 1995; Kuan & Chau, 2001).

In Jamaica, the importance of SMEs for economic growth and reducing unemployment was explicitly recognised within the ICT initiatives of the Jamaican national strategic plan. Eight small and medium-sized companies in Jamaica were investigated by the authors in an attempt to gauge the impact that interventions had on their adoption and use of Internet technologies. All companies had participated in initiatives jointly funded by the Government of Jamaica and an international donor agency. The two initiatives examined: the New Economy project (NEP) and the Trade Development project (TDP) were aid-supported with funding from the United States Agency for International Development (USAID) and the European Union (EU) respectively. Both NEP and TDP were geared specifically towards the micro, small, and medium sized enterprises sector. The eight companies undertook ICT related activities as a direct result of their participation in the schemes.

The focus was mainly on the adoption and use of Internet technologies. In most cases the companies had Internet access but were only engaging in low complexity activities such as e-mail and had basic or no Web sites. By participating in the scheme, they benefited from funding to update their Web sites and make them e-commerce ready.

Technical support was received from consultants provided under the initiative to aid these companies in the implementation process. It is important to note that these initiatives were demand driven rather than support led; this gave the organizations the power to decide on the activities that would be most beneficial to their company and that supported their business strategy.

Evidence from the companies examined in Jamaican situation suggests that although the motivation for the adoption of Internet technology was not explicitly stated as strategic, there were perceptions of potential benefits that could be harnessed by implementing the technology. One of the most common benefits perceived was cheaper and more efficient communication via e-mail with business partners and customers, especially with overseas business. The idea of being easily accessible to potential business partners and customers within the region was another appealing factor for the implementation of the technology. After the intervention, all eight companies were using broadband technology, purchased online and used the Internet extensively for research purposes to source new suppliers, to improve production processes and also to keep abreast of international competitors.

The existing IT infrastructure within the companies varied ranging from relatively advanced to very basic. Only two of the companies had specialist ICT personnel in-house, while the others relied on outside personnel. Although for all of the companies Internet adoption was seen as highly desirable if they wanted to be competitive, there were external factors that dampened their eagerness to make it a 'way of life' in their organizations. For example, the lack of proper legislation for e-commerce, such as electronic signatures, had made it impossible for some to proceed into more complex e-business activities. Similarly, the lack of support and services geared towards SMEs from banks and other financial institutions also impacted negatively on these companies' ability to offer services such as online selling. Of the eight companies, only two are selling fully online and both companies have had to partner with overseas companies in order to offer this service. Despite these shortcomings the evidence is clear that the direct intervention through targeted ICT initiatives has resulted in successful and sustained benefits in all of the eight sampled companies.

The link between direct intervention and positive outcomes is also evidenced in the cases of community-based initiatives. For example, the ICT Training for disadvantaged youth is an aid-supported initiative, implemented by the Jamaica Sustainable Development Network Programme (JSDNP), funded by United Nations Development Programme (UNDP) and Microsoft Jamaica. The project aims to provide ICT skills training 600 marginalized youths between the ages 16-25. Likewise, the Container project, funded by United Nations Educational, Scientific, and Cultural Organization (UNESCO), is another community-based, aid-supported initiative that exemplifies positive outcomes as a result of direct intervention in Jamaica.

FUTURE ISSUES

For the future GOJ faces two main issues. Firstly, a relatively high national debt means the GOJ has limited capacity to service loan commitments arising from these collaborative initiatives with international development partners. Even though a portion of the funds is given as grants, the significant portion of the budget needed to implement the initiative represents a loan. It is not unusual for developing countries to fund economic development with debts but in the case of Jamaica where the national debt is 131.48% of GDP in 2006, this will become an issue. The second main issue is the deployment of ICT to rural communities. Again this is a situation faced by other developing countries and can be more problematic for some more than others. For example, in the case of India and China where large and widely dispersed populations are two factors that contribute to the digital divide between urban and rural areas. While these factors may be less problematic for Jamaica, there are still issues especially regarding access to and use of the Internet. These manifest in problems of infrastructure, ICT education, and personal financial constraints. GOJ will start to address this broad issue through the e-learning project, which will provide community access points, and the Wifi and WiMax policy initiatives. These are positive moves but the limitation on committed resources and the setbacks associated with stakeholders' participation mean that the issue of the digital divide is likely to be longstanding.

IMPLICATIONS AND CONCLUSION

A cohesive national plan for IT has been beneficial to Jamaica. The Global Information Technology Report for 2004/2005 ranked Jamaica 49th out of 104 countries in terms of networked readiness. This ranking is the highest achieved by an English-speaking Caribbean country and is exceeded only by the United States, Canada, Chile, and Brazil from the Americas. For the first time, Jamaica was included in the 2005 e-readiness rankings produced by the Economic Intelligence Unit of the Economist; the country was ranked 41st and according to the report Internet penetration has nearly tripled since the liberalization of the telecommunications sector in 2000 (Paulwell, 2005). There has been significant movement since market liberalization of telecommunications.

Twenty-two new licenses were issued in 2002 for Internet companies, and the number of Internet hosts rose from approximately 300 in 1999 to 1,472 in 2000–2003. Currently, the focus of the government is on becoming an outsourcing location to the United States and Europe and the main thrust has been to invest in call centres. There have been favorable reviews from site selection companies and Jamaica is considered a viable option for near shore outsourcing for the United States. There have been claims by the Government that they are not able to meet the demand for companies wanting to establish call centres due to a lack of space. Since 2000, when the strategic plan was introduced, the above progress represents a strong performance. Similarly, the ICT initiatives aimed at the smaller firms to enhance their prospects have been successful, although the scale of these initiatives has been modest.

The main problem, however, has been the failure to create significant numbers of jobs. Millions of dollars were lost under the INTEC fund when one of the main beneficiary organizations, Netserv Communication Jamaica Limited, was placed in receivership. The aim of the government was to create 40,000 jobs in the information and communications technology sector, mainly through call centres operations but this target was not remotely reached. There are operators who started businesses during the period 2001–2003 but abandoned the sector. While mobile telephony has increased dramatically, the diffusion of the Internet has been sluggish. Supporting sectors such as financial institutions have been slow in their decisions to offer services to small and medium-sized companies and citizens that could foster e-commerce. Companies have had to seek solutions from overseas providers to circumvent the challenges faced by this lack of support. Finally, the government has been slow in passing the various pieces of legislation needed to support e-business.

The stability of the political environment in the last 16 years in Jamaica has helped the implementation of national policy for ICT. The current ruling party is the one that sanctioned the original policy and the portfolio belongs to the same Minister who developed and mobilized it. Also helpful, and arising from this relative stability, has been the cross Government integration for the national strategic plan. For example, the Ministry of Education and Culture developed its own information technology policy, which directly complemented the ICT plan. In 2006 it is safe to assume that the strategic ICT vision for Jamaica remains unchanged, but the difficulties of implementation are better appreciated.

More generally, the Jamaican experience shows that a comprehensive national strategy for ICT can help harness the benefits to be gained from the technology and thereby promote economic development. Such a strategy is operationalized through specific initiatives and it is here where the implementation difficulties arise. Jamaica has had both good and poor experiences. ICT is a facilitating technology for other sectors and helping these sectors, such as SMEs, can work but it requires effective partners with knowledge of the technology and the sector. Merely providing a budget for infrastructure is not a sufficient requirement. Similarly liberalization of the telecommunications sector of itself in Jamaica has not led to ubiquitous access and other mechanisms are likely to be needed.

For other developing countries there are lessons from the Jamaican experience with ICT planning. The value of a nationally agreed plan, the importance of integration across government departments, and the need for carefully planned initiatives to implement policy are well evidenced in this six-year retrospective reflection. So too is the importance of partnerships, both local and international, to finance or progress the initiatives. Regarding implementing national ICT policies, other countries should seek to adopt best practices but be mindful of the fact that it is not a matter of one-size that fits all. In particular, it is important for countries to exercise care and caution when using policy implementation to achieve certain results under market conditions. The range of stakeholders; their interests and agendas can produce unintended outcomes. In Jamaica, the failure to create sufficient jobs is an example. Developing countries should seek to become parties to initiatives from international donor agencies that support and are aligned with national policies for growth and socio-economic development, and try locally to develop initiatives that are characterized by local knowledge and by ICT technical and managerial competence.

POLICY INITIATIVES IN JAMAICA AND THEIR IMPACT ON THE DEVELOPMENT OF INFORMATION AND COMMUNICATION TECHNOLOGY

Future Research

Through the example of Jamaica, this chapter has demonstrated the relevance of national policy to ICT development and adoption. In the main, this has produced tangible benefits to individual companies and to the Jamaican economy generally. This is in line with the mounting evidence of the economic benefits from IT. It is now accepted that virtually all the increase in US productivity in the past 20 years can be attributed to IT (Brynjolfsson & Hitt, 2003). Within this context some suggestions for the nature of future research in this area can be explored.

There is little doubt that in the context of developing countries more work is needed to understand the contribution of national policy in developing ICT's potential. A recent UN report on the information economy (UNCTAD, 2006) reviews this area and identifies key issues across a range of advanced and developing economies. In terms of further policy research in Jamaica the task is to the same as in other developing economies, namely to assess the efficiency and effectiveness of existing policies. There are two considerations. Firstly, as the Jamaican example illustrates ICT policy should not exist in isolation. Within the broad economic goals a portfolio of policies is needed to encourage and facilitate ICT adoption and usage that between them cover telecommunications, security, human resources (capacity building), ICT related trade, and investment, innovation building. Secondly, implementation of policy is the critical task. For this to happen, the roles of stakeholders within the institutional framework need to be clear and prioritised. A policy related research agenda for Jamaica, and in other developing countries, would include a critical review of existing arrangements against a conceptualised model of necessary activities.

Beneath the broad policy level, research is needed on the needs of specific sectors, transport, finance, retail, trade, and so forth, and in relation to firm size. Although there are common requirements between sectors, such as basic ICT skills, there will be advanced skills which are more sector specific. These are not well understood and the ability of specific sectors to provide that training will vary. The balance between on job training and national provision needs further attention. This is one example but similar issues arise between sector stakeholders in respect of technology provision such as Web services.

Finally, there needs to be more research into the factors affecting the adoption of technology from a business model perspective at the firm level. The provision of technology and related services, for example dial-up and then broadband, does not guarantee the adoption and effective use of ICT. The advent of widespread e-business practices has required many new intermediaries to service collaborative applications such as supply chains and to facilitate payments. Each provider of such services has their own requirements and this has led to uneven adoption of best practices.

REFERENCES

Annan, K. (2000). We the peoples: The role of the United Nations in the 21st century. *Millennium Report of the Secretary General of the United Nations*. Retrieved April 20, 2006, from http://www.un.org/millennium/sg/report/summ.htm

Brown, D., & Lockett, N. (2004). Potential of critical e-applications for engaging SMEs in e-business: a provider perspective. *European Journal of Information Systems, 13*(1), 21-34.

Brynjolfsson, E., & Hitt, L. M. (2003). Computing productivity: Firm-level evidence. MIT Sloan Working Paper No. 4210-01, available at SSRN: http://ssrn.com/abstract=290325

Chwelos, P., Benbasat, I., & Dexter, A. (2001). Research report: Empirical test of an EDI adoption model. *Information Systems Research, 12*(3), 304-321.

Damsgaard, J., & Lyytinen, K. (2001). The role of intermediating institutions in the diffusion of electronic data interchange (EDI): How industry associations intervened in Denmark, Finland and Hong Kong. *The Information Society, 17*, 195-210.

DTI (2002). *A government action plan for small business*. Retrieved August 18, 2005, from http://www.sbs.gov.uk/content/7-strategies/sbs_evidence.pdf

European Union (2005). *eEurope 2005 Action Plan*. Retrieved August 28, 2005, from http://www.eeurope2005.com

Gilbert, A. L. (1992). A transaction costs model of international information technology transfers: The dynamics of intelligence and of information control. In Palvia, Palvia, & Zigli (Eds.), *The global issues technology management* (pp. 403-426). Hershey, PA: Idea Group Publishing.

GOJ (2002). *Five-year strategic information technology plan for Jamaica*. Retrieved November 1, 2004, from http://www.mct.gov.jm/gojitplanrevisedversionmarch20021.pdf

GOJ (2003). *Electronic commerce (Transaction) policy*. Retrieved February 12, 2005, from http://www.mct.gov.jm/electronictransactionspolicy-april2003ps1submitted.pdf

GOJ (2006). *Transformation of the ICT sector: Overview of ICT development in Jamaica.* Retrieved July 27, 2006, from http://www.mct.gov.jm/transformationoftheicsector2.pdf

Henriksen, H.Z., & Andersen, K.V. (2004). Diffusion of e-commerce in Denmark: An analysis of institutional intervention. *Knowledge, Technology & Policy, 17*(2), 63-81.

Iacovou, C., Benbasat, I., & Dexter, A. (1995). Electronic data interchange and small organizations: Adoption and impact of technology. *MIS Quarterly, 19*(4), 465-485.

ITU (2003). *World Telecommunication Development Report.* Retrieved June 14, 2006, from http://www.itu.int/itud/ict/publications/wtdr_03/material/chap4_wtdr2003_e. pdf

King, J.L., Gurbaxani, V., Kraemer, K.L., McFarlan, F.W., Raman, K.S., & Yap, C.S. (1994). Institutional factors in information technology innovation. *Information Systems Research, 5*(2), 139-169.

Kraemer, K., Gurbaxani, V., & King, J.L. (1992). Economic development, government policy, and the diffusion of computing in Asia-Pacific countries. *Public Administration Review, 52*(2), 146-156.

Kuan, K. K. Y., & Chau, P. Y. K. (2001). A perception-based model for EDI adoption in small businesses using a technology-organization-environment framework. *Information & Management, 38*(8), 507-521.

Madon, S. (2000). The Internet and socio-economic development. *Information Technology and People, 13*(2), 85-101.

Mehrtens, J., Cragg, P.B., & Mills, A.M. (2001). A model of Internet adoption by SMEs. *Information & Management, 39*(3), 165-176.

Montealegre, R. (1999). A temporal model of institutional interventions for information technology adoption in less-developed countries. *Journal of Management Information Systems, 16*(1), 207-232.

Pace, N. (2006). *China surpasses U.S. in Internet use.* Retrieved July 12, 2006, from http://www.forbes.com/2006/03/31/china-internet-usage-cx_nwp_0403china.html

Palvia, P., Palvia, S., & Roche, E. (Eds.). (2002). *Global information technology & electronic commerce: Issues for the new millennium.* Marietta, GA: Ivy League Publishing

Paulwell, P. (2005). *Sectoral budget presentation 2005/2006.* Kingston: Ministry of Commerce, Science & Technology (with Energy).

Poon, S., & Swatman, P. (1999). An exploratory study of small business Internet commerce issues. *Information & Management, 35*(1), 9-18.

Sadowski, B., Maitland, C., & van Dogen, J. (2002). Strategic use of the Internet by small and medium-sized companies: An exploratory study. *Information Economics and Policy, 14*(1), 75-93.

Silva, L., & Figueroa E. (2002). Institutional intervention and the expansion of ICTs in Latin America: The case of Chile. *Information Technology & People, 15*(1), 8.

Statistical Institute of Jamaica (2006). Retrieved July 26, 2006, from http://www.statinja.com

Steinberg, J. (2003). Information technology and development beyond "either/or". *The Brookings Review, 21*(2), 45-48.

Twist, J. (2004). *China leads world on broadband.* Retrieved July 12, 2006, from http://news.bbc.co.uk/1/hi/technology/3699820.stm

UNCTAD (2006). *Information Economy Report 2006: The Development Perspective.* UN New York and Geneva.

Walsham, G. (2001). *Making a world of difference*. IT in a Global Context. Chichester: John Wiley & Sons.

World Bank (2001). *Bridging the digital divide in the Americas*. Retrieved April 20, 2006, from http://lnweb18.worldbank.org/External/lac/lac.nsf/e4ce3b63837a57ef85256 a37004a8dac/ea45f9696e38b03585256a370052bcaf/$file/English%20.pdf

World Bank (2006). *2006 Information & Communications for Development (IC4D): Global trends and policies*. Retrieved July 27, 2006, from http://web.worldbank.org/wbsite/external/topics/informationandcommunicationandtechnologies/contentmdk:20831214~pagepk:210058~pipk:210062~thesitepk:282823,00.html

ADDITIONAL READINGS

Anderson, K., Bjorn-Anderson, N., & Dedrick, J. (2003). Government initiatives creating a demand driven e-commerce approach: The case of Denmark. *Information Society, 19*(1), 95-105.

Avgerou, C. (2000). IT and organizational change: An institutionalist perspective. *Information Technology and People, 13*(4), 234.

Avgerou, C. (2003). The link between ICT and economic growth in the discourse of development. In M. Korpela, R. Montealegre, & A. Poulymenakou (Eds.), *Organizational information systems in the context of globalisation* (pp. 372-387). Kluwer: Academic Publishers.

Baliamoune-Lutz, M. (2003). An analysis of the determinants and effects of ICT diffusion in developing countries. *Information Technology for Development 10,* 151-169.

Brown, D., & Lockett, N. (2001). Engaging SMEs in e-commerce: The role of intermediaries within eClusters. *Electronic Markets, 11*(1), 52-58.

Castells, M. (2000). *The rise of the network society*. Oxford: Blackwell Publishers Inc.

Daamsgaard, J., & Lyytinen, K. (2001). The role of intermediating institutions in the diffusion of electronic data interchange (EDI): How industry associations intervened in Denmark, Finland and Hong Kong. *The Information Society, 17*, 195-210.

DiMaggio, P.J., & Powell, W.W. (1983). The iron cage revisited: Institutional isomorphism and collective rationality in organizational fields. *American Sociological Review, 48*(2), 147-160.

DiMaggio, P. (Ed.). (2001). *The twenty-first-century firm: Changing economic organization in international perspective*. New Jersey: Princeton University Press.

Edwards, S. (2002). Information technology and economic growth in developing countries. *Challenge, 45*(3), 19-43.

Gibbs, J.L., & Kraemer, K.L. (2004). A cross-country investigation of the determinants of the scope of e-commerce use: An institutional approach. *Electronic Markets, 14*(2), 124-137.

Hanna, N. (2003). *Why national strategies are needed for ICT-enabled development*. World Bank Staff Working Paper. Washington D.C.

Harris, R. (2004). Information and Communication Technologies for Poverty Alleviation. E-Primers for the Information Economy, Society and Polity, UNDP-APDIP.

Henriksen, H., & Andersen, K. (2004). Diffusion of e-Commerce in Denmark: An analysis of institutional intervention. *Knowledge, Technology & Policy, 17*(2), 63-81.

Huggins, R. (2000). *The business of networks: Inter-firm interaction, institutional policy and the TEC experiment*. Ashgate Publishing Ltd.

IICD (2005). The TIC Bolivia Country Programme: The impact of IICD support for poverty reduction and development using ICTS 2000-2004. International Institute for Communication and Development.

Madon, S. (2000). The Internet and socio-economic development. *Information Technology and People, 13*(2), 85-101.

Martinez-Frias, J. (2003). The importance of ICTs for developing countries. *Interdisciplinary Science Reviews, 28*(1), 10-14.

Odera, M., & Madon, S. (1993). Information *Technology Policies and Applications in the Commonwealth Developing Countries*. London: Commonwealth Secretariat.

Odedra-Straub, M. (Ed.). (1996). *Global Information Technology and socio-economic development*. Nashua, N. H: Ivy League Publishing.

Orlikowski, W.J., & Barley, S.R. (2001). Technology and institutions: What can research on information technology and research on organizations learn from each other? *MIS Quarterly, 25*(2), 145-165.

Pohjola, M. (2002). The new economy: Facts, impacts and policies. *Information Economics and Policy, 14*, 133-144.

Rogers, E.M. (2003). Diffusion of innovations. *Free Press*.

Rovere La, R. L. (1998). Diffusion of information technologies and changes in the telecommunications sector. The case of Brazilian small and medium-sized enterprises. *Information Technology & People, 11*(3), 194.

Sumit, R. (2005). *Globalisation, ICT and developing nations: Challenges in the information age*. London: Sage Publications.

Sayed, E.H., & Westrup, C. (2003). Egypt and ICTs bring national initiatives, global actors and local companies together. *Information Technology and People, 16*(1), 93-110.

Scott, W.R. (2001). *Institutions and organizations*. Thousand Oaks, CA: Sage.

Steinmueller, W. (2001). ICTs and the possibilities for leapfrogging by developing countries. *International Labour Review, 140*(2), 193-210.

UNCTAD (2006). *Information Economy Report 2006*. United Nations publication, New York and Geneva. Retrieved from http://www.unctad.org/ecommerce

Walsham, G., & Sahay, S. (2006). Research on information systems in developing countries: Current landscape and future prospects. *Information Technology for Development, 12*(1), 7-24.

Chapter XVII
Digital Government Development Strategies:
Lessons for Policy Makers from a Comparative Perspective

Yu-Che Chen
Iowa State University, USA

Richard Knepper
Indiana State University, USA

INTRODUCTION

Digital government has a better chance of success if it follows a well-articulated and sound strategy. Failures of digital government projects in terms of cost overrun, delay in delivery, and problems in implementation are the rule rather than the exception (Heeks, 1999). Having a sound strategy is considered as the first and most important step in securing the success of information technology projects (Fletcher, 1999). More importantly, a strategic master plan helps align information and communication technology investment and resource allocation with the objectives of organizations. If done properly, a strategic plan can help public organizations realize the full potential of an information technology investment.

The key question that guides this study is: How can a national digital government strategy be better designed and developed? As reflected in the research question, this chapter takes a normative perspective of policy and institutional design with emphasis on policy makers. A large number of countries around the world have recognized the importance of information and communication technology for staying competitive in the information age. Moreover, it is recognized that a national strategy is the critical first step in building an information and communication infrastructure and allocating the necessary critical resources to build digital government. Some European countries, Asian economies, and North and South American nations, have some form of a national digital government plan, such as UK Online, e-Japan, and e-Korea.

However, there is a shortage of research-based frameworks for guiding the development of a national strategy. This chapter fills this gap by offering policy makers a comprehensive framework for developing national digital government strategies. Moreover, the authors further examine the experiences of Poland and Taiwan to illuminate the proposed framework and to provide insights into specific considerations for designing and improving digital government through strategic planning.

BACKGROUND

Digital government refers to the use of information and communication technology to better the relations between government and its employees, citizens, businesses, nonprofit partners, and other agencies by enhancing the access to and delivery of government information and services.[1] The terms of digital government and e-government are used interchangeably in this chapter because, in the authors' view, they cover the same set of activities as outlined in the definition of digital government. Digital government strategic plans are national plans to guide the development of a country's efforts to deliver information and service via digital means.

An overarching national digital government plan is less studied than individual digital government projects. Most of the critical success factors come out of studies of individual IT projects at state and local levels (Heeks, 1999; Dawes et al., 1997). Since national strategic plans provide the framework in which most digital government projects are prioritized, designed, and implemented, it is important to study those. There is lack of a research-based guidance on strategic planning for digital government at the national level and this study aims to fill the gap by drawing from three streams of literature to develop a framework for developing a national digital government strategic plan: digital government, institutionalism, and strategic information system planning. Each body of literature, as reviewed below, points out some key factors for the development of a conceptual framework. By choosing Taiwan and Poland, which are at markedly different points in their implementation of digital government plans, the authors believe that they can more rigorously test the validity and salience of the framework.

Digital Government

The broad literature on digital government points out the importance of political forces, institutions, and the all-encompassing notion of digital government to include citizen participation and electronic production of public services. In addition to these broad environmental factors, this body of literature offers a set of critical success factors that should be an integral part of national digital government strategies, including building management/rational decision-making capacity, establishing national information and technology infrastructure, developing IT-related human resources, protecting privacy and security, and managing government information resource management.

In addition, political forces and institutional settings play a significant role in national digital government efforts. Digital government plans and projects, unlike other private sector information technology projects, need to be attentive to political considerations and governmental processes at all levels of government (Bozeman and Bretschneider, 1986; Rocheleau, 2003). In the international arena, countries need to harmonize their digital government plans with regional associations that they aspire to be associated with.

For example, current and future European Union members look for EU e-governments plan for guidance on what components need to be included in their digital government plan to take advantage of the resources available at the regional level. A number of unique political challenges face development of digital government at the domestic level. These include the election cycle, budget requirements, politics of intergovernmental information-sharing, competing policy objectives, and existing civil service systems.

Moreover, legal and institutional incentives and constraints shape the design and use of information technology in the public sector (Fountain, 2001; Landsbergen and Wolken, 2001). For example, privacy laws may govern how information is collected and shared among public agencies. Institutional incentives are critical for soliciting cooperation from individual public employees and agencies as a whole to share information and resources. Moreover, security law allows government to win the trust of citizens and business to conduct transactions with government without the fear of information fraud or identity theft.

Citizens and their participation also constitute an important environmental element of digital government (Marchionini, 2003; UN/ASPA, 2002; Schedler and Scharf, 2001). When an active civil society and well-established information and communication infrastructure is in place, an e-government strategy that is more responsive to citizen needs is more likely to be implemented.

Keeping these environmental conditions in mind, the broad digital government literature has pointed out some critical elements needed in a sound national plan. First, rational/administrative capacity is one critical factor for digital government, as various studies of e-government projects around the world have suggested (Heeks, 1999). This is a precondition for any successful implementation of national digital government strategy. Second, national governments need to facilitate citizen participation in establishing an information and communication (hereafter referred to as ICT) infrastructure. As highlighted in the United Nations' report on an e-government survey, accessibility to e-government services depends on the information and communication infrastructure of a country (UN/ASPA, 2002). Since building ICT infrastructure usually has an economy of scale, national government is best suited for conducting its design and development.

Third, the training of IT related personnel should also be included in a national plan. The building of human infrastructure for the use of information technology is as critical as building the physical one. This is a particularly severe challenge at the local level in the United States and probably for most countries a challenge at all levels of government. The International City/County Management Association has conducted two e-government surveys of local government and has consistently ranked the lack of IT personnel as the number two barrier to e-government (ICMA 2000, ICMA 2002).

Fourth, the importance of launching a national campaign for the protection of privacy and security has been increasingly recognized by countries around the world. Cyber-crime, such as identity theft and child pornography, has greatly affected many people's lives. The protection of privacy and security is also critical for winning citizens' trust in government and for them to conduct transactions online (Edmiston, 2003). Lastly, the management of government information resources has been discussed as a critical component for a national digital government plan (Fletcher, 2003). Government information is the underlying content base on which information and transaction services are created. National government is in a good position to set a common standard for data format and information sharing protocols.

Institutionalism

Institutionalism has contributed to the study of national digital government strategies in a number of ways.[2] First, it points out the importance of national administrative and legal institutions. Therefore, a sound national digital government plan should have a component on building an administrative and legal institution. Second, it explains the ways in which institutions may shape the design and implementation of digital government projects. Specific programs can be developed when there is a good understanding of how incentive structure works in various national and subnational context. Third, institutionalism stresses the link between institutional arrangements and performance. A comprehensive national strategy should have a mechanism to track performance. Lastly, for a design of an institution or policy, it is important to address the physical environment (Ostrom, 1990; Farris and Tang, 1993).

National institutions are critical for the success or failure of national digital government efforts. The comparison between various nations' economic performance has articulated the importance of developing national institutions (North, 1990). Countries with similar resources may perform quite differently if one has efficient and rational national institutions and another is plagued with corruption and mismanagement. Institutions are broadly defined as rules that govern relationships in the specific domain of human interchange, such as the rule of law and democratic governance. In the digital government domain, privacy and security laws are fundamental for developing trust for citizens, businesses, and other governmental agencies to conduct transactions with one another. The health of national digital government institutions is in part judged by the effectiveness of national government in making rational decisions and implementing policies. The lack of rationality in government decision-making has undermined digital government projects in many developing countries (Heeks, 1999). A comprehensive digital government strategy should include the effort to build rational decision-making if so lacking.

Institutions shape not only the selection of digital government projects but also their design and implementation (Fountain, 2001). For example, a national identity card is permissible in Singapore but raises heated debates in the United States, which are due to the difference in legal protection requirements for privacy. The role of institutions is particularly pronounced when a cross-country comparison is concerned. As a result, a national strategic framework needs to factor in the role of institutions.

The link between institutions as an incentive structure and performance is another important insight (Ostrom, 1990; Ostrom et al., 1993). For example, an IT training program can only be successful when the perceived expected benefit of participation outweighs the costs involved. A national digital government plan needs to think through the costs and benefits. Performance is critical for linking incentives and outcomes (North, 1990; Ostrom et al., 1993). A national digital government strategy needs to have performance measures specified for the evaluation and modification of strategies.

The last important insight that institutionalism has for digital government is that it is critical to match policy and institutions to address the conditions of the physical information and communication infrastructure. Any strategy and policy developed needs to be sensitive to the unique challenge of each individual country's ICT infrastructure.

Strategic Information System Planning

Strategic information system planning is another key body of literature that provides insights into the development of a national digital government strategy. A national strategic plan, if implemented properly, can anticipate and address the political, institutional, economic, and technical context in which a national digital government strategy is formulated. The strategic information system planning literature states the important components of strategic planning, the role of context, and specific issues for public sector strategic information system planning.

Strategic planning literature offers guidelines on the essential components of an effective strategy. An effective strategy has a statement of strategic goals, an articulation of policies that affect the achievement of strategic goals, and more detailed programs to carry out the strategy (Mintzberg & Quinn, 1991). Similarly, an effective digital government strategy should at least include explicitly stated strategic goals. For example, the State of Minnesota has articulated its goals as to create a technologically literate public and to both create and maintain an efficient and effective state government. Moreover, that strategy should list relevant policies that limit or facilitate the deployment of such strategies as well as specific programs to achieve these strategic goals. Information system (IS) planning adds a technical component to strategic planning in general. In particular, strategic IS planning looks at the inventory of information technology resources both inside an organization and in the marketplace.

Strategic IS planning literature has pointed out the importance of scanning societal, organizational, and managerial issues before formulating a strategy. These organizational and managerial issues are usually more difficult than the technical ones. Taking organizational issues into account before drafting strategy increases the fit between strategy and organization.

Public sector strategic information system planning needs to consider several unique characteristics of government. First, government usually has multiple and usually competing objectives (Bozeman & Bretschneider, 1986). A public sector strategic IS plan needs to take that into account. Second, the planning horizon needs to correspond with election cycles, so that the planning horizon is usually shorter than the one advocated by the private sector (Guy, 2000; Caudle, Gorr, & Newcomer, 1991). Lastly, stakeholders' involvement is regarded as critical for the successful implementation of an IS strategic plan (Bryson & Alston, 1996). However, a balance needs to be made between timeliness and extent of stakeholder involvement.

Figure 1. A conceptual framework for the development of a national digital government strategy

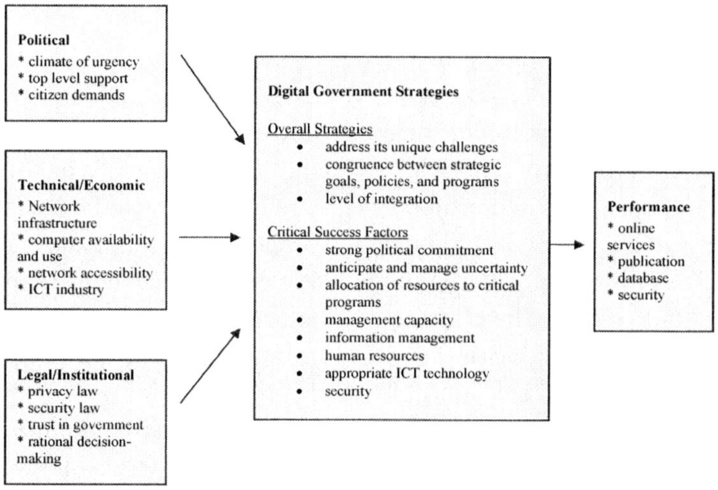

AN ANALYTICAL FRAMEWORK

The framework developed below draws from the insights offered by the three streams of literature as identified by their respective opening paragraphs presented in the previous section. As shown in Figure 1, these insights are organized into digital government strategies (center), environmental conditions (left), and performance results (right). Each group of factors is grounded either in these bodies of literature or some empirical evidence.

Four principles govern the development of the framework. First, a good strategic plan needs to address the unique challenges posed by the environmental circumstances facing a national digital government effort. What matters is the extent to which a strategy addresses its environmental conditions rather than setting ambitious goals. Second, there exists a set of principles and success factors, which, if done correctly, will increase the likelihood of success. Third, the framework acknowledges the evolutionary nature of digital government efforts. A strategy also needs to adapt to and possibly shape the environment that a national government is in. Lastly, performance of digital government is an important element of the framework to trace the effect of strategies on outcomes.

Environmental Factors

Political Considerations

Three political drivers, as shown in the top left box of Figure 1, set the stage for the development of a national digital government strategy. International political pressure may have an impact on a strategic plan. For example, a national digital government plan can be used as a precondition for a membership in an international organization, or for external help.

Another driver for a national digital government plan is the policy agenda of administration. A country's leaders may recognize the need for digital government as being part of a coherent strategy to stay competitive in the information era. They may view digital government as a means to revitalize public services (Snellen, 2000). The digital government plan could be supported by national legislature and the executive branch.

Citizen demand is another driver for digital government (Kamarck & Nye, 2002). Citizens, after accustomed to the convenience and efficiency accompanied by e-commerce, will probably demand that government do the same. Grass-roots demand involves both electronic voting and civic participation in the government decision-making process. This is an important area of development for a digital government plan to pay attention to (Marchionini et al., 2003). From a management perspective, electronic democracy should be integrated into the core mission of digital government. It should be noted that the extent to which civic participation may shape a national digital government plan depends on the power of democratic institutions in shaping public policy.

Technical/Economic Conditions

Several technical/economic conditions, as shown in Figure 1, are relevant for the design and formulation of a national digital government strategy. A good strategy needs to first assess the current condition as the first step to developing a path to the desired results. The national information and communication technology infrastructure is the first critical element for assessment. In particular, the UN/ASPA report has pointed out the importance of network availability (UN/ASPA, 2002). The availability of a high-speed network for both government and private use is an important indicator of the maturity of an infrastructure.

Second, computer availability and use determines whether the information and communication infrastructure is utilized. Citizens and businesses need to have computers and Internet service providers for them to get access to government information and services online. This is usually measured by the number of computers per capita. Wide disparity is seen among countries around the world. For example, in the United States, more than half of population has a computer whereas in Brazil less than 5 percent of the population has one (UN/ASPA, 2002).

Third, the distribution of use and accessibility among various groups also shapes national digital government strategies. Digital divide is usually the term used to capture this issue. Different socio/economic groups and demographics have unique needs and uses for digital government.

Lastly, the existence of a viable domestic IT industry is beneficial to the design and implementation of a national digital government strategy. This factor has a positive spill over effect on the affordability of computers, the establishment of an information and communication network, and the viability of building e-commerce applications for government use.

Legal and Institutional Issues

Two prominent legal and institutional issues pertaining to digital government are privacy and security (see Figure 1). These two issues have been identified by various levels of U.S. Government as important (ICMA, 2000; 2002; GAO, 2001). Privacy laws govern the use and distribution of information by government and businesses. For example, US Privacy Act dictates the protection of sensitive individual information from improper distribution. An exception can only be made when there is strong public interest such as public safety or national security. The EU has also developed directives to specify policies for the protection of individual data. The protection of privacy aims to foster consumer confidence and encourage e-commerce.

Security is a major concern when government information systems store vital information about individuals and government units. Unauthorized access to critical information could have devastating consequences. Social security information may be stolen as a result of identity theft. Individuals' lives are shattered as a result. Countries around the world started to experience security problems associated with increasingly computerized government information systems. The U.S instituted the Computer Security Act of 1987 to spell out principles for computer security.

A national digital government strategy needs to address both privacy and security to win the trust and willingness of citizens and businesses to submit personal information and conduct transactions online. If privacy is not adequately protected to facilitate private and business transactions with government online, a national digital government strategy needs to include efforts to pass relevant laws and regulations. A national digital government strategy should also address gaps in the area of security. Some countries have launched the Public Key Infrastructure Initiative and other security measures to address this issue.

As institutionalist literature suggests, administrative capacity is one of the key elements for enhancing performance. As a governmental effort, digital government also relies on the institutional capacity of government to be efficient and rational. If such capacity does not exist, a national digital government strategy needs to make this a high priority and incorporate it into its plan.

Digital Government Strategies

A successful national digital government strategy should be able to appropriately address the political, institutional, social, and technical challenges outlined above as environmental factors. As suggested by the review of relevant literature, digital government strategies have two components (see Figure 1). The first component is overall strategies that a country should follow to address its environmental challenges. Since each country faces unique challenges, an assessment of the overall strategy should focus on the extent to which these challenges are addressed. Having a congruent strategic plan and striving for a higher level of integration of electronic services are two additional indicators of a sound strategic plan.

The second component includes critical factors for the success of a national digital government strategy. These factors include strong political support, management of uncertainties, allocation of resources, development of management capacity, the use and quality of managed information, the development of human resources, an appropriate level of technology, and the protection of privacy and security. Each factor will be detailed in the following section.

Overall Strategies

A successful digital government strategy needs to have the following three characteristics. First, the strategy needs to address its unique environmental conditions as articulated in institutionalism and strategic planning. For instance, if a country is still behind in setting up the network infrastructure, then this should receive the highest priority. This is consistent with the strategic planning literature. The UN/ASPA survey of digital government around the world also supports the need for addressing its social, political, and economic composition.

Second, strategic goals, policies, and programs should be congruent with one another. The internal consistency between goals, policies, and program as three integral components of a strategic plan are the preconditions for successful implementation. For instance, a national objective of bringing access to all citizens should have policy and programs that address digital divide issues. The congruence can be seen when policies and programs support the outlined strategic goals.

Another common characteristic is the integration of various digital government infrastructure components and services. The integration of government networks and infrastructure is the foundation for seamless e-government services. Such integration usually involves a common platform for all government services and protocol for information and resource sharing. This is also a sign of maturity in the development of digital government (Layne & Kim, 2001).

Critical Success Factors

A good strategic plan addresses the constraints of political authority. A good plan mobilizes the support for digital government, particularly high level support. In addition, it will reflect citizen demands and other political considerations. Another critical success factor of a national strategic plan is the anticipation and management of political and technological uncertainties.

Government is subject to the turn-over of political appointees and elected officials. Changes in the course of policies are commonplace. A strategic plan needs to identify and anticipate those political uncertainties. Another source of uncertainty is technology. A leading technology today may be obsolete tomorrow. As a result, the cost projection for information technology may be incorrect.

Moreover, a strategic plan needs to allocate necessary resources to critical programs. One main challenge of digital government initiatives is finding the resources. A plan without committed resources is very likely to fail. One main resource that most countries are short of is skilled IT personnel. A strategic plan needs to address the recruiting and training of IT staff for critical programs.

Quality and security control of government information resources is at the center of any successful national digital government strategic plan. Ensuring the quality and security of information is the necessary groundwork for disseminating government information and providing government services online.

Rational decision-making and implementation capacity is central to a national strategic plan. This institutional capacity affects both the design and implementation of a digital government plan. This is the foundation for much successful deciionmaking at national and local levels. With this capacity, a national government is able to identify what is lacking in its political, economic/technical, and institutional conditions and develop a plan to address those.

Performance

The performance of digital government usually has multiple dimensions (see Figure 1). The main indicators include the availability of online services, the number of government publications online, the access to government databases for information retrieval, the level of security control, and the extent to which a national information and communication infrastructure is developed.

Online government services range from the simple provision of information to client-centered transaction environments. Most developing countries are capable of delivering static information and forms for a number of standard services over the Internet. Many governments are beginning to offer a limited range of services via the Internet. The most advanced online government portals provide services, such as filing for benefits and renewing identification cards, which are organized by clients' needs, rather than by organizational structures. This client-centered model for a transactional portal requires a high degree of both horizontal and vertical cooperation between agencies and levels of government, or an intermediary agent that aggregates disparate functions into a single interface. Public service portals and government information systems require a database containing information relevant to public services and functions, and these systems must ensure security and privacy for their users. These include such data as records of vehicles and insurance, social security, medical, and demographic information. One of the most difficult pieces in any information system is to ensure that it provides privacy and security to its users. Secure information systems require thorough investigation and testing at implementation and long-term maintenance.

In order for public information systems to be used by clients, and not just public administrators, a healthy telecommunications infrastructure must be in place to convey information to clients. In order to provide access to as many citizens as possible, many government plans emphasize the development of high-speed networks and public access points, as well as the stimulation of Internet service provider markets. These development plans include links to the global network infrastructure, fast connections between metropolitan area networks and research institutions, and the provision of network infrastructure to public places such as post offices and public transportations stations.

ILLUSTRATIVE CASES: A COMPARISON BETWEEN TAIWAN AND POLAND

The utility of the proposed framework (see Figure 1) is illustrated with the analysis of Taiwan's and Poland's digital government strategies. When guided by the framework, policy makers are able to identify the strengths and weaknesses of digital government strategies. The results of analysis are presented below with summary tables.

Poland and Taiwan were selected because each country is at a different point along the developmental curve. Selecting cases with significant differences in context and outcomes provide theoretical insights (Yin, 1994; Miles & Huberman, 1994). Moreover, it assists with the assessment of applicability of the framework in a wide variety of situations. By investigating two cases with different characteristics, this chapter identifies success factors that are common to both situations, investigates which factors seem to be isolated to specific developmental processes, and provide policy recommendations.

Performance of National Digital Government Strategies

The result of implementing existing digital government plans varies between Taiwan and Poland. According to a 2003 report comparing countries around the world (West, 2003), Taiwan was ranked number 5 among one hundred and sixty-nine countries surveyed in terms of its digital government performance. In contrast, Poland ranked 44 among them. Taiwan out performed Poland in all five main areas examined in the report: online services, online publications, use of databases, privacy policy, and security policy. Taiwan was particularly advanced in making publications online and employment of integrating databases for services.

Poland's progress in the implementing its plan has been slow, although some areas are showing success. The number of ministry, regional, and local government websites continues to grow. Many online services, such as tax services and an online system for entering complaints and requests for information, are still in pilot stages. However, databases and internal systems for bureaucratic functions are being introduced, and some programs to ensure the safe transmission of data are in place. A large part of Poland's plan focuses on the reengineering and replacement of old civil service information systems. Poland appears to be focusing on developing a strong internal structure before opening that structure to its citizens.

Analysis of Environmental Factors

Taiwan's overall political environment is in favor of a digital government strategy. Taiwan's political authority has recognized the importance of becoming a "green silicon island." The government's commitment to a digital island is seen in the emphasis it places in e-government as an important piece of its overarching national development strategic plan—the sixth national economic developmental plan (Lin, 2002). Top level government officials have touted the current achievement in the area of information technology and hope to sustain growth in this area. Citizens are used to the convenience provided by e-commerce. As a result, they tend to demand similar services from their government.

The information and communication infrastructure in Taiwan is relatively more mature than most of the emerging economies. In September 2000, Internet users have grown to 5.94 million, about 25 percent of the total population. The communication infrastructure has both domestic and international components. In December 2000, the connection between major cities in Taiwan has reached 45 Gbps, a 29000-fold increase over the 1.544 Mbps in April 1995. A direct international connection has been established between Taiwan's major trading partners such as United States, Japan, Hong Kong, and Singapore. In terms of broadband service coverage, ADSL services covered 97 percent of Taiwan in 2000.

Moreover, Taiwan has a vibrant IT manufacturing industry and aspires to further develop network related industries. In 2000, Taiwan was the fourth IT hardware manufacturing country in the world with the total value of approximately 23 billion dollars. Taiwan has over 60% of the global market share in products such as keyboards, hubs, and motherboards. The National Science Council, a Taiwanese counterpart to the National Science Foundation in the United States, has aimed to promote collaboration among industry, academia, and government to promote wireless communications and next generation Internet technologies (Yo et al., 2001).

For the past few years, Taiwan has made efforts toward formulating laws and policies in the areas of digital transactions and security. The legal framework governing electronic signature to facilitate electronic commerce and electronic transactions with government was in place before the year 2002. The revision of existing laws on digital property rights, regulations on fair trade in electronic commerce, protection of online consumer transactions, and online banking practices was planned to be underway in 2002 (Executive Yuan, 2001). Digital security efforts center on the establishment of new offices inside policy departments to handle computer crime and creation of a new reporting system. Specific attention is paid to the prevention of computer crime and problems with pornography (Yo et al. 2001). Privacy has not been specifically discussed as part of the e-Taiwan initiative.

Poland has only recently started to embark upon market and ICT infrastructure development, and requires extensive infrastructure development and education of public administrators and citizens in order to ensure the success of its initiatives. Poland is on the way to joining the European Union and needs to rapidly adopt EU practices and standards in order to become a member state. The political motivation for the implementation of a strategic development plan is strong. The ePolska plan also emphasizes the benefit of ICT for business and government effectiveness. At the same time Polish citizens are increasingly frustrated with government inefficiency and irrational processes, which can benefit (or possibly suffer) from the use of ICT and reengineering. In contrast to Taiwan, however, citizens are used to doing business personally and have little experience with government on the Internet. Some statistics find that only 4% of Polish Internet users access public administration websites (Taylor Nelson Sofres & Center for Public Opinion Research, 2002). Providing Polish citizens with channels to access government services must precede the adoption of those services.

Information technology infrastructure in Poland is still maturing, and many citizens are still without some means of Internet access. Currently about 2 million of Poland's 38 million citizens are listed as Internet subscribers (SourceOECD, 2001). Poland has one major telecommunications company, Telekomunikacja Polska S.A. (TPSA), which is undergoing a lengthy privatization and market rationalization process, and this company provides a subscription-free dial-up Internet service to every telephone client, as well as subscription-based ISDN and broadband services. Broadband subscriptions are much less frequent, around 1000 subscribers (SourceOECD, 2001). Poland's metropolitan area networks are connected by a single (originally academic) network, and there are three connections to the European backbone. Poland's information and communications technology industry is small; there are few Polish companies producing new technology. The relatively high price of computers has kept computer ownership low: twenty-three percent of households have computers, and computer ownership is concentrated in the 25% of households with the highest income (Main Statistical Office, 2002).

The Polish government is struggling to improve its practices and make administrative procedures more rational, as well as to maintain citizens' privacy and security. Laws against unauthorized access and computer and telecommunications fraud are part of the criminal code, although police forces may not be well equipped to deal with computer crime. Ensuring identity and reliability of documents on the Internet in order to regulate e-commerce has been a concern for some time, generating a project to create an electronic signature that has been working for several years. The Ministry of Internal Affairs and Administration is working to rationalize Poland's personal identification system. Implementation of policy initiatives has been difficult so far; Poland's scarce resources have limited the introduction of new programs, although a number of plans have been drafted. Engendering trust in these new programs may be difficult, even if Poland succeeds in providing access for all citizens. Public trust in the government is low, and government services are often seen as inefficient and operating without rationality, and new initiatives may not be better. As an example, a program to allow business owners to report incomc tax with their computer met with difficulties that severely limited the program's efficacy. Digital government offers Poland the chance to rationalize administrative decision-making and make government services easier to deal with by eliminating problems such as multiple collections of data and paper-based archives.

Assessing Digital Government Strategies

Both countries address three overall strategy areas well. Both the ePolska and e-Taiwan plans fit well with external concerns, fitting their e-government development plan well into an overall strategy. Both plans also clearly delineate relationships between goals and policy measures, and initiatives clearly move towards overarching goals. Taiwan and Poland are also able to integrate those measures effectively. Multiple initiatives are dependent upon each other for eventual success.

Poland and Taiwan both seem to be facing the challenges effectively, but both development plans have weaknesses that require attention. The two countries focus on network infrastructure and have a strong commitment to address universal access issues, as well as high-level support and standardization. However, both plans require extensive interagency cooperation, which is often difficult to secure. Poland and Taiwan also score low in data quality and training areas, as well as in the area of security. In many of the areas Taiwan performs better, owing to a strong existing infrastructure and support for the measures in the plan.

Following the proposed framework, the analysis of digital government strategies includes two phases. The first phase involves the overall fitness of a digital government strategy to address external challenges, the congruence

Table 1. Strengths and weaknesses of the overall digital government strategies of e-Polska (2003) and e-Taiwan (2001)

	Poland		**Taiwan**	
Criteria	**Strengths**	**Weaknesses**	Strengths	**Weaknesses**
Meeting external challenges	*catch up to EU standards * addressing digital divide	* goals are set too high, timelines too short	* efficiency * addresses legal and institutional issues * G2G, G2B, G2C * use of multiple channels * strong ICT infrastructure and IT industry * addressing digital divide	* unable to scan external changes
Congruence between goals, policy, and programs	*Three long range goals * aligned with eEurope initiatives * creating standards	* large infrastructure investment * lack of reengineering	* clear overall strategies * clear measurable goals * clear role of common platform * process reengineering	* lack of training
Level of integration	*Portal (Wrota Polski) * Central Administrative Database * unification of identification schemes	* Agency cooperation difficult * Inertia	* portal * common service platform * one-stop services	* lack of organizational strategy * lack of resource commitment

between strategic goals, policy, and programs, as well as the level of integration among different components of the strategy. Table 1 illustrates the main findings. The second phase discusses the strengths and weaknesses of specific areas of the digital government strategies by comparing Taiwan with Poland. A summary table for this section is Table 2.

Assessing the Overall Digital Government Strategy

In addressing external concerns, Poland ranks slightly behind Taiwan. Each country stresses competition as a driving force behind improvement, as well as digital divide issues, and both focus on relationships between government, business, and citizens (G2G, G2B, and G2C). Each country's plan recognizes the gains in efficiency and efficacy that e-government offers.

The Polish strategy notes several gaps in ownership access equipment between Poland and its neighbors (ePolska, 2003) and accordingly sets goals for increasing citizens' access. Measures for success in the ePolska plan include the number of stationary and mobile phones, computers, and the number of homes with broadband access. Access-related goals also include the construction of public Internet kiosks and provisioning schools with computers and high-speed Internet access, as well as the creation of a tax credit or other economic incentive for computer owners. Another part of the plan calls for improving Poland's telecommunications market. By stimulating price-and service-based competition, the plan aims to improve Internet access prices and service. Poland's initiatives also include a number of public-private partnerships with foreign firms, especially in education.

Impending EU accession is a driving factor in Poland's development plans and serves to tie together a number of initiatives, in order to bring Poland into compliance with EU requirements, as well as to prepare for participation in the Schengen Information System. Poland's portal will include a "Polish Gateway to Europe" which will increase the efficiency of requests for EU financial aid both before and after accession. The structure of Poland's plan follows the foundation laid out in the eEurope plan, dividing digital government development into providing access for all, training and education of citizens and public administrators, and the creation of resources and applications.

However, Poland's strategy has a two-fold weakness, which gives rise to its relatively poor fit between government strategy and external environment. First, the goal set for building communication infrastructure and improving network access is overly aggressive since most of the goals are modeled after eEurope without giving consideration to feasibility. Second, Poland seems to rely heavily on technological solutions when human capital should receive more attention. For example, there is an emphasis on telework and telemedicine while the fundamental issue of training of personnel is not properly addressed.

In contrast, Taiwan scores relatively high in addressing its environmental conditions. Its assessment of the feasibility of its digital government plan seems to be reasonable. This is evident in the achievement of key objectives ahead of schedule in the first three years of e-government efforts from 1997-2000. Moreover, the e-government strategy included partnership with Taiwan's vibrant IT manufacturing industry to quickly establish and upgrade the ICT infrastructure. In its e-government plan, Taiwan understands the need for multiple channels of service delivery to address the problem of the digital divide. For instance, in the area of health care service, electronic service is integrated into a broader strategy of service delivery rather than being the only service delivery channel. The e-Taiwan plan also covers some aspects of the legal and institutional framework such as electronic transactions and digital signatures necessary for e-government to function.

Nonetheless, Taiwan's digital government strategy also has its weaknesses. The common problem that Taiwan shares with Poland is the lack of a proactive mechanism for environmental scanning. Due to fast-changing technology and political conditions, a good strategic plan must constantly scan changes in environment and make adjustments accordingly. However, no section of the digital government plan has been devoted to that.

With regard to each country's plans in terms of long-term strategic goals, the two countries score close to each other. Each country defines clear strategic goals in its plan, and Poland's plan identifies clear goals and their relation to wider targets. The ePolska plan also calls for the development of overarching standards, such as common file formats, applications, protocols, and a system for document exchange. The primary goal of Taiwan's e-government plan is to provide seamless online service to citizens, businesses, and government employees. Specific policies are formulated to achieve a secure and reliable information exchange environment, reduction in paperwork, and online access for all government units. This is supported by programs for establishing a government-wide service network with a common platform supporting various public service applications, setting web sites for all national agencies, putting application forms online, and e-routing for government (Research Development and Evaluation Commission, 2001).

Both countries still have some flaws in the area of strategic goals that prevent them from receiving high marks. As part of its preparation for EU membership, Poland is reforming a number of other areas as well, from agriculture to public health to transportation. Financial and administrative support for IT reforms may be overshadowed by demands in other areas. In order to marshal the funds necessary to implement the ePolska plan, leaders will have to ensure that there is enough impetus year-by-year to make sure that continuous funding supports plan initiatives, rather than leaving them without financial support in the middle of implementation.

Other areas in the framework are also lacking. Process reengineering is not mentioned in the Polish plan, and this could be fatal to the execution of the plan as a whole if public administrators hold to the old way of conducting business, rather than adopting new practices. Taiwan's challenges include the deficit problem that limits the resources available for e-government projects. Another related documented deficiency is the lack of training for government employees to establish web sites and move services online.

Turning to the integration of different components of a national digital government strategy, both countries have a modest plan. The key element of their plan is a unified portal for their services. The ePolska plan calls for the creation of a "Gateway to Poland" which will include services and content for Polish citizens as well as the goal of promoting Poland to other countries. Poland will create a unified data model, which will ensure that databases for personal identification, vehicle registration, and taxes are interoperable and rational. Poland also proposes the unification of its various identification schemes. These schemes are strongly ingrained at the organizational level and replacing multiple systems will prove to be a challenge.

Taiwan has identified one-stop shopping in the form of a portal as an objective in its strategy. This portal is supported by a common service platform that integrates all government services. In the year 2000, 61 percent of government offices had local area networks to connect computers at the office. Seventy-four percent of government offices had Internet connections while 47 percent of total government offices had a web site. These numbers are expected to reach 100 percent by the end of 2002, according to Taiwan's e-government plan (Research Development and Evaluation Commission, 2001).

Both countries are likely to meet problems with interagency cooperation. Interagency cooperation has been identified as one main barrier to seamless e-government. In Taiwan's digital government strategy, there is a lack of specific reference to any systematic way of addressing potential barriers to interagency sharing of resources and information. A council at the cabinet level is required to facilitate interagency cooperation. However, no acknowledgement of the need for a high-level coordinating agency is in the plan and a description of such a coordinating body is not yet available. The ePolska plan also lacks a high-level coordinating agency to oversee collaboration among various agencies to deliver seamless e-government services.

Individual Critical Success Factors

Concerning the success factors that we have identified in the framework, Taiwan seems to be ahead in several areas, but both strategies have

Table 2. Strengths and weaknesses of the specific areas of digital government strategies of e-Polska (2003) and e-Taiwan (2001)

	Poland		**Taiwan**	
Specific Areas	**Strengths**	**Weaknesses**	Strengths	Weaknesses
Political commitment	* Strong support * Reorganization	* Low funding, * Lack of follow-through	* Strong support	
Uncertainty management	* Leaves technology choices open * Does not dictate broad solutions	* Vulnerable to uncertainty		* Vulnerable to change
Resource allocation		* Funds not allocated	* Large allocation of funds	
Management capacity	*Periodic evaluation	*No oversight	* Tied to NICI * Periodic evaluation	
Information management (use & quality)	*Creating information resources * Standards and common systems	*No use of existing information	* Creating information resources * Use of existing information	* Lack of quality assurance
Human resources	*Training for disadvantaged groups	* lack of trained professionals * Lack of training programs	* Training emphasized in all areas	* Lack of training of government officials
ICT Technology	*Focus on disadvantaged groups * efforts to broadband access	* Market rationalization far off	* maturity of network technology * broadband access	
Security	* Some training * Emergency response	* Lack of standards * still under development	* PKI * Vulnerability scanning * Emergency response * Training	* Still developing

some common weak points. These critical success factors and their assessment for Taiwan and Poland are summarized in Table 2. Each factor will be discussed separately.

Both strategies have a fair amount of political commitment behind them. Taiwan's e-government strategy is part of digital Taiwan, a major component of the national development plan, the blue-print for Taiwan, which is evidence of major political recognition of the importance of having a digital Taiwan. Implementation of the plan will come directly from Taiwan's Executive Yuan, the primary executive branch of government. Another sign of support is numerous statements on the important role that information and communication will play in securing Taiwan's competitiveness in the global market.

Poland's plan comes from the State Committee for Scientific Research, although it has some high-level backing and input from important ministries. Unfortunately, initiatives in the Polish strategy may not be not followed through, and a lack of resources has severely reduced the number of initiatives implemented on time. The ePolska plan outlines agencies responsible for the execution of plan initiatives as well as target completion dates, but it does not define sources for funding of individual initiatives and it appears that government branches are responsible for securing their own funding for initiatives.

Neither plan seems to adequately prepare for unexpected developments. Digital government needs to be particularly sensitive to technological changes and possible social changes. In a strategic plan, these changes should be considered as risks and a plan to manage risks should subsequently be in place. Poland makes no mention of any ways to adapt to possible technological change or change in political regime. In its e-government plan Taiwan also makes no reference to possible risks and the management of those risks. As a result, both countries score low in risk management.

Taiwan performs better in resource allocation. Strong resource commitment can be seen in the NT $36.6 billions (approximately one billion US dollars) allocated for e-Taiwan over 2003-2008 under the framework of the six-year development plan. Institutional resources are also available. For instance, the National Information and Communication Initiative (NICI) was formed to outline the blueprint for e-Taiwan with e-government as a major component. In contrast, Poland's economy is going through a recession and government resources are scarce. Implementation of individual measures requires the presence of a strong leader to gather and protect funds in an environment where resources are often contested. Funding sources are not identified for individual initiatives in the Polish plan, nor does there appear to be a funding source for the plan overall.

In terms of management capacity, Taiwan's strategy seems to be well-grounded. The e-Taiwan plan undergoes periodic evaluation, and planning for the strategy was executed under the oversight of the National Information and Communication Initiative (NICI). The existence of technocrats who serve a rational decision-making capacity has been documented. Further evidence of management capacity is the comprehensive and specific nature of Taiwan's e-government plan. Specific measures are used to gauge the progress of e-government. For instance, according to the plan, the use of e-mail by government employees needs to reach 100 percent by 2003, compared to 33 percent use in 1999. Poland's plan also has a periodic evaluation which sets multiple milestones for completion and also outlines the responsible agency or office for each initiative in the plan. The measures for success are open-ended, and there are no set levels to be reached for each measure. For example, success is measured by the number of homes with broadband connections, but there is no number or percentage that represents success or failure of initiatives related to providing access to citizens. In addition, there is no facility for oversight by either government or citizens, or for assessment by external bodies such as the European Union.

With regard to information resource management, the two countries have different problems and goals. Taiwan concentrates on the need for more information resources in its own language as well as the use of existing informational resources. As stipulated in the e-government plan (Research Development and Evaluation Commission, 2001), content needs to be further developed to meet the needs of citizens and businesses. Another area of information resource management emphasized in Taiwan is the use of geographic information systems and health information systems. This is accomplished by information sharing between governments. However, one weakness of the current information resource management effort is the lack of a concerted effort in ensuring information quality. Since information quality is the most fundamental issue for information resources, missing it has serious consequences.

In contrast, the Polish strategy focuses on the creation of content and new resources, as well moving the flow of information from paper to computer for initiatives such as a Polish Internet Library, medical databases, or encouraging electronic trade. The ePolska plan also mandates the creation of standards for data quality and a national data model with majority of public bodies adopting the data model as one of the measures for success. Poland makes no mention of converting existing older records to an electronic format, but initiatives to create an electronic document exchange service for government bodies are included in the plan.

Both Taiwanese and Polish plans mention a lack of trained professionals and the difficulty of training citizens to ensure that they are able to use new services. The ePolska plan mandates a minimum level of computer and Internet skills for K-12 students, certifiable by examination. The Polish plan also encourages cooperative partnerships between schools and firms such as Cisco and Intel, and emphasizes the educational use of IT in all subject areas, rather than just in computer science classes. The Polish plan addresses issues of access by creating training programs for adults and those outside the school system, including special programs for the unemployed, disabled and those in rural areas.

Taiwan's e-government plan has recognized that the lack of training for government employees was one main drawback of the first phase of the e-government effort from 1998-2000. To address such a deficiency, the second phase of e-government effort from 2001-2004 has proposed to include the use of online e-government transactions as part of performance measures for government employees. Training and e-learning are also part of the overall strategy. The overall e-Taiwan plan has proposed a "Knowledge and Learning Logistic Center" that helps improve both quality and access to knowledge. This begins by addressing the need of various industries and then extends to the general public for a range of educational opportunities.

Taiwan is much further ahead in the use and application of ICT, and demonstrates an excellent understanding of implementing and managing networks. Taiwan has emphasized having a broadband connection to 6 million households in an island of 22 million people. To increase its capacity for managing domains on the Internet, Taiwan has also proposed to establish an IPv6 framework. In government, a common communication platform has been developed called the government service network and this lays the foundation for information exchange and seamless service delivery. The Polish plan identifies a number of essential factors for reaching a similar level of development. The Polish telecommunications market requires strong intervention in order to ensure that businesses and citizens get the level of access they require at fair prices, and the ePolska plan details a number of milestones on the way to providing broadband access to schools and citizens. ePolska also focuses on ways to provide Internet access to the disadvantaged. There is also an initiative to create PIONIER, a high-speed optical academic network linking research universities, which will increase linkages across the country.

Security is an area that both Taiwan and Poland emphasize. Currently, Taiwan has done a better job than Poland in addressing security issues for its digital government. Taiwan has set out to protect its critical information infrastructure. For example, a center has been set up to track computer crimes. Another main initiative is the use of a Public Key Infrastructure. Government has planned to have security policy and measures in place for at least 50 percent of main government offices by 2007. Training for computer security should reach more than 5 percent of government employees during the same period. However, the goals do not seem to be aggressive enough to address the potential threats properly. The ePolska plan focuses on the propagation of best practices, assuming that users will implement measures on their own. The ePolska plan also places emphasis on the protection of intellectual property. The Polish government intends to implement multiple security levels for communication, such as authentication and digital signatures, as well as an advanced authentication method for every citizen to use: the Integrated Packet of Personal Documents. Finally, ePolska calls for the proper qualification of judges and legal staff in order to prosecute electronic crime effectively.

Between the two countries, Taiwan appears to be in a stronger position, with good political and financial support for the plan as well as sound initiatives. Poland's plan is well in line with the objectives outlined by eEurope, but lacks the financial support of Taiwan's plan. Both ePolska and eTaiwan have well-reasoned initiatives that span a number of developmental dimensions, and concentrate on areas that are central to continuing long-term development. Viewing each country through the framework outlined in this chapter allows comparison of two development strategies that at first glance may seem wildly disparate. With Taiwan's existing advanced infrastructure and strong political and financial commitment to the plan, it might seem that Taiwan's plan is far ahead of where Poland could hope to be. By examining the two plans through the analytical framework, however, areas where the two plans diverge and correspond can be seen. Both plans could use more risk management, as well as increased training and security measures, and both plans have strengths in the area of information management, management capacity, security and use of technology. Each country can benefit from the other's strengths by incorporating initiatives that the other plan contains, and other national digital government strategies can also act on the differences seen here to ensure success.

DEVELOPMENT AND TRENDS

Examining these development strategies yields a number of implications for public administrators in other countries. In order for a country to see the benefits offered by digital government, development must be coordinated strategically with broad goals outlined in a plan. Both Poland and Taiwan's development plans make use of initiatives that cut across agencies and depend upon successive stages. A strategic plan allows a government to pick out large scale goals and outline the steps required to reach that goal, which may take a number of years. Without such a plan, implementation of individual measures will occur haphazardly, without coordination of efforts. Resource allocation will be carried out on a case-by-case basis, resulting in gaps in implementation. Common data models and application models are impossible to implement without some top-level planning.

As these plans and others like them are implemented, a number of concerns still remain. Security and privacy will continue to be important for any government activity, and needs for secure systems and private means of communication will only increase. A system designed to be secure today will undoubtedly have some vulnerabilities months or years down the road. Secure systems and privacy safeguards are a matter of constant maintenance and testing, as no program, tool, or protocol is perfect, and often only time brings security holes to light. In order for digital government systems to be secure, these systems will not only need to be built with security in mind to protect against the exploits used today, but also constantly updated and audited to defend against future attacks.

Another concern unlikely to disappear is the issue of the "digital divide". Despite countries' best efforts to provide as many access channels as possible for citizens, some citizens will remain out of reach, either because they are somehow denied access by disability or situation, or because they choose traditional means of interacting with government. Governments have a responsibility to constantly improve accessibility for the disadvantaged, provide training for citizens to allow them to effectively use access channels, and create systems that are both effective and easy to use. One way of addressing the digital divide is to make services easy, effective, and appealing enough to encourage citizens to seek out the means to use them.

In order for services to become and remain effective, government institutions will need to be constructed and maintained to reflect the current environment. Governments need up-to-date privacy and security laws, as well as legal professionals capable of making informed decisions. By creating a body of policy and law that is conversant with technological issues, governments need to increase their ability to adapt to changes and be more prepared for future digital government initiatives. Having strong and current institutions allows a government to support continuous reform. Governments that do not have institutions that take technological change into account will be forced to reengineer these institutions in order to move forward.

In keeping with the idea of maintenance, both of security and government institutions, government entities need to maintain flexibility with regard to their environment and act correspondingly. Constant assessment of changing capacity, such as resources, expertise, and time, which dictate the courses of action open to public administrators, is necessary to be prepared for changes. While implementing digital government initiatives is an exercise in closing the gaps between the current and desired situations, it is possible for these gaps to change. In order to be sure that initiatives are successfully implemented, public administrators must therefore constantly scan the environment as well as internal capabilities in order to see what changes may be necessary. It is critical for a strategy to be flexible in order to adapt initiatives to best fit changing gaps.

Further research can improve this analytical framework in a number of ways. This examination of two countries is rather limited, and a multi-country study would undoubtedly refine the elements of the framework, and highlight some areas that have been left unexplored. The choice of Poland and Taiwan uses subjects that are quite distant, in terms of culture, development, and geography, and a regional study of Asian states or Central European states may serve to better ground the framework with a range of subjects that are closer together, and improve the rating system for each of the framework's elements. Increasing the number of subjects for study will also bring greater variety to the elements of the framework, and can serve as a cross-pollination of initiatives, as similar situations and success factors for more cases are compared.

A future study might follow digital government strategies in a longitudinal study of plans from drafting to implementation to completion and maintenance. Such a view of strategies put into action can expose factors that the plans themselves do not contain, such as social networks that cause governments to behave differently than expected, or non-rational behavior that obstructs implementation, or problems related to the public-private partnership aspects that are part of many strategies. Long-term studies will also serve to better identify which factors are most important for success. Furthermore, the contrast between issues in building initial capacity for putting strategy into place versus flexibility in adapting to environmental changes may be highlighted, showing if governments with more capacity or more flexibility are more successful in implementing strategy.

CONCLUSION AND RECOMMENDATIONS

The framework of national digital government strategy is useful in incorporating various building blocks for the success of digital government. This framework addresses the environmental conditions necessary for various advances in digital government, including political, institutional, technical, and economic ones. A less inclusive approach is likely to cause problems. Moreover, this proposed framework provides a list of key factors for a successful digital government endeavor. As the above comparison has demonstrated, if these factors are not attended to at the national level, implementation and performance are likely to be problematic.

Moreover, the proposed framework offers several general guidelines for developing national digital government strategies. First, the broad political, technical, economic, and institutional conditions are of critical importance for the development of a digital government strategy. Second, it is the role of a national digital government strategy to address its unique setting and give priority to its areas of critical need. Third, various components of a national digital government strategy should complement each other. One productive way of looking at the integration and interoperability is the portal concept and the network infrastructure that supports it.

However, it should be acknowledged that this framework does have some limitations. The framework could be complex and the interactions between its various elements may not be fully understood. Moreover, the policy recommendations that it provides are largely normative in nature. This framework, although aiming to be comprehensive, may not be able to account for all possible initiatives embodied in existing digital government plans.

The analysis of Taiwan's and Poland's digital government strategies has yielded specific recommendations for both countries. For Poland, more emphasis needs to be placed on the development of information and communication infrastructure. More reasonable goals need to be set to take into account its resource issues. Poland also needs stronger political support for the implementation of its ePolska plan. This political support may come in the form of establishing a higher central office to oversee the ePolska plan or in the form of more resource allocation. Management of information and risk is another area for potential improvement for Poland. The conversion of existing information resources to an electronic format would be useful. A risk management plan that anticipates changes in technology and political and social conditions is also critical. Training and security are another two areas for Poland to improve upon. Currently, training is one of the main barriers to establishing and utilizing e-government services. Training should be conducted for users in the rural areas as well as government employees. Security is in its infancy for Poland. Specific measures need to be taken to further develop this area of capacity.

Taiwan is lacking in the mechanism to conduct a scan of environmental factors as outlined in the proposed framework. This is critical for the government's ability to align its overall strategy to its environment. A specific section on building this mechanism should be included in the next version of the plan. The other area that needs further emphasis is interagency cooperation. This is at least one factor that needs to be explicitly dealt with. The current e-government plan is limited in its ways of addressing interagency cooperation when turf fighting is likely to become a barrier to information exchange and resource sharing. Risk management capacity also needs to be further developed for Taiwan. In the Taiwanese e-government plan, there seems to be no explicit mentioning of any particular risk. The discussion has only focused on lessons learned from the first phase of e-government. The training of both government employees and the general public needs to be strengthened. Some progress has been made, but the real implementation may be challenging. The last area for improvement is security. In particular, security training seems to be the weakest link of the comprehensive security policy. Until 2007, only less than 5 percent of government employees are expected to get computer security training.

This chapter is the first significant step to formulate a comprehensive framework for the development and implementation of national digital government strategic plans. The knowledge of such a comprehensive framework can be further advanced in two ways. First, the proposed framework can be applied to more countries to test its comprehensiveness and utility. Second, more research can be done to link a national strategy to e-government performance. As a result, a more comprehensive framework will provide even more detailed guidance for the development and implementation of national digital government strategies.

REFERENCES

Bozeman, B., & Bretschneider, S. (1986). Public management information systems: Theory and prescription. *Public Administration Review, 46* (Special Issue), 475-487.

Bryson, J. M., & Alston, F. K. (1996). *Creating and implementing your strategic plan: A workbook for public and nonprofit organizations.* San Francisco: Jossey-Bass Publishers.

Caudle, S. L., Gorr, W. L., & Newcomer, K.E. (1991). Key information systems management issues for the public sector. *MIS Quarterly, 15*(2), 171-188.

Dawes S., Pardo T., Green D., Mclnemey C., Connelly D., & DiCatireno A. (1997). *Tying a sensible knot: A practical guide to state-local information systems.* Albany, NY: Center for Technology in Government.

Edmiston, K. (2003). State and local e-Government: Prospects and challenges. *American Review of Public Administration, 33*(1), 20-45.

Executive Yuan. (2001, December 26). *National Information and Communication Infrastructure Development Plan.* Taipei, Taiwan: Executive Yuan.

Ferris, J. M., & Tang, S.-Y. (1993). The new institutionalism and public administration: An overview. *Journal of Public Administration Research and Theory, 3*(1), 4-10.

Fletcher, P. D. (2003). The realities of the paperwork reduction Act of 1995: A government-wide strategy for information resources management. In D. Garson (Ed.), *Public information technology: Policy and management issues* (pp. 74-93). Hershey, PA: Idea Group Publishing.

Fletcher, P. D. (1999). Strategic planning for information technology management in state government. In G. D. Garson (Ed.), *Information technology and computer applications in public administration: Issues and trends* (pp. 81-98). Hershey, PA: Idea Group Publishing.

Fountain, J. (2001). *Building the virtual state: Information technology and institutional change.* Washington, DC: Brookings Institution Press.

General Accounting Office. (2001, July 11). *Electronic government: Challenges must be addressed with effective leadership and management.* Washington, DC: General Accounting Office. GAO-01-959T.

Guy, M. E. (2000). Public management. In J. M. Shafritz (Ed.), *Defining public administration*. (pp. 166-168). Boulder, CO: Westview Press.

Heeks, R. (1999). *Reinventing government in the information age: International practice in IT-enabled public sector reform*. London; New York: Routledge.

International City/County Management Association. (2001). 2000 Electronic Government Survey Dataset [Data file]. Washington, DC: International City/County Management Association.

International City/County Management Association. (2002, August). 2002 Electronic Government Survey Dataset. Washington, DC: International City/County Management Association.

Kamarch, E. C., & Nye, J. S. (2002). *Governance.com: Democracy in the information age*. Cambridge, MA: Visions of Governance in the 21st Century Brookings Institution Press.

Landsbergen, D. Jr. G. Wolken. (2001). Realizing the promise: Government information systems and the fourth generation of information technology. *Public Administration Review, 61*(2), 206-220.

Layne, K., & Lee, J. (2001). Developing fully functional e-Government: A four stage model. *Government Information Quarterly, 18*(2), 122-136.

Lin, K.H. (2002, June 5). Cabinet to pour NT$36 billion into 'e-Taiwan' Project. *Taiwan News*.

Main Statistical Office. (2002). Household situation in 2002 From the Results of the Household Budget Survey. Retrieved July 15, 2003, from http://www.stat.gov.pl/serwis/nieregularne/gosp_dom2002.htm

Marchionini, G., Samet, H., & Brandt, L. (2003). Digital government. *Communications of the ACM, 46*(1), 25-27.

Miles, M., & M. Huberman. (1994). *Qualitative data analysis*. Thousand Oaks, CA: SAGE Publications.

Mintzberg, H., & Quinn, J. B. (1991). *The strategy process*. Englewood Cliffs, NJ: Prentice-Hall.

North, D. C. (1990). *Institutions, institutional change, and economic performance*. Cambridge, MA: Cambridge University Press.

Ostrom, E. (1990). *Governing the commons: The evolution of institutions for collective action*. Cambridge: Cambridge University Press.

Ostrom, E., Schroeder, L., & Wynne, S. (1993). Analyzing the performance of alternative institutional arrangements for sustaining rural infrastructure in developing countries. *Journal of Public Administration Research and Theory, 3*(1), 11-45.

Research Development and Evaluation Commission (RDEC). (2001). *E-government implementation plan*. Taipei, Taiwan: Research Development and Evaluation Commission, Executive Yuan.

Rocheleau, B. (2003). Politics, accountability, and governmental information systems. In D. Garson (Ed.), *Public information technology: Policy and management issues* (pp. 20-52). Hershey: Idea Group Publishing.

Schedler, K., & Scharg, M.C. (2002). *Exploring the interrelations between electronic government and the new public management: A managerial framework for electronic government*. Paper presented at APPAM 2002 Conference.

Snellen, I. (2000). Public service in an information society. In G. Peters & D. J. Savoie (Eds.), *Governance in the twenty-first century: Revitalizing the public service*. Montreal & Kingston: Canadian Centre for Management Development.

SourceOECD Telecommunications Database. (n.d). Retrieved July 15, 2003, from http://www.

sourceoecd.org/content/templates/el/el_location_ivt.htm?comm=telecomm

State Committee for Scientific Research. (2003). Strategy for Information Technology Development of the Republic of Poland – ePolska.

Taylor Nelson Sofres & Center for Public Opinion Research. (2002). *Government online: A national perspective*. Annual Country Report: Poland.

United Nations. DPEPA & ASPA. (2002). *Benchmarking e-government: A global perspective: assessing the progress of the UN member states*. New York: United Nations American Society for Public Administration.

West, D. (2003). *Global e-Government.* Providence, RI: report published by author.

Yin, R. (2003). *Case study research: Design and methods* (3rd ed.) Thousand Oak, CA: SAGE Publications.

Yo, C. -D., Tsai, W. -C., & Tien, D. (2001). National Information Infrastructure Development in R.O.C. Industrial Forum.

ENDNOTES

1 This definition is adopted from the General Accounting Office's (2001) report on e-government.

2 Institutions are sets of rules that constrain and guide behaviors (North, 1990; Ostrom, 1990). These rules manifest themselves in laws, regulations, governing structures, patterns of behavior, etc.

This work was previously published in Electronic Government Strategies and Implementations, edited by W. Huang, K. Siau, & K.K. Wei, pp. 394-420, copyright 2005 by IGI Publishing, formerly Idea Group Publishing (an imprint of IGI Global).

Compilation of References

(EU) Commission Regulation (EC) No874/2004 of 28 April 2004 (OJ L 162 30.04.2004)

(EU) Council Regulation (EC) No 44/2001 of 22 December 2000 on jurisdiction and the recognition and enforcement of judgments in civil and commercial matters

(EU) Directive 98/34/EC

(EU) Directive on Electronic Commerce 2000 (Directive 2000/311EC of the European Parliament and Council)

(EU) European Brussels Convention on Jurisdiction and Recognition and Enforcements of Judgments (1968).

(EU) Regulation (EC) No733/2002 of 22 April 2002, of the European Parliament and of the Council (OJ L 113, 30.04.2002)

(UK) Defamation Act 1996

(UK) Electronic Commerce (EC Directive) Regulations 2000

(US) Digital Millenium Copyright Act (DMCA), 1998

(US) Online Copyright Infringement Liability Limitation Act of 1998

Accounting Education Change Commission (1990). *Objectives of education for accountants (Position Statement Number 1)*. Bainbridge Island:AECC.

Adamsson, V., & Puukka, J. (2004). Ihme, Yrittäjätarinoita Loimaan seudulta. Turun ammattikorkeakoulun oppimateriaaleja 13. Turku.

Adera, E., & Camara, A. (2003). Opportunities and Challenges for Community Development. Preface. In Ramata Malo Thioune (Ed.), *Information and Communication Technologies for Development in Africa: Vol. 1*. International Development Research Centre, Canada, 2003. ISBN 2-86978-114-8.

Ahituv, N., Neumann, S., & Zviran, M. (1989). Factors affecting the policy for distributing computing resources, *MIS Quarterly, 13*(4), 389-401.

AIMC (2006). *Estudios General de Medios*. Madrid: AIMC.

Alabau, A. (2005). The European Union and its e-government development policy. *Vodafone Spain Foundation*, (p.17).

Alajoutsijärvi, K., Möller, K., & Rosenbröijer, C.J. (1999). Relevance of focal nets in understanding the dynamics of business relationships. *Journal of Business-to-Business Marketing, 6*(3), 3-35.

Alter, A. (2001). Thinking out loud: Interview with JoAnn Heissen, *CIO INSIGHT*, December.

American Society of Association Executives Foundation (2001). *People make it work: The social fabric of community on the Web*. Retrieved April 20, 2006, from http://ww.asaenet/org/foundation

Amit, R., & Zott, Ch. (2001). Value creation in e-business. *Strategic Management Journal, 22*, 493-520.

Amutio-Gómez, M.A. (2003). The construction of a Pan-European e-government: The IDA Programme. *UPGRADE, the European Journal for the Informatics Professional, 4*(2), 52-62.

Anand. A., & Uppal, M. (2002). *Engendering management and regulation of ICTs*. Prepared for the UN/IN-

STRAW Virtual Seminar Series on Gender and ICTs Seminar Two: Women and ICTs Enabling and Disabling Environments. July 15-26, 2002. Retrieved July 26, 2002, from http://www.uninstraw.org/en/research/gender_and_ict/virtual_seminars.html

Anderson v. New York Telephone Co., 35 NY 2 d 746 (NY,1974).

Anderson, A.R., & Jack, S.L. (2002). The articulation of social capital in entrepreneurial networks: A glue or lubricant? *Entrepreneurship and Regional Development, 14*, 193-210.

Anderson, D.L., & Lee H.L. (2000). The Internet-enabled supply chain: From the first click to the last mile. *Achieving supply chain excellence through technology* (pp. 15-22). Montgomery Research Inc.

André and Jean-Pierre Vasarhelyi v. Guiness World Records Ltd., No. 01/19480 Tribunal de Grande Instance de Paris [2003] ECDR 14.

Annan, K. (2000). We the peoples: The role of the United Nations in the 21st century. *Millennium Report of the Secretary General of the United Nations.* Retrieved April 20, 2006, from http://www.un.org/millennium/sg/report/summ.htm

Antonia-Belika Koubareli v. Evangelos Volotas and Forthnet S.A. No. 1639/2001, Athens Court of First Instance ECDR 19 (2003).

Anyaegbunum, Mefalopulus, & Moetsabi(1998). Participatory Rural Communication Appraisal Methodology (PRCA): The application of participatory approaches to communication programme design and implementation for sustainable human development. In D. Richardson & L. Paisley (Eds.), *The first mile of connectivity*. Rome: FAO.

Argyris, C. (1980). Some limitations of the case method: Experiences in a management development programme. *Academy of Management Review, 5*(2), 21-34.

Arun, S., Heeks, R., & Morgan, S. (2004). *Researching ICT-based enterprise for women in developing countries: A livelihoods perspective.* Women's ICT-Based Enterprise for Development project. Manchester, UK: IDPM, University of Manchester. Retrieved June 2006, from http://www.womenictenterprise.org

Ashby, W. R. (1956). *Introduction to cybernetics.* New York, John Wiley & Sons.

Aslin, H.J., & Brown, V.A. (2004). *Towards whole of community engagement.* Canberra: ACT Murray-Darling Basin Commission.

Avgerou, C., & Walsham, G. (Eds.). (2000). *Information technology in context: Studies from the perspective of developing countries.* London, Ashgate (2000) ISBN 0 7546 1480. From: Sein, M. K., & Harindranath, G. (2004). Conceptualizing the ICT artifact: Towards understanding the role of ICT in national development. *The Information Society, 20*, 15-24.

Bagla, P. (2005). Booming computer sector seen as a mixed blessing. *Science, 310*, 1754. Retrieved from www.sciencemag.org

Baharul Islam, K. M. (2003, May 6). Information age government: Success stories of online land records & revenue governance from India. *Executive Summary, United Nations Economic and Social Council.*

Baker R.E., Simon J.R., & Bazeli, F.P. (1986). An assessment of learning style preferences of accounting managers. *Issues in Accounting Education, 1*(1), 1-12.

Bakos, Y. (1998). Towards friction-free markets: The emerging role of electronic marketplaces on the Internet. *Communications of the ACM. 41*(8), 35-42.

Balit, S. (1999). *Voices for change: Rural women and communication.* Rome: FAO.

Banta, T., & Associates (2002). *Building a scholarship of assessment.* San Francisco: Jossey-Bass.

Barker, K. (2002). *Canadian recommended e-learning assessment guidelines (CanRegs).* Retrieved May 20, 2004, from the LICEF Research Center Web site http://www.licef.teluq.uquebec.ca/fr/ pdf/CanREGs%20Eng.pdf

Barnett, N. J., & Crowther, D. (1998). Community identity in the 21st century: A postmodernist evaluation of local government structure. *International Journal of Public Sector Management, 11*(6/7), 425-439.

Barney, J. B. (1991). Firm resources and sustained competitive advantage. *Journal of Management, 17*, 99-120.

Beardsley, S.C., Bray, W.D., & van Rooijen, M.C.W. (1995). The great European multimedia gamble. *McKinsley Quarterly ,3.*

Beer, A., Maude, A., & Pritchard, B. (2003). *Developing Australia's regions: Theory and practice*. Sydney NSW: UNSW Press.

Beer, S. (1994). *Beyond dispute: The invention of team syntegrity.* John Wiley & Sons.

Beishuizen, J. J. (1992). Studying a complex knowledge domain by exploration or explanation. *Journal of Computer Assisted Learning, 8*(2), 104-117.

Benjamin Franklin Institute of Global Education (2001). *Distance education: Quality assurance initiative.* Retrieved May 5, 2004, from http://www.academyWeb.com/DEQA

Benjamin, R., & Wigand, R. (1995). Electronic markets and virtual value chain on the information superhighway. *MIT Sloan Management Review, 36*(2), 62-72.

Bennet, R. (2003). Using electronic assessment to measure student performance. *Issue Brief, NGA Center for Best Practices,*

Bennett, B.R. (2001). How the Internet will help large-scale assessment reinvent itself. *Education Policy Analysis Archives, 9*(5).

Bennett, M.J. (n.d.). Towards ethnorelativism: A developmental model of intercultural sensitivity. In R. M. Paige (Ed.), *Cross-cultural orientation: New conceptualizations and applications* (pp. 27-70). New York: University Press of America.

Bennett, R. (1998). *Reinventing assessment.* Princeton: Educational Testing Service.

Berezousky v. Michaels, 1 WLR 1004 (2000).

Berezousky v. Michaels, 1 WLR 1004, 1024 (2000).

Berezousky v. Michaels, 1 WLR 1004,1032 (Lord Hope, 2000).

Berlin Communiqué (2003). Bologna Process Berlin 2003, Realising the European Higher Education Area. Communiqué of the Conference of Ministers Responsible for Higher Education in Berlin on 19 September.

Bettelheim, N. (January 2006). Personal jurisdiction and the Internet: Cyber differences shed new light on existing conflicts. *Journal of Internet Law, 22.*

Bhatnagar, S., & Schware, R. (2000). *Information and communication technology in rural development.* Case Studies from India. Washington, DC: World Bank Institute.

Birley, S. (1985). The role of networks in the entrepreneurial process. *Journal of Business Venturing, 1*, 107-117.

Blanke, H.T. (1998). Librarianship and public culture in the age of information capitalism. In R.N. Stichler & R. Hauptman (Eds.), *Ethics, information and technology: Readings* (pp. 184-199). Jefferson, NC: MacFarland & Company.

Bloom, B. S. (1956). *Taxonomy of educational objectives: Cognitive domain.* London: Longmans.

Blumenfeld, E., Carone, P.J., & Jain, S. (January 1999). *Journal of Internet Law,* 20.

Born, G. B. (1992). Reappraisal of the extraterritorial reach of U.S. Law. *Law and Policy in International Business, 24*, 1.

Bourn, J. (2002, April 4). Better public services through e-government. *Report by the Comptroller and Auditor General HC 704-III Session, National Audit Office.*

Boynton, A. C. (1993). Achieving dynamic stability through information technology, *California Management Review, 35*(2), 58-77.

Bozeman, B., & Bretschneider, S. (1986). Public management information systems: Theory and prescription. *Public Administration Review, 46* (Special Issue), 475-487.

Braintech v. Kostiuk (1999) 171 D.L.R. (4th) 46 (B.C.C.A).

Breen, M. (1999). Counterrevolution in the infrastructure: A cultural study of techno-scientific impoverishment. In L.J. Pourciau (Ed.), *Ethics and electronic information in the 21st century* (pp. 29-45). West Lafayette, IN: Purdue University Press.

Brennan, J., & Shah, T. (2000). *Managing quality in higher education: An international perspective on institutional assessment and change.* Buckingham: Open University Press.

Briggs, J. (February 2006). *64 billion impressions per month for Google and Yahoo PPC adverts.* Retrieved July 18, 2006, from http://othermedia.com/industry/64-billion-impressions-per-month-for-google-and-yahoo-ppc-adverts,158,TA.html

Bringing Markets Closer in the United Republic of Tanzania. (2006, February). *IFAD Update.* 2006: 2. Retrieved May 15, 2006, from http://www.ifad.org/newsletter/update/2/6.htm

British Association for Open Learning. (2002). *The development, implementation and use of the BAOL Quality Mark: A report to the Department for Education and Skills.* Retrieved May 5, 2004, from http://www.baol.co.uk/PDF/qmrepdfes.pdf

British Broadcasting Corporation (2002). *Annual Report and Accounts 2001/2002.* London: BBC.

British Broadcasting Corporation (2003a). *Annual Report and Accounts 2002/2003.* London: BBC.

British Broadcasting Corporation (2003b). *Market Impact Assessment of BBC's Online Service.* London: BBC.

British Broadcasting Corporation (2004a). *Annual Report and Accounts 2003/2004.* London: BBC.

British Broadcasting Corporation (2004b). *Building Public Value. Renewing the BBC for a Digital World.* London: BBC.

British Broadcasting Corporation (2004c). *Report of the Independent Review of BBC Online.* London: BBC.

British Broadcasting Corporation (2005). *Annual Report and Accounts 2004/2005.* London: BBC.

British Internet Publishers' Alliance (1999). *BIPA Submission to the Davies Review Panel.* London: BIPA.

Brown, C. V., & Magill, S. L. (1994). Alignment of the IS function with the enterprise: Toward a model of antecedents, *MIS Quarterly, 18*(4), 371-403.

Brown, C. V., & Magill, S. L. (1998). Reconceptualizing the context-design issue for the information systems function, *Organization Science*, *9*(2), 176-194.

Brown, D., & Lockett, N. (2004). Potential of critical e-applications for engaging SMEs in e-business: a provider perspective. *European Journal of Information Systems, 13*(1), 21-34.

Brown, J.R., Goma, G., & Gazzard, S. (1999). *Assisting student learning using Web-based assessment: an overview of the WebMCQ system.* Poster presentation at the 15th Annual Conference of the Australasian Society for Computers in Learning in Tertiary Education. Retrieved December 01, 2002, from http://www.webmcq.com/

Brown, M.D., Armon, O., Ploeger, L., & Traynor, M. (2005, Dec). Secondary liability for inducing copyright infringement, MGM v. Grokster Infringement-Prevention and Product Design. *Journal of Internet Law, 21.*

Brown, R. (2000). The new UK quality framework. *Higher Education Quarterly, 54*(4), 323-342.

Brusilovsky, P., & Miller, P. (1999). *Web-based testing for distance education.* New York: Kogan Page.

Brynjolfsson, E., & Hitt, L. M. (2003). Computing productivity: Firm-level evidence. MIT Sloan Working Paper No. 4210-01, available at SSRN: http://ssrn.com/abstract=290325

Bryson, J. M., & Alston, F. K. (1996). *Creating and implementing your strategic plan: A workbook for public and nonprofit organizations.* San Francisco: Jossey-Bass Publishers.

Budhiraja, R. (2003). *Electronic governance-A key issue in the 21st century.* Retrieved from http://mit.gov.in

Buenger, V., Daft, R. L., Conlon, E. J., & Austin, J. (1996). Competing values in organizations: Contextual influences and structural consequences, *Organization Science, 7*(5), 557-576.

Bull, J. (1997). *A glimpse of the future in computer-assisted assessment in higher education.* London: Kogan Page.

Burgoyne, J., & Stuart, R. (1977). Implicit learning theories as determinants of the effect of management development programmes. *Personnel Review, 6*(2), 5-14.

Burstein, J., Kukich, K., Wolff, S., Chi Lu, Chodorow, M., Braden-Harder, L., & Harris, M. (1998). *Automated scoring using a hybrid feature identification technique.* Retrieved March 21, 2003, from http://www.webmcq.com/

Burstein, J.C., Wolff, & Lu, C. (1999). *Using lexical semantic techniques to classify free-responses. The depth and breadth of semantic lexicons.* London: Pergamon Press.

Burt, R. S. (1980). Models of network structure. *Annual Review of Sociology, 6*, 79-141

Bustamante, E. (2003). *Comunicación y cultura en la era digital. Industrias, mercados y diversidad en España.* Barcelona: Gedisa.

Cabral, L., Domingue J., Motta, E., Payne, T., & Hakimpour, F. (2004). Approaches to Semantic Web services: An overview and comparisons. *Computer Science, 3053*, 225-239.

Caliyurt, K. T., & Crowther, D. (2006). The necessity of fraud education for accounting students: A research study from Turkey. *Social Responsibility Journal, 2*(3/4), 322-328.

Campbell v. MGN Ltd. [2004] UKHL 22; [2004] 2 WLR 1232.

Capacity dot org (2004). *Embedding ICT in development*, Issue 23, October 2004, ISSN 1566-6603.

Carnevale, D. (2000). Assessing the quality of online courses remains a challenge and educators agree. *Chronicle of Higher Education,* p. A59.

Carnevale, D. (2002). Accreditors offer views on distance programs. *Chronicle of Higher Education,* p. A36.

Carson, L., & Gelber, K. (2001). *Ideas for community consultation: A discussion on principles and procedures for making consultation work.*Sydney NSW: NSW Department of Urban Affairs and Planning.

Carter, C., & Grieco, M. (2000). New deals, no wheels: Social exclusion. tele-options and electronic ontology. *Urban Studies, 37*(10), 1735-1748.

Carvalho, M. M., Laurindo, F. J. B., Pessôa, M. S. De P. (2003). Information technology project management to achieve efficiency in Brazilian companies. In S. Kamel, *Managing globally with information technology* (pp. 260-271). Hershey, PA: IGI Global.

Carvin, M., & Sermin A. (2006). *From the Ground Up: Evolution of the Telecentre Movement.* Ottawa, CN: IDRC. Retrieved July 2006, from http://www.telecenter.org

Case of the Barcelona Traction Company (Belgium v Spain), ICJ Rep. 3, 47 (1970).

Castells, M. (1996). *The rise of network society.* MA: Blackwell, Cambridge.

Castells, M. (2000). *The information age: Economy, society, and culture* (Volume I). The Rise of the Network Society. 2nd edition. Oxford: Blackwell.

Castells, M. (2001). Epilogue: Informationalism and the network society. In P. Himanen (Ed.), *The hacker ethic and the spirit of the information age* (pp. 155-178). London: Secker & Warburg.

Caudle, S. L., Gorr, W. L., & Newcomer, K.E. (1991). Key information systems management issues for the public sector. *MIS Quarterly, 15*(2), 171-188.

Caudron, S. (2001). Evaluating e-degrees. *Workforce, 80*, 44-48.

CELAC, *Collecting and Exchanging of Local Agricultural Content Program in Uganda.* Retrieved July 15, 2006, from www.celac.or.ug/

Certo, S. C. (1997). *Modern management*. NJ: Prentice Hall International.

Certo, Samuel C., & Peter, J. P. (1995). *Strategic management; Concepts and applications*. Chicago: Austen Press/ Irwin.

Chandrashekhar, R. (2005, January 14). e-Governance: Unlocking India's latent potential. *ASSOCHAM–World Bank 2nd Summit*

Chapman, A. (2002). *Abraham Maslow's hierarchy of needs motivational model*. Retrieved May 11, 2006, from http://www.businessballs.com/

Chapman, G., & Rotenberg, M. (1995). The national information infrastructure: A public interest opportunity. In D.G. Johnson & H. Nissenbaum (Eds.), *Computers, ethics & social values* (pp. 628-644). Upper Saddle River: Prentice Hall.

Chopra, P., & Lata, S. (2004). E-governance-Localization issues. *Vishva Bharat @ tdil*. Retrieved from www.mit.gov.in/tdil/Oct_2004/egov-li-16.pdf

Church of Spiritual Technology, Religious Technology Centre, New Era Publications International APS v. Dataweb B.V. & 21 Others, No. 99/1040 (Court Appeal of The Hague) ECDR 25 (2004).

Chwelos, P., Benbasat, I., & Dexter, A. (2001). Research report: Empirical test of an EDI adoption model. *Information Systems Research, 12*(3), 304-321.

Cilia, M., Bornhövd, C., & Buchmann, A. P. (2005). Event handling for the universal enterprise. *Information Technology and Management, 6*, 123-148.

Ciske, M.M., & Webb, M.C. (2005, September). Suing the downloading doe: Post Verizon ISP liability and infringemnet prosecution under the DMCA. *Journal of Internet Law, 3*.

Clark, C., Cavanaugh, N., Brown, C. V., & Sambamurthy, V. (1997). Building change-readiness IT capabilities: Insights from the Bell Atlantic experience. *MIS Quarterly, 21*(4), 425-455.

Clark, T. D. (1992). Corporate systems management: An overview and research perspective, *Communications of the ACM, 35*(2), 61-75.

Cleary, T. S. (2001). Indicators of quality. *Planning for Higher Education, 29*(3), 19-28.

Clegg, S.R. (1989). *Frameworks of power*. London: Sage.

Cleve, A., Henrard J., & Hainaut, J-L. (2005). Co-transformations in information system reengineering. *Science Direct (Electronic Notes in Theoretical Computer Science), 137*, 5-15.

Coates, B. E. (2004). E-entrepreneurship in a disadvantaged community: Project EdNet in California. *International Journal on E-Learning, 3*(3), 25-31.

Cochrane, P. (1995). The virtual university; *Business of Education, 5*, 17-18.

Cole, G. (1994). Learning with computers. *Accountancy* (UK), *113*, 1209.

Collins, M. (2005). *The law of defamation and the Internet* (2nd ed.). Oxford: University Press.

Collis, B. (1999). *Designing for differences: Cultural issues in the design of WWW-based course-support sites*. BJET.

Collis, B., & Moonen, J. (2001). *Flexible learning in a digital world: Experiences and expectations*. London: Kogan Page.

Collis, B., & van der Wende, M. (2002). *Models of technology and change in higher education: An international comparative survey on the current and future use of ICT in higher education*. Cheps, Center for Higher Education Policy Studies.

Collis, D. J., & Montgomery, C. A. (1995). Competing on resources: strategy in the 1990s. *Harvard Business Review, 73*(4), 118-128.

Concurso Autobiografía de Mujeres.(1997). Electronic Conference. Santiago: RIMISP. Retrieved June 1999, from http://www.fidamerica.cl/seccion.php?seccion=156

Contenti, M., Termini, A., Mecella, M., & Baldoni R. (2003). An e-service-based framework for inter-administration cooperation. *LNAI, 2645*, 13-24.

Convention on Cybercrime (2001)

Cooper, S., Crowther, D., Davies. M., & Davis, E.W. (2001). *Shareholder or stakeholder value.* London: CIMA.

Cornwall, A. (2001). *Making a difference? Gender and participatory development.* Working Paper 378. Brighton, UK: IDS. Retrieved March 2002, from http://server.ntd.co.uk/ids/bookshop/details.asp?id=597

Cornwall, A., & Gaventa, J. (2001). *From users and choosers to makers and shapers.* Working Paper 127. Brighton, UK: IDS. Retrieved March 2002, from http://server.ntd.co.uk/ids/bookshop/details.asp?id=618

Council of the State-Owned Media Reform (2005). *Report on the Reform of the State-Owned Public Media.* Madrid: Secretaría de Estado de Comunicación.

Cox, A., Sanderson, J. E., & Watson, G. (2001) Supply chains and power regimes: Toward and analytic framework for managing extended networks of buyer and supplier relationships. *The Journal of Supply Chain Management, 37*(2), 28-35.

Cranfield University (2006). Performance prism measuring stakeholder value. Available at http://www.valuebasedmanagement.net/methods_performance_prism.html

Crede, A., & Mansell, R. (1998). *Knowledge Societies in a Nutshell.* Information Technologies for Sustainable Development. IDRC, Ottawa, Canada.

Cross, J., Earl, M., & Sampler, J. L. (1997). Transformation of the IT function at British Petroleum. *MIS Quarterly, 21*(4), 401-423.

Croteau, D., & Hoynes, W. (2001). *The Business of Media. Corporate Media and the Public Interest.* Thousand Oaks: Pine Forge Press.

Crowther, D. (2000). Corporate reporting, stakeholders and the Internet: Mapping the new corporate landscape. *Urban Studies, 37*(10), 1837-1848.

Crowther, D. (2002a). *The importance of corporate social responsibility.* Escola Superior de Tecnologia e Gestao da Guarda. Working paper no 05/02.

Crowther, D. (2002b). *A social critique of corporate reporting.* Aldershot: Ashgate.

Crowther, D. (2002c). *Creating shareholder value.* London: Spiro.

Crowther, D. (2003). Social responsibility and the marketisation of higher education. *Staff and Educational Development International, 7*(1), 23-40.

Crowther, D., & Carter, C. (2002). Legitimitating irrelevance: Management education in higher education institutions. *International Journal of Education Management, 16*(6), 268-278.

Crowther, D., & Davies, M. (1998). Integrating Multimedia technology into a management accounting curriculum: an evaluation of effectiveness in the learning process. In *Proceedings of the 9th CTI-AFM Conference* (pp. 85-93).

Crowther, D., & Davila Gomez, A. M. (2006). Stress in the back office. In *Proceedings of India-the processing Office to the World,* Kochi, India (pp. 27-38).

Crowther, D., Davies, M., & Barnett, N. (1997). Evaluating computer based courseware: A stakeholder approach. In *Proceedings of the 8th CTI-AFM Conference* (pp. 32-39).

Crowther, D.E.A., & Davies, M. L. (1996). Multimedia, mesmerism and teaching strategy; *CALECO 1995 Conference Proceedings.* University of Bristol, 11-17.

CTC Association Web site. Retrieved October 16, 2006, from http://www.ctca.net.au

CTC@NSW Web site. Retrieved July 20, 2004, from http://ctc.nsw.gov.au

Cummings, S., et al. (Eds.). (1999). *Women's information services and networks.* Amsterdam: KIT Press and Oxfam GB.

Cummins, G.R. (April 2003). Global liability risks emerge in cyberspace. *Journal of Internet Law, 1.*

Currie, W. (2000). *The global information society*. Chichester: John Wiley & Sons.

Curtin, R. (2001). *Promoting youth employment through information and communication technologies (ICT): Best practices examples in Asia and the Pacific*. Report prepared for ILO/Japan Tripartite Regional Meeting on Youth Employment in Asia and the Pacific, Bangkok.

Cyert, R. M., & March, J. G. (1963). *A behavioral theory of the firm*. Englewood Cliffs, NJ: Prentice Hall, Inc.

D'Aveni, R. A. (1994). *Hypercompetition: Managing the dynamics of strategic maneuvering*. New York: Free Press.

D'Aveni, R. A. (1999). Strategic supremacy through disruption and dominance, *Sloan Management Review*, *40*(3), 127-135.

D'Aveni, R. A. (2002). Competitive pressure systems: Mapping and managing multimarket contact. *Harvard Business Review*, Fall.

Daft, R. L. (1998). *Organization theory and design* (6th ed.). Cincinnati, OH: South Western College.

Dahlgren, P. (2002). Reconfiguring Civic Culture in the Evolving Media Milieu. In J. Corner & D. Pels (Eds.), *Media and political style: Essays on representation and civic culture*. London. Sage.

Dahlman, C.J., Routti, J., & Ylä-Anttila, P. (Eds.). (2005). A prepublication version of the Overview of the book *Finland as a Knowledge Economy: Elements of Success and Lessons Learned*, published in January 2006. The International Bank for Reconstruction and Development / The World Bank, Washington, 2005.

Damsgaard, J., & Lyytinen, K. (2001). The role of intermediating institutions in the diffusion of electronic data interchange (EDI): How industry associations intervened in Denmark, Finland and Hong Kong. *The Information Society*, *17*, 195-210.

Dash, S. S., Ray, D., & Surinder, K. (2005). Developing interoperable information system. *e-gov*, *1*(2), 10-13. Retrieved from www.csdms.in/egov/march05.pdf

Davenport, S., Carr, A., & Bibby, D. (2002). Leveraging talent: Spin-off strategy at industrial research. *R&D Management*, *32*(3), 241-254.

Davidson, S.J., & Bryant, D.M. (2001, November). The right of privacy: Historical development, international discord, and the interface with intellectual property law. *Journal of Internet Law*, *8*.

Davies, B., & Ellison, L. (2003). *The new strategic direction and development of the school*. London: Routledge Falmer.

Davies, M. L., & Crowther, D.E.A. (1995). The benefits of using multimedia in higher education: Myths and realities. *Active Learning*, *3*, 3-6.

Davis, J.P. (1961). *Corporations: A study of origin and development of great business combinations and of their relation to the authority of the state*. New York: Capricorn Books.

Davis, N. J., Fensel, D. & Richardson, M. (2004). The future of Web services. *BT Technology Journal*, *22*(1), 118-130.

Dawes S., Pardo T., Green D., Mclnemey C., Connelly D., & DiCatireno A. (1997). *Tying a sensible knot: A practical guide to state-local information systems*. Albany, NY: Center for Technology in Government.

De George, R.T. (1999). *Business ethics* (5th ed.). Upper Saddle River, NJ: Prentice Hall.

De Jong, T., & Ferguson-Hessler, M. (1996). Types and qualities of knowledge. *Educational Psychologist*, *31*(2).

De Weaver, L., & Ellis, A. (2006). The CTC @NSW Program: Achievements and ongoing challenges. *International Journal of Education and Development using ICT*, *2*(2),15-26.

De Weaver, L., & Lloyd, D. (2005, August). *The language of community engagement in a regional and indigenous context*. Paper presented at International Conference on Engaging Communities. Brisbane: QLD. Government of the State of Queensland.

Denegar, A.P. (December 2001). ISP liability for third-party defamation in Canada: Adopting a US-style statutory regime. *Journal of Internet Law, 18*.

Department of Communications, Information Technology & the Arts Web site. *Communications for Business Report*. Retrieved November 30, 2006, from http://www.dcita.gov.au/communications_for_business/funding_programs__and__support/clever_networks

Department of Communications, Information Technology and the Arts (2004, February). *Regional Policy Research Forum Report*. Canberra: ACT.

Department of Local Government and Regional Development (2004). *Planning for the future report*. Perth, WA: Government of Western Australia.

Deschamps, N. (1995). *Product juggernauts*. Harvard Business School.

Dickenson, J. (1997). *Using WebMCQ for formative and summative assessment*. Proceedings of university science teaching and the Web workshop. Sydney: Uniserve Science. Retrieved December 12, 2002, from http://www.webmcq.com/

Digital Equip. Corp. v. Altavista Tech. Inc., 960 F. Supp. 456 (D. Mass. 1997).

Dillenbourg, P. (1999). Introduction: What do you mean by 'collaborative learning'? In P. Dillenbourg (Ed.), *Collaborative learning-cognitive and computational approaches* (pp. 1-19). Oxford: Pergamon.

Douglas v. Hello! Ltd., No.3, EWCA Civ. 595 (2005).

Drury, R. R. (1998). The regulation and recognition of foreign corporations: Responses to the Delaware Syndrome. *Cambridge Law Journal, 165*.

DTI (2002). *A government action plan for small business*. Retrieved August 18, 2005, from http://www.sbs.gov.uk/content/7-strategies/sbs_evidence.pdf

Duke of Brunswick v. Harmer, 14 QB 185 (1849).

Earl, M. (2000). Local lessons for global businesses. In D. A. Marchand, T. H. Davenport, & T. Dickson (Ed.), *Mastering information management*, London: Financial Times/Prentice Hall.

EC (2004). *Specific programme for research, technological development and demonstration: Integrating and strengthening the European research Area*. 2004 Work Programme (SP1-10), European Commission, 2004.

EC (2005). *Specific programme for research, technological development and demonstration: Integrating and strengthening the European research Area*. 2005 Work Programme (SP1-10), European Commission, 2005.

EC (2006). *Specific programme for research, technological development and demonstration: Integrating and strengthening the European research Area*. 2006 Work Programme (SP1-10), European Commission, 2006.

ECOM International Relations Group announced its 2002 Survey on the State of E-commerce in Other Countries. (2003). ECOM Newsletter, 34.Retrieved from http://www.ecom.jp/ecom_e/latest/newsletter_no34.htm

ECOSOC (2006). Informal background note on possible options for consideration by ECOSOC in the reform of the CSTD.

Edmiston, K. (2003). State and local e-Government: Prospects and challenges. *American Review of Public Administration, 33*(1), 20-45.

Edmundson, A. (2004). *The cross-cultural dimensions of e-learning*. PhD Dissertation, Walden University

Ein-Dor, P., & Segev, E. (1982). Organizational context and MIS structure: Some empirical evidence. *MIS Quarterly, 6*(3), 55-68.

El Sawy, O., Malhotra, A., Gosain, S., & Young, K. (1999). IT-intensive value innovation in the electronic economy: Insights from Marshall industries. *MIS Quarterly, 23*(3), 305-335.

Escobar, A. (2000). Notes on networks and anti-globalization social movements. Available at http://www.unc.edu/depts/anthro/faculty/fac_pages /escobarpaper.html

Ess, C. (1996). The political computer: Democracy, CMC, and Habermas. In C. Ess (Ed.), *Philosophical perspectives on computer-mediated communication* (pp. 197-230). Albany: State University of New York Press.

Estelle Smet-Hallyday v. Valentin Lacambre, ECC 439 (1999).

Euromarket Designs Incorporated v. Peters, ETMR 1025 (Jacob J., 2000).

Europa News (2004, September). *Belgium close to solving electronic ID cards distribution deadlock.*

European Commission (2004). The European e-Business Report. *A Portrait of e-Business in 10 Sectors of the EU Economy.* European Commission, Enterprise Directorate General, Enterprise Publications.

European Union (2005). *eEurope 2005 Action Plan.* Retrieved August 28, 2005, from http://www.eeurope2005.com

Executive Yuan. (2001, December 26). *National Information and Communication Infrastructure Development Plan.* Taipei, Taiwan: Executive Yuan.

Fagin, B. (2000). Liberty and community online. In R.M. Baird, R. Ramsower, & S.E. Rosenbaum (Eds.), *Cyberethics - Social and moral issues in the computer age.* New York: Prometheus Books.

Fairclough, N. (2000). Multiliteracies and language: Orders of discourse and intertextuality. In Cope & Kalantis (Eds.), *Multiliteracies: Literacy learning and the design of social futures* (pp. 162-181). South Yarra,Vic: Macmillan Publishers Australia Pty Ltd.

Fang, Z. (2002). E-government in digital era: Concept, practice, and development. *International Journal of The Computer (The Internet and Management), 10*(2), 1-22.

Fayol, H. (1970 edition). Administration industrielle et générale: prévoyance, organisation, commandement, coordination et contrôle, extrait du *Bulletin de la société de l'industrie minérale.* Paris: Dunod.

Feeny, D. F., & Willcocks, L. (1998). Core IS capabilities for exploiting information technology, *Sloan Management Review, 39*(3), 9-21.

Feeny, D. F., Edwards, B. R., & Earl, M. (1989). *Complex organizations and the information systems function,* Oxford Institute for Information Management, Oxford, RDP 87/7.

Feka, V., Xouris, D., & Tsiotras, G. (1997). Mapping strategic groups: An international example. *Journal of Business & Industrial Marketing, 12*(1), 66-75.

Ferris, J. M., & Tang, S.-Y. (1993). The new institutionalism and public administration: An overview. *Journal of Public Administration Research and Theory, 3*(1), 4-10.

Fidler, B. (2002). *Strategic management for school development.* London: Paul Chapman Publishing.

Filipczak, B. (1995). Different strokes: Learning styles in the classroom. *Training, 32*(3), 43-48.

Flamm, K. (1987). *Targeting the computer: Government support and international competition.* Washington, DC: Brookings Institution.

Fletcher, P. D. (1999). Strategic planning for information technology management in state government. In G. D. Garson (Ed.), *Information technology and computer applications in public administration: Issues and trends* (pp. 81-98). Hershey, PA: Idea Group Publishing.

Fletcher, P. D. (2003). The realities of the paperwork reduction Act of 1995: A government-wide strategy for information resources management. In D. Garson (Ed.), *Public information technology: Policy and management issues* (pp. 74-93). Hershey, PA: Idea Group Publishing.

Forestier, E., Grace, J., & Kenny, C. (2002). Can information and communications technologies be pro-poor? *Telecommunications Policy, 26*(11), 623-646.

Foucault, M. (1977). *Discipline and punish.* trans A Sheridan. London: Penguin.

Fountain, J. (2001). *Building the virtual state: Information technology and institutional change.* Washington, DC: Brookings Institution Press.

Freeman, R. E. (1984). *Strategic management: A stakeholder approach.* Boston: Pitman.

Fried, E. (June 2003). Issues and challenges presented by defamation on the Internet. *Journal of Internet Law, 3.*

Fryman, H. (2005). Data migration: Lowering cost and risk to move to new systems and applications. *Informatica, What Works, 19*(5). Retrieved from http://www.tdwi.org/Publications/WhatWorks/display.aspx?id=7524

Furrer, O., & Thomas, H. (2000). The rivarly matrix: Understanding rivarly and competitive dynamics. *European Management Journal, 18*(6).

Gagne, R., Briggs, L., & Wager, W. (1992). Designing instructional systems. In *Principles of instructional design*, Wadsworth.

Galbraith, J. R. (1973). *Designing complex organizations.* MA: Addison-Wesley.

Galbraith, J. R. (1974). Organization design: An information processing view. *Interfaces, 4*(3), 28-39.

Galbraith, J. R. (1987). Organizational design. In J. W. Lorsch (Ed.), *Handbook of organizational behavior*, Englewood Cliffs, Prentice Hall, Inc.

Galbraith, J. R. (1994). *Competing with flexible lateral organizations.* MA: Addison-Wesley Pub.Co.

Galbraith, J. R., & Lawler, E. E. (1993). *Organizing the future: The new logic of managing complex organizations.* San Francisco: Jossey-Bass.

García Mestanza, J., & Fernández, V. (2002). El valor estratégico de los contenidos en el mercado de la televisión de pago. Una perspectiva de recursos y capacidades. Boletín económico de ICE, 2.719.

Garcia, C. (1999). Weaving webs of unity: The experience of Asia and the Pacific. In S. Cummings, et al. (Eds.), *Women's information services and networks.* Amsterdam: KIT Press and Oxfam GB.

Gates, B., & Hemingway, C. (1999). *Business @ the speed of thought.* Warner Books.

Gayeski, D. (1993, April). De-mystifying multimedia. *IABC Communication World,* 27-32.

Gazzard, S. (1998). *ComDesign principles for next wave software: The development of the WebMCQ system.* Proceedings of the Fifteenth Annual Conference of the Australasian Society for Computers in Learning in Tertiary Education. Retrieved November 23, 2002, from http://www.webmcq.com/

Gemunden, H. G., & Ritter, T. (1997). Managing technological networks: The concept of network competence. In H. G. Gemunden, T. Ritter, & A. Walter (Eds.), *Relationships and networks in international markets* (pp. 294-304). Oxford: Elsevier Science.

General Accounting Office. (2001, July 11). *Electronic government: Challenges must be addressed with effective leadership and management.* Washington, DC: General Accounting Office. GAO-01-959T.

Gershwin Publishing Corp. v. Columbia Artists Management Inc., 443 F. 2d 1159, 1162 (2d Cir. 1971).

Ghosh, S. (1998). Making business sense of the Internet. *Harvard Business Review*, March-April, 126-35.

Gibbs, G. (1992). *Improving the quality of student learning.* Bristol: Technical & Educational Services.

Gilbert, A. L. (1992). A transaction costs model of international information technology transfers: The dynamics of intelligence and of information control. In Palvia, Palvia, & Zigli (Eds.), *The global issues technology management* (pp. 403-426). Hershey, PA: Idea Group Publishing.

Gittler, A. (1999). Mapping women's global communications and networking. In W. Harcourt (Ed.), *Women@Internet.* London: Zed.

Godfrey v. Demon Internet Ltd. QB 201 (Morland J., 2001).

GOJ (2002). *Five-year strategic information technology plan for Jamaica.* Retrieved November 1, 2004, from http://www.mct.gov.jm/gojitplanrevisedversion-march20021.pdf

GOJ (2003). *Electronic commerce (Transaction) policy.* Retrieved February 12, 2005, from http://www.mct.gov.

jm/electronictransactionspolicyapril2003ps1submitted.pdf

GOJ (2006). *Transformation of the ICT sector: Overview of ICT development in Jamaica.* Retrieved July 27, 2006, from http://www.mct.gov.jm/transformationoftheicsector2.pdf

Goodyear, P. (1999). Pedagogical frameworks and action research in open and distance learning. *European Journal of Open and Distance Learning.*

Gore, A. (1995). Global information infrastructure. In Johnson & Nissenbaum (Eds.), *Computers, ethics & social values.* Upper Saddle River: Prentice Hall.

Gosnell, G. R. (September 1999). Internet-related intellectual property issues. *Journal of Internet Law, 23.*

Government of Japan (2005, February). *IT Policy Package 2005: Towards the Realization of the World's Most Advanced IT Nation.* The 29th Meeting of the IT Strategic Headquarters.

Gowthman, K., Mustafa, K., & Khan, R. A. (2005). Legacy source code migration to .NET architecture. *Journal of Computer Science, 1*(1), 45-52.

Grabher, G. (Ed.). (1993). *The embedded firm. On the socio-economics of industrial networks.* London: Routledge.

Grabow, B., Drüke, H., & Siegfried, C. (2004). Factors for success for local community e-government. *German Journal of Urban Studies, 45*(2). Retrieved from http://www.difu.de/ index.shtml?/publikationen/dfk/en/04_2/04_2_kubicek_wind.shtml

Grace, J., Kenny, C., & Qiang, C. with J. Liu and T. Reynolds. (2001). *Information and communication technologies and broad-based development: A partial review of the evidence* (Draft). Washington, DC: World Bank.

Graham, S. (1988). The end of geography or the explosion of place? Conceptualizing space, place and information technology. *Progress in Human Geography, 22*(2), 165-185.

Grainger, G. (1999). *Broadcasting Coregulation and Public Good.* The 1999 Spry Memorial Lecture. Retrieved July 8, 2006, from www.com-umontreal.ca/spry/spry-gg-lec-html

Grassroots Caucus Recommendations to the WSIS Plan of Action. (2005). Retrieved November 2005, from http://www.grassroots-caucus.org

Griffith, D.A., & Harvey, M. G. (2004). The influence of individual and firm level social capital of marketing managers in a firm's global network. *Journal of World Business, 39,* 244-254.

Groth, L. (1999). *Future organizational design: The scope for the IT-based enter*prise. Chichester: John Wiley & Sons.

Grundén, K. (2005, September 13). Implementation of "Vision of E-Government 2007" at the county administration of Sweden. *eGovernment Workshop '05 (eGOV05),* Brunel University, West London.

Grupe, F. H., & Connolly, F .W. (1995). Grownups are different: Computer training for adult learners. *Journal of Systems Management, 46*(1), 58-64.

Gupta, M. P., & Sahu, G. P. (2005). Need fruits of e-governance? *e-gov., 1*(2), 6-9.

Gurbaxani, V., King, J. L., Kraemer, K.L., Mc Farlan, F.W.,Raman, K.S., & Yap, C.S. (1990). Institutions and international diffusion of information technology. In J.I DeGross, M. Alavi, & H. Oppelland (Eds.), *Proceedings of the Eleventh International Conference on Information Systems* (pp. 97-98). Copenhagen.

Gury-Rosenblit, S. (1999). *Distance and campus universities: Tensions and interactions. A comparative study of five countries.* Unesco-Pergamon.

Guy, M. E. (2000). Public management. In J. M. Shafritz (Ed.), *Defining public administration.* (pp. 166-168). Boulder, CO: Westview Press.

Hafkin, N. (2002). *Gender issues in ICT policy in development countries.* An Overview. Prepared for the UNDAW Expert Group Meeting on "Information and communication technologies and their impact on and use as an instrument for the advancement and empowerment of women. Seoul, Republic of Korea. Retrieved from http://www.un.org/womenwatch/daw/egm/ict2002/documents.html

Hafkin, N. (2006). Women, gender and ICT statistics and indicators. In N. Hafkin & S. Huyer (Eds.), *Cinderella or cyberella? Empowering women in the knowledge society*. Bloomfield, CT: Kumarian Press.

Hafkin, N., & Huyer, S. (2006). *Cinderella or cyberella? Empowering women in the knowledge society*. Bloomfield, CT: Kumarian Press.

Håkansson, H., & I. Snehota (Eds.). (1995). *Developing relationships in business networks*. London: Routledge.

Håkansson, H., & Waluszewski, A. (2002). *Managing technological development. IKEA, the environment and technology*. Routledge Advances in Management and Business Studies. London: Routledge.

Hak-Su, K. (2003). Forward. *Regional Road Map towards an Information Society in Asia and the Pacific, Economic and Social Commission for Asia and the Pacific*. United Nations, New York.

Halinen, A., & Törnroos, J.Å. (1998). The role of embeddedness in the evolution of business networks. *Scandinavian Journal of Management*, *14*(3), 187-205.

Halinen, A., Salmi, A., & Havila,V. (1999). From dyadic change to changing business networks: An analytical framework. *Journal of Management Studies*, *36*(6), 779-94.

Hall, E. (1977). *Beyond culture*. Anchor Press/Doubleday.

Handa, S. (February 2004). Online defamation: Broadcasting and republication in Canada. *Journal of Internet Law, 18*.

Hanna, N. (1994). *Exploiting information technology for development: A case study of India*. World Bank Discussion paper 264. Washington, DC: World Bank.

Hannafin, M. J., & Land, S. M. (1997). The foundations and assumptions of technology-enhanced learning environments. *International Journal of Learning and Cognition, 25*(3).

Haraway, D. (1990) [1985]. A manifesto for cyborgs: Science, technology, and socialist feminism in the 1980's. In L. Nicholson (Ed.), *Feminism/postmodernism*. New York, NY: Routledge.

Haraway, D. (1994). A game of cat's cradle: Science studies, feminist theory, cultural studies. *Configurations, 2*(1), 59-71.

Harcourt, W. (Ed.). (1999). *Women@Internet*. London: Zed Books.

Haulmark, M. (2002). *Accommodating cultural differences in a web-based distance education course: A case study*. In the 9th Annual International Distance Education Conference, Texas A&M University

Hausman, D.M., & McPherson, M.S. (1996). *Economic analysis and moral philosophy*. Cambridge: Cambridge University Press.

Hayat, A., Posch, R., & Leitold, H. (2005). *Identifying obstacles in moving towards an interoperable electronic identity management system*. IDABC Conference.

Hayek, F. A. von (1994). *The road to serfdom*. Fifthieth Anniversary Edition. Chicago: The University of Chicago Press.

Healy, M., & Iles, J. (2002). The impact of information and communications technology on managerial practices: The use of codes of conduct. In Alvarez, et al. (Eds.), *The transformation of organisations in the information age: Social and ethical implications. Proceedings of the sixth ETHICOMP Conference*, Lisbon, Portugal. Lisbon: Universidade Lusiada.

Heard, B. (1997). A correlation of mutiple-choice and essay assessment measures. *Research in Education, 56,* 73-84.

Hearn, G., Kimber, M., Lennie, J., & Simpson, L. (2004). *ICTS and regional sustainability: A critique and a way forward*. Paper presented at CIRN Conference and Colloquium, Prato, Italy.

Heeks, R. (1999). *Reinventing government in the information age: International practice in IT-enabled public sector reform*. London; New York: Routledge.

Heeks, R., Arun, S., & Morgan, S. (2004). *Researching ICT-based enterprise for women in developing countries: An enterprise perspective.* Women's ICT-Based Enterprise for Development project. Manchester, UK: IDPM, University of Manchester. Retrieved June 2006, from http://www.womenictenterprise.org

Heeks, R., Arun, S., & Morgan, S. (2005). *Researching women's ICT-based enterprise for development: Methods, tools and lessons from fieldwork*. Women's ICT-Based Enterprise for Development project. Manchester, UK: IDPM, University of Manchester. Retrieved June 2006, from http://www.womenictenterprise.org/

Heide, J.B., & John, G. (1995). Measurement issues in research on interfirm relations. In K. Möller & D. Wilson (Eds.), *Business marketing: An interaction and network perspective* (pp. 531-554). Boston: Kluwer Academic Publishing.

Heizer, R. (1999). *Operations management.* Prentice-Hall.

Henderson, J. C. (1990). Plugging into strategic partnerships: The critical IS connection. *Sloan Management Review, 31*(3), 7-18.

Henderson, L. (1996). *Instructional design of interactive multimedia: A cultural critique*. Educational Technology Research & Development.

Henriksen, H.Z., & Andersen, K.V. (2004). Diffusion of e-commerce in Denmark: An analysis of institutional intervention. *Knowledge, Technology & Policy, 17*(2), 63-81.

Hetherington, K., & Law, J. (2000). After networks. *Environment and Planning D: Society and Space, 18*(2), 127-132.

Hetherington, K., & Lee, N. (2000). Social order and the blank figure. *Environment and Planning D: Society and Space, 18*(2), 169-184.

Hirschheim, R., & Klein, H.K. (1994). Realizing emancipatory principles in information systems development: The case for ETHICS. In *MIS Quarterly, 18*(1), 83-109.

Hitt, M. A., Keats, B. W., & DeMarie, S. M. (1998). Navigating the new competitive landscape: Building strategic flexibility and competitive advantage in the 21st century, *Academy of Management Executive, 12*(4), 22-42.

Hodgkinson, S. L. (1996). The role of the corporate IT function in the federal IT organization. In M. Earl (Ed.), *Information management: The organizational dimension* (pp. 247-268). Oxford, Oxford University Press.

Hofstede, G. (2001). Culture's consequences. *Comparing values, behaviors, institutions and organizations across nations* (2nd ed.). Sage.

Hogan, J., & Greene, A.M. (2002). E-collectivism and e-commerce: On-line action and on-line mobilisation. In L. Holmes, D. M. Hosking, & M. Grieco (Eds.), *Organising in the information age.* Aldershot: Ashgate.

Hogg G. (1997). About multiple choice questions, multiple choice testing: Green Guide, 16, *The Higher Education Research and Development Society of Australasia.* Cambeltown: Associated Press.

Holmes, L., & Grieco, M. (1999, June). *The power of transparency: The Internet, e-mail and the Malaysian political crisis.* Paper presented to Asian Management in Crisis Conference, Association of South East Asian Studies, UNL.

Holmqvist, M. (2003). A dynamic model of intra- and interorganizational learning. *Organization Studies, 24*(1), 93-121.

Hölttä, P. (1998a). *Lisäselvitys Palmu-projektin tuotekehityshakemukseen*. 30.4.1998. Helsingin Puhelin Oyj. Helsinki, Finland.

Hölttä, P. (1998b). *Palmu-projektin väliraportti ajalta 1.1.1998–30.9.1998.* Helsingin Puhelin Oyj. Helsinki, Finland.

Hooker, M. (1997). The transformation of higher education. In Diana Oblinger & Sean C. Rush (Eds.), *The learning revolution.* Boston: Anker Publishing Company, Inc. Retrieved May 5, 2004, from http://horizon.unc.edu/projects/seminars/Hooker.asp

Houldsworth, E. (1994). Multimedia in management education: Theoretical framework and methodological choices. *Henley Management College working paper* no. HWP9420.

HPY Research (1998). *Palmu-project. 5.2.1998. Palmu-projektin kick-off tilaisuus.* Helsingin Puhelin Oyj. Helsinki, Finland.

Huber, G.P., & Van de Ven, A. H. (Eds.). (1995). *Longitudinal field research methods. Studying processes of organizational change.* USA: SAGE Publications Inc.

Huges, TK. (1996). An introduction to latent semantic analysis. *Discourse Processes, 25*(2), 259-284.

Huyer, S. (1997). *Supporting women's use of information technologies for sustainable development.* Manuscript prepared for the Gender and Sustainable Development Unit.

Huyer, S. (2006). Understanding gender equality and women's empowerment. In N. Hafkin & S. Huyer (Eds.), *Cinderella or cyberella? Empowering women in the knowledge society.* Bloomfield, CT: Kumarian Press.

Iacovou, C., Benbasat, I., & Dexter, A. (1995). Electronic data interchange and small organizations: Adoption and impact of technology. *MIS Quarterly, 19*(4), 465-485.

ICT Riga (2006, June). *Ministerial Declaration.* Latvia.

IMF (2001, October). *World Economic and Financial Surveys, World Economic Outlook: The Information Technology Revolution.* International Monetary Fund, Washington, ISBN 1-58906-073-3.

Inoue, A. (1996). *Developing and integrating a Web-based quiz into the curriculum.* Proceedings of WebNet97, World Conference of the WWW and Internet. Toronto: Association for the Advancement of Computing in Education.

International City/County Management Association. (2001). 2000 Electronic Government Survey Dataset [Data file]. Washington, DC: International City/County Management Association.

International City/County Management Association. (2002, August). 2002 Electronic Government Survey Dataset. Washington, DC: International City/County Management Association.

Introna, L. (2001). Workplace surveillance, privacy, and distributive justice. In R.A. Spinello & H.T. Tavani (Eds.), *Readings in cyberethics.* Sudbury, MA: Jones and Bartlett.

Ischy, F., & Simoni, O. (2002). Representations as factor of organizational change. In Alvarez, et al. (Eds.), *The transformation of organisations in the information age: Social and ethical implications. Proceedings of the Sixth ETHICOMP Conference*, Lisbon, Portugal. Lisbon: Universidade Lusiada.

Isomäki, M. & Jäntere, K. *PALMU working draft. Laajakaistainen DSL Liittymä–Business-mallien ja niihin liittyvien teknisten toteutustapojen vertailu.* Helsinki, Finland.

IT Governance Institute (2001). Board Briefing on IT Governance. Retrieved from http://www.itgovernance.org

ITU (2003). *World Telecommunication Development Report.* Retrieved June 14, 2006, from http://www.itu.int/itud/ict/publications/wtdr_03/material/chap4_wtdr2003_e. pdf

ITU (2004, September). *The Portable Internet.* ITU Internet Reports 2004, Switzerland. ISBN 92-61-10741-9.

ITU (2005). *World Telecommunications Development Report 2005*, ITU, Geneva.

ITU (2006). *World Telecommunication/ICT Indicators*, ITU, Geneva.

Jakubowicz, K.(2002). *We know it when we see it? Public Service Broadcasting: Definitions, Descriptions and Policy Dilemmas.* Changing Media-Changing Europe

Working Papers, 2. Retrieved July 5, 2006, from http:/info.lut.ac.uk/research/changing.media/K.J.-Paper.html

Jameel (Yousef) v. Dow Jones & Co. Inc. EWCA Civ. 75 (2005).

Jarillo, J. C. (1993). *Strategic networks: Creating the borderless organization*. Bodmin, Cornwall: MPG Books Ltd.

Jean-Louis Costes v. Ministere Public,) ECC 451 (Cour D'Appeal, Paris, 2000).

Jenkins, G. (2002). Lessons from the trenches of e-Gov: Part 2. *Ubiquity: An ACM IT magazine and forum*. Retrieved from http://www.acm.org/ubiquity/views/g_jenkins_2.html

Johnson, D. G. (2000). Democratic values and the Internet. In D. Langford (Ed.), *Internet ethics* (pp. 181-196). London: McMillan.

Johnson, D.G. (2001). *Computer ethics* (3rd ed.). Upper Saddle River, NJ: Prentice Hall.

Johnson, G., & Scholes, K. (2001). *Exploring corporate strategy: Text and cases*. FT Prentice Hall.

Joia, L. A. (2005). A framework for developing regional e-government capacity building networks. *MIT Press Journal, 2*(4), 61-73.

Juntunen, A. (2005), *The emergence of a new business through collaborative networks: a longitudinal study in the ICT sector*. Acta Universitatis oeconomicae Helsingiensis. Printed in HSEPrint. A-series, no 256, ISSN 1237-556X. ISBN: 951-791-957-3/1.

Kaasinen, K. (2001). *Describing business concept based on partnership network*. Thesis. Lappeenranta University of Technology.

Kahn, K. B., & McDonough, F. (1997). An empirical study of the relationships among co-location, integration, performance, and satisfaction, *Journal of Production Innovation Management, 14*, 161-178.

Kahney, H. (1993). *Problem solving*. Buckingham: Open University Press.

Kalim, R., & Lodhi, S. (2004). Digital economy and its impact on employment in the developing countries. *The ICFAI Journal of Applied Economics, 3*(3), 61-73.

Kamarch, E. C., & Nye, J. S. (2002). *Governance.com: Democracy in the information age*. Cambridge, MA: Visions of Governance in the 21st Century Brookings Institution Press.

Kaplan, R. S., & Norton, D. P. (1992). The balanced scorecard: Measures that drive performance. *Harvard Business Review, 70*(1), 71-79.

Kaplan, R. S., & Norton, D. P. (1996). *The balanced scorecard: Translating strategy into action*. Boston: Harvard Business School Press.

Kaplan, R. S., & Norton, D. P. (1996). Using the balanced scorecard as a strategic management system. *Harvard Business Review, 74*(1), 75-85.

Kaplan, R., & Norton, D. (1992). The balanced scorecard: Measures that drive performance, *Harvard Business Review*, January-February, 71-79.

Kaplan, R., & Norton, D. (1993). Putting the balanced scorecard to work. *Harvard Business Review*. September-October, 134-147.

Kaplan, R., & Norton, D. (1996). *The balanced scorecard*. Boston: Harvard Business School Press.

Kaplan, R., & Norton, D. (2001). *The strategy-focused organisation*. Boston: Harvard Business School Press.

Kaplan, R., & Norton, D. (2004). *Strategy maps*. Boston: Harvard Business School Press.

Kember, D. (2000). *Misconceptions about the learning approaches, motivation and study practices of Asian students*. Higher Education, Kluwer Academic Publishers.

Kenis, P., & Knoke, D. (2002). How organizational field networks shape interorganizational tie-formation rates. *Academy of Management Review, 27*(2), 275-293.

Kester, G.H. (1998). Access denied: Information policy and the limits of liberalism. In R. N. Stichler & R.

Hauptman (Eds.), *Ethics, information and technology: Readings* (pp. 207-230). Jefferson, NC: MacFarland & Company.

Kettunen, J. (2002). Competitive strategies in higher education. *Journal of Institutional Research, 11*(2), 38-47.

Kettunen, J. (2004). The strategic evaluation of regional development. *Assessment and Evaluation in Higher Education, 29*(3), 357-368.

Khandwalla, P. N. (1976). The techno-economic ecology of corporate strategy. *Journal of Management Studies, 13*(1), 62-75.

Kim, C., & Galliers, R. D. (2004). Toward a diffusion model for Internet systems. *Internet Research, 14*(2), 155-166.

King v. Lewis [2004] EWCA Civ. 1329; [2004] I.L.Pr. 31.

King, G. (1994). *Developing adaptive tests for school children.* NJ: Lawrence Erlbaum Associates.

King, J. L. (1983). Centralized versus decentralized computing: Organizational considerations and management options, *Computing Surveys, 15*(4), 319-349.

King, J.L., Gurbaxani, V., Kraemer, K.L., McFarlan, F.W., Raman, K.S., & Yap, C.S. (1994). Institutional factors in information technology innovation. *Information Systems Research, 5*(2), 139-169.

Kingsbury, G.G., & Hauser, C. (2004, April). *Computerized adaptive testing and No Child Left Behind.* Northwest Evaluation Association

Kling, R. (2000). Learning about information technologies and social change: The contribution of social informatics. *The Information Society, 16*(3), 217-232.

Klischewski, R. (2004). *Information integration or process integration? How to achieve interoperability in administration.* Retrieved from http://is.guc.edu.eg/uploads/egov2004/ klischewski.pdf

Kochha, S., & Dhanjal, G. (2005). E-governance report card. *Yojana, 49*, 60-65.

Kohl, U. (2004). The rule of law, jurisdiction and the Internet. *International Journal of Law and Information Technology, 365*, 12.

Kolb, D. A. (1984). *Experiential learning: Experience as the source of learning and development.* NJ: Englewood Cliffs.

Kompf, M. (2001). ICT could be the death knell of professoriate as we know it. *CAUT-ACPPU Bulletin Online.* Retrieved May 5, 2004, from http://www.caut.ca/english/bulletin/2001_sep/commentary.asp

Kothari, S., & Bansal, R. (2005). Innovative thinking in government. *Yojana, 49*, 34-39.

Kraemer, K., & Dedrick, J. (1994). Payoffs from investments in information technology: Lessons from the Asia-Pacific region. *World Development, 22*(12), 1921-1931.

Kraemer, K., Gurbaxani, V., & King, J.L. (1992). Economic development, government policy, and the diffusion of computing in Asia-Pacific countries. *Public Administration Review, 52*(2), 146-156.

Kreikebaum, H. (1996). *Grundlagen der Unternehmensethik.* Stuttgart: Schaeffer-Poeschel

Kuan, K. K. Y., & Chau, P. Y. K. (2001). A perception-based model for EDI adoption in small businesses using a technology-organization-environment framework. *Information & Management, 38*(8), 507-521.

Kuhn, T.S. (1996). *The structure of scientific revolutions.* 3rd edition. Chicago; London: The University of Chicago Press.

Kumar, K. (2005, February). Strategic framework for effective e-governance. *PC Quest*, 68-69.

Kunstelj, M., & Vintar, M. (2004). Evaluating the progress of e-government development: A critical analysis. *Information Polity, 9*(3-4), 131-148.

Lairson, T. D. (1999). Rethinking the "course" in and online world. *Campus-wide Information Systems, 16*(5). Retrieved May 5, 2004, from http://fox.rollins.edu/~tlairson/online.html

Lambe, P. (2002). *The autism of knowledge management.* Retrieved from www.straitsknowledge.com

Lamizet, B., & Weymouth, T. (Eds.). (1996). *Markets and Myths: Forces for change in the European media.* London: Longman.

Landsbergen, D. Jr. G. Wolken. (2001). Realizing the promise: Government information systems and the fourth generation of information technology. *Public Administration Review, 61*(2), 206-220.

Langdon, C. S. (2003). Information systems architecture styles and business interaction patterns: Toward theoretic correspondence. *Information Systems and e-Business Management, 1*, 283-304.

Larson, A. (1991). Partner networks: Leveraging external ties to improve entrepreneurial performance. *Journal of Business Venturing, 6*, 173-188.

Larsson, R., Bengtsson, L., Henriksson, K., & Sparks, J. (1998). The interorganizational learning dilemma: Collective knowledge development in strategic alliances. *Organization Science, 9*(3), 285-305.

Latour, B. (1988). *The pasteurization of France*. Cambridge, MA: Harvard University Press.

Latour, B. (1999). On recalling ANT. In J. Law & J. Hassard (Eds.), *Actor network and after.* Oxford: Blackwell Publishers.

Laurillard, D., Swift,B., & Darby, J. (1993). Academics' use of courseware materials: A survey. *Association for Learning Technology Journal, 1*(1), 4-12.

Law, J. (1992). *Notes on the theory of the actor network: Ordering, strategy and heterogeneity.* Centre for Science Studies and the Department of Sociology. Lancaster, UK: Lancaster University. Retrieved August 28, 2001, from http://www.lancs.ac.uk/fss/sociology/papers/law-notes-on-ant.pdf

Law, J. (1999). Introduction: Monsters, machines and sociotechnical relations. In Law & J. Hassard (Eds.), *Actor network theory and after.* Oxford: Blackwell Publishers.

Law, J. (2000a). *Objects, spaces, others, draft manuscript.* Centre for Science Studies and Department of Sociology, Lancaster University, Lancaster. Retrieved August 28, 2001, from http://tina.lancs.ac.uk/sociology/soc027jl.html

Law, J. (2000b). *Networks, relations, cyborgs: On the social study of technology*, (Draft). Centre for Science Studies and the Department of Sociology, Lancaster University. Retrieved August 28, 2001, from http://www.comp.lancs.ac.uk/sociology/soc042jl.htm

Law, J., & Hassard, J. (1999). *Actor network theory and after.* Oxford: Blackwell Publishers.

Lawrence, P.R., & Lorsch, J.W. (1969). *Developing organizations: Diagnosis and action.* Addison-Wesley Publishing Company Inc.

Lawrence, P.R., & Lorsch, J.W. (1967). Differentiation and integration in complex organizations, *Administrative Science Quarterly, 12*(1), 1-47.

Lawson, B. (2002). *How designers think: The design process demystified.* Architectural Press.

Layne, K., & Lee, J. (2001). Developing fully functional e-government: A four stage model. *Government Information Quarterly, 18*, 122-136.

Layne, K., & Lee, J. (2001). Developing fully functional e-Government: A four stage model. *Government Information Quarterly, 18*(2), 122-136.

Learner, E., & Levinsohn, J. (1995). International trade theory: The evidence. In *Handbook of Internal Economics, Vol. 3.* Amsterdam: North Holland.

Léautier, F.A. (2005). Forward. In Dahlman, C.J., Routti, J., & Ylä-Anttila, P. (Eds.) A prepublication version of the Overview of the book *Finland as a Knowledge Economy: Elements of Success and Lessons Learned,* published in January 2006. The International Bank for Reconstruction and Development / The World Bank, Washington, 2005.

Lechner, C., & Dowling, M. (2003). Firm networks: External relationships as sources for the growth and competitiveness of entrepreneurial firms. *Entrepreneurship & Regional Development, 15*, 1-26.

Leddo, H.J. (1996). *Computer-based features of the junior astronomy course at the University of Sydney.* Proceedings of University Science Teaching and the Web Workshop. Sydney: Uniserve Science.

Lee, C., Lee, K., &. Pennigs, J.M. (2001). Internal capabilities, external networks and performance: A study on technology-based ventures. *Strategic Management Journal, 22*, 615-640.

Lee, N.M., & Brown, S. D. (1994). Otherness and the actor-network: The undiscovered continent. *American Behavioral Scientist, 37*(6), 772-790.

Lehmus, P. (1999). *Strategiasuunnitelma 2000-2003. Arena.* Helsingin Puhelin Oyj, Liiketoimintaprojektit. 14.4. (1999). Helsinki, Finland.

Lévy, P. (1997). *Cyberculture.* Paris: Editions Odile Jacob.

Liebl, F. (1999). What system have you announced? (Impressionen aus einer Ökonomie der Ankündigung). In Liebl (Ed.), *e-Conomy - Management und Ökonomie in digitalen Kontexten.* Marburg: Metropolis Verlag.

Lim, B.B.L., Sun, Y., & Vila, J. (2004). Incorporating WS-security into a Web services-based portal. *Information Management & Computer Security, 12*(3), 206-217.

Lin, K.H. (2002, June 5). Cabinet to pour NT$36 billion into 'e-Taiwan' Project. *Taiwan News.*

Litho, P. (2006, June). *ICTs, 'empowerment' and women living in rural Uganda.* Paper presented at Women, ICTs and Development Workshop. Manchester, UK: IDPM, University of Manchester.

Littlejohn, A. (1999). The virtual university as a conceptual model for faculty change and innovation. *Journal of Interactive Learning Environments, 7*(2/3), 209-226.

LLL: An Internet Learning Support Service for Linking Local Learners. (n.d.) Retrieved July 15, 2006, from http://www.linkinglearners.net/

Local Access Global Reach: Five Years of Tasmanian Communities Online 1998- 2002. Retrieved May 31, 2005, from http://www.education.tas.gov.au/tco

Lorsch, J. W., & Lawrence, P. R. (1970). *Studies in organization design.* Georgetown, Ontario: Irwin-Dorsey Limited.

Loutchansky v. The Times Newspapers Ltd., EWCA Civ. 1805 (2001).

Lucas, F.B. (1999). *A radio broadcasting model for rural women and farm households.* Prepared for FAO, Regional Office for Asia and the Pacific.

Luftman, J., & Brier, T. (1999). Achieving and sustaining business-IT alignment, *California Management Review, 42*(1), 109-122.

Luke, A., & Gilbert, P. (1993). *Literacy in contexts: Australian perspectives and issues.* St. Leonards, NSW: Allan & Unwin.

Lundgren, A. (1991). *Technological innovation and network evolution. The emergence of industrial networks.* EFI, Stockholm School of Economics, Stockholm.

Lunney v. Prodigy Services Co., 250 AD 2d 230 (1998).

Lynda Lacoste (née Heineman) v. Multimania Production SA, E.C.C. 22 (2001).

MacLean, D., Souter, D., Deane, J., & Lilley, S. (2002). *Louder voices: Strengthening developing country participation in international ICT decision-making.* London: Commonwealth Telecommunications Organization and Panos Institute.

Madon, S. (1994). Designing information system for development planning. Alfred Waller, Oxon. (From : Sein, M. K. and Harindranath, G. (2004). Conceptualizing the ICT artifact: Towards understanding the role of ICT in national development. *The Information Society, 20*,15-24.

Madon, S. (2000). The Internet and socio-economic development. *Information Technology and People, 13*(2), 85-101.

Main Statistical Office. (2002). Household situation in 2002 From the Results of the Household Budget Survey. Retrieved July 15, 2003, from http://www.stat.gov.pl/serwis/nieregularne/gosp_dom2002.htm

Malone, T. W., & Crowston, K. (1994). The interdisciplinary study of coordination, *ACM Computing Surveys, 26*(1), 87-118.

Malone, T. W., & Rockart, J. F. (1991). Computers, networks, and the corporation. *Scientific America, 265,* 128-136.

Mano, C. (2001). Não sei se caso ou compro uma bicicleta. *Revista Exame,* i. 735.

Mansell, R., & When, U. (Eds.). (1998). *Information Technology for Sustainable Development.* Oxford: Oxford University Press.

Mansen, J. (1999). *Innovations in computerised assessment.* NJ: Lawrence Erlbaum Associates.

March, J. G. (1991). Exploration and exploitation of organizational learning, *Organization Science, 2,* 71-87.

March, J.G., & Simon, H.A. (1958). *Organizations* (2nd ed.). [s.l.], Wiley.

Marchand, D. A. (2000). Hard IM choices for senior managers. In D. A. Marchand, T. H. Davenport, & T. Dickson (Ed.), *Mastering information management* (pp. 295-300). London: Financial Times/Prentice Hall.

Marchand, D. A. (2000). How to keep up with the hypercompetition. In D. A. Marchand, T. H. Davenport, & T. Dickson (Ed.), *Mastering information management* (pp. 120-126). London: Financial Times/Prentice Hall.

Marchionini, G., Samet, H., & Brandt, L. (2003). Digital government. *Communications of the ACM, 46*(1), 25-27.

Maskell, D., & Robinson, I. (2001). *The new idea of a university.* London: Haven Books.

Massey, D. (1994). *Space, place, and gender.* Minneapolis: University of Minnesota Press.

Maul, G. P., & Spotts, D. S. (1993b). A comparison of computer-based training and classroom instruction. *Industrial Engineering, 25*(2), 25-27.

Maul, G.P., & Spotts, D.S. (1993a). Developing computer-based instructional courses. *Industrial Management, 35*(6), 9-11.

McCalman, J. (2003). What can we do for corporate nomads? IT and facilities management. In Joia (Ed.), *IT-based management: Challenges and solutions* (pp. 130-142). Hershey: Idea Group Publishing.

McGonagle, T. (2001). *Changing Aspects of Broadcasting: New Territories and New Challenges.* IRIS Plus, Issue 2001-10.

McQuail, D. & Siune, K. (Eds.). (1998). *Media Policy. Convergente, Concentration and Commerce.* Euromedia Research Group. London: Sage.

Medjahed, B., Rezgui, A., Bouguettaya, A., & Ouzzani, M. (2003, January/February). Infrastructure for e-government Web services. *IEEE Internet Computing,* 58-65.

Meeks, B. N. (2000). Better democracy through technology. In Baird, Ramsower, & Rosenbaum (Eds.), *Cyberethics - Social and moral issues in the computer age.* New York: Prometheus Books.

Mehrtens, J., Cragg, P.B., & Mills, A.M. (2001). A model of Internet adoption by SMEs. *Information & Management, 39*(3), 165-176.

MGM v. Grokster ,125 S.Ct. 2764 (2005).

MIC (2004). Ministry of Internal Affairs and Communications. *MIC Communications News. 15*,12.

MIC (2005). Ministry of Internal Affairs and Communications, *MIC Communications News.16*,1.

Miles, M., & M. Huberman. (1994). *Qualitative data analysis.* Thousand Oaks, CA: SAGE Publications.

Miller, D. (1988). Relating Porter's business strategies to environment and structure: Analysis and performance implications, *Academy of Management Journal, 31*(2), 280-308.

Miller, D., & Friesen, P. H. (1978). Archetypes of strategy formulation, *Management Science, 24*(9), 921-933.

Milliron, M. D., & Miles, C. L. (2000). Education in a digital democracy: Leading the charge for learning about, with and beyond technology. *EDUCAUSE Review*, 50-62. Retrieved May 5, 2004, from http://www.educause.edu/pub/er/erm00/articles006/ perm0064.pdf

Milmo, P., & Rogers, W.V.H. (Ed.). (2004). *Gatley on Libel and Slander 10th edition.* London: Sweet & Maxwell. *Dow Jones & Co. Inc v Gutnick* [2002] HCA 56.

Ministry of Education (1999). Koulutuksen ja tutkimuksen tietostrategia sekä toimeenpanosuunnitelma 2000-2004. Helsinki: Ministry of Education.

Ministry of Education (2003a). Koulutuksen ja tutkimuksen tietoyhteiskuntaohjelma 2004-2006. Helsinki: Ministry of Education.

Ministry of Education (2003b). Koulutus ja tutkimus vuosina 2003-2008, Kehittämissuunnitelma. Helsinki: Ministry of Education.

Mintzberg, H. (1979). *The structuring of organizations.* Prentice-Hall: Englewood Cliffs.

Mintzberg, H., & Quinn, J. B. (1991). *The strategy process.* Englewood Cliffs, NJ: Prentice-Hall.

Mol, A., & Law, J. (1994). Regions, networks and fluids: Anaemia and social topology. *Social Studies of Science, 24*(4), 641-671.

Möller, K., & Svahn, S. (2006). Role of knowledge in value creation in business nets. *Journal of Management Studies, 43*(5), 985-1007.

Möller, K., & Halinen, A. (1999). Business relationships and networks: Managerial challenge of network era. *Industrial Marketing Management, 28*, 413-427.

Möller, K., & Svahn, S. (2003). Managing strategic nets: A capability perspective. *Marketing Theory, 3*(2), 201-226.

Möller, K., Rajala, A., & Svahn, S. (2005). Strategic business nets–Their types and management. *Journal of Business Research*, 1274-1284.

Montealegre, R. (1999). A temporal model of institutional interventions for information technology adoption in less-developed countries. *Journal of Management Information Systems, 16*(1), 207-232.

Moon, M. J., & Norris, D. F. (2005). Does managerial orientation matter? The adoption of reinventing government and e-government at the municipal level. *Information Systems Journal, 15*, 43-60.

Morgan, S. (2006, June). *Studying gender, ICTs and development using new institutionalism.* Paper presented at Women, ICTs and Development Workshop. Manchester, UK: IDPM, University of Manchester.

Morgan, S., Heeks R., & Arun, S. (2004). *Researching ICT-based enterprise for women in developing countries: A gender perspective.* Manchester, UK: IDPM, University of Manchester. Retrieved July 2006, from http://www.womenictenterprise.org

Morse, K. (2003). *Does one size fit all? Exploring asynchronous learning in a multicultural environment.* JALN.

MPT (2003, December). *Info-Communication Country Report in Vietnam, Ministry of Posts and Telematics (MPT).* The 29th Asian Info-communication Council conference, Tokyo, Japan.

Muchlinski, P. (1999). *Multinationals and the law* (2nd ed.). Oxford: Blackwell Publishing.

Muller, B. (2005). The European interoperability framework: An industry perspective. *European Review of Political Technologies, 3.* Retrieved from www.politech-institute.org/review/articles/MULLER_Benoit_volume_3.pdf

Musgrave, S.J. (2005). Community portals –The UK experience a false dawn over the field of dreams? *The Journal of Community Informatics, 1*(2), 32-44.

Nadamato, S. (2006, June). *Gender and ICTs: Empowerment of women in telecentres in Jamaica.* Paper presented at Women, ICTs and Development Workshop. Manchester, UK.

Nadler, D., & Tushman, M. L. (1998). *Competing by design: The power of organizational architecture.* Oxford: Oxford University Press.

Nahapiet, J., & Ghoshal, S. (1998). Social capital, intellectual capital, and the organizational advantage. *Academy of Management Review, 23*, 242-266.

Nath, V. (2001). Empowerment and governance through information and communication technologies: Women's perspective. *Intl. Inform. & Libr. Rev., 33*, 317-339.

Nath, V. (2006). Empowerment of women through ICT-enabled networks. In N. Hafkin & S. Huyer (Eds.), *Cinderella or cyberella? Empowering women in the knowledge society.* Bloomfield, CT: Kumarian Press.

National Assembly for Wales (2001). *Information and Communication Technology for Education.* Education and Life-Long Learning Committee. Final Report, Cardiff, UK.

Nelson, L. (1999). Collaborative problem solving In *Instructional: Design Theories and Models.*

Noble, D. F. (2001). *Digital diploma mills: The automation of higher education.* New York: Monthly Review Press.

Nonaka, I., Toyama, R., & Konno, N. (2001). SECI, Ba and leadership: A unified model of dynamic knowledge creation. In Nonaka & Teece (Eds.), *Managing industrial knowledge. Creation, transfer and utilization* (pp. 13-43). London: Sage.

Norman, D. (1996). Design as practiced. In *Winograd—Bringing design to software.* Addison Wesley.

Normann, R., & Ramírez, R. (1993). From value chain to value constellation. Designing interactive strategy. *Harvard Business Review, 71*(4), 65-77.

North, D. C. (1990). *Institutions, institutional change, and economic performance.* Cambridge, MA: Cambridge University Press.

NSW Department of Commerce & Office of Information Technology (June 2004), *CTC Final Report,* Sydney: NSW.

OECD (1987). *Information technology and economic prospects. Information, Computer and Communications Policy Series.* ICCP No. 12. Paris: OECD.

OECD (1988). *New technologies in the 1990s: A socio-economic strategy.* Paris: OECD.

OECD (2004). *The economic impact of ICT: measurement, evidence and implications.* Paris: OECD Publications Service.

OECD (2005). *Financing ICTs for Development: Efforts of DAC Members,*

OII (2005, May). *The Internet in Britain.* The Oxford Internet Survey (OxIS), Oxford Internet Institute, University of Oxford, Oxford.

Olukoshi, A., & O'Neil, M. (2003). In Ramata Malo Thioune (Ed.), *INFORMATION AND COMMUNICATION TECHNOLOGIES FOR DEVELOPMENT IN AFRICA: Vol.1., Opportunities and Challenges for Community Development.* International Development Research Centre, Canada. ISBN 2-86978-114-8.

Ostrom, E. (1990). *Governing the commons: The evolution of institutions for collective action.* Cambridge: Cambridge University Press.

Ostrom, E., Schroeder, L., & Wynne, S. (1993). Analyzing the performance of alternative institutional arrangements for sustaining rural infrastructure in developing countries. *Journal of Public Administration Research and Theory, 3*(1), 11-45.

Pace, N. (2006). *China surpasses U.S. in Internet use.* Retrieved July 12, 2006, from http://www.forbes.com/2006/03/31/china-internet-usage-cx_nwp_0403china.html

Palanisamy, R. (2004). Issues and challenges in e-governance planning. *Electronic Government, 1*(3), 253-272.

Paletz, D. L. (2000). Advanced information technology and political communication. In Baird, Ramsower, & Rosenbaum (Eds.), *Cyberethics - Social and moral issues in the computer age* (pp. 285-287). New York: Prometheus Books.

Palmu (1999). *Palvelujen Hallinta Mobiilissa Internetissä.* Helsingin Puhelin Oyj.13.1.1998. Helsinki, Finland.

Palmu-kokousmuistio. 13.1. (1999). Helsingin Puhelin Oyj. Helsinki, Finland.

Palmu-projektisuunnitelma HPY .(1999). Helsingin Puhelin Oyj. Helsinki, Finland.

Palvia, P., Palvia, S., & Roche, E. (Eds.). (2002). *Global information technology & electronic commerce: Issues for the new millennium.* Marietta, GA: Ivy League Publishing

Park, S. H. (1996). Managing an interorganizational network: A framework of the institutional mechanism for network control. *Organization Studies, 17*(5), 795-824.

Parker, M., Benson, R., & Trainor, H. (1988). *Information economics: Linking business performance to information technology.* NJ: Prentice Hall, Englewood Cliffs.

Parker, M., Trainor, H., & Benson, R. (1989). *Information strategy and economics: Linking business performance to information technology.* NJ: Prentice Hall.

Parker, M.M., & Benson, R.J. (1989). Enterprisewide information management: State-of-the-art strategic planning. *Journal of Information Systems Management,* 14-23.

Parker, M.M., Ribbers, P.M., & Parker, S.R. (1997). *Critical IT Issues for Business Transformation.* Handboek BIK, D 1005-1, Tilburg University.

Parolini, C. (1999). *The value net: A tool for competitive strategy.* UK: John Wiley & Sons

Pascarella, E. T. (2001). Identifying excellence in undergraduate education: Are we even close? *Change,* 19-23.

Patel, R., & Parmentier, M.J. (2005). The persistance of traditional gender roles in the information technology sector: A study of female engineers in India. *Information Technologies and International Development, 2*(3), 29-46.

Paulwell, P. (2005). *Sectoral budget presentation 2005/2006.* Kingston: Ministry of Commerce, Science & Technology (with Energy).

Penrose, E. (1959). *The theory of growth of the firm.* London: Basil Blackwell.

Pentland (2003). Panel contribution to: Ackerman, Mark, Pentland, Brian T., Qureshi, Sajda & Yakura, Elaine K. (2003): Visual Elements in the Discourse on Information Technology. In: Wynn, Eleanor; Whitley, Edgar; Myers, Michael D. & DeGross, Janice (Eds.) (2003): *Global and Organizational Discourse About Information Technology*: Ifip Tc8/Wg8.2 Working Conference on Global and Organizational Discourse About Information Technology, Barcelona, Spain. Dordrecht: Kluwer Academic Publishers.

Peppard, J., & Ward, S. (1999). Mind the gap: Diagnosing the relationship between the IT organization and the rest of the business. *Journal of Strategic Information Systems, 8*, 29-60.

Pereira, P.L.S., Carvalho, M. M., & Laurindo, F.J.B. (2004). Competitive mapping in a Brazilian food industry. In *EUROMA2004 - European Operations Management Association International Conference, 2004, Fontainebleau. Proceedings of EurOMA2004* (Vol. 1, pp. 247-256).

Peterson, R., & Ribbers, P.M.A (2002) *Service Level Agreements: Noodzakelijk en toereikend?* .ego. Retrieved from http://www.sbit.nl/ego/archief.php

Peterson, R.R. (1998). Successful exploitation of information technology in financial services. The role of IT governance. In *Proceedings of the International Conference on Information Systems (Doctoral Consortium)*, Helsinki, Finland.

Peterson, R.R. (2000). *Emerging capabilities for IT governance: Exploring stakeholder perspectives in Financial Services.* Conference Proceedings European Conference on Information Systems 2000, Vienna, Austria.

Peterson, R.R. (2000). *Examining ambidextrous designs for information governance: Evidence from a longitudinal comparative case study in Financial Services.*

Conference proceedings of the European Conference on IT Evaluation 2000, Dublin, Ireland.

Peterson, R.R. (2001). *Configurations and coordination for global information governance: Complex designs in a transnational European Context.* Proceedings of the 34th HICSS Conference, Hawaii.

Peterson, R.R. (2001). *Information governance: An empirical investigation into the differentiation and integration of strategic decision-making for IT.* Tilburg University, The Netherlands

Peterson, R.R., O'Callaghan, R., & Ribbers, P.M.A. (2000). *Information technology governance by design.* Conference proceedings of the International Conference on Information Systems 2000, Brisbane, Australia.

Peterson, R.R., Smits, M., & Spanjers, R. (2000). *Exploring IT-enabled networked organizations in health care: Emerging practices and phases of development.* Conference Proceedings the European Conference on Information Systems 2000, Vienna, Austria.

Pettigrew, A.M. (1997). What is a processual Analysis? *Scandinavian Journal of Management, 13,* 337-348.

Phipps, R. A., & Merisotis, J. P. (2000). *Quality on the line: Benchmarks for success in Internet-based education.* Retrieved May 5, 2004, from the Institute for Higher Education Policy Web site http://www.ihep.com/Pubs/PDF/Quality.pdf

Pigg, K., & Crank, L.D. (2003). *Rural economic development: The impact of information and telecommunications technology.* Paper presented at 5th International Information Technology in Regional Areas Conference, Caloundra, Qld.

Plowman, I., Ashkanasy, N.M., Gardner, J., & Letts, M. (2003). Main Report, *Innovation in rural Queensland: Why some towns thrive while others languish.* Brisbane, QLD: University of Queensland.

Pond, W. K. (2002). Distributed education in the 21st century: Implications for quality assurance. *Online Journal of Distance Learning Administration, 5(2).* Retrieved May 5, 2004, from http://www.westga.edu/%7Edistance/ojdla/summer52/pond52.html

Poon, S., & Swatman, P. (1999). An exploratory study of small business Internet commerce issues. *Information & Management, 35*(1), 9-18.

Popov, A. R. (May 2005). The extraterritorial reach of the Lanham Act over trademarks on the Internet: Adopting the Lanham Act to global e-commerce or an unjustified extension of US laws? *Journal of Internet Law, 19.*

Porter, M. (1990). *The competitive advantage of nations.* London: Macmillan.

Porter, M. (1996). What is Strategy? *Harvard Business Review,* November-December, 61-78.

Porter, M. E. (1980). *Competitive advantage.* Harvard Business Press.

Porter, M. E. (2001, March). *Strategy and the Internet.* Harvard Business Review.

Porter, M.E. (1979). How competitive forces shape strategy. *Harvard Business Review,* 137-145.

Porter, M.E. (1987). From competitive advantage to corporate strategy. *Harvard Business Review,* May-June.

Porter, M.E. (2001). Strategy and the Internet. *Harvard Business Review*, 63-78.

Postman, N. (1992). *Technopoly - The surrender of culture to technology.* New York: Vintage Books.

Powell, T. C. (1992). Organizational alignment as competitive advantage, *Strategic Management Journal, 13,* 119-134.

Prahalad, C.K., & Hamel, G. (1990). The core competence of the corporation. *Harvard Business Review,* 79-91.

Prosser, D. (1953). Interstate publication. *Michigan Law Review, 959*, 51.

PTI, April 5, 2005. www.agencyfaqs.com, April 5, 2005. Indo-Asian News Service. Retrieved April 3, 2005, from http://www.infochangeindia.org/Itanddltop.jsp?section_idv=9#1917 as accessed on 22/08/2005

Quinn, J. B. (2001). Services and technology: Revolutionizing higher education. *EDUCAUSE,* 28-37. Retrieved May 5, 2004, from http://www.educause.edu/ir/library/pdf/erm0141.pdf

Quinn, R. E., & Rohrbaugh, J. (1983). A spatial model of effectiveness criteria: Towards a competing values approach to organizational analysis, *Management Science, 29*(3), 363-377.

Quinn, R. E., DeGraff, J., & Thakor, A. (2000). Creating shareholder value - and dispelling some myths. In T. Dickson (Ed.), *Mastering strategy.* London: Pearson Education Limited.

Rabechini Jr, R., & Carvalho, M.M. (2003) O perfil das competências em equipes de projeto. Revista de Administração de Empresas—RAE Eletrônica—FGV.

Rabinowitz, S., & Brandt, T. (2001). *Computer-based assessment: Can it deliver on its promise?* Vancouver: Kogan Page.

Radloff, J. (1999). Celebrating the power, diversity and strength of African women's networks. In S. Cummings, et al. (Eds.), *Women's information services and networks.* Amsterdam: KIT Press and Oxfam GB.

Raghavan, G. V., & Nair, V. S. M. (2004). e-Governance: Leveraging IT for reinventing the government. *Manorama Yearbook 2004* (pp. 130-138).

Rammelo, S. (2001). *Corporations in private international law: A European perspective.* Oxford: University Press.

Rappa, M. (2002). Business models on the Web. Managing the digital enterprise. Retrived from http://digitalenterprise.org/models/models.html

Re Enforcement of an English Injunction, ILPr 320 (1997).

Re the Law to Promote Confidence in the Digital Economy, 1 CMLR 14 (French Conseil Constitutionnel Constitutional Court, 2005).

Reagle, J.M. Jr. (1996). Trust in electronic markets. In *First Monday* (1:2). Available www.firstmonday.dk

Reese, R. A. (2005, December). The problems of judging young technologies. *Journal of Internet Law, 1.*

Reich, B. H., & Benbasat, I. (1996). Measuring the linkage between business and information technology objectives. *MIS Quarterly, 20*(1), 55-81.

Reigeluth, C. (1999). What is instructional-design theory and how is it changing? In *Instructional—Design Theories and Models.*

Reigeluth, C., & Moore, J. (1999). Cognitive education and the cognitive domain. In *Instructional- Design Theories and Models.*

Reilly, K. (2002). *Government ICT use in Central America... a view from civil society.* Unpublished Manuscript. Retrieved March 2002, from http://katherine.reilly.net/publications.html

Remenyi, D. (2000). The elusive nature of delivering benefits from IT investment. *Electronic Journal of Information Systems Evaluation,* 1. Retireved from http://www.iteva.rug.nl/ejise/index.html

Remenyi, D., & Bannister, F. (Eds.). (2003). In *Proceedings of the European Conference on e-Government,* Trinity College Dublin.

Research Development and Evaluation Commission (RDEC). (2001). *E-government implementation plan.* Taipei, Taiwan: Research Development and Evaluation Commission, Executive Yuan.

Research Program: Engendering ICT Policy—Call for Grant Proposals, WSIS Gender Caucus. Retrieved March 2005, from http://wsis.ecommons.ca/node/view/502

Review of Recent Trends of ODA and its Contribution, Report to the UN Task Force on Financial Mechanisms for ICT for Development (TFFM), OECD Publications, France, 2005.

Revista Exame (2000). Na terra do açaí. São Paulo Nov.

Ritter, T. (1999). The networking company. *Industrial Marketing Management, 28*(5), 497-506.

Robbins, S. P. (1994). *Essentials of organizational behavior.* London: Prentice Hall International.

Robinson, J. (1999). *Computer-assisted peer review.* New York: Kogan Page.

Rocheleau, B. (2003). Politics, accountability, and governmental information systems. In D. Garson (Ed.), *Public information technology: Policy and management issues* (pp. 20-52). Hershey: Idea Group Publishing.

Rockart, J. F. (1988). The line takes the leadership: IS management in a wired society. *Sloan Management Review, 29*(4), 57-64.

Rockart, J. F., Earl, M., & Ross, J. W. (1996). Eight imperatives for the new IT organization, *Sloan Management Review, 38*(1), 43-55.

Rocks, S., Gilmore, Audrey & Carson, David (2005). Developing strategic marketing through the use of marketing networks. *Journal of Strategic Marketing, 13*, 81-92.

Rogers, C.R. (1951). *Client centred therapy.* London: Constable & Co.

Rogers, C.R. (1961). *On becoming a person.* London: Constable & Co.

Rosenthal, R. E. (April 2003). From ZIPPO to ALS: Jurisdiction over IP violations on the Internet. *Journal of Internet Law, 12.*

Ross, J. W., Beath, C. M., & Goodhue, D. L. (1996). Develop long-term competitiveness through IT assets. *Sloan Management Review, 38*(1), 31-42.

Rossman, P. (2005). *TRANSFORMING HUMAN SOCIETY: The Future of Higher (Lifelong) Education: A Vision of Job Training and Education for Everyone on the Planet during this Century, Planning For all Worldwide, a Holistic View.* Retrieved June 12, 2006, from http://ecolecon.missouri.edu/globalresearch/

Rubinoff, D. (2003). *Campesina leaders, life histories and ICT for development: Cybernetworking and rural women in Central America.* Unpublished Doctoral Dissertation, University of Colorado-Boulder.

Rubinoff, D. (2005). Life histories in cyberspace: Life writing as a development tool for rural women. *Geoforum, 36*(1), 59-75.

Rumelhart, D.E., & Norman, D.A. (1983). Representation in memory. In R.C. Atkinson, R. J. Herrnstein, G. Lindsey, & R. D. Luce (Eds.), *Handbook of experimental psychology.* London: Wiley & Sons.

Rumelt, R. F. (1974).*Strategy, structure, and economic performance in large industrial corporations.* Boston: Harvard University Press.

Rundle, M. (April 2006). Beyond Internet governance: The emerging international framework for governing the networked world. *Journal of Internet Law, 1.*

Russell, B. (1975). *Power.* London: Routledge.

Russell, M., & Haney, W. (2000). The gap between testing and technology in schools. *Statements, 1,* 2, National Board on Educational Testing and Public Policy, January 2000.

Rustad, M.L., & Koenig, T. H. (May 2006). Harmonizing Internet law: Lessons from Europe. *Journal of Internet Law, 3.*

Ryan, Nader, J.R., & Lihan, T.P. (2003, Sept). Internet-based exercise of personal jurisdiction. *Journal of Internet Law, 12.*

S.A. Groupe Progres v. Syndicat National Des Journalistes, ECC 8 (Cour D'Appeal, Lyon France 2001).

Sableman, M. (June 1999). Business liabilities on the Internet. *Journal of Internet Law, 1.*

Sadowski, B., Maitland, C., & van Dogen, J. (2002). Strategic use of the Internet by small and medium-sized companies: An exploratory study. *Information Economics and Policy, 14*(1), 75-93.

Salmi, P. (1999). *Helsinki Arena 2000-esittelymateriaali.* Helsingin Puhelin Oyj. Helsinki, Finland.

Sambamurthy, V., & Zmud, R.W. (1999). Arrangements for information technology governance: A theory of multiple contingencies. *MIS Quarterly, 23*(2), 261-290.

Sambamurthy, V., & Zmud, R.W. (2000). Research commentary: the organizing logic for an enterprise's IT activities in the digital era: A prognosis of practice and a call for research, *Information Systems Research, 11*(2), 105-114.

Sankaran, S., Dick, B., Passfield, R., & Swepson, P. (2001). *Effective change management using action learning and action research.* Lismore, NSW: Southern Cross University Press.

Sauer, C., & Yetton, P. W. (1997). The right stuff: An introduction to new thinking about IT management. In C. Sauer, P. W. Yetton, & Associates (Eds.), *Steps to the future: Fresh thinking on the management of IT-based organizational transformation* (pp. 1-25). San Francisco: Jossey-Bass Publishers.

Scaranello, B.M. (2004). Análise da Concorrência: Um Estudo de Caso do Setor de Bebidas. Trabalho de Formatura. Escola Politécnica da USP. Orientadora: Marly Monteiro de Carvalho. São Paulo.

Schedler, K., & Scharg, M.C. (2002). *Exploring the interrelations between electronic government and the new public management: A managerial framework for electronic government.* Paper presented at APPAM 2002 Conference.

Scheier, R. (2001). Central intelligence: Johnson & Johnson case study. *CIO INSIGHT,* December.

Schiller, D. (1999). *Digital capitalism: Networking the global market system.* Cambridge, MA / London: MIT Press.

Schwarz, G., & Watson, B. M. (2005). The influence of perceptions of social identity on information technology-enabled change. *Group & Organization Management, 30*(3), 289-318.

Scouller, A. (1994). Test writing on computers: an experiment comparing student performance on tests conducted via computer and via paper and pencil. *Education and Policy Analysis Archives, 5*(30), 21-25.

Seidl-Hohenveldern, I. (1987). *Corporations in and under international law.* Cambridge: Grotius Publications Ltd.

Sein, M. K., & Ahmad, I. (2001, June 4-6). A framework to study the impact of information and communication technologies on developing countries: The case of cellular phones in Bangladesh. In *Proceedings of the BIT World International Conference*, Cairo, Egypt.

Sein, M. K., & Harindranath, G. (2004). Conceptualizing the ICT artifact: Towards understanding the role of ICT in national development. *The Information Society, 20,* 15-24.

Selinger, M. (2004). Cultural and pedagogical implications of a global e-learning programme. *Cambridge Journal of Education.*

Sen, A. (1987). *On ethics and economics.* New York: Basil Blackwell.

Senge, P. M. (1990). *The fifth discipline.* New York: DoubleDay.

Seoul-Gyeonggi Declaration on the Equal Participation of Women in the Information Society (2005). Retrieved October 2005, from http://women.or.kr/wsisforum/

SGAE: *Anuario SGAE de las Artes Escénicas, Musicales y Audiovisuales.* Madrid: SGAE.

Sharma, R. R., & Sharma, L. (2003). Use of information technology in education: Some issues and challenges. In B. Shah, Pub., K. Mahal, & A. (Eds.), *Information technology and education: Vision and prospects* (pp. 44-57).

Shepherd, A. (1998). *Sustainable rural development.* London: Macmillan Press Ltd.

Shevill v. Presse Alliance SA,AC 959 (1996).

Shin, N. (2003). Productivity gains from IT's reduction of coordination costs. In Shin (Ed.), *Creating business value with information technology: Challenges and solutions* (pp. 125-145). Hershey, PA: Idea Group Publishing.

Silander, P., & Koli, H. (2003). *Oppimisaihioista oppimisprosessiin.* Helsinki: Finn Lectura Oy.

Silva, L., & Figueroa E. (2002). Institutional intervention and the expansion of ICTs in Latin America: The case of Chile. *Information Technology & People, 15*(1), 8.

Simon, H.A., & Barnard, C.I. (1961). *Administrative behavior: A study of decision-making processes in administrative organization*. New York: The Macmillan Company

Sims, R., Dobbs, G., & Hand, T. (2002). Enhancing quality in online learning: Scaffolding planning and design through proactive evaluation. *Distance Education, 23*(2), 135-148.

Simson, E. M. v. (1990). The centrally decentralized IS organization. *Harvard Business Review, 68*(4), 158-160.

Simson, E. M. v. (1995). The recentralization of IT. *Computerworld, 1*(11), 2-7.

Siochru, Sean O., & Girard, Bruce (2005). *Community-based Networks and Innovative Technologies: New models serve and empower the poor.* UNDP, New York, 2005.

Sly, L., & Rennie, L. (1999). Computer managed learning as an aid to formative assessment in higher education. In S. Brown, J. Bull, & P. Race (Eds.), *Computer-assisted assessment in higher education*. London: Kogan Page.

Snellen, I. (2000). Public service in an information society. In G. Peters & D. J. Savoie (Eds.), *Governance in the twenty-first century: Revitalizing the public service.* Montreal & Kingston: Canadian Centre for Management Development.

Soan (2006). *Digital sector comes into sharp focus.* Retrieved July 12, 2006, from http://english.vietnamnet.vn/tech/2006/06/582289/

SourceOECD Telecommunications Database. (n.d). Retrieved July 15, 2003, from http://www.sourceoecd.org/content/templates/el/el_location_ivt.htm?comm=telecomm

Spinello, R. (2000). *Cyberethics: Morality and law in cyberspace*. London: Jones and Bartlett.

Stallman, R. (1995). Are computer property rights absolute? In Johnson & Nissenbaum (Eds.), *Computers, ethics & social values*. Upper Saddle River: Prentice Hall.

Star, S. L. (1991). Power, technologies and the phenomenology of conventions: On being allergic to onions. In J. Law (Ed.), *A sociology of monsters? Essays on power, technology and domination, Sociological Review Monograph*. London, Routledge.

State Committee for Scientific Research. (2003). Strategy for Information Technology Development of the Republic of Poland – ePolska.

Statistical Institute of Jamaica (2006). Retrieved July 26, 2006, from http://www.statinja.com

Steinberg v. Pritchard Englefield (A Firm), EWCA Civ. 824 (2005).

Steinberg, J. (2003). Information technology and development beyond "either/or". *The Brookings Review, 21*(2), 45-48.

Steward, F., Conway, S., Mahay, V., Yearley, S., Bailey, P., Garcia, C., Joly, P., Lemarie, J., & Hansen, A. (2000). *Consumer and sectoral networks in the societal management of food technology.* Final report to the EC Environment and Climate Research Programme, Contract ENV4-CT97-0695.

Stichler, R. N. (1998). Ethics in the information market. In Stichler & Hauptman (Eds.), *Ethics, information and technology: Readings*. Jefferson, NC: MacFarland & Company.

Stichler, R. N., & Hauptman, R. (Eds.). (1998). *Ethics, information and technology: Readings*. Jefferson, NC: MacFarland & Company.

Strathern, M. (1996). Cutting the network. *Journal of the Royal Anthropological Institute, 2*(3), 517-535.

Tacchi, J., Slater, D., & Hearn, G. (2003). *Ethnographic participation research*. New Delhi: UNESCO Regional Bureau for Communication and Information. Retrieved July 15, 2006, from http://portal.unesco.org/ci/en/ev.php-URL_ID=16619&URL_DO=DO_TOPIC&URL_SECTION=201.html

Tavakolian, H. (1989). Linking the information technology structure with organizational competitive strategy: A survey. *MIS Quarterly, 13*(3), 309-317.

Taylor Nelson Sofres & Center for Public Opinion Research. (2002). *Government online: A national perspective.* Annual Country Report: Poland.

Taylor, S. (1995). *Handbook of modern item response theory.* New York: Springer-Verlag.

Teece, D. J., Pisano, G., & Shuen, A. (1997).Dynamic capabilities and strategic management. *Strategic Management Journal, 18*(7), 509-533.

Tenhovuori, T. (1996). *Helsinki Arena 2000–palveluverkko. Tavoitesuunnitelma Helsinki Arena 2000.* Helsingin Puhelin Oyj. 2.12.1996. Helsinki, Finland.

Teo, T. S. H., & King, W. R. (1999). An empirical study of the impacts of integrating business planning and information systems planning. *European Journal of Information Systems, 8*, 200-210.

Thanh, P.X. (2003, March). *ICT Policy Report.* Vietnam, Paper presented at the Interregional Seminar and Regional Demonstration Workshop on Accessible Information and Communications Technologies and Persons with Disabilities, Manila, Philippines.

The Blind Men and the Elephant. (1963). McGraw-Hill Book Company.

The Gazette of India (2005, June 21). Extraordinary Part II, Section I. Mnistry of Law and Justice (Legislative Department). Govt. of India. New Delhi.

The Role of ICT in Building Communities and Social Capital (2005, January). *A discussion paper.* Access Branch Department of Communications, Information Technology and the Arts.

Thioune, R.M. (Ed.) (2003). *Information and Communication Technologies for Development in Africa, Volume 1.* Opportunities and Challenges for Community Development, IDRC, Ottawa, Canada, 2003.

Thomas, J. (2004). Strategically transforming governance for the new global economy. *Strategies for Competitiveness in the New Global Economy.* Presented at the 4th International Conference on Quality Management, Teheran, Iran. Retrieved at www.qmconf.com/Docs/0097.pdf

Thompson, C. (2003). *The Mission for Public Service Broadcasting in a competitive digital environment.* Retrieved July 6, 2006, from www.bbc.co.uk/pressoffice/speeches/stories/thomsoncaroline_oxfordmediaconv.shtml

Thompson, G., Frances, J., & Levačić, R., & Mitchell, J. (Eds.). (1994). *Markets, hierarchies & networks: The coordination of social life.* Thousand Oaks, CA: SAGE Publications.

Thompson, J. (1967). *Organizations in action: Social sciences bases of administrative theory.* New York: McGraw-Hill Book Co.

ThuHue, N. (2005, July). *Internet governance for socio-economic development.* Case study from Vietnam, a case study in GIPI, Geneva.

Tooley, J. (2000). *Reclaiming education.* London: Cassell.

Torre, N. P., & Ponce de la (2003). e-Government and the "network society". *UPGRADE, the European Journal for the Informatics Professional, 4*(2), 14-20. Retrieved from http://www.upgrade-cepis.org

Treacy, M., & Wiersema, F. (1993). Customer intimacy and other value disciplines. *Harvard Business Review, 71*(1), 84-93.

Treacy, M., & Wiersema, F. (1995). *The discipline of market leaders.* MA: Addison-Wesley.

Treacy, M., & Wiersma, F. (1995). *The discipline of market leaders: Choose your customers, narrow your focus. Dominate your market.* Reading, MA: Addison-Wesley.

Trompenaars, F. (1997). *Riding the waves of culture: Understanding cultural diversity in business.* Nicholas Brealey Publishing Ltd.

Turban, E., King, D., Lee, J., Warkentin, M., & Chung, H.M. (2002). Electronic commerce: A managerial perspective. Upper Saddle River, NJ: Prentice Hall.

Twist, J. (2004). *China leads world on broadband.* Retrieved July 12, 2006, from http://news.bbc.co.uk/1/hi/technology/3699820.stm

Uganda-Africa (2005). Retrieved June 06, 2006 from http://www.ungana-afrika.org/resources/techplanning/tech_planning_guide.pdf

Ulaga, W., & Eggert A. (2005). Relationship value in business markets: The construct and its dimensions. *Journal of Business-to-Business Marketing, 12*(1), 73-99.

Ulrich, P., Chacko, J.G., & Says, P, (2004). Overview of ICT Policies and e-strategies in the Asia-Pacific Region. In Sayo, P., Chacko, J.G., & Pradhan, G. (Eds.), *ICT Policies and e-Strategies in the Asia-Pacific: A critical assessment of the way forward*, Elsevier, New Delhi. ISBN: 81-8147-513-5.

UN (2003). *Regional Road Map towards an Information Society in Asia and the Pacific, Economic and Social Commission for Asia and the Pacific*, United Nations, New York.

UN (2005). *United Nations for Sustainable Development, Expert Group Meeting on Indicators of Sustainable Development*, New York.

UN (2006). *The Digital Divide Report: ICT Diffusion Index 2005*, United Nations Conference on Trade and Development, United nations, New York and Geneva.

UN General Assembly (2000). *United nations Millennium Declaration 55/2,* available at http://www.un.org/millennium/declaration/ares552e.htm), para. 20.

UNCTAD (2006). *Information Economy Report 2006: The Development Perspective*. UN New York and Geneva.

UNDP (2001). *Human Development Report 2001: Making New technologies Work for Human Development*, Oxford University Press, New York.

UNDP (2004, May). *Local e-Government in FYR Macedonia 2004.* ICT Assessment Report of Municipalities, Skopje, UNDP FYR Macedonia.

UNESCAP (2006). Retrieved February 15, 2007, from http://www.unescap.org/pmd/03_tc/pmd_dp_tc_ra_ictps.asp

United Nations Development Programme (2001). *Human Development Report 2001: Making New Technologies work for human development*. Oxford: University press.

United Nations Development Programme: Human Development Report (2005). Retrieved from http://hdr.undp.org

United Nations. DPEPA & ASPA. (2002). *Benchmarking e-government: A global perspective: assessing the progress of the UN member states*. New York: United Nations American Society for Public Administration.

US v. Aluminium Co. of America,148 F. 2d 416 (2d Cir. 1945).

Uzzi, B. (1997). Social structure and competition in interfirm networks: The paradox of embeddedness. *Administrative Science Quarterly, 42*, 35-67.

Van de Ven, A.H., & Poole, M. S. (1990). Methods for studying innovation development in the Minnnesota Innovation Research Program. *Organization Science, 1*(3) 313-335.

Van der Heijden, K. (1996). *Scenarios. The art of strategic conversation*. New York: John Wiley & Sons.

Van Eecke, P., & Rombouts, D. (October 2005). The EU domain: The sun is rising. *Journal of Internet Law, 26.*

Van Eecke, P., & Truyens, M. (2006, March). Recent events in privacy legislation. *Journal of Internet Law, 22.*

Van Grembergen, W. (2002). Introduction to the minitrack: IT governance and its mechanisms. In *Proceedings of the 35th Hawaii International Conference on System Sciences (HICSS).* IEEE.

Van Grembergen, W., & Saull, R. (2001). Information technology governance through the balanced scorecard. In *Proceedings of the 34th Hawaii International Conference on System Sciences (HICSS).* IEEE.

Van Grembergen, W., & Van Bruggen, R. (1997). Measuring and improving corporate information technology through thc balanced scorecard technique. In *Proceedings of the 4th European Conference on the Evaluation of Information Technology* (pp. 163-171). Delft.

Vanhaverbeke, V. (2001). Realizing new regional core competencies: Establishing a customer-oriented SME network. *Entrepreneurship & Regional Development, 13*, 97-116.

Venkatraman, N. (2000). Five steps to a dot-com strategy: How to find your footing on the web. *Sloan Management Review, 41*(3), 15-28.

Verma, N., Kalra, S., & Mohandas, V. S. (2005). Assessing citizen centricity of government websites. *e-gov (www.csdms.in/egov/july05.pdf), 1*(4), 29-31.

Vitale, M. (2001). *The dot.com legacy: Governing IT on Internet time*. Working Paper, Information Systems Research Center, University of Houston, Houston, Texas.

Von Bertalanffy, L. (1968). *General system theory: Foundations, development, applications*. New York: George Braziller.

Walsham, G. (2001). *Making a world of difference.* IT in a Global Context. Chichester: John Wiley & Sons.

Warren, S.D., & Brandeis, L.D. (1890). The right to privacy. *Harvard Law Review, 4*, 5.

Wastell, D.G. (2003). Organizational discourse as a social defense: Taming the tiger of electronic government. In Wynn, Whitley, Myers, & DeGross (Eds.), *Global and Organizational Discourse About Information Technology: Ifip Tc8/Wg8.2 Working Conference on Global and Organizational Discourse About Information Technology*, Barcelona, Spain. Dordrecht: Kluwer Academic Publishers.

Weckert, J. (2000). What is new or unique about Internet activities? In Langford (Ed.), *Internet ethics*. London: McMillan.

Weick, K. E., & Roberts, K. H. (1993). Collective mind in organizations: Heedful interrelating on flight decks. *Academy of Science Quarterly, 38*, 357-381.

Weill, P., & Broadbent, M. (1998). *Leveraging the new infrastructure: How market leaders capitalize on information technology*. Harvard Business School Press.

Weill, P., & Vitale, M. (2001). *Place to space. Migrating to eBusiness models.* Harvard Business School Press.

Weill, P., & Vitale, M. (2002). What IT infrastructural capabilities are needed to implement e-business models. *MIS Quarterly Executive, 1*(1), 17- 34.

Weill, P., & Woodam, R. (2002). *Don't just lead, govern: Implementing effective IT governance.* CISR Working Paper No. 326. Sloan School of Management, Cambridge, Massachusetts.

Weiser, M., & Molnar, K. K. (1996). Advanced telecommunications infrastructure policies - A comparative analysis. In *Proceedings of the Americas Conference on Information Systems.*

Weizenbaum, J. (1976). *Computer power and human reason*. San Francisco: W. H. Freeman and Company.

Welty, B., & Becerra-Fernandez, I. (2001). Managing trust and commitment in collaborative supply chain relationships. *Communications of the ACM* (6), 67-73.

Wenger, E. (1999). *Communities of practice. Learning, meaning, and identity.* CUP.

Wernerfelt, B.(1984). A resource-based view of the firm. *Strategic Management Journal, 5*, 171-180.

West, D. (2003). *Global e-Government.* Providence, RI: report published by author.

West-Burnham, J. (1994). Strategy, policy and planning. In T. Bush & J. West-Burnham (Ed.), *The principles of educational management*. Harlow: Longman.

Wheale, J. (1991). *Generating income for educational institutions: A business planning approach.* London: Kogan Page.

Whetherbe, J. (2001). *Achieving the high-performance, information-based networked organization.* Working Paper, Information Systems Research Center, University of Houston, Texas.

Whittington, C.D., & Sclater, N. (1998). Building and testing a virtual university. *Computers and Education, 30*(1/2), 41-47. Retrieved February 21, 2003, from http://i.am/davewhittington/

Widrick, S. M., Mergen, E., & Grant, D. (2002). Measuring the dimensions of quality in higher education. *Total Quality Management, 13*(1), 123-131.

Wiio, J. (2004). *From the Information Society to a Society of Information: Public Service Digital Television as a Player in the Knowledge Society of the Future*, RIPE@2004. Paper presented at Mission, Market and Management. Public Service Broadcasting and the Cultural Commons, Copenhagen and Arhus, Denmark.

Willcocks, L. (1996). *Investing in information systems: Evaluation and management.* London: Chapman & Hall.

Willcocks, L.P., Feeny, D.F., & Islei, G. (1997). *Managing IT as a strategic resource.* London: McGraw-Hill Companies.

Wilson, G., & Heeks, R. (2000). Technology, poverty and development. In T. Allen & A. Thomson (Eds.), *Poverty and development: Into the 21st century* (pp. 403-424). Oxford University Press.

Wimmer, M., & Krenner, J. (2001). An integrated online one-stop government platform: The eGOV Project. In *Proceedings Hofer, Chroust IDIMT-2001 9th Interdisciplinary Information Management Talks* (pp. 329-337) (ISBN 3-85487-272-0).

Winner, L. (2000). Cyberlibertarian myths and the prospects for community. In Baird, Ramsower, & Rosenbaum (Eds.), *Cyberethics - Social and moral issues in the computer age.* New York: Prometheus Books.

Wollstenholme, R. (1995). *Understanding small rural communities and the consultation process.* Armidale, NSW: The Rural Development Centre, University of New England.

Women and ICTs: Creating Global Transformation. (2005, June). Conference Proceedings. Retrieved July 2005, from http://portal.acm.org/toc.cfm?id=1117417&coll=portal&dl=ACM&type=proceeding

Wood, P.R. (1995). *Comparative financial law.* London: Sweet & Maxwell.

World Bank (2001). *Bridging the digital divide in the Americas.* Retrieved April 20, 2006, from http://lnweb18.worldbank.org/External/lac/lac.nsf/e4ce3b63837a57ef85256a37004a8dac/ea45f9696e38b03585256a370052bcaf/$file/English%20.pdf

World Bank (2005). *Project Appraisal Document and Project Information Document*, World Bank, Retrieved July 12, 2006, from http://web.worldbank.org/external/projects/main?pagePK=104231&.htm

World Bank (2006). *2006 Information & Communications for Development (IC4D): Global trends and policies.* Retrieved July 27, 2006, from http://web.worldbank.org/wbsite/external/topics/informationandcommunicationandtechnologies/contentmdk:20831214~pagepk:210058~pipk:210062~thesitepk:282823,00.html

World Bank (2006). *Korea as a Knowledge Economy: Evolutionary Process and Lessons Learned, The World Bank,* Washington, 2006.

World Bank. (2005, September 14). *Gender and the World Summit on the Information Society.* [Conference Streaming Video]. Washington, DC: World Bank. Retrieved March 2006, from http://web.worldbank.org/WBSITE/EXTERNAL/TOPICS/EXTGENDER/

Yagüe, M. I., Maña, A., & Lopez, J. (2005). A metadata-based access control model for web services. *Internet Research, 15*(1), 99-116.

Yahoo Inc. v. La Lique Contre le Racisme et L'Antisemitisme.169 F. Supp. 2d 1181 CND (Cal. 2001).

Yin, R. (2003). *Case study research: Design and methods* (3rd ed.) Thousand Oak, CA: SAGE Publications.

Yo, C. -D., Tsai, W. -C., & Tien, D. (2001). National Information Infrastructure Development in R.O.C. Industrial Forum.

Yoon, S-H. (1996). Power online: A post-structuralist perspective on computer-mediated communication. In C. Ess (Ed.), *Philosophical perspectives on computer-mediated communication* (pp. 171-196). Albany: State University of New York Press.

Young v. New Haven Advocate,315 F. 3d 256 (4th Cir. 2002).

Zachman, J.A. (1987). A framework for information systems architecture. *IBM Systems Journal, 26*(3), 276-292.

Zakaria, N., & Gould, E. (2001). *Applying cross-cultural theory to instructional design*. Geneseo.

Zakzewski, M. (1999). A linguistic perspective on multiple choice questioning. *The Higher Education Research and Development Society of Australasia Newsletter.*

Zerdick, A., et al. (2001). *European Communication Councel Report: Die Internet-Ökonomie: Strategien für die digitale Wirtschaft.* 3rd edition Berlin, Heidelberg: Springer.

Zmud, R. W., Boynton, A. C., & Jacobs, G. C. (1986). The information economy: A new perspective for effective information systems management, *Database,* Fall, 17-23.

Zollo, M., & Winter, S.G. (2002). Deliberate learning and the evolution of dynamic capabilities. *Organization Science*, *13*(3), 339-351.

About the Contributors

Engr. Md. **Hakikur Rahman**, PhD, is the national project coordinator of SDNP Bangladesh, a global initiative of UNDP since December 1999. He is also acting as the secretary of South Asia Foundation Bangladesh. Before joining SDNP he worked as the director, Computer Division, Bangladesh Open University. He has written and edited several books on computer education for the informal sector, distance education, and research. He is the founder-chairperson of Internet Society Bangladesh, editor of monthly *Computer Bichitra*, founder-principal of ICMS Computer College, head examiner (computer), Technical Education Board, executive director of the BAERIN Foundation, and involved in establishing an IT based distance education university in Bangladesh.

* * *

Munir Ahmed, born in Sialkot, studied in Murray College, Sialkot, Institute of Statistics, Punjab University, Lahore, and taught at Punjab University, Karachi University, Karachi, and Michigan Tech. University, Houston, Michigan, USA. He did his post graduate degree at Aberdeen University, Aberdeen, UK and PhD from Iowa State University, Iowa, USA. He is the author of more than 150 research papers published in national and international journals. He is currently working as rector, National College of Business Administration & Economics and professor of statistics in the School of Business Administration and Economics. His area of research is statistics, population, data neural network, and various management sciences.

David Brown is chair of strategy and information systems and director of the Lancaster China Management Centre at Lancaster University Management School, UK. He is also a fellow professor at Renmin University, PRC. His research interests have two separate but linked strands. Firstly, the strategic studies including strategic information systems and e-business, and secondly the application of these ideas internationally, especially in Asia. Current research includes e-business and SMEs, e-science, and e-business and innovation; all funded by Research Council and EU grants. He has published widely on information systems, e-business, and transition in China.

Jamilur Reza Choudhury is the vice chancellor of the BRAC University in Bangladesh. Obtaining his PhD from the University of Southampton in 1968, he headed the Civil Engineering Department of the Bangladesh University of Engineering and Technology during 1978-1979 and 1981-1983; and was dean of the faculty of Civil Engineering from 1983-1985. He served as an advisor with the rank of a minister of the Non-Party Caretaker Government in 1996. During his 43 years of professional career he has served many corporations, institutions, and organizations in the capacity of chairman and director of their governing councils. He has numerous publications in civil engineering and IT. His contribution to the National ICT Task Force is notable through his many policy initiatives and guidance on information technology of the country.

David Crowther (BA, MBA, MEd, PhD, DSocSc, DSc, PGCE, FCMA, CPFA, MCMI; professor of corporate social responsibility, De Montfort University, UK) worked as an accountant, systems specialist, and general manager for 20 years. After a number of years in the financial services sector, including a spell in which he set up and ran a store credit card scheme, he decided to leave the business world and become an academic. In 1994 he joined Aston University as a lecturer in accounting and there obtained a PhD for research into corporate social performance and reporting. He is the author of 18 books and several hundred articles in academic, business and professional journals, and edited book collections. He has also spoken widely at conferences and seminars and acted as a consultant to a wide range of organisations.

Allan H. Ellis is an associate professor in the School of Commerce and Management at Southern Cross University. He supervises post-graduate research students in Masters, DBA, and PhD programs. His research interests are in using computer networks for communication, learning, and training. He chairs the AusWeb conference series (http://ausweb.scu.edu.au) and is president of the IW3C2 the Geneva-based group that runs the International World Wide Web conference series (http://www.iw3c2.org).

Emmanuel Eze received his BSc (Hons.) degree in computer science from The University of Nigeria in 1999. He is presently concluding his PhD in computer science at the Centre for Internet Computing, University of Hull, UK that commenced in 2003. He has also been providing IT consultancy service to a variety of companies in the area of Internet-based software technologies. His research interests include multimedia semantics, multimedia ontology, contextual knowledge engineering, Internet technologies, and the building of software architectures. He is a member of BCS, ACM, and IEEE Computer Society and has published papers in leading conferences and journals.

Devan Govender is currently working as an e-learning consultant in the Institute for Curriculum and Learning Development (ICLD) at the University of South Africa (Unisa). Before joining Unisa in 2002, he was a lecturer at the University of Durban Westville in the Department of Education Technology where he taught various modules within the field of education technology to both masters and honours students. He completed his doctoral thesis in 2002, which focussed on technology enhanced teaching and learning in South African higher education.

Arla Juntunen received her master's degree in administrative information systems at the University of Helsinki and a PhD in marketing at the Helsinki School of Economics (HSE), Finland. Arla is a post-doc researcher and senior advisor in Finland's Ministry of the Interior. Her research interests focus on strategic management, business networks, B2B marketing, and managerial capabilities.

Suleman Aziz Lodhi is a system analyst with MSc in physics and computer programming expertise in high-level languages. He joined the Public Health Engineering Department (Pakistan) in 1995 as assistant director MIS and worked on community development projects. It was during this time that he developed a keen interest in community development and sustainability issues. He later joined secretariat of a premier trade association and headed its economic research cell. Presently, he is working with academia after completing his PhD in knowledge management. His present research interests include knowledge economy, community development, and social network analysis.

Helena Mitev (MBus (Prof Acc.), MSc (Econ), ASA CPA, EAA, AAA) is a lecturer in the School of Accounting and Corporate Governance at the University of Tasmania, Australia. Her teaching and research are informed by her international business experience. She has traveled across three continents working as a company executive officer and an auditor. She has taught undergraduate and postgraduate auditing courses. Her specialist teaching areas include auditing, organizational risk management, and management accounting. She is pursuing a PhD in the area of international accounting education, which is part of her wider research interest in corporate social responsibility and corporate governance. She is a member of the Association of Certified Fraud Examiners, AAA, and EAA.

Onyeka Kingsley Osuji is a postgraduate law research student and teaching assistant at The University of Manchester, UK. A 2003 Shell Centenary/FCO Chevening Scholar at Oxford University, he also has law degrees from the Universities of Oxford and Nigeria. He is both a barrister and solicitor of Nigeria, and a solicitor of England and Wales. Mr Osuji's has interest on law and business issues. He has written papers and delivered conference presentations on issues such as such as removal of company directors, aviation financing, multinational enterprises, corporate governance, and corporate social responsibility.

Donna D. Rubinoff received her doctorate in geography from the University of Colorado-Boulder and teaches there in the Departments of Geography and Women's Studies. She also holds master's degrees in planning and landscape architecture from the University of Virginia and worked in that field for 15 years. Dr. Rubinoff's interests are in development, especially ICT4D, sustainable development, and gender and development. She has conducted research in Central America and is currently working on land use planning and development in Rwanda. She has received funding from the NSF and is a Woodrow Wilson fellow in Women's Studies.

Juan Luis Manfredi Sánchez is PhD in communication sciences. He has both degrees in journalism and history from the University of Seville (Spain), as well both MBA in audiovisual companies and public administration. He is a recognized expert in public service broadcasting and he has written several books and scientific articles about the transformation of the television markets in the information age. He is the dean of the School of Humanities, Social and Communication Sciences at the SEK University (Segovia, Spain).

Lalitsen Sharma is a senior lecturer in Department of Computer Science, University of Jammu, India. He has obtained master of science in mathematics and master of computer applications from Guru Nanak Dev University, Amritsar, India. He has submitted PhD thesis in May 2006. He has been teaching to post graduate students in computer applications for 10 years. He is a life member of Indian Science Congress Association, Institute of Electronic and Communication Engineers, India and National HRD Network, India. His specialization is in data communication and network, Internet and WWW, and data structures. He has attended a number of national conferences and workshops. He has also organised workshops and acted as a member of organising committee to organise national conferences.

A graduate of the University of the West Indies, Kingston, Jamaica and Cambridge University, United Kingdom, **Sheryl Thompson** is in the final stages of her PhD at Lancaster University Management School (LUMS). Prior to undertaking doctoral studies at LUMS, she worked in IT for various organisations in Jamaica. Her research interests include e-business, ICT diffusion and adoption in developing countries, IT strategy, and e-business adoption by small and medium-sized enterprises.

Lynne De Weaver is currently conducting research for her PhD at Southern Cross University as well as serving as their marketing manager. Prior to joining the University, she was the Northern Rivers Regional Community coordinator for the CTC@NSW program. She also worked in Sydney as creative director for Clemenger Direct, Brann Direct, and LINTAS. Clients included Primus Telecommunications, Ozemail, IBM Australia, Microsoft Australia, Australia Post, the Commonwealth Government, and the NSW Department of State & Regional Development (DSARD). She is the author of the books, *Marketing for E-Business in Australia* and *How to Use Direct Marketing in Your Business.*

Index

T

U

W